Foundations of Optimization

PRENTICE-HALL INTERNATIONAL SERIES
IN INDUSTRIAL AND SYSTEMS ENGINEERING

W. J. Fabrycky and J. H. Mize, Editors

BEIGHTLER, PHILLIPS, AND WILDE *Foundations of Optimization, 2nd ed.*
BLANCHARD *Logistics Engineering and Management*
BROWN *Systems Analysis and Design for Safety*
BUSSEY *The Economic Analysis of Industrial Projects*
FABRYCKY, GHARE, AND TORGERSEN *Industrial Operations Research*
FRANCIS AND WHITE *Facility Layout and Location: An Analytical Approach*
GOTTFRIED AND WEISMAN *Introduction to Optimization Theory*
KIRKPATRICK *Introductory Statistics and Probability for Engineering, Science, and Technology*
MIZE, WHITE, AND BROOKS *Operations Planning and Control*
MUNDEL *Motion and Time Study: Improving Productivity, 5th ed.*
OSTWALD *Cost Estimating for Engineering and Management*
SIVAZLIAN AND STANFEL *Analysis of Systems in Operations Research*
SIVAZLIAN AND STANFEL *Optimization Techniques in Operations Research*
THUESEN, FABRYCKY, AND THUESEN *Engineering Economy, 5th ed.*
TURNER, MIZE, AND CASE *Introduction to Industrial and Systems Engineering*
WHITEHOUSE *Systems Analysis and Design Using Network Techniques*

Foundations of Optimization

second edition

CHARLES S. BEIGHTLER
University of Texas

DON T. PHILLIPS
Texas A & M University

DOUGLASS J. WILDE
Stanford University

PRENTICE-HALL, INC., Englewood Cliffs, N. J. 07632

Library of Congress Cataloging in Publication Data

Beightler, Charles S (date)
Foundations of optimization.

First ed. published in 1967 by D. J. Wilde and C. S. Beightler.
Includes bibliographies and index.
1. Mathematical optimization. I. Phillips, Don T. joint author. II. Wilde, Douglass J., joint author. III. Wilde, Douglass J. Foundations of optimization. IV. Title.
QA402.5.W47 1979 519.7 78-12386
ISBN 0-13-330332-2

Editorial/production supervision and interior design by Barbara A. Cassel
Manufacturing buyer: Gordon Osbourne

Printed in the United States of America

10 9 8 7 6 5 4 3 2 1

PRENTICE-HALL INTERNATIONAL, INC., *London*
PRENTICE-HALL OF AUSTRALIA PTY. LIMITED, *Sydney*
PRENTICE-HALL OF CANADA, LTD., *Toronto*
PRENTICE-HALL OF INDIA PRIVATE LIMITED, *New Delhi*
PRENTICE-HALL OF JAPAN, INC., *Tokyo*
PRENTICE-HALL OF SOUTHEAST ASIA PTE. LTD., *Singapore*
WHITEHALL BOOKS LIMITED, *Wellington, New Zealand*

To Pat, Candy, and Jane;
for
the Next Generation

Contents

2 *INDIRECT METHODS: THE DIFFERENTIAL VIEWPOINT* 11

3 *LINEAR PROGRAMMING, SENSITIVITY ANALYSIS, AND INTEGER PROGRAMMING* 59

Preface to the Second Edition

This book, the second edition of *Foundations of Optimization*, includes some significant changes from the original work published in 1967 by Douglass Wilde and Charles Beightler. Coauthored by Don T. Phillips, this edition preserves the fundamental objectives of the first in unifying a large body of optimization techniques in a practical, yet rigorous, presentation. Since the first edition was so well received, we have delayed making any revisions; now, however, the steady advances made in this field during the past dozen years have necessitated a rather extensive updating and rewriting of the original work. In addition, we have corrected numerous minor errors that have been brought to our attention by many colleagues, students, and research workers. Each chapter now stands alone; the entire text can be covered in a two-semester sequence, or as three separate quarterly courses using the chapters on linear programming, nonlinear optimization, and dynamic programming.

The purpose of Chapter 1 is to place optimization theory in its proper historical perspective, and to summarize the key developments that have led to the current state of the art. Chapter 2 satisfies two basic objectives. First, it introduces a basic set of mathematical procedures which are necessary in solving mathematical programming problems. Second, it introduces the differential approach as embodied in the fundamental concept of constrained derivatives, which then forms the cornerstone of the many algorithms described in the later chapters.

Chapter 3 is an extensive treatment of linear programming, which is now independent of the nonlinear development on which it was based in the first edition. After a thorough coverage of the simplex method, duality theory, sensitivity analysis, parametric programming, and the dual-simplex algorithm, transportation and assignment problems are covered in detail. The chapter concludes with a brief but solid presentation of integer programming. The emphasis throughout this chapter, as in the remainder of the book, is on practical problem-solving techniques. Chapter 4 extends the coverage of linear systems to encompass nonlinear unconstrained optimization. Both direct elimination and direct climbing procedures are presented, covering most of the material in Wilde's very successful treatise, *Optimum Seeking Methods*. In addition, many new results and algorithms have been included in this updated chapter.

Constrained nonlinear optimization is covered in rigorous fashion in Chapter 5. Constrained derivatives are employed to handle nonlinear inequality constraints, and new proofs of sufficiency conditions are given here for the first time. Also, new sections have been added on penalty function methods, generalized Lagrange multiplier techniques, and nonlinear duality theory. The growing new field of geometric programming is covered in Chapter 6, including the basic derivations of the novel duality theory on which this important technique depends, and generalized signomial programming. In addition, new sections have been added on functional substitutions, sensitivity analysis, and the solution of the general nonlinear programming problem.

Chapter 7 is a detailed treatment of dynamic programming for both serial and nonserial systems. Also, new sections now deal with infinite stage systems, multiple state variables, and the algorithms for branch and loop absorption and compression.

The authors owe a debt of gratitude to Ury Passy, M. Avriel, Ron Dembo, A. Fiacco, H. Everett, and G. McCormick for their work as reflected in this revised edition. Also, special thanks are reserved for A. Ravindran, J. J. Solberg, Byron Gottfried, and Joel Weismann for individual contributions from prior work in the field. There are many other people who have contributed to this book either directly or indirectly, too numerous to acknowledge. Each will recognize his own contribution, and for that we are grateful.

CHARLES S. BEIGHTLER

DON T. PHILLIPS

DOUGLASS J. WILDE

Preface to the First Edition

Scattered bits of isolated knowledge have been organized in this book into a compact, unified theory of optimization. Dealing as it does with achieving the best—maximum gain or minimum loss—in a rational manner, optimization theory naturally holds great interest for the practical professions of engineering, economics, administration, and operations research. Its development over the centuries by architects, physicists, politicians, merchants, astronomers (and astrologers), clerics, and philosophers gives optimization a colorful history and a claim to be considered a branch of mathematics, for most of its contributors are posthumously called *mathematicians*. Yet no one recognized this body of work as "optimization theory" until the middle of the twentieth century, when high-speed computers implemented forgotten procedures of the past and stimulated research on new methods. Spectacular advances followed, producing a massive, jargon-filled literature on "linear, nonlinear, and dynamic programming," as well as on the "maximum principle" and "modern control theory." At first glance, these diverse developments appear to share only their goal of achieving an optimum. But one can, having once mastered their different languages, discern many ideas common to all of them. This book abstracts the concepts underlying the various procedures, constructing from them a definitive theory of optimization. The subject can then take its place among such other theoretical fundamentals of applied science as thermodynamics, mechanics, differential equations, and probability theory. With its foundations identified and its framework strengthened, optimization theory should be able to support even greater achievements in the future.

The authors have accumulated the material in the book during five years of teaching optimization to graduate and advanced undergraduate students of engineering and operations research at Stanford University and the University of Texas. The principal mathematics required is understanding of the differential (not integral) calculus. Matrix algebra is deliberately avoided except where it can be used without confusion for abbreviation. Although, for economy of thought, many abstractions are presented, they are always illustrated with detailed industrial examples based on the authors' experience as practicing engineers and consultants. Clarity and plausibility take precedence over formal logic, although the authors have not hesitated to be rigorous when important fine points are at stake or when an informal treatment would consume too much space. The material takes between three and six semester hours to cover, depending on the preparation and quality of the professor and students. Rather than devote a separate course to optimization, a faculty may prefer to introduce various aspects of it into existing design and analysis courses, using the book for reference rather than as a text.

Readers already expert in optimization will notice several concepts developed here for the first time. The "constrained derivatives" and a novel definition of states and decisions are the key ideas unifying the theory. They lead to improved ways of handling constrained optimum-seeking problems and give novel conditions for nonconvex programming. The generalization of geometric programming, itself a very new topic, has not even appeared in the literature as we go to press. This accomplishment of Passy's, as well as Avriel's block search method, come almost directly from their Ph.D dissertations. In addition to this original material, many results are assembled here for the first time in book form. Among these are several optimum-seeking methods, the use of functional diagrams in nonserial systems, and idealized industrial examples of sensitivity analysis, the decomposition principle, and automatic control.

Besides describing quantitative technical advances, the book tries to distill qualitative guides for making decisions in the sort of ill-defined, hurried situation occurring so often in practice. The authors have also attempted to place the subject in historical perspective in order to help the reader understand the motives and accomplishments of the pioneers of optimization. To this end each chapter has its own bibliography, giving as far as possible the original sources cited in the text.

Over the half decade that the ideas for this book have been germinating and maturing, the authors have been influenced by many colleagues, among them R. Aris, R. J. Buehler, E. D. Crandall, A. Harkins, F. Horn, L. G. Mitten, G. M. Nemhauser, D. F. Rudd, M. E. Thomas, and C. F. Wood. The U.S. National Science Foundation has supported not only research upon which some of the results are based, but also a course on optimization theory for chemical engineering professors in August 1965. The participants were excellent critics for the preliminary edition. To all these people and institutions, the authors offer their sincere gratitude.

DOUGLASS J. WILDE

CHARLES S. BEIGHTLER

Foundations of Optimization

Optimization and Optimism

1

Decide not rashly. The decision made
Can never be recalled. The gods implore not,
Plead not, solicit not; they only offer
Choice and occasion, which once being passed
Return no more. Dost thou accept the gift?

HENRY WADSWORTH LONGFELLOW,
"Masque of Pandora,"
Tower of Prometheus on Mount Caucasus

Man's longing for perfection finds expression in the theory of *optimization.* It studies how to describe and attain what is Best, once one knows how to measure and alter what is Good or Bad. Normally, one wishes the most, or *maximum,* and the least, or *minimum.* The word *optimum,* or maximum, meaning "best," is synonymous with "most" or "maximum" in the former case, and with "least" or "minimum" in the latter. *Optimum* has become a technical term connoting quantitative measurement and mathematical analysis, whereas "best" remains a less precise word more suitable for everyday affairs. The technical verb *optimize,* a stronger word than "improve," means to achieve the optimum, and *optimization* refers to the act of optimizing. Thus *optimization theory* encompasses the quantitative study of optima and methods for finding them.

This book is intended to introduce the theory of optimization to students of engineering, economics, and administration, as well as of the physical, mathematical, and social sciences. Since optimization involves finding the best way to do things, it has obvious applications in the practical world of production, trade, and politics, where sometimes small changes in efficiency spell the difference between success and failure.

Although many phases of optimization theory have been known to mathematicians for centuries, the tedious and voluminous computations required prevented their practical application. The development of rapid, inexpensive, automatic computers in the middle of the twentieth century has not only made these older methods attractive, but also encouraged much new research on optimization. This book surveys these diverse developments, old and new, and tries to unify them into a single cohesive theory.

There is, however, more to optimization theory than a set of numerical recipes for finding optima. By studying various optimization techniques, each suitable for different quantitative, if idealized, situations, one often discerns fairly general decision rules appropriate to problems not entirely mathematically describable. This can develop decisiveness through skill in recognizing the proper form of an optimal solution even when a problem is not completely formulated in mathematical terms. It can also nurture appreciation of the value of the information needed to describe a system well enough for it to be optimized. Even when the path to the ideal optimum is blocked or obscured, optimization theory often shows how existing conditions can be improved.

This introductory chapter discusses, in nonmathematical language, the role of optimization theory both in synthesis—design and decision making—and in analysis—understanding how the world behaves. Citing successful applications of optimization to such varied endeavors as city planning, statistics, optics, mechanics, astronomy, economics, and chemistry suggests its possibilities in formulating new decision rules and natural laws. As a matter of historical interest, the mutual influence of optimization and the philosopher-mathematician Leibniz is recounted. In closing the chapter with Voltaire's witty demolition of Leibniz's "philosophical optimism," we expose the limitations of optimization theory while emphasizing its legitimate unrealized potentialities.

1-01 SYNTHESIS: OPTIMAL DESIGN AND DECISION

Today many important decisions are made by choosing a quantitative measure of effectiveness and then optimizing it. Deciding how to design, build, regulate, or operate a physical or economics system ideally involves three steps: First, one should know, accurately and quantitatively, how the system variables interact. Second, one needs a single measure of system effectiveness expressible in terms of the system variables. Finally, one should choose those values of the system variables that yield optimum effectiveness. Thus optimization and choice are closely related.

The first step, knowledge of the system, is of paramount importance, for it is here that the decision maker brings to bear his/her professional skill and training as an engineer or operations analyst. There is little point in optimizing a model that does not describe what is truly happening in the system. Hence most of the effort expended on an "optimization" study will in practice be devoted to understanding the system and describing it quantitatively in terms of tables, graphs, computer programs, or mathematical equations. It seems better, therefore, to add optimization theory to professional training in existing disciplines than to develop optimization specialists unable to comprehend the systems to be optimized. Optimization theory must reinforce, rather than supplant, the present professions.

Since the second step, finding a measure of system effectivness, often involves value judgment, it is usually either trivially simple or practically impossible to accomplish. In many physical and economic systems, the measure is obviously profit, cost, or efficiency, but a social or political system may have such conflicting goals that optimization cannot be carried out at all. Even when the type of measure is clear, it is not always easy to express its quantitative dependence on the system variables. Yet one must try to obtain this information if the fruits of optimization are to be tasted, and one advantage of optimization studies is their way of making economic information valuable enough to be gathered. It is not easy for professionals to agree on a unique measure of effectiveness, but in the words of Confucius, "Those whose courses are different cannot lay plans for one another."

Only upon finishing the first two steps, requiring knowledge and value judgment, can one proceed to apply the theory developed in this book. Optimization is decisive because it narrows down the possible choices to one—the best one. Moreover, it often yields information about the sensitivity of optimum conditions to fluctuations and uncertainties in the original system description. For these reasons no rational decision-making process is really complete without optimization.

Let us consider examples of the three-step decision-making process just described. The earliest and most poetic is given by Virgil, whose legendary Queen Dido procured for the founding of Carthage the largest area of land that could be surrounded by the hide of a bull. From the hide she made a rope which she arranged in a semicircle with the ends against the sea. Her queenly intuition told her that this half circle had the largest possible area for the perimeter given, a fact conjectured by Archimedes (187–212 B.C.) but not proved for over two millennia. Many ancient cities are, in fact, circular, probably to minimize the length of the city walls needed to enclose the city's fixed area.

A famous decision problem of the late Renaissance was to design the *brachistochrone* (Greek for "shortest time"), a slide down which a frictionless object would slip in the least possible time. Galileo guessed that it should be a circular arc, but here unaided intuition failed, and in 1694 Johann Bernoulli proved it to be cycloidal.

More recent examples of using optimization to make decisions involve invention of plausible or mathematically convenient measures of effectiveness. Statisticians fit curves to experimental data by Gauss's "method of least squares," in which one minimizes the sum of the squared deviations between curve and data. Wiener's extension of this idea is used by control engineers to design control systems and electronic filters that minimize the time integral of the squared error. Engineers and economists commonly employ minimum cost or maximum profit as decision-making criteria, and it is to such problems that this book principally addresses itself.

1-02 PHILOSOPHICAL OPTIMISM: THE BEST OF ALL POSSIBLE WORLDS

In 1710 Leibniz coined the word "optimum" in his *Theodicy*: *Essays on the Goodness of God, the Freedom of Man and the Origin of Evil.* Although his theological and metaphysical conclusions are not appropriate subjects for discussion here, his line of reasoning illustrates well the role of optimization in both synthesis and analysis. Leibniz, who continually sought philosophical truths through mathematics, speculated on the nature of the world and on its creation, or in our words, its *synthesis*. He writes:

> There is an infinitude of possible worlds, among which God must needs have chosen the best, since he does nothing without acting in accordance with supreme wisdom. Now this supreme wisdom, united to a goodness that is no less infinite, cannot but have chosen the best. . . . As in mathematics, when there is not maximum or minimum, everything is done equally or . . . nothing at all is done: so it may be said . . . that if there were not the best (optimum) among all possible worlds, God would not have produced any.

Thus does Leibniz draw his conclusions from his premises, following the three steps given in the preceding section for making rational decisions. For the first step, knowledge, he postulates infinite wisdom; for the second, value judgment, he assumes infinite goodness. The optimization step calls for exhaustive evaluation of all possibilities, a procedure made plausible, if intellectually uninteresting, by the premise of infinite wisdom. Leibniz reduced his analysis of the nature of the world to a study of how it might have been synthesized, a strategy employed, as we shall see, by many later scientists in their studies of nature and man. Notice that he felt it necessary to attempt to prove the existence of an optimum. In modern optimization theory, an existence proof is also sometimes essential. We see, then, in Leibniz's theological speculations the rudiments of decision theory, a description of phenomena in terms of an optimum principle and an existence proof. These patterns will be found throughout this exposition of optimization theory.

The cheerful doctrine that we live in the best of all possible worlds became known as *philosophical optimism*, its adherents being called *Optimists*. They perverted Leibniz's ideas into an excuse for impotent fatalism, since it seemed to them a waste of time to try to improve that which was already optimum. Voltaire, not at all satisfied with the state of the eighteenth-century world, lampooned them so successfully in *Candide* that philosophical optimism did not live to see the French Revolution. But the effect of Leibniz's point of view on the scientists of his day was powerful, as shown in the next section, where it is seen reflected in elegant formulations of natural laws and economic principles.

1-03 ANALYSIS: OPTIMUM PRINCIPLES

The renowned mathematician Leonhard Euler (1707–1783) appears to have been a philosophical optimist, having written: "Since the fabric of the world is the most perfect and was established by the wisest Creator, nothing happens in this world in which some reason of maximum or minimum would not come to light." Such a sweeping generalization is difficult to accept, even from so great an authority as Euler. Yet this idea has produced many strikingly simple formulations of certain complex laws of nature.

In attempting to use optimization theory for analyzing natural behavior, one reverses the order of the three steps for rational decision making. Knowledge about a system is deduced by assuming that it behaves so as to optimize some given measure of effectiveness. Thus the system behavior is completely specified by identifying the criterion of effectiveness and applying optimization theory to it. This approach is known as describing nature in terms of an *optimum principle*.

The earliest optimum principles concerned the behavior of light. Around 100 B.C., Heron of Alexandria asserted that light travels between two points by the shortest path. This minimum-distance principle leads to two experimentally verifiable facts: first, that light rays are straight lines unless they are reflected or refracted; second, that rays leave a reflecting surface at the same angle at which they strike it. The more general principle of Fermat (1657), that light travels between two points in the least time rather than the least distance, generates Snell's refraction law without contradicting Heron's principle.

It was in the heyday of philosophical optimism that laws of mechanics were first formulated in terms of minimum principles. Maupertius' *least-action principle*, strongly defended by Euler, led Lagrange to invent the "kinetic potential." Even after the demise of optimism, Gauss, in 1829, stated a "principle of least restraint" from which could be deduced the equality of internal to external forces in statics.

Light and mechanics were brought together by a single minimum principle conceived by W. R. Hamilton, who was, appropriately, the Astronomer Royal of Ireland (1834, 1835). From Hamilton's single principle could be obtained, by optimization, all the optical and mechanical laws then known. It remains one of the foundations of wave mechanics (Schrödinger, 1926) and relativity (Einstein, 1916).

Many laws of chemistry and thermodynamics are summarized compactly by saying that a system in equilibrium has minimum "free energy" (Gibbs, 1875–1878). But efforts to find optimum principles have not been confined entirely to the exact sciences. In 1776, for example, Adam Smith tried to explain complex economic phenomena of eighteenth-century England in terms of the "economic man," who always acts to maximize his personal profit.

1-04 MAXIMS

From Aesop to George Ade, men have deduced rules for conducting their affairs from fables and parables. These stories, abstractions of the real world, strip away mundane details in order to focus attention on the situation under study. At the end of each fable is a moral, a short summary of the lesson to be learned. Although the moral may not always be valid when taken out of context, recollection of the fable helps one remember the qualifications placed on the conclusions.

In a sense, the problems studied in this book are mathematical fables, being simplified abstractions of real situations. Although these problems are principally intended to illustrate optimization techniques, one cannot help drawing more general conclusions from the results. In fact, one reason for studying optimization theory is the insight it gives into how one should react to certain circumstances. It would be pertinent, therefore, to conclude each fable with a moral, or more appropriately, a *maxim*, since this word means "the most important sentence" (Webster).

But the authors, being engineers rather than poets or philosophers, have not always been able to discern the maxims in the text, much less phrase them in suitable language. Since the mathematical variables populating these fables do not lend themselves to imagery as well as Aesop's talking animals, we have been forced to draw on the existing literature for our maxims. This has not always been successful, for the precision and power of mathematics permit analysis of situations far more complicated than is possible by qualitative methods. The reader would be wise then to look for maxims where we may not have noticed them, and to improve upon those we have chosen. Optimization still awaits its Poor Richard.

1-05 THE AURA OF OPTIMIZATION

Having explored the development of optimization both historically and philosophically, it is now beneficial for us to disengage from the rationale of philosophy to the uncompromising structure of optimization theory. In doing so, we will attempt to place the scientific foundations of optimization in their proper perspective with rational decision making. We will first discuss the qualitative and then the quantitative implications of optimization theory, summarizing the objectives of optimum decision making. We will begin our discourse by tracing Leibniz's word "optimum" back to pre-Roman times to show how much the ancients shared our contemporary enthusiasm for the fruits of optimization.

Although the mathematics of optimization rarely goes back further than the Age of Reason, wanting the best is an ancient desire. Leibniz based his coined word "optimum" on the Latin *optimus*, meaning "best." *Optimus* contains the name of Ops, the Sabine goddess of agricultural abundance introduced into Rome in the eighth-century B.C. (Varro, Duruy). From her name come the English words "opulence" and "copious," reflecting her later status as the Roman divinity of wealth. In Imperial times her temple in the Forum contained the Roman treasury. Wealth became the symbol of power, the rich aristocracy of Rome being known as the *optimates*, a name still used at Oxford University as a title for outstanding scholars. Thus "most" came to mean "best," and Jupiter took the surname *Optimus Maximus*.

Let us not forget the sources of plenty. Earlier Latin words deriving from Ops were *opus* (work) and *opera* (works), from which the English words "operations" and "operator" are descended. The fruits of Ops are not gathered without labor.

Optimization is only the last of three steps needed to reach a rational decision. The first two, description of the system and adoption of a measure of effectiveness, are absolute prerequisites for the third. Therefore, in taking the time to apply optimization theory, one cannot neglect more conventional engineering and economic phases of a problem without risking ultimate failure. Optimization theory should be regarded not as an isolated specialty to be applied only by detached consultants, but rather as a valuable addition to the existing professional knowledge of the practicing economist, operations analyst, engineer, or administrator. In most industrial problems the work expended on defining the decision problem mathematically, gathering reliable data, and agreeing on objectives far exceeds the effort needed for mathematical optimization. True, a decision without optimization is as unfinished as an arch without a keystone. But optimization, like a keystone, is only a small part of the total structure and consequently cannot compensate for shoddy workmanship or faulty materials elsewhere in the project.

Surprisingly, the main justification for attempting to optimize a complicated decision, say an engineering design, may be that it motivates good modeling and accurate economic estimates. Unless a project is to be optimized, there is little advantage in describing the system carefully. But as soon as optimization enters the picture, everybody must become quantitative, and consequently more meticulous, rigorous, and scientific. A good way to rekindle an engineer's waning interest in the scientific basis of his profession is to teach him optimization theory, for getting his optimization plans to work usually forces him to examine closely the hypotheses and assumptions that go into the system description. Even after the optimization has been completed, one can use sensitivity analysis to assess in economic terms the value of reducing uncertainties in the model. This could help managers decide where to expend further research effort. Thus the precise mathematical nature of optimization theory imposes on a project a certain discipline which focuses effort into the channels most valuable economically.

Although current interest in optimization theory depends on its applicability to mathematical decision problems, its most permanent contribution may well be qualitative rather than quantitative. Often one can identify the structure of a problem without going into the numerical details, and this done, discern the character of the decision to be made.

Once the rationale of optimization is clearly understood, quantitative situations give insight into the underlying structure of rational decisions. This understanding not only guides the fact-finding phase of a study, but also helps one decide wisely even when time does not permit gathering all the information needed. For in the words of Syrus, "The opportunity is often lost by deliberating."

1-06 CANDIDE: OPTIMISM AND OPTIMISTS

Philosophical optimism was so badly distorted by the followers of Leibniz that Voltaire was perhaps merciful when he put it out of its misery in 1759 with his *Candide*, subtitled Optimism. Leibniz meant that people should not despair when confronted by bad luck, but should continue striving for improvement, a point of view that modern "optimists" can accept. Yet the disciples of Leibniz, while parroting his technical terms, missed his main point and settled into gloomy inaction, maintaining that although things were bad, they could not be made better.

A similar fate could overtake modern optimization theory, although for different reasons. As in the eighteenth century, some of its advocates tend not only to bury ideas in impenetrable jargon but also to use knowledge of optimization theory as an excuse for neglecting their professional training and not getting the information needed for the first two steps of the decision process. The remedies are clear: plain speech and hard work. There is, moreover, a tendency toward uncontrolled enthusiasm which leads to exaggerated claims about what particular optimization techniques can do. This, together with the use of incorrect results, could bring upon optimization theory the same undeserved disrepute that Voltaire hung on Leibniz's philosophical optimism.

On the last page of *Candide*, Dr. Pangloss, Voltaire's caricature of a philosophical optimist, says to Candide:

> There is a chain of events in this best of all possible worlds; for if you had not been turned out of a beautiful mansion at the point of a jackboot for the love of Lady Cunegonde, and if you had not been involved in the Inquisition, and had not wandered over America on foot, and lost all those sheep you brought from Eldorado, you would not be here eating candied fruit and pistachio nuts. "That's true enough," said Candide; "but we must go and work in the garden."

Let us follow Candide's advice and proceed to the task of exploring the choice of optimization. Perhaps we can then make our own meager parts of this "best of all possible worlds" a little better.

BIBLIOGRAPHY

ADE, GEORGE, *Fables in Slang* (Grosset & Dunlap, New York, 1899).

AESOP, *The Fables of Aesop*, by J. Jacobs (Macmillan, London, 1854).

ARCHIMEDES, cited in Cajori, p. 370.

BALL, W. W. R., *A Short Account of the History of Mathematics* (Macmillan, London, 1888).

BATEMAN, H., in *A Collection of Papers in Memory of Sir William Rowan Hamilton* (Scripta Mathematica, New York, 1945).

BELL, E. T., *The Development of Mathematics* (McGraw-Hill, New York, 1940).

BERNOULLI, JOHANN, cited in Cajori, p. 217.

CAJORI, F., *A History of Mathematics* (Macmillan, New York, 1919).

CONFUCIUS, *Analects*, Book XV, Chap. 31, J. Legge, trans., p. 169.

DAVIES, O. L., *The Design and Analysis of Industrial Experiments* (Oliver & Boyd, London, 1956).

DURUY, V., *History of Rome and of the Roman People*, Vol. I, W. J. Clarke, trans. (Estes and Lauriat, 1894), pp. 145–146.

EINSTEIN, ALBERT, *Sitzungber* (Acad. Wissenschaften, Berlin), **42** (1916), 1111–1116.

EULER, LEONHARD, cited in Polya, p. 121.

———, *Methodus inveniendi lineas curvas maximi minimive proprietate gaudentes* (1744), cited in Cajori, p. 232.

——— (1966), cited in Cajori, p. 251.

FERMAT, PIERRE DE, *Oeuvres* (1657), **1**, 170–173; **2**, 354, 457.

FRANKLIN, BENJAMIN, *Poor Richard's Almanack* (*The Sayings of Poor Richard*) (1733–1758), T. H. Russell, ed.

GALILEO, *Dialogues* (1630), cited by Bell, p. 351.

GAUSS, CARL FRIEDRICH, *Werke*, **4** (Göttingen, 1821), cited in Davies, p. 578.

——— (1829), cited in Bell.

GIBBS, J. WILLARD, "On the equilibrium of heterogenous substances," *Trans. Conn. Acad.*, **3** (October 1875–May 1876), 108–248; (May 1877–July 1878), 343–524.

GRIMAL, P., *Dictionnaire de la mythologie greque et romaine* (Presses Universitaires de France, Paris, 1963), p. 329.

HAMILTON, W. R., *Roy. Soc. Lond. Trans.* (1834), 247–308, cited in Bateman and Bell, p. 348.

———, *Roy. Soc. Lond. Trans.* (1835), 95–144, cited in Bateman and Bell, p. 348.

HEILBRONER, R. L., *The Worldly Philosophers* (Simon and Schuster, New York, 1953).

LAGRANGE, JOSEPH LOUIS, cited in Bell, p. 347.

LEGGE, J., *The Chinese Classics*, Vol. 1 (Trübner and Co., London, 1861).

LEIBNIZ, GOTTFRIED WILHELM, Letter to Father Bouvet (c. April 1702), cited in Wilhelm, p. 217.

———, *Theodicy: Essays on the Goodness of God, the Freedom of Man and the Origin of Evil* (1710), E. M. Huggard, trans. (Yale Univ. Press, New Haven, Conn., 1952), p. 128.

MAUPERTUIS, P. L. M., cited in Bell, p. 370.

POLYA, G., *Induction and Analogy in Mathematics*, Vol. I: *Of Mathematics and Plausible Reasoning* (Princeton Univ. Press, Princeton, N.J., 1954).

SCHRÖDINGER, E., *Annalen Physik*, s. 4, **79**, 361–376 (1926), cited in Bateman, p. 52.

SMITH, ADAM, *The Wealth of Nations* (Modern Library, New York, 1937); see also Heilbroner.

SYRUS, *Maxims* cited in *Hoyt's Quotations*.

VARRO, MARCUS TERENTIUS (116–27 B.C.), *De Lingua Latina* (Kent, London, 1938), pp. 64, 74.

VIRGIL, *Aeneid*, Vol. I, pp. 364–368, cited in Aris, p. 2.

VOLTAIRE, *Candide, or Optimism* (1759), John Butt, trans. (Penguin Books, Middlesex, England, 1947).

WILHELM, HELMUT, *Leibniz and the I-Ching* (1943), pp. 205–219.

Indirect Methods: The Differential Viewpoint

2

There are many paths to the top
of the mountain, but the view there
is always the same.

CHINESE MAXIM, QUOTED BY H.L. MENCKEN

The maxim expresses, in allegorical speech, the business of this chapter. For "paths to the top of the mountain," read "optimization methods"; for "view," substitute "mathematics"; for "there," insert "at the optimum." Then the statement becomes, "There are many optimization methods, but the mathematics at the optimum is always the same"—clumsily prosaic, but to the point.

Whatever route a mountain climber takes, he recognizes the peak when he arrives there. Equally important, he can tell when he is not at the top and must therefore continue his climb. These simple facts, intuitively evident where mountains are concerned, have mathematical analogs which provide not only a precise description of an optimum, but also a way to unify the entire subject of optimization theory. For the feature common to almost all optimization techniques is their continual progress by successive betterment. Since a well-conceived method will not overlook any possibilities for improvement, it cannot stop before reaching the optimum, which, by definition, is where further progress is impossible. This point of view exposes a strong resemblance between various methods which may at first glance appear quite different. Its unifying power permits us, in this single volume, to survey optimization in its entirety.

Our zeal for economy of thought must not, however, lead us to oversimplification. Despite strong similarities, the diverse optimization schemes have many intriguing differences and special characteristics. After all, differently shaped mountains require different climbing strategies, and one must base his approach not only on the obstacles in his path, but also on the climate surrounding the mountain and on the equipment at hand. Similarly, in selecting a specific optimization technique one should take account not only of the mathematical topography, with its computational glaciers, cliffs, and crevasses, but also of the intellectual climate surrounding the problem and the computing facilities available. After this chapter describes the similarities between the methods, the rest of the book will develop the differences.

The technical part of the chapter begins with a simple optimization problem involving a hypothetical manufacturing plant. As far as possible, all general concepts developed are demonstrated in concrete terms by application to this problem. Next, optima are described mathematically and distinctions made between local and global maxima, minima, suprema, and infima. After *feasibility* has been defined, a brief discussion of optimization by total enumeration clears the agenda for a survey of more efficient, and hence more interesting, methods. Ideas dating back over three centuries to Kepler and Fermat are developed for extension in future chapters to problems solved only since World War II. The unifying powers of the concepts of "solution," "decision," and "constrained derivative" permit straightforward derivation of most of the optimization methods described in the rest of the book. Thus we find simple mathematical expression for the unchanging view from the mountain top, where all climbers ultimately meet.

2-01 DESIGNING A HYPOTHETICAL MANUFACTURING PLANT

The generalities to be developed will, whenever possible, be illustrated by applying them to a particular decision problem involving design of a factory. To prevent our bogging down in technological and computational detail, the hypothetical plant will be oversimplified to the point of fiction. Imagine it intended to produce 10^7 lb/year of a certain chemical, using the five pieces of equipment shown in Fig. 2-1: a main compressor, a chemical reactor, a separator, a recirculating compressor, and a mixer. Raw-material gases are brought up to operating pressure x_1 (atmospheres), mixed with reused gas at the same pressure, and passed through a reactor where the gases are partly

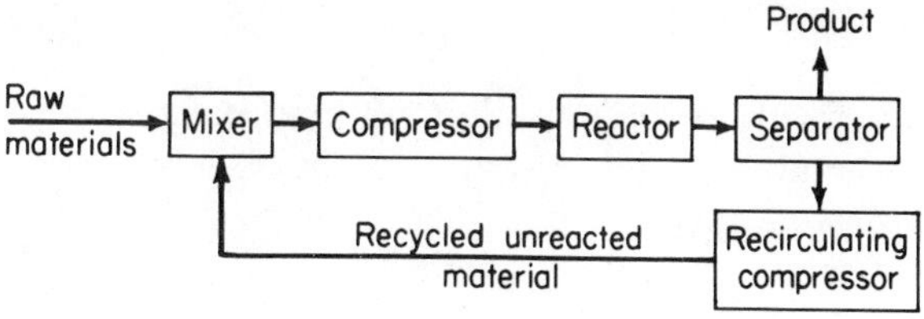

Figure 2-1. Hypothetical chemical plant.

converted into product. Then a separator removes the product for sale, leaving the unreacted gases to be sent back to the mixer by the recirculating compressor. The fraction converted to product in the reactor relates directly to the ratio of recirculated unreacted gas to raw material entering the process. This second process variable is called the *recycle ratio*, denoted x_2.

Now that the first step of the decision process—acquiring knowledge of the system—has been completed, it remains to develop a measure of effectiveness before optimization is possible. In this case we choose to minimize the total annual cost, including direct operating expenses, such as power cost, as well as capital expenditure amortized over the life of the process. Procedures for doing this are not obvious, and the reader interested in studying them would find any book on engineering economics valuable. Here we shall omit these important considerations and merely list, in Table 2-1, the costs as functions of the two operating variables for each piece of equipment. The *objective function*, y, to be minimized will be taken as the sum of all costs which depend on the process variables, x_1 and x_2.

$$y \equiv 1000x_1 + 4 \times 10^9 x_1^{-1} x_2^{-1} + 2.5 \times 10^5 x_2 \tag{2-1}$$

Notice that the objective function does not include the cost of the mixer because of its independence of x_1 and x_2. Although this fixed amount of \$$10^4$ affects the total cost, adding this quantity does not influence the optimal pressure x_1 and recycle ratio x_2, which is what we seek.

TABLE 2-1

Equipment Annual Costs for Hypothetical Plant

Equipment	*Annual cost* (\$)
Main compressor	$1000x_1$
Mixer	10^4
Reactor	$4 \times 10^9/[x_1x_2]$
Separator	10^5x_2
Recirculating compressor	1.5×10^5x_2

2-02 DEFINITIONS AND USEFUL RELATIONS

Before proceeding with the optimization, we must define precisely what we mean by it. In general the scalar *objective function*, y, depends on n real scalar *independent variables* $x_1, x_2, \ldots, x_n$, often assembled for abbreviation into an n-component column vector or point $\mathbf{x}$.

$$\mathbf{x} \equiv \begin{pmatrix} x_1 \\ x_2 \\ \cdot \\ \cdot \\ \cdot \\ x_n \end{pmatrix} = (x_1, x_2, \ldots, x_n)^T \tag{2-2}$$

In the example, $n = 2$. Vectors are always set in **boldface** type, and the superscript T denotes transposition. When particular numbers are assigned to the components of $\mathbf{x}$ the resulting vector will sometimes be called a *policy* or a *design*. The fact that y is a function depending on $\mathbf{x}$ often is expressed by writing $y(\mathbf{x})$.

In practice many conceivable policies are physically impossible, illegal, unsafe, or known in advance to be uneconomical. In the hypothetical chemical plant, negative values of pressure x_1 and recycle ratio x_2 are not physically possible, and let us assume that safety codes prohibit pressures higher than 2200 atm. Suppose, moreover, that industry practice rules out recycle ratios greater than 8. The remaining points, which satisfy the inequalities

$$0 \leq x_1 \leq 2200 \tag{2-3}$$

$$0 \leq x_2 \leq 8 \tag{2-4}$$

are said collectively to form the problem's *feasible region* $\mathfrak{F}$. This terminology is applicable to all problems. When, as in this case, all inequalities also admit the possibility of strict equality, $\mathfrak{F}$ is a *closed* region, since it contains all its boundary points [the line segments $(0, x_2)$, $(x_1, 0)$, $(2200, x_2)$, and $(x_1, 8)$ satisfying also Eqs. (2-3) and (2-4)]. If any of the inequalities are strict, which is really the situation here since absolute zero pressure is not achievable, then parts of the boundary of $\mathfrak{F}$ are outside $\mathfrak{F}$, and $\mathfrak{F}$ is said to be *open*. For reasons that will soon be apparent, problems should be formulated with closed feasible regions whenever possible.

Consider a particular point $\mathbf{x}^*$ in $\mathfrak{F}$ such that the value of the objective function there is less than at any other point in the region.

$$y(\mathbf{x}^*) < y(\mathbf{x}) \qquad \text{for all } \mathbf{x} \neq \mathbf{x}^* \text{ in } \mathfrak{F} \tag{2-5}$$

Then $y(\mathbf{x}^*)$, abbreviated y^* (read "y-star"), is the *minimum* of y, and $\mathbf{x}^*$, called the *minimizing* (or *minimal*) *policy*, is unique. There is more than one minimal policy, but still only one minimum, when Eq. (2-5) is replaced by

$$y(\mathbf{x}^*) \leq y(\mathbf{x}) \qquad \text{for all } \mathbf{x} \neq \mathbf{x}^* \text{ in } \mathfrak{F} \tag{2-6}$$

and Eq. (2-6) may be written

$$y^* \equiv \min_{\mathbf{x} \in \mathfrak{F}} [y(\mathbf{x})] \tag{2-7}$$

read "y^* is defined as the minimum of y for all feasible $\mathbf{x}$ ($\mathbf{x} \in \mathfrak{F}$ is read '$\mathbf{x}$ is an element of $\mathfrak{F}$')." When $\mathbf{x}^*$ is on the boundary of $\mathfrak{F}$, y^* is called a *boundary minimum;* otherwise, it is an *interior minimum.*

There are fairly ordinary situations in which no minimum exists. For example, suppose that one wishes to minimize the value of

$$y(x_1) = 2x_1 + 3$$

where $\mathfrak{F}$ is the *open* region defined by

$$0 < x_1 \leq 1$$

Then $y(0) < y(x_1)$ for all x_1 in $\mathfrak{F}$, but 0 itself is not in $\mathfrak{F}$. Furthermore, there is no point in $\mathfrak{F}$ where y is less than at every other point, for one can always find a better point by moving closer to the origin without actually reaching it.

To handle this situation, consider any point x_1^-, not necessarily in $\mathfrak{F}$, which gives a value of y lower than at any point in $\mathfrak{F}$.

$$y(x_1^-) < y(x_1) \qquad \text{for } x_1 \text{ in } \mathfrak{F} \tag{2-8}$$

Such a point x_1^- is called a *lower bound* for $y(x_1)$; let $\mathfrak{F}^-$ be the set of all such lower bounds, in this case all nonpositive real numbers. Now consider the *greatest lower bound* (glb), which is the point x_1^i such that

$$y(x_1^i) \geq y(x_1^-) \qquad \text{for } x_1^- \text{ in } \mathfrak{F}^- \tag{2-9}$$

In this example, $x_1^i = 0$, and in general $y(\mathbf{x}^i)$ is called the *infimum* of y, written

$$y(\mathbf{x}^i) \equiv \inf_{\mathbf{x} \in \mathfrak{F}} [y(\mathbf{x})] \equiv y^i \tag{2-10}$$

An infimum, or greatest lower bound, always exists when y is real, even though a minimum may not (Birkhoff and Mac Lane, p. 90).

To avoid worry about such fine points, use a closed region, for in this case a theorem of Weierstrass guarantees the existence of a minimum if the region is bounded and the objective function continuous. Thus, if $\mathfrak{F}$ is closed (strict equality possible), then

$$y^* \equiv \min_{\mathbf{x}} [y(\mathbf{x})] = \inf_{\mathbf{x}} [y(\mathbf{x})] \equiv y^i \tag{2-11}$$

Notice that "$\mathbf{x} \in \mathfrak{F}$" has been replaced by "$\mathbf{x}$" alone, which will be our practice from now on whenever it is clear that $\mathbf{x}$ must be feasible.

The *maximum* of y and the corresponding *maximizing* (or *maximal*) policy, unique or not, is defined in the same manner, except that all inequalities must be reversed. Asterisks are also used to designate these quantities, since the context of any given problem should make it clear whether maximization or minimization is desired. Thus we shall write

$$y(\mathbf{x}^*) \equiv y^* \equiv \max_{\mathbf{x}} [y(\mathbf{x})]$$

in maximization problems. The *least upper bound* (lub), called the *supremum*, is the maximization analog to the infimum, and Weierstrass's theorem holds for maxima as well as for minima. To cover optimization in general, we shall call y^* the *optimum* and $\mathbf{x}^*$ the *optimal policy*, writing

$$y(\mathbf{x}^*) \equiv y^* \equiv \operatorname*{opt}_{\mathbf{x}} [y(\mathbf{x})] \tag{2-12}$$

Consider now a feasible point $\mathbf{x}^*$ and the set of points $\mathbf{x}$ in a *feasible neighborhood* of $\mathbf{x}^*$. These are points satisfying not only the inequalities describing $\mathfrak{F}$, but also the additional condition

$$0 < |\mathbf{x} - \mathbf{x}^*| < \epsilon \tag{2-13}$$

where

$$|\mathbf{x} - \mathbf{x}^*| \equiv \left[\sum_{j=1}^{n} (x_j - x_j^*)^2\right]^{1/2} \tag{2-14}$$

Thus ϵ is the radius of an n-dimensional spherical open region centered at $\mathbf{x}^*$ and containing the points $\mathbf{x}$ and $\mathbf{x}^*$. The feasible neighborhood will be denoted $\mathfrak{N}$. If there exists a feasible neighborhood containing $\mathbf{x}^*$ such that

$$y(\mathbf{x}^*) \leq y(\mathbf{x}) \qquad \text{for } \mathbf{x} \text{ in } \mathfrak{N} \tag{2-15}$$

then $y(\mathbf{x}^*)$ is called a *local minimum* and $\mathbf{x}^*$ a *locally minimum policy*. We shall write

$$y^* \equiv y(\mathbf{x}^*) \equiv \operatorname*{lmin}_{\mathbf{x}} [y(\mathbf{x})] \tag{2-16}$$

The notation "$\operatorname*{lmin}_{\mathbf{x}}$" means that $\mathbf{x}$ is restricted to a feasible neighborhood of $\mathbf{x}^*$. This new notation is needed to preserve the important distinction between the global minimum, which is what we usually want, and a local minimum, which by present optimization techniques is what we usually get. The concepts of local infimum, maximum, supremum, and optimum are derived in a similar manner, using the notion of feasible neighborhood. When employing existing optimization methods, one must prove (or hopefully assume) in advance that there is only one peak, or at least that one is not climbing the wrong mountain. An objective function with a unique local, and hence global, optimum is said to be *unimodal.*

Certain elementary relations between maxima and minima are of interest. If b is positive, and a arbitrary, then if $\mathbf{x}^*$ minimizes $y(\mathbf{x})$, it also minimizes $a + by(\mathbf{x})$, and

$$a + by(\mathbf{x}^*) = \min_{\mathbf{x}} [a + by(\mathbf{x})] \tag{2-17}$$

Moreover, it maximizes $a - by(\mathbf{x})$, and

$$a - by(\mathbf{x}^*) = \max_{\mathbf{x}} [a - by(\mathbf{x})] \tag{2-18}$$

Proofs, which involve manipulation of the defining inequalities, are left as exercises (Exercise 2-1). The mathematical moral is twofold: (1) that additive constants and positive factors do not affect the location of the optimum; and (2) that minimization techniques can be used in maximization problems (and vice versa) simply by changing the sign of the objective function.

2-03 EXHAUSTIVE ENUMERATION

It may be that but a finite number of policies need be considered. Suppose in the hypothetical chemical plant that only designs with pressures in integral numbers of atmospheres 1, 2, . . . , 2199, 2200, and tenths of a recycle ratio 0.1, 0.2, . . . , 7.9, 8.0 are acceptable. Then there are in all (2200)(80) = 176,000 possibilities, a finite, if large, number. The method, if it can be called one, of *exhaustive enumeration*, is simply to evaluate the objective function for every case and pick out the optimum directly. Theoretically simple but practically tedious, this technique, also called informally the *brute force approach*, is appropriate only when the number of cases is small compared to the speed of the computation facilities available. Even with today's high-speed computers, one is justified in asking for better procedures than this.

2-04 THE DIFFERENTIAL APPROACH

Fortunately, there are methods for finding a peak without mapping the entire mountain. The first inkling that there might be a better way came at the beginning of the seventeenth century, when Johannes Kepler noticed in the midst of his astronomical calculations that differences between successive values of a dependent variable, computed at equally spaced values of the independent variable, tended to vanish near an optimum. A generation later, Pierre de Fermat developed this hint into a method for finding interior optima of continuous functions of a single variable. Since our approach resembles his in some ways, it is instructive to demonstrate its practical strength and logical weakness with a simple example.

Suppose in the hypothetical chemical plant that the recycle ratio x_2 is fixed at unity, defining a new objective function y_1 of a single variable x_1.

$$y_1 \equiv 1000x_1 + 4 \times 10^9 x_1^{-1} + 2.5 \times 10^5$$

Fermat would seek the point x_1^* where the value of the objective function is the same as at a nearby point $x_1^* + \partial x_1$, where ∂x_1 is a small, and for Fermat, ill-defined, quantity known in the seventeenth century as a *virtual displacement* and in the eighteenth as a *differential*. It should be read "differential x." Equating the two values and performing the customary algebraic simplifications gives

$$y(x^* + \partial x_1) = y(x_1^*)$$

$$= \begin{cases} 1000(x_1^* + \partial x_1) + 4 \times 10^9 (x_1^* + \partial x_1)^{-1} + 2.5 \times 10^5 \\ 1000x_1^* + 4 \times 10^9 (x_1^*)^{-1} + 2.5 \times 10^6 \\ 1000(x_1^*)^2 \partial x_1 + 1000x_1^*(\partial x_1)^2 = 4 \times 10^9 \partial x_1 \end{cases} \tag{2-19}$$

Here Fermat canceled a factor ∂x_1 from each side of Eq. (2-19), a permissible operation only as long as ∂x_1 is not zero.

$$1000(x_1^*)^2 + 1000x_1^*(\partial x_1) = 4 \times 10^9 \tag{2-20}$$

His next move was completely illogical; he set the differential to zero, obtaining by accident the correct equation for the location of the optimum.

$$1000(x_1^*)^2 = 4 \times 10^9 \tag{2-21}$$

whence

$$x_1^* = \pm 2000 \tag{2-22}$$

In the hypothetical plant only the positive root is feasible, since only positive pressures are physically meaningful.

Before developing the advantages of this approach, namely that it is simple and usually gives correct answers, let us emphasize its logical inconsistencies and show how to avoid them. Fermat's method immediately aroused the suspicions of the clear-thinking Descartes, mainly because of the vagueness of its original description. The logical flaw, which remained concealed when Newton, in 1669, and Leibniz, in 1675, used differentials in their invention of the infinitesimal calculus, was finally brought to light by an Irish clergyman-philosopher with a theological ax to grind. George Berkeley, Bishop of Cloyne, was annoyed by mathematicians who had criticized certain contradictions in the religious dogma of the times. His revenge was sweet; in his *Discourse addressed to an infidel mathematician* he exposed the inconsistency in having the differential ∂x_1, assumed nonzero in the beginning of the derivation, vanish conveniently as soon as it gets in the way. Berkeley's taunt drove the eighteenth-century mathematicians to seek a rigorous basis for the calculus, but it was not until 1821, when Cauchy defined the concept of limit, that the contradiction was finally removed. The correct way to pass from Eq. (2-20) to Eq. (2-21) is to remark that as the differential ∂x_1 *approaches* zero, the term $1000x_1^*(\partial x_1)$ approaches zero *as a limit*.

$$\lim_{\partial x_1 \to 0} [1000x_1^*(\partial x_1)] = 0 \tag{2-23}$$

In any rigorous demonstration this must be proved, not assumed, a task not always as easy as in this simple example.

Despite Bishop Berkeley's rebuke, arguments involving differentials—what we shall call the *differential approach*—were used with great success throughout the eighteenth century. Euler manipulated differentials with no qualms whatever, extending Fermat's method to functions of many variables, while deriving most of the formulas of modern calculus. Later (1799) Lagrange tried to put the calculus on a rigorous basis using series expansions, a project that contributed greatly to optimization theory, even though it did not achieve its original goal. This ancient, although not always honorable, differential approach will, because of its clarity and simplicity, be our guide through the maze of optimization theory. Almost all optimization methods, despite the diverse mathematics of their inventors, can be analyzed successfully from the differential point of view. By using it we decrease the amount of mathematics needed to master the subject and bring optimization theory within the understanding of anyone who has studied the differential (not even integral) calculus. This includes undergraduate students of engineering, economics, and the physical, mathematical, and social sciences, for whom this book is written. But we would be derelict in our duties if we used the differential approach blindly, unaware of its dangers. Therefore, we provide references to formal proofs in the literature substantiating any results derived by manipulating differentials. And to instill a sense of caution in the reader, we shall also describe incorrect results derived by differential methods insufficiently tempered with rigor. Not even the great Lagrange was immune to subtle errors of this sort, and they are still being committed today.

2-05 THE CLASSICAL INDIRECT METHOD

Optimization techniques can, for convenience, be divided into two classes: direct and indirect methods (Edelbaum). Direct methods, which start at an arbitrary point and proceed stepwise toward the peak by successive improvement, will be described later, leaving the rest of the chapter for development of the indirect methods. Fermat's classical method is indirect because it ultimately involves solving an equation rather than searching for an optimum. This works because the root of the equation is also the location of the optimum. Because they often pick out an optimum without examining any nonoptimal points, indirect methods are very effective when they can be applied.

Since Fermat's method does not generalize easily to functions of several variables, it behooves us to introduce the more advanced notion of *derivative.* In modern terminology, the *first derivative of y with respect to x_1*, designated $\partial y/\partial x_1$, is defined by

$$\frac{\partial y}{\partial x_1} \equiv \lim_{\partial x_1 \to 0} \left[\frac{y(x_1 + \partial x_1) - y(x_1)}{\partial x_1}\right] \tag{2-24}$$

Actually it is more common to write the derivative as dy/dx_1 when there is only one independent variable, the notation $\partial y/\partial x_1$ being reserved for multivariable situations. Here we depart from this tradition, dating back to Leibniz, in order to save the letter d for later use as a mnemonic symbol for a *decision.* In the example, the first derivative, interpreted geometrically as the slope of the curve of y versus x_1, depends on x_1 and is given by

$$\frac{\partial y}{\partial x_1} = 1000 - 4 \times 10^9 x_1^{-2} \tag{2-25}$$

In 1714 the elder Johann Bernoulli remarked that the first derivative vanishes at an optimum

$$\left(\frac{\partial y}{\partial x_1}\right)^* = 0 \tag{2-26}$$

where the asterisk indicates that the derivative is evaluated at $x_1 = x_1^*$. This is, in the notation of the differential calculus, the observation made over a century earlier by Kepler. The reader can verify in the example that combination of Eqs. (2-25) and (2-26) gives $x_1^* = \pm 2000$, confirming Fermat's result.

When there is more than one independent variable the appropriate generalization is the set of *n first derivatives* $\partial y/\partial x_j$ $(j = 1, 2, \ldots, n)$, each defined as the function obtained by differentiation with respect to x_j alone, all other independent variables being held constant.

$$\frac{\partial y}{\partial x_j} \equiv \lim_{\partial x_j \to 0} \left[\frac{y(x_1, \ldots, x_{j-1}, x_j + \partial x_j, x_{j+1}, \ldots, x_n) - y(\mathbf{x})}{\partial x_j} \right] \tag{2-27}$$

In the original example involving two independent variables, the first partial derivatives at any point (x_1, x_2) are

$$\frac{\partial y}{\partial x_1} = 1000 - 4 \times 10^9 (x_1)^{-2}(x_2)^{-1} \tag{2-28}$$

$$\frac{\partial y}{\partial x_2} = 2.5 \times 10^5 - 4 \times 10^9 (x_1)^{-1}(x_2)^{-2} \tag{2-29}$$

Taylor showed that if $\partial \mathbf{x}$ is a column vector of small displacements

$$\partial \mathbf{x} \equiv (\partial x_1, \ldots, \partial x_n)^T \tag{2-30}$$

then the difference between $y(\mathbf{x} + \partial \mathbf{x})$ and $y(\mathbf{x})$, abbreviated ∂y, is given at any point $\mathbf{x}$ by the series expansion

$$\begin{aligned} \partial y &\equiv y(\mathbf{x} + \partial \mathbf{x}) - y(\mathbf{x}) \\ &= \left(\frac{\partial y}{\partial x_1}\right) \partial x_1 + \cdots + \left(\frac{\partial y}{\partial x_n}\right) \partial x_n + 0(\partial \mathbf{x}^2) \\ &= \sum_{j=1}^{n} \left(\frac{\partial y}{\partial x_j}\right) \partial x_j + 0(\partial \mathbf{x}^2) \end{aligned} \tag{2-31}$$

where $0(\partial \mathbf{x}^2)$ represents an infinity of terms, each involving products of at least two differentials. Lagrange, who was the first to write the remainder as $0(\partial \mathbf{x}^2)$, neglected it in the computations to follow; Euler ignored it entirely. A rigorous proof would need to show that as ∂x_j approaches zero, the ratio $0(\partial \mathbf{x}^2)/\partial x_j$ vanishes in the limit. Let us assume this has been done and consider an analogous expression involving *finite* displacements Δx_j.

$$\Delta y = \sum_{j=1}^{n} \left(\frac{\partial y}{\partial x_j}\right) \Delta x_j \tag{2-32}$$

This equation may be written in abbreviated vector (or inner) product form as

$$\Delta y = \nabla y \, \Delta \mathbf{x} \tag{2-33}$$

for finite variations, or

$$\partial y = \nabla y \, \partial \mathbf{x} \tag{2-34}$$

for infinitesimal ones. Here ∇y is the row vector of first partial derivatives

$$\nabla y \equiv \left(\frac{\partial y}{\partial x_1}, \ldots, \frac{\partial y}{\partial x_n}\right) = \frac{\partial y}{\partial \mathbf{x}} \equiv y_x \tag{2-35}$$

called the *gradient of y* for reasons to be made clear in Chapter 4.

Conditions that must hold at an interior local optimum $\mathbf{x}^*$ are derived by combining its definition with the Taylor expansion in any open feasible neighborhood $\mathfrak{N}$ containing $\mathbf{x}^*$. Suppose, to be definite, that $\mathbf{x}^*$ is a local *minimum*. Then in the limit

$$\sum_{j=1}^{n} \left(\frac{\partial y}{\partial x_j}\right)^* \partial x_j = \partial y = y(\mathbf{x}^* + \partial \mathbf{x}) - y(\mathbf{x}^*) \geq 0 \tag{2-36}$$

for all possible perturbations ∂x_j. This implies that every partial derivative must vanish

$$\left(\frac{\partial y}{\partial x_j}\right)^* = 0; \qquad j = 1, \ldots, n \tag{2-37}$$

for if on the contrary any $(\partial y/\partial x_j)^* \neq 0$, then any perturbation $\partial x_j'$ with sign opposite from that of $(\partial y/\partial x_j)^*$ would give a negative total change

$$\left(\frac{\partial y}{\partial x_j}\right)^* \partial x_j' < 0$$

Holding all other perturbations at zero would generate a better point in every feasible neighborhood of y^*, contradicting the assumption that y^* was a local minimum. The same form of argument can be used to prove that Eq. (2-37) must also hold at a local maximum. In summary, a necessary condition that $\mathbf{x}^*$ be optimum is that the gradient vanish there.

$$\nabla y(\mathbf{x}^*) \equiv \nabla y^* = \mathbf{0} \tag{2-38}$$

where $\mathbf{0}$ is the n-component null vector. In geometric terms, the tangent plane is horizontal at an optimum.

Johann Bernoulli guessed this result while settling a dispute between Huygens and Chevalier Renau (chief marine engineer to Louis XIV) on the best sail angle and tack for ships, given the direction of the wind relative to the course. But in his zeal to show that a landlocked Swiss mathematician could solve sailing problems too subtle for colleagues in seafaring nations, he did not bother to prove his results. The methods described are therefore attributed to Euler, even though the latter's "proof" is marred by his unrigorous casting away of the higher-order differentials.

Euler's classical indirect method, applied to the two-variable example, would be to solve the nonlinear equations resulting when the algebraic expressions for the first partial derivatives are equated to zero.

$$1000 - 4 \times 10^9 (x_1^*)^{-2}(x_2^*)^{-1} = 0 \qquad (2\text{-}39)$$

$$2.5 \times 10^5 - 4 \times 10^9 (x_1^*)^{-1}(x_2^*)^{-2} = 0 \qquad (2\text{-}40)$$

The reader can verify that the only feasible solution is $x_1^* = 1000$, $x_2^* = 4$. Thus the classical indirect method reduces the original optimization problem to the solution of simultaneous equations, usually nonlinear. If the optimum is not interior, or the equations are too difficult to solve, other methods must be employed. Contours of the function are shown in Fig. 2-2.

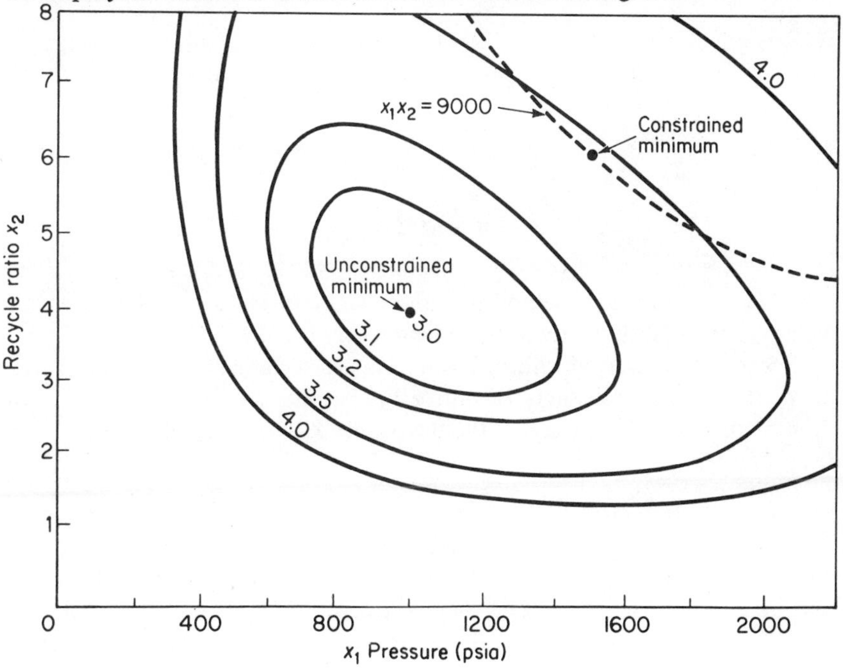

Figure 2-2. Objective function contours.

2-06 SOLVING NONLINEAR EQUATIONS

Since the classical indirect method eventually involves solving simultaneous nonlinear equations, let us consider the well-known method conceived by Newton and refined by Raphson for doing just this. Let the first partial derivatives, which are scalar functions of the vector $\mathbf{x}$, be abbreviated by

$$\frac{\partial y}{\partial x_j} \equiv y_j'(\mathbf{x}); \qquad j = 1, \ldots, n \qquad (2\text{-}41)$$

so that Eq. (2-38) becomes

$$y'_j(\mathbf{x}^*) = 0 \tag{2-42}$$

Let $\mathbf{x}_k$ be a trial point, and let $(\partial y'_j/\partial x_p)_k$ be the n first partial derivatives of the functions y'_j, evaluated at the point $\mathbf{x}_k$. These, of course, are *second* partial derivatives of the original objective function y, and they can also be abbreviated

$$\left(\frac{\partial y'_j}{\partial x_p}\right)_k \equiv \left(\frac{\partial^2 y}{\partial x_j\, \partial x_p}\right)_k \equiv y''_{jp}(\mathbf{x}_k) \tag{2-43}$$

For the hypothetical chemical plant, the second partial derivatives are, at $\mathbf{x}^*$,

$$\frac{\partial^2 y}{\partial x_1^2} \equiv y''_{11} = 8 \times 10^9 (x_1^*)^{-3}(x_2^*)^{-1} = 2$$

$$\frac{\partial^2 y}{\partial x_1\, \partial x_2} = y''_{12} = y''_{21} = 4 \times 10^9 (x_1^* x_2^*)^{-2} = 250$$

$$\frac{\partial^2 y}{\partial x_2^2} = y''_{22} = 8 \times 10^9 (x_1^*)^{-1}(x_2^*)^{-3} = 125{,}000$$

The Newton–Raphson method uses these second derivatives to estimate the values of the first derivatives in the neighborhood of $\mathbf{x}_k$. Then the new point $\mathbf{x}_{k+1}$ where the predicted values all vanish is selected as a better approximation to $\mathbf{x}^*$. The procedure is iterated until all y'_j become acceptably small. Expansion of the y'_j in Taylor series about $\mathbf{x}_k$ gives $\mathbf{x}_{k+1}$ as the solution to

$$y'_j(\mathbf{x}_{k+1}) = 0 = y'_j(\mathbf{x}_k) + \sum_{p=1}^{n} y''_{jp}(\mathbf{x}_k)(x_{p,k+1} - x_{pk}) \tag{2-44}$$

Since the $y'_j(\mathbf{x}_k)$ and $y''_{jp}(\mathbf{x}_k)$ are all known, these n linear equations have only the n coordinates $x_{p,k+1}$ of $\mathbf{x}_{k+1}$ as unknowns.

It is convenient to express all this in matrix notation, which is introduced here formally without any theoretical justification (see Birkhoff and Mac Lane, Chaps. 7–10, for a deeper treatment of matrices). Let the n^2 second partial derivatives be arranged in an array with n rows and n columns called the *Hessian matrix* $\mathbf{H}$.

$$\mathbf{H}(\mathbf{x}_k) \equiv \mathbf{H}_k = \begin{pmatrix} y''_{11} & y''_{12} & \cdots & y''_{1n} \\ y''_{21} & y''_{22} & \cdots & y''_{2n} \\ \cdot & \cdot & & \cdot \\ \cdot & \cdot & & \cdot \\ \cdot & \cdot & & \cdot \\ y''_{n1} & y''_{n2} & \cdots & y''_{nn} \end{pmatrix} \equiv \frac{\partial^2 y}{\partial \mathbf{x}^2} \equiv y_{xx} \tag{2-45}$$

By the rules of matrix multiplication, Eq. (2-44) can be written, after transposition of terms involving $\mathbf{x}_{k+1}$ to the left and changing signs,

$$\mathbf{H}_k \mathbf{x}_{k+1} = \mathbf{H}_k \mathbf{x}_k - (\nabla y_k)^T \tag{2-46}$$

where $\nabla y_k \equiv \nabla y(\mathbf{x}_k)$ is the gradient computed at $\mathbf{x}_k$. If the n linear equations in the n unknown components of $\mathbf{x}_{k+1}$ are linearly independent, then the solution to Eq. (2-46) is unique and can be represented formally by

$$\mathbf{x}_{k+1} = \mathbf{x}_k - \mathbf{H}_k^{-1}(\nabla y_k)^T = \mathbf{x}_k - y_{xx}^{-1} y_x^T \tag{2-47}$$

Here $\mathbf{H}_k^{-1}$ is the unique n by n *inverse matrix*, which has the property that

$$\mathbf{H}_k^{-1}\mathbf{H}_k = \mathbf{H}_k \mathbf{H}_k^{-1} = \mathbf{I} \tag{2-48}$$

where $\mathbf{I}$ is the unit diagonal matrix having diagonal elements unity and all others zero. Equation (2-47) is obtained by multiplying all vectors in Eq. (2-46) on the left by $\mathbf{H}_k^{-1}$ and using Eq. (2-48) for simplification. The significance of all this is not yet clear; suffice it to say that it will be useful in the future to have the matrices $\mathbf{H}_k$ and $\mathbf{H}_k^{-1}$, which we have seen must be calculated anyway when the Newton–Raphson method is employed.

The Newton–Raphson method, when it converges at all, may not find the proper root. Consider Fig. 2-3, which plots y_1' as a function of the single variable x_1. The Hessian "matrix" is in this case a single number equal to the slope y_{11}'' of the curve; its "inverse" is $1/y_{11}''$. The set of points 1, 2, 3, 4, demonstrates how the method works, or, for this unfortunate start, does not work. The reader may verify that any search starting to the left of **b** will always

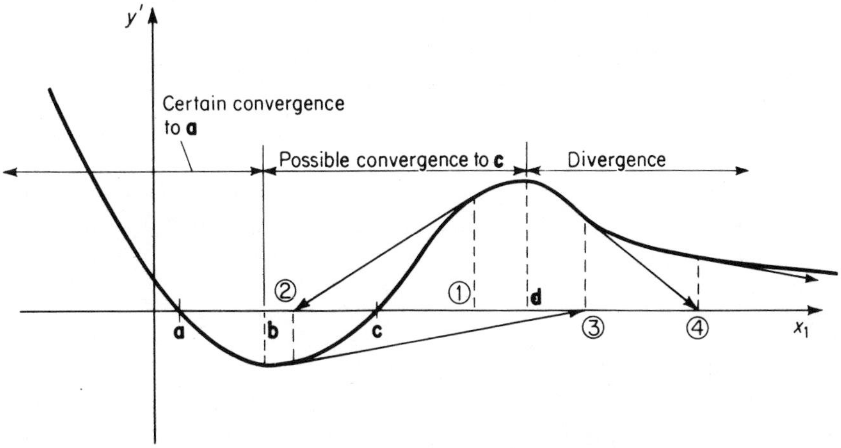

Figure 2-3. Newton–Raphson procedure.

converge to the root at **a**. Starts to the right of **d** never converge, while between **b** and **d** a search converges either to **c** or to **a**, or else diverges.

2-07 STATIONARY POINTS

The vanishing of the first derivatives is a necessary, but by no means sufficient condition for a point to be a local minimum, for the tangent is also horizontal at a maximum and at an inflection point or a flat place. Any point where the gradient vanishes, Eq. (2-38), is called a *stationary point*, and we shall now define the various kinds so that minima can be distinguished among them.

Figure 2-4 displays a minimum of a function of one or two variables; by changing the sign of the objective y the points become maxima. A set of adjacent local minima (maxima), shown in Fig. 2-5, forms a *valley* (*ridge*), a situation that arises when the equality sign in Eq. (2-17) is satisfied for some $\mathbf{x} \neq \mathbf{x}^*$ in the neighborhood of $\mathbf{x}^*$. Figure 2-6 shows *saddle points* $\mathbf{x}^0$ where Eq. (2-17) is violated for some points $\mathbf{x}$ in every feasible neighborhood of the stationary point $\mathbf{x}^0$. Henceforth, any stationary point will be denoted $\mathbf{x}^0$ until it has been definitely proved to be an optimum, when it can be positively identified by $\mathbf{x}^*$. We seek then *sufficient* conditions for a minimum.

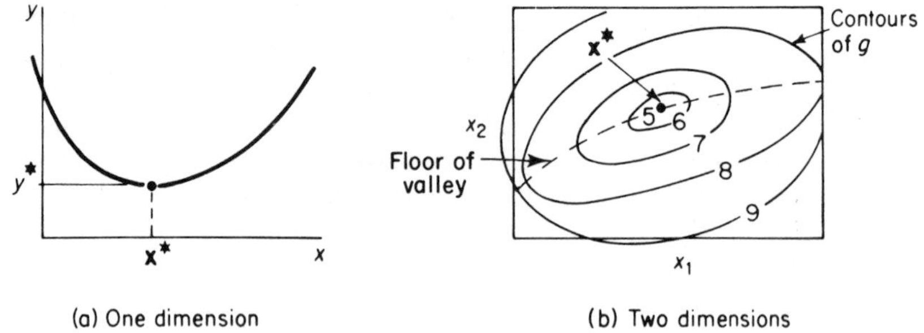

Figure 2-4. Minima.

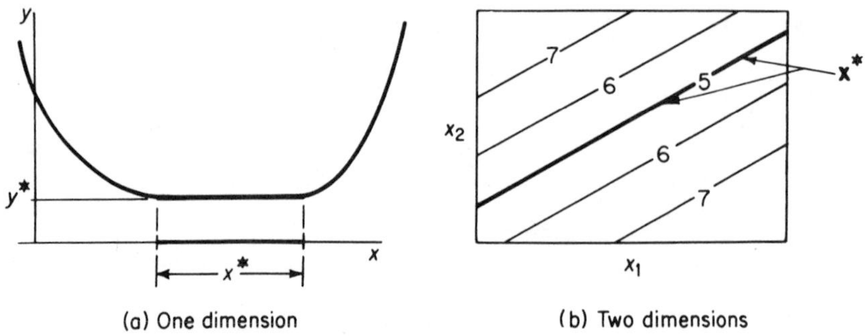

Figure 2-5. Valleys.

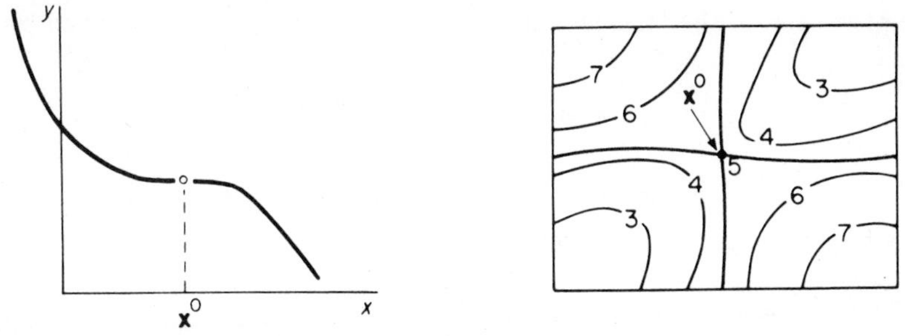

(a) One dimension (b) Two dimensions

Figure 2-6. Saddle points.

Maclaurin first gave rules for distinguishing between various kinds of stationary points in one dimension, and Lagrange extended his ideas to the multidimensional case. Let $y(\mathbf{x})$ be expanded about $\mathbf{x}^0$ in a Taylor series exhibiting the second-order terms explicitly.

$$
\begin{aligned}
\partial y^0 &\equiv y(\mathbf{x}^0 + \partial\mathbf{x}) - y(\mathbf{x}^0) \\
&\equiv \sum_{j=1}^{n} \left(\frac{\partial y}{\partial x_j}\right)^0 \partial x_j + \frac{1}{2} \sum_{j=1}^{n} \sum_{k=1}^{n} \left(\frac{\partial^2 y}{\partial x_j\, \partial x_k}\right)^0 \partial x_j\, \partial x_k + 0(\partial x^3)
\end{aligned} \tag{2-49}
$$

or more compactly in terms of the gradient vector ∇y and Hessian matrix $\mathbf{H}$,

$$\partial y^0 = \nabla y^0(\partial\mathbf{x}) + \tfrac{1}{2}(\partial\mathbf{x})^T\mathbf{H}^0(\partial\mathbf{x}) + 0(\partial x^3) \tag{2-50}$$

Here the superscript zero reminds us that all quantities are evaluated at the stationary point, and $0(\partial x^3)$ represents terms in the expansion of degree three and higher. For all $\partial\mathbf{x}$ such that

$$(\partial\mathbf{x})^T\mathbf{H}^0(\partial\mathbf{x}) \neq 0 \tag{2-51}$$

it is true that

$$\lim_{\partial x \to 0} \left[\frac{0(\partial x^3)}{(\partial\mathbf{x})^T\mathbf{H}^0(\partial\mathbf{x})}\right] = 0 \tag{2-52}$$

so that $0(\partial x^3)$ will be neglected in subsequent arguments when Eq. (2-51) holds. By Eq. (2-38) the gradient vanishes, and Eq. (2-50) simplifies to

$$\partial y^0 = \tfrac{1}{2}(\partial\mathbf{x})^T\mathbf{H}^0(\partial\mathbf{x}) \tag{2-53}$$

Combination of this with previous definitions involving ∂y gives the conditions for identifying the various kinds of stationary points in terms of calculable properties of the Hessian matrix $\mathbf{H}^0$, which in the example is given by

$$\mathbf{H}(\mathbf{x}^0) = \begin{pmatrix} 2 & 250 \\ 250 & 125{,}000 \end{pmatrix} \tag{2-54}$$

The scalar $(\partial\mathbf{x})^T\mathbf{H}^0(\partial\mathbf{x})$ is called a *differential quadratic form*, and it may be written in terms of the second partial derivatives of y as

$$(\partial\mathbf{x})^T\mathbf{H}(\partial\mathbf{x}) = \sum_{j=1}^{n}\sum_{k=1}^{n} y''_{jk}\,\partial x_j\,\partial x_k \tag{2-55}$$

In the example

$$\partial y^0 = (\partial x_1)^2 + 250(\partial x_1)(\partial x_2) + 62{,}500(\partial x_2)^2 \tag{2-56}$$

If this quadratic form is positive for all possible $\partial\mathbf{x}(\neq \mathbf{0})$, then the form (and $\mathbf{H}$) is said to be *positive-definite;* this is the sufficient condition for $\mathbf{x}^0$ to be a local minimum. At local maxima $\mathbf{H}$ is *negative-definite.* In valleys (ridges) the form is nonnegative (nonpositive), vanishing for $\partial\mathbf{x}$ along the valley (ridge), in which case $\mathbf{H}$ is said to be *semidefinite.* At saddlepoints, the quadratic form can be positive, negative, or zero, depending on the value of $\partial\mathbf{x}$ chosen; $\mathbf{H}^0$ is called *indefinite* under these circumstances. It remains only to show how to find perturbations $\partial\mathbf{x}$ which improve y near $\mathbf{x}^0$. If there are none, the point is an optimum.

In the example this can be done most easily by Lagrange's method of completing the square (1759). Thus, by adding and subtracting $(125\partial x_2)^2$ to Eq. (2-56) and rearranging, one obtains a sum of two perfect squares.

$$\begin{aligned}\partial y^0 &= [(\partial x_1)^2 + 250(\partial x_1)(\partial x_2) + 15{,}625(\partial x_2)^2] \\ &\quad + (62{,}500 - 15{,}625)(\partial x_2)^2 \\ &= (\partial x_1 + 125\partial x_2)^2 + 46{,}875(\partial x_2)^2\end{aligned} \tag{2-57}$$

Since this must be positive for every value of $\partial\mathbf{x} \neq \mathbf{0}$, we are assured that the point is a local minimum. If the signs of the squares had all been negative, the point would have been identified as a maximum. Similarly, if any signs are different, $\mathbf{x}^0$ is a saddle. For instance, if $\partial y^0 = (\partial x_1 + 125\partial x_2)^2 - 46{,}875(\partial x_2)^2$, then $\partial y^0 > 0$ as long as $\partial x_2 = 0$, and $\partial y^0 < 0$ whenever $\partial x_1 = -125\partial x_2$. The next three sections generalize this procedure to any number of variables.

If any of the coefficients of the squares vanish, more powerful methods are needed to test the character of the stationary point. Suppose, for example, that $\partial y^0 = (\partial x_1 + 125\partial x_2)^2 + 0(\partial x_2)^2$. Then for any $\partial x_2 \neq 0$ and $\partial x_1 = -125\partial x_2$, $\partial y^0 = 0$. Since for any other choice of ∂x_1, ∂y^0 is strictly positive, one might jump to the conclusion that $\mathbf{x}^0$ is in a valley. Although y_{xx} must be semidefinite whenever $\mathbf{x}^0$ is in a valley, the converse is not always true, since the valley itself may curve up or down. The vanishing of the second-order terms in these directions means that the third- and higher-order terms cannot be neglected, for if Eq. (2-51) does not hold, the limit in Eq. (2-52) may not be zero. Hence the theory developed so far does not give positive identification of a stationary point when the Hessian matrix is semidefinite.

Things are simple when there is but one independent variable, for if the second derivative $\partial^2 y/\partial x_1^2$ (the one-dimensional Hessian) vanishes, the third derivative must also be zero if $\mathbf{x}^0$ is an optimum. This leaves the fourth derivative; if positive, $\mathbf{x}^0$ is a minimum; if negative, $\mathbf{x}^0$ is a maximum; if zero, the question is still unsettled and one must examine the fifth derivative; and so on. Thus, in 1742, Maclaurin could assert with truth that $\mathbf{x}^0$ is a minimum (maximum) if and only if the lowest order nonvanishing derivative is positive (negative) and of even order. In 1797 Lagrange propounded an analogous principle for multivariable functions; namely that if the lowest-order nonvanishing differentials are positive (negative)-definite and of even order, then $\mathbf{x}^0$ is a minimum (maximum). This myth was not exploded for almost a century, when Peano (in Genocchi and Peano) exhibited an elementary counterexample (see Exercise 2-4) having a *curved* valley. Lagrange's spurious principle, based on an improper discarding of higher-order differentials, died hard because of its plausibility, but a rigorous and more difficult theory due to Scheefer and Stolz has finally replaced it. (Since such anomalous situations do not arise enough in practice to justify exposition of the Scheefer–Stolz theory here, the interested reader should consult Hancock's book for a complete discussion.) Let us take Lagrange's error as a warning of the dangers lurking in nonlinear problems, especially when there are several independent variables. Sections 2-12 and 2-14 treat such questions in more detail.

2-08 DIAGONALIZATION

Consider the quadratic form

$$q \equiv \mathbf{x}'\mathbf{Q}\mathbf{x} = \sum_{j=1}^{N} \sum_{n=1}^{N} x_j q_{jn} x_n \tag{2-58}$$

where $\mathbf{x}$ is an N element vector, and $\mathbf{Q} \equiv (q_{jn})$ is an N by N symmetric nonsingular matrix. There exist nonsingular linear transformations (Birkhoff and Mac Lane, pp. 248–250) such that, in terms of the N variables z_h, the quadratic form

$$\hat{q} = \sum_{h=1}^{N} d_{hh} z_h^2 \tag{2-59}$$

becomes a sum of perfect squares. Let $\mathbf{T}$ be the N by N matrix of such a transformation, and let $\mathbf{D}$ be the N by N matrix associated with the new quadratic form $\hat{q}$. Then

$$\mathbf{z} = \mathbf{T}\mathbf{x} \tag{2-60}$$

and since $\mathbf{T}$ is nonsingular,

$$\mathbf{x} = \mathbf{T}^{-1}\mathbf{z} \tag{2-61}$$

Equations (2-58) and (2-61), together with the definition of $\mathbf{D}$, give

$$\hat{q} = \mathbf{z}'(\mathbf{T}^{-1})'\mathbf{Q}\mathbf{T}^{-1}\mathbf{z} = \mathbf{z}'\mathbf{D}\mathbf{z} \tag{2-62}$$

Since $\hat{q}$ is a sum of perfect squares, it follows that $\mathbf{D}$ is a *diagonal* matrix of elements d_{hh}, and therefore this transformation process is called the *diagonalization* of the coefficient matrix $\mathbf{Q}$.

Diagonalization is widely used by operations analysts to simplify systems arising in optimization, control, and statistics problems. The standard diagonalization technique consists of finding the eigenvalues and eigenvectors of $\mathbf{Q}$, reducing it to canonical form (Birkhoff and Mac Lane, pp. 266–268). This *congruence* transformation is appealing theoretically because of its uniqueness and orthogonality, but finding it requires factoring an Nth-degree polynomial followed by solving N *sets* of N linear equations in N unknowns. All this labor is necessary in certain important problems in vibrations, fluid mechanics, and quantum theory where only the congruence transformation is physically meaningful. But most operations analysis problems can be diagonalized with much less effort by using a *similarity* rather than a congruence transformation.

Lagrange's transformation, which involves completing the square, is described in many algebra texts (Birkhoff and Mac Lane, p. 270); less familiar is the knowledge that this transformation can be obtained by the standard *Gauss elimination* procedure for solving linear equations (Beightler and Wilde). First, we describe Lagrange's method constructively and then show that the operations involved are exactly the same as those used in Gaussian elimination. Then, we present an algorithm for inverting the transformation, and show that it is possible to diagonalize any quadratic form systematically, using a standard computation routine which takes even less effort than solving a *single* set of N linear equations in N variables.

2-09 *COMPLETING THE SQUARE (LAGRANGE'S TRANSFORMATION)*

Consider the quadratic form

$$q = x_1^2 + 2x_1x_2 + 4x_1x_3 + 3x_2^2 + 2x_2x_3 + 5x_3^2 \tag{2-63}$$

We begin the process of completing the square by expressing as a perfect square all terms containing x_1 as a factor:

$$q = (x_1 + x_2 + 2x_3)^2 + 2x_2^2 - 2x_2x_3 + x_3^2 \tag{2-64}$$

The next step consists of writing as a perfect square all terms containing x_2 as a factor:

$$q = (x_1 + x_2 + 2x_3)^2 + 2(x_2 - \tfrac{1}{2}x_3)^2 + \tfrac{1}{2}x_3^2 \tag{2-65}$$

For a quadratic form consisting of N variables, completing the square involves a repetition of the foregoing process for $x_1, x_2, \ldots, x_{N-1}$. Since here $N = 3$, we have finished. From the form of Eq. (2-65), we are led to make the following changes of variable:

$$z_1 = x_1 + x_2 + 2x_3, \qquad z_2 = x_2 - \tfrac{1}{2}x_3, \qquad z_3 = x_3 \tag{2-66}$$

so that Eq. (2-63) may now be written as

$$\hat{q} = z_1^2 + 2z_2^2 + \tfrac{1}{2}z_3^2 \tag{2-67}$$

The foregoing procedure may now be generalized; let N be the order of $\mathbf{Q}$, and let the constants $q_{jn}^{(h)}$ be generated recursively from the q_{jn} as follows:

$$q_{jn}^{(o)} \equiv q_{jn}; \qquad j, n = 1, \ldots, N \tag{2-68}$$

$$q_{hn}^{(h)} \equiv \frac{q_{hn}^{(h-1)}}{q_{hh}^{(h-1)}}; \qquad n = 1, \ldots, N \tag{2-69}$$

$$q_{jn}^{(h)} \equiv q_{jn}^{(h-1)} - q_{jn}^{(h-1)}q_{hn}^{(h)}; \qquad j = h+1, \ldots, N; \quad n = h, \ldots, N \tag{2-70}$$

$$q_{jn}^{(h)} \equiv q_{jn}^{(h-1)}; \qquad j = 1, \ldots, h-1; \quad n = 1, \ldots, N \tag{2-71}$$

$$q_{jn}^{(h)} \equiv 0; \qquad j = h, \ldots, N; \quad n = 1, \ldots, h-1 \tag{2-72}$$

If any $q_{hh}^{(h-1)}$ is zero, but some other diagonal element $q_{jj}^{(h-1)}$ is nonzero, then renumber the variables so as to interchange the indices h and j. Let m be the smallest number for which $q_{jj}^{(m)} = 0$ for all $j = m+1, \ldots, N$. Define the linear change of variable:

$$z_h = x_h + \sum_{n=h+1}^{N} q_{hn}^{(h)}x_n; \qquad h = 1, \ldots, m \tag{2-73}$$

Then

$$\sum_{j=1}^{N}\sum_{n=1}^{N} x_j q_{jn}^{(h-1)} x_n = q_{hh}^{(h-1)}z_h^2 + \sum_{j=h+1}^{N}\sum_{n=h+1}^{N} x_j q_{jn}^{(h)} x_n; \qquad h = 1, \ldots, m \tag{2-74}$$

Equation (2-73) may be used inductively to prove that

$$q(z_1, \ldots, z_m, x_{m+1}, \ldots, x_N) = \sum_{h=1}^{m} q_{hh}^{(h-1)}z_h^2 + \sum_{j=m+1}^{N}\sum_{n=m+1}^{N} x_j q_{jn}^{(m)} x_n \tag{2-75}$$

If $m = N$, then comparison of Eqs. (2-59) and (2-74) shows that the $q_{hh}^{(h-1)}$ are the elements of the diagonal matrix $\mathbf{D}$:

$$d_{hh} = q_{hh}^{(h-1)} \tag{2-76}$$

Moreover, the elements t_{jn} of the corresponding transformation matrix $\mathbf{T}$ are, by Eqs. (2-60) and (2-73),

$$t_{jn} = q_{jn}^{(j)}, \text{ for } j = 1, \ldots, n-1 \tag{2-77a}$$

$$t_{jn} = 1, \quad \text{for } j = n \tag{2-77b}$$

$$t_{jn} = 0, \quad \text{for } j = n+1, \ldots, N \tag{2-77c}$$

From Eqs. (2-58), (2-60), (2-62), (2-76), and (2-77), we have

$$q = \mathbf{x}'\mathbf{Q}\mathbf{x} = \mathbf{z}'\mathbf{D}\mathbf{z} = \mathbf{x}'\mathbf{T}'\mathbf{D}\mathbf{T}\mathbf{x} \tag{2-78}$$

whence

$$\mathbf{Q} = \mathbf{T}'\mathbf{D}\mathbf{T} \tag{2-79}$$

If $m < N$, and the $q_{jn}^{(m)}$ all vanish for $j, n > m$, then m is the rank of $\mathbf{Q}$, evidently a singular matrix. In this case, set $d_{hh} \equiv 0$ and $t_{jn} \equiv 0$ for all h, $j = m + 1, \ldots, N$. Then Eqs. (2-78) and (2-79) still hold even though $\mathbf{T}$ is singular. If $m < N$ but there is a nonvanishing $q_{jn}^{(m)}$ for some $j, n > m$, then it can be shown (Beightler and Wilde) that $\mathbf{Q}$ is an indefinite matrix. As a consequence of this fact, if $\mathbf{Q}$ is definite, or semidefinite, then $q_{jn}^{(m)} = 0$ for all $j, n > m$, and m is the rank of $\mathbf{Q}$. Therefore, in such cases, the process described in Eqs. (2-68) through (2-79) will always accomplish the diagonalization of $\mathbf{Q}$.

2-10 GAUSSIAN ELIMINATION

We now show that Lagrange's diagonalization algorithm may be expressed compactly in the format of the widely known Gauss elimination technique for solving linear equations. Consider the application of this method to the hypothetical set of equations represented by

$$\mathbf{Q}\mathbf{x} = \mathbf{0}$$

where $\mathbf{0}$ (zero) is a null column vector of N elements. To begin, the elements q_{1n} of the first row of $\mathbf{Q}$ are divided by q_{11}, the *first pivot*, which is equivalent to solving the first equation for x_1. This result is then used to eliminate x_1 from the remaining $N - 1$ equations. When this is done, the coefficient matrix $\mathbf{Q}^{(1)}$ for the new equations has for elements the $q_{jn}^{(1)}$ of Eqs. (2-69),

(2-70), (2-71), and (2-72), specialized for $h = 1$. Notice that

$$q_{11}^{(1)} = 1 \quad \text{and} \quad q_{j1}^{(1)} = 0; \qquad j = 2, 3, \ldots, N$$

Repeating this cycle of operations, that is, eliminating x_2, gives a new coefficient matrix $\mathbf{Q}^{(2)}$ in which the elements are the $q_{jn}^{(2)}$ of Eqs. (2-69), (2-70), (2-71), and (2-72), for $h = 2$. The second pivot is $q_{22}^{(1)}$, which by Eq. (2-76) is d_{22}, the second diagonal element of $\mathbf{D}$. After N iterations, the unit upper triangular matrix $\mathbf{Q}^{(N)}$ is obtained, and this is identical with the desired transformation matrix $\mathbf{T}$. The N pivots $q_{hh}^{(h-1)}$ must be recorded, since they form $\mathbf{D}$, but otherwise the diagonalization procedure involves only the ordinary elimination phase of the Gauss reduction. Alternatively, one may use the Crout reduction (Kunz, pp. 220, 226), a compact version of the Gauss elimination which avoids writing down the intermediate matrices $\mathbf{Q}^{(1)}$ through $\mathbf{Q}^{(N-1)}$.

The coefficient matrix for the quadratic form (2-63), Section 2-09, is

$$\mathbf{Q} = \begin{pmatrix} 1 & 1 & 2 \\ 1 & 3 & 1 \\ 2 & 1 & 5 \end{pmatrix}$$

Using Gauss elimination, we find the successive matrices:

$$\mathbf{Q}^{(1)} = \begin{pmatrix} 1 & 1 & 2 \\ 0 & 2 & -1 \\ 0 & -1 & 1 \end{pmatrix}$$

$$\mathbf{Q}^{(2)} = \begin{pmatrix} 1 & 1 & 2 \\ 0 & 1 & -\frac{1}{2} \\ 0 & 0 & \frac{1}{2} \end{pmatrix}$$

$$\mathbf{Q}^{(3)} \equiv \mathbf{T} = \begin{pmatrix} 1 & 1 & 2 \\ 0 & 1 & -\frac{1}{2} \\ 0 & 0 & 1 \end{pmatrix}$$

$$\mathbf{D} = \begin{pmatrix} 1 & 0 & 0 \\ 0 & 2 & 0 \\ 0 & 0 & \frac{1}{2} \end{pmatrix}$$

Using the foregoing transformation matrix $\mathbf{T}$ in Eq. (2-60) produces the Eq. (2-66) change of variable needed in Lagrange's transformation; the preceding diagonal matrix $\mathbf{D}$, when used in Eq. (2-62), yields the value of $\hat{q}$ given by Eq. (2-67). Since the diagonal elements of $\mathbf{D}$ are all positive, we know that $\mathbf{Q}$ is positive-definite.

2-11 QUASI-NEWTON METHODS

The Newton–Raphson method is appropriate when the curvature matrix y_{xx} of second derivatives is easily computed, as when the objective is made up of power functions. Imagine, however, a situation where second derivatives are difficult or impossible to evaluate. When this happens, y_{xx} can in principle be constructed from gradient measurements, taken at as many points as there are independent variables. Since the early part of a search for an optimum should involve the gradient method, it is reasonable to use the available gradient information to generate a Newton–Raphson step.

Define a quadratic approximation $a(\mathbf{x})$ to the objective function $y(\mathbf{x})$ by

$$y(\mathbf{x}) \approx a(\mathbf{x}) \equiv \mathbf{L}^T\mathbf{x} + \tfrac{1}{2}\mathbf{x}^T\mathbf{Q}\mathbf{x}$$

where $\mathbf{L}$ is an unknown N-vector and $\mathbf{Q}$ an unknown N^2 matrix. The gradient is $a_x = \mathbf{L}^T + \mathbf{x}^T\mathbf{Q}$, or $a_x^T = \mathbf{L} + \mathbf{Q}\mathbf{x}$ in transposed form. At a base point $\mathbf{x}_0$ and at N other points $\mathbf{x}_n$ $(n = 1, \ldots, N)$ nearby, let the gradients (y_x) be evaluated and set equal to the gradients of the approximation there. Then for $n = 1, \ldots, N$,

$$(y_x^T)_n - (y_x^T)_0 \sim (a_x^T)_n - (a_x^T)_0 = \mathbf{Q}(\mathbf{x}_n - \mathbf{x}_0)$$

Arrange these into columns of an N^2 *gradient difference matrix*

$$\mathbf{G} \equiv ((y_x^T)_1 - (y_x^T)_0, \ldots, (y_x^T)_N - (y_x^T)_0)$$

and similarly form the N^2 *excursion matrix*

$$\mathbf{X} \equiv (\mathbf{x}_1 - \mathbf{x}_0, \ldots, \mathbf{x}_N - \mathbf{x}_0)$$

Then $\mathbf{G} = \mathbf{QX}$, and the matrix $\mathbf{Q}$, which approximates y_{xx}, could be found from

$$\mathbf{Q} = \mathbf{GX}^{-1}$$

The Newton–Raphson step $\Delta\mathbf{x}$, which can be taken from any of the $N + 1$ points, is therefore the solution to $\mathbf{Q}\,\Delta\mathbf{x} = \mathbf{GX}^{-1}\,\Delta\mathbf{x} = -y_x^T$. Since this would involve solving simultaneous equations after inversion of $\mathbf{X}$, the explicit solution is actually easier to obtain, since it would involve only the inversion of $\mathbf{G}$.

$$\Delta\mathbf{x} = -\mathbf{XG}^{-1}y_x^T \tag{2-80}$$

If the points $\mathbf{x}_1, \ldots, \mathbf{x}_N$ are obtained as part of an iterated gradient search, then the objective function takes its best value at the last point measured, say $\mathbf{x}_N$. The Newton–Raphson step from $\mathbf{x}_N$ may then be regarded as an acceleration step which, converging immediately to the optimum if y is indeed quadratic, overcomes the oscillating tendency of the iterated gradient method near the optimum. More realistically, when y is not quadratic, this step will merely produce another point and associated gradient measurement to replace the worst point and gradient in the old set. Then a new NR step with a better matrix **Q** can be calculated from this newest point. As the optimum is approached and the approximation improves, this procedure will converge rapidly.

Matrix inversion can be avoided by choosing **X** to be the N^2 unit matrix **I**. Then $\mathbf{G}_I$, the corresponding gradient difference matrix, will be the approximate second derivative matrix **Q**, since

$$\mathbf{G}_I \equiv \mathbf{QI} = \mathbf{Q}$$

Then the only computation is the solution of the linear equations for the Newton–Raphson step $\Delta\mathbf{x}$ in

$$\mathbf{G}_I \, \Delta\mathbf{x} = -y_x^T \tag{2-81}$$

Unlike the iterated gradient method, this scheme would not produce continued improvement, so it is best used near the end of a search when improvement is difficult.

One can imagine many ways to locate the points, and there is in fact a substantial literature on this subject. These "parallel tangents," "conjugate," or "quasi-Newton" methods, often identified by the names of their inventors (Davidon, Shah, Buehler, Kempthorne, Fletcher, Powell, and Broyden) are widely available as effective computer codes.

2-12 SINGULAR VALLEYS

Consider now the effects of completely or nearly singular curvature matrices y_{xx}. This situation can be troublesome because analysis higher than second order may be needed. For this reason the topic is usually omitted from optimization texts, singularity being regarded as an event too rare to justify detailed discussion. But analysts cannot overlook this possibility, which will be shown to be caused by certain styles of mathematical modeling that on first glance would seem not only harmless, but even quite natural to the careful engineer. Mishandled, this situation can give glacially slow or even false convergence; correctly analyzed, it can yield elegant simplifications and insights. "Danger strikes the sooner it is despised," said Publius Syrus (ca. 50 B.C.), so prudence dictates a thorough study of the perverse subject of the singular curvature of valleys, ridges, and saddles. Resulting will be not just awareness of the problems, but positive ways of taking advantage of the situation.

For notational purposes, denote the gradient vector and the Hessian matrix as y_x and y_{xx}. When referring to a particular function $y_k(\mathbf{x})$, the notation will be y_{kx} and y_{kxx} for the gradient vector and the Hessian matrix respectively.

First, a little geometry. Near a local minimum, where the Hessian y_{xx} is positive definite, contours are concentric ellipsoids in x-space, as shown in Fig. 2-4(b) for $N = 2$. When the contours are far apart in some directions but close together in others, the objective is said to have a *valley*, as shown in Fig. 2-4(b). (Near a maximum, contours of this shape would form a *ridge*.) As long as y_{xx} remains positive definite, y will curve upward away from the minimum, although along the floor of the valley this increase is slight.

The contours may open enough to become parallel, in which case y_{xx} is positive *semi*definite, the contours becoming parallel lines as shown in Fig. 2-5(b). There is no change in y along the floor of the valley, and the minimum is not unique. From an economic viewpoint, this is good, because there is more than one set of values of the problem variables that will achieve the minimum. Yet a valley, with its alternative minima, can confuse the uninitiated, whose numerical optimization techniques obsessively seek improvement where none is possible or of practical importance.

Valleys occur when unimportant or even irrelevant variables are in the problem. This can happen when a model is made overly sophisticated to avoid leaving anything out. It can also occur when there is a hidden physical or mathematical relationship, such as a mass balance, linking the variables. And in least-squares regression problems, valleys appear when the data do not properly cover the region of interest. One lesson of this section is how to avoid creating models with valleys. Another related one is how to recognize and cope with any valleys that have crept into the formulation.

First consider the quadratic function

$$y_1(\mathbf{x}) = -4x_1 + 2x_2 + 4x_1^2 - 4x_1x_2 + x_2^2$$

Differentiation gives

$$y_{1x} = (-4 + 8x_1 - 4x_2,\ 2 - 4x_1 + 2x_2); \qquad y_{1xx} = \begin{pmatrix} 8 & -4 \\ -4 & 2 \end{pmatrix}$$

Stationary points $\mathbf{x}^0$ satisfy $(y_{1x})^0 = \mathbf{0}^T$, but the two scalar equations resulting are linearly dependent since $\partial y_1/\partial x_1 = -2\partial y_1/\partial x_2$. Hence there is an infinity of stationary points along the straight line $2x_1^0 - x_2^0 = 1$. At every stationary point, or for that matter, at every point, the curvature matrix y_{1xx} is the same, and so

$$\begin{aligned} \partial y_1 &= \partial \mathbf{x}^T(y_{1xx})\partial \mathbf{x} = 8(\partial x_1)^2 - 8(\partial x_1)(\partial x_2) + 2(\partial x_2)^2 \\ &= 2(2\partial x_1 - \partial x_2)^2 \geq 0 \qquad \text{for } \mathbf{x} \neq \mathbf{0} \end{aligned}$$

Hence y_1 increases in all directions from a stationary point except those for which $2\partial x_1 - \partial x_2 = 0$, where $\partial y_1 = 0$ to second order. But these directions exactly correspond to the line of stationary points, as shown earlier. This coincidence happens because y_1 is quadratic in $\mathbf{x}$, so the second-order Taylor expansion is exact, there being no terms higher than second order. This is a manifestation of the

> ***Quadratic Unconstrained Sufficiency Theorem:*** *If $y(\mathbf{x})$ is quadratic and $(y_{xx})^0$ is positive semidefinite at a stationary point $\mathbf{x}^0$, then $\mathbf{x}^0$ is a local minimum, along with all points $\mathbf{x}^0 + \mathbf{v}$, where $\mathbf{v}$ satisfies $(y_{xx})^0\mathbf{v} = \mathbf{0}$.*

This theorem, whose proof was sketched informally before its statement, defines the *valley vectors* $\mathbf{v}$. Being orthogonal to $(y_{xx})^0$, they point along the valley in the directions where the first and second derivatives of y vanish.

The Newton–Raphson procedure detects a valley nicely. Let $\mathbf{x}_0 = (1, 2)^T$ be a starting point for the Newton–Raphson method in the example. Then $(y_{1x})_0 = (-4, 2)$, and the Newton–Raphson step $\Delta\mathbf{x}$ is the set of all solutions to $(y_{1xx})_0\Delta\mathbf{x} = -(y_{1x})_0$, which gives the two equations: $8\Delta x_1 - 4\Delta x_2 = 4$ and $-4\Delta x_1 + 2\Delta x_2 = -2$. Since the second equation is a multiple of the first, any solution to one will solve the other. Thus the solutions are not unique, and the components of $\Delta\mathbf{x}$ are related by $2\Delta x_1 - \Delta x_2 = 1$. The new point (or in this case points) is given by $(\mathbf{x})_1 = (\mathbf{x})_0 + \Delta\mathbf{x}$, so $2[(x_1)_1 - 1] - [(x_2)_1 - 2] = 1$, or $2(x_1)_1 - (x_2)_1 = 1$. The points satisfying this were shown earlier to be *all* the stationary points along the floor of the valley, so in a sense NR generates the entire valley in one step, although only when y is quadratic.

When, as in the example, a perfectly flat valley can be identified, the number of independent variables can be reduced by a linear transformation. Thus let $\hat{x}_1 \equiv 2x_1 - x_2$. Then $y_1(\mathbf{x}) = -4x_1 + 2x_2 + 4x_1^2 - 4x_1x_2 + x_2^2 = -2(2x_1 - x_2) + (2x_1 - x_2)^2 = -2\hat{x}_1 + \hat{x}_1^2 = y(\hat{x}_1)$. Hence the objective can be written as a function of the single variable $\hat{\mathbf{x}}$, and minimized with respect to it alone. Often such a variable can be given a physical or economic interpretation lending insight into the system under study.

When the objective function is of higher than second degree, any valleys that occur will be curved. For instance, consider the fourth-degree function $y_2(\mathbf{x}) \equiv x_1^4 - 2x_1^2x_2 + x_2^2$. Gradient and Hessian are, respectively, $y_{2x} = (4x_1^3 - 4x_1x_2, -2x_1^2 + 2x_2) = 2(x_1^2 - x_2)(2x_1, -1)$ and

$$y_{2xx} = \begin{pmatrix} 12x_1^2 - 4x_2 & -4x_1 \\ -4x_1 & 2 \end{pmatrix}$$

Here the possible presence of a valley is signaled by the scalar factor $x_1^2 - x_2$ in the gradient, which consequently vanishes if and only if this factor is zero, since the vector $(2x_1, -1)$ cannot vanish. Hence an infinity of stationary points $\mathbf{x}^0$ lie on the parabola $(x_1^2 - x_2)^0 = 0$. This equation can be used to eliminate x_2 from the Hessian at any stationary point.

$$(y_{2xx})^0 = \begin{pmatrix} 12x_1^2 - 4x_1^2 & -4x_1 \\ -4x_1 & 2 \end{pmatrix}^0 = 2\begin{pmatrix} 4x_1^2 & -2x_1 \\ -2x_1 & 1 \end{pmatrix}^0$$

Since the first row equals $-2(x_1)^0$ times the second, $(y_{2xx})^0$ is singular at every stationary point $\mathbf{x}^0$. Completing the square gives $\partial\mathbf{x}^T(y_{2xx})^0\,\partial\mathbf{x} = 2[4(x_1^0)^2(\partial x_1)^2 - 4(x_1^0)\,\partial x_1\,\partial x_2 + (\partial x_2)^2] = 2[2(x_1)^0\,\partial x_1 - \partial x_2]^2 \geq 0$ for all $\partial\mathbf{x} \neq \mathbf{0}$. Hence $(y_{2xx})^0$ is positive semidefinite at every stationary point $\mathbf{x}^0$. However, this does *not* prove that every such point is a minimum, because y_2 is not quadratic, and so the terms of order higher than second in the Taylor expansion are not negligible in the valley directions where the first- and second-order parts vanish.

Away from the valley of stationary points, y_{2xx} is not singular, so the Newton–Raphson method would eventually converge to one of the stationary points. The only difficulty would be that near the valley, roundoff errors would be magnified by the near singularity of y_{2xx}. Suppose, however, that $\mathbf{x}_0 \equiv (1, 1)^T$ has been located as a stationary point by some direct search method such as Newton–Raphson. There the direction $2\partial x_1 - \partial x_2 = 0$ is tangent to the valley, the valley vector being any scalar multiple of $(1, 2)^T$. That is, for k real, $\mathbf{v}^T = k(1, 2)$. Consider a move along the valley, that is, let $\Delta\mathbf{x} = \mathbf{v}$. Then $\Delta y_2 = y_2(\mathbf{x}_0 + \Delta\mathbf{x}) - y_2(\mathbf{x}_0) = (1 + k)^4 - 2(1 + k)^2(1 + 2k) + (1 + 2k)^2 = [(1 + 2k)^2 - (1 + 2k)]^2 = k^4 > 0$ for $k \neq 0$. Therefore, y_2 will strictly increase along a *straight* line in the valley direction. In fact, y_2 increases along *every* straight line through $(1, 1)^T$. Yet there are points in the neighborhood which have the same value as at $(1, 1)^T$, a paradox explained by the curvature of the valley itself. Thus second-order methods cannot always detect a valley because of the neglected higher-order terms.

As for a straight valley, the equation of a curved valley can be used to simplify the problem by reducing the number of variables. In the example, let $\hat{x}_2 \equiv x_1^2 - x_2$. Then $y_2(\mathbf{x}) = x_1^4 - 2x_1^2x_2 + x_2^2 = (x_1^2 - x_2)^2 = (\hat{x}_2)^2 = y_2(\hat{x}_2)$. This transformation has reduced the degree of the problem to 2, so all second-order methods used on it would be exact. Any ambiguity in the solution now results strictly from inverting the transformation, which is a good reason *not* to invert it. That is, one should work with $\hat{x}_2$ instead of x_1 and x_2.

Valleys sloping gently away from a unique minimum can also occur. For example, consider the sum of the first objective y_1 with 100 times the second objective y_2. $y_3 \equiv y_1 + 100y_2 = -4x_1 + 2x_2 + 4x_1^2 - 4x_1x_2 + 101x_2^2 - 200x_1^2x_2 + 100x_1^4$. Previous analysis of the component functions y_1 and y_2 showed that their gradients both vanish at the unique solution of $(2x_1 - x_2)^0 = 1$ and $(x_1^2 - x_2)^0 = 0$, which is $\mathbf{x}^0 = (1, 1)^T$. Hence this point is the only stationary point for the new composite function. Analysis of the Hessians also showed them to be positive semidefinite at $(1, 1)^T$, both of their quadratic forms vanishing in the same direction $k(1, 2)^T$. Therefore, the Hessian $(y_{3xx})^0$ of the composite function y_3 is also positive semidefinite at $\mathbf{x}^0$. Yet the methods developed so far are not adequate to prove that $\mathbf{x}^0$ is a minimum. Section 2.14 will give a slightly modified function having the same second-order character, but which is a saddle rather than a minimum.

First-order search procedures like the gradient method experience great difficulty with such functions because the slope is very gradual on the floor of the valley. Newton–Raphson performs well, however.

Neglecting relatively small portions of an objective function, which is equivalent to ignoring components of a system contributing little to the cost, can lead to important simplifications, greater insight, and even results of theoretical importance. Thus neglecting y_1 would lead to the analysis already made of the dominant part $100y_2$ and would suggest the nonlinear transformation $\hat{x}_2 = x_1^2 - x_2$. This indicates that all the points in the flat, curved valley where $\hat{x}_2 = 0$, or, equivalently, where $x_1^2 = x_2$, are very good solutions, an important thing for an analyst to realize.

As usual, constructing a lower bound makes things even more precise. Minimizing the neglected part $y_1 = -2\hat{x}_1 + \hat{x}_1^2$ gives $y_1 \geq y_1^* = -1$, from which it follows that $y_3 \equiv y_1 + 100y_2 \geq -1 + 100y_2$. But $y_2 = \hat{x}_2^2$ is clearly bounded below by zero, so $y_3 \geq -1$. Thus if the analyst or designer wishes, for some reason, to set $x_1 = 2$, the minimization of $100y_2$ would put x_2 at $(2)^2 = 4$, where the neglected portion $y_1(2, 4)$ equals zero. Then the total objective for this design has the value zero, only one unit of objective function greater than the lower bound of -1.

For deeper insight into the behavior of the objective, one can express it in terms of the new variables $(\hat{x}_1, \hat{x}_2)^T \equiv \hat{\mathbf{x}}^T$, the exact objective being $y_3(\hat{\mathbf{x}}) = -2\hat{x}_1 + (\hat{x}_1)^2 + 100(\hat{x}_2)^2$. This is easily minimized, being quadratic in $\hat{\mathbf{x}}$. The derivatives are

$$\frac{\partial y_3}{\partial \hat{\mathbf{x}}} = (-2 + 2\hat{x}_1, 200\hat{x}_2) \quad \text{and} \quad \frac{\partial^2 y_3}{\partial \hat{\mathbf{x}}^2} = \begin{pmatrix} 2 & 0 \\ 0 & 200 \end{pmatrix}$$

The unique solution for the stationary point is therefore $\hat{\mathbf{x}}_\dagger = (1, 0)^T$, and the Hessian is positive definite everywhere. Hence this point is the global minimum, $y_3(\hat{\mathbf{x}})$ being quadratic. Any ambiguity in the original $\mathbf{x}$-space is due entirely to difficulties in inverting from $\hat{\mathbf{x}}$ back to $\mathbf{x}$: $x_1^* = (1 \pm (1 - (\hat{x}_1 - \hat{x}_2^2))^{1/2})^* = 1$; $(x_2)^* = (2 - \hat{x}_1 \pm 2(1 - (\hat{x}_1 - \hat{x}_2^2))^{1/2})^* = 1$.

It happens that this particular function y_3 resembles a rather famous one notorious as a test for direct search methods. Shown in Fig. 2-7, the function $y = (1 - x_1)^2 + 100(x_2 - x_1^2)^2$ has come to be called "Rosenbrock's Banana" from the name of its creator and the shape of the curved valley's contours. A common test for a numerical optimization procedure is

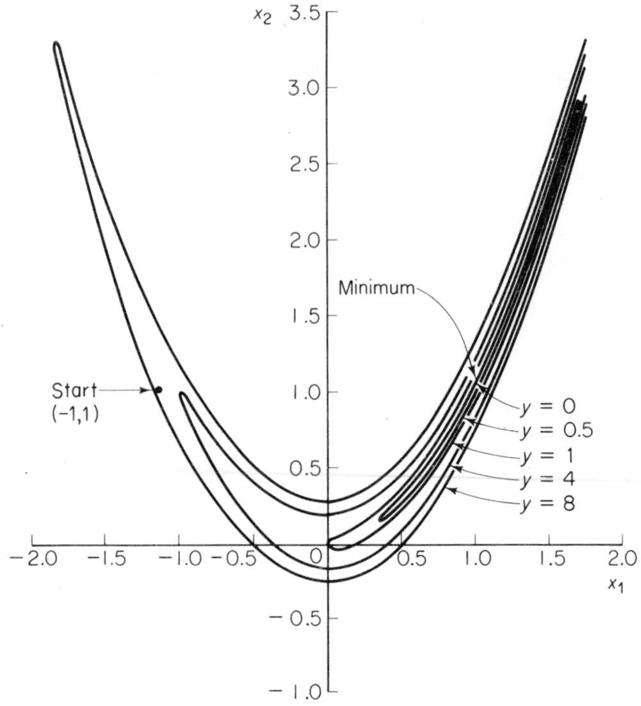

Figure 2-7. Rosenbrock's curved valley.

to start it at $(-1, 1)^T$ and count the number of iterations needed to find the minimum at $(1, 1)^T$ (see Chapter 4). The gradient method will not even move away from $(-1, 1)^T$, and the other techniques using first derivatives or differences take from 100 to 200 iterations to reach the minimum when they converge at all. The most effective direct search procedure was found to be a Newton–Raphson method adapted to the special form of the function, which is a sum of squared functions. This second-order scheme, discovered by Gauss in his early nineteenth-century development of the method of least squares, only took two or three iterations, on the way making wide excursions from the valley floor. Gauss's method is described in the next section because of its importance for curve fitting and statistical correlation of experimental data, a topic of interest to the progressive designer. But of course all this computation can be avoided, and deeper insight gained into the function, when the analysis of the present section is applicable.

2-13 LEAST SQUARES

The method of least squares of Gauss and Legendre works with *errors* $e_i(\mathbf{x})$, differences between the ith result of a set of I experiments and a prediction based on variables $\mathbf{x}$ to be selected so as to make the errors small. When N, the number of adjustibles, equals I, the number of experiments, then each error can be driven to zero in principle. But when there are more measurements than variables ($I > N$), as in most experimental work, one wants the value of $\mathbf{x}$ minimizing $y = \sum e_i^2$, the sum of squared errors.

Let $\mathbf{e}(\mathbf{x})$ be the column vector of errors, expressed as functions of $\mathbf{x}$, so that the sum of squares can be written

$$y = \mathbf{e}^T\mathbf{e}$$

This can be viewed as a quadratic form with unit matrix $\mathbf{I}$ so that $y = \mathbf{e}^T\mathbf{I}\mathbf{e}$. Then differentiation with respect to $\mathbf{e}$ gives $y_e = 2\mathbf{e}^T$, and so the chain rule yields

$$y_x = \frac{\partial y}{\partial \mathbf{e}}\frac{\partial \mathbf{e}}{\partial \mathbf{x}} = 2\mathbf{e}^T\mathbf{e}_x$$

In transposed form, $y_x^T = 2\mathbf{e}_x^T\mathbf{e}$. To obtain an approximation to y_{xx}, the curvature matrix, Gauss treated $\mathbf{e}_x$ as constant while differentiating y_x^T with respect to $\mathbf{x}$ to get $y_{xx} = 2\mathbf{e}_x^T\mathbf{e}_x$. Application of the Newton–Raphson formula then gives the Newton–Raphson step $\Delta\mathbf{x}$ as the solution of

$$\mathbf{e}_x^T\mathbf{e}_x\,\Delta x = -\mathbf{e}_x^T\mathbf{e} \qquad (2\text{-}82)$$

The remarkable fact that a second-order method is generated from first derivatives alone results from the special sum of squares form and Gauss's clever approximation, which is exact when the errors are all linear functions of $\mathbf{x}$, a situation the statisticians call "linear regression." Gauss's procedure, devised to compute orbital parameters for the planets from astronomical measurements, can correlate test data even when the underlying model is nonlinear. For a detailed discussion of this topic, including a numerical example, see Chapter 4.

2-14 HIDDEN SADDLES

Let us return now to the discussion of singular stationary points. If the objective function y curves upward in some directions and downward in others in the neighborhood of a stationary point, the point picturesquely is called a *saddle*. There the tangent plane is horizontal ($y_x = \mathbf{0}^T$) and the Hessian y_{xx} is indefinite. More generally, points where y curves both above and below the tangent plane, meaning that y_{xx} is indefinite, have hyperbolic contours as shown in Fig. 2-6(b). Notice the two valleys sloping down from the saddle, the two ridges sloping upward, and the two lines (four directions) along which y does not change to second order. Quadratic functions will always have as many double valleys and ridges as there are negative terms in the diagonalized quadratic form. Hence a saddle signals multiple minima.

First-order methods, or any others which descend strictly at each step, will by their nature avoid saddles, as shown in Fig. 2-8. Second-order methods such as Newton–Raphson may, unless precautions are taken, converge to a saddle, since they are intended to seek out stationary points. Directly solving the gradient equations $y_x = \mathbf{0}^T$ can also lead to saddles.

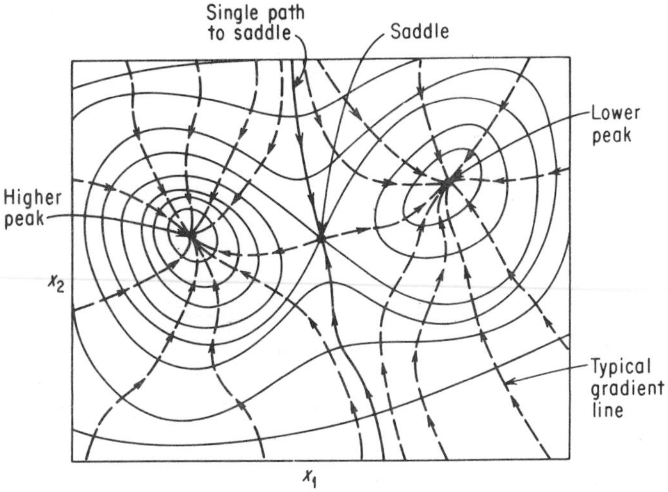

Figure 2-8. Gradient lines in the presence of a saddle.

Any stationary point whose curvature is not already known should always be tested to see if it is a saddle. If it is, then the search for the minimum is not over, and in fact several directions may need exploration. This section shows how to detect these directions and warns against the anomalies of functions having terms of higher than second degree.

As an example, consider the quadratic function

$$y_4 = 2x_1^2 - 8x_1x_2 + 1.5x_2^2 + 4x_1 + 5x_2$$

constructed to have a unique stationary point at the base case $(1, 1)^T$.

$$(y_x)^0 = (4x_1 - 8x_2 + 4,\ 3x_2 - 8x_1 + 5)^0 = (0, 0)$$

The curvature matrix y_{xx} is indefinite at x^0 (and everywhere else)

$$\begin{aligned} \tfrac{1}{2}\partial\mathbf{x}^T y_{xx} \partial\mathbf{x} &= 4(\partial x_1)^2 - 8\partial x_1 \partial x_2 + 3(\partial x_2)^2 \\ &= (2\partial x_1 - 2\partial x_2)^2 - (\partial x_2)^2 = \partial y \end{aligned}$$

Hence y decreases in every direction from x^0 where $4(\partial x_1 - \partial x_2)^2 < (\partial x_2)^2$ and increases where this inequality is reversed. Hence x^0 is a *saddle*. The directions of the contours at x^0 are obtained by using the identity $a^2 - b^2 = (a - b)(a + b)$ to the diagonalized quadratic form and setting it to zero. Hence

$$(\partial y)^0 = 4(\partial x_1^0 - \partial x_2^0)^2 - (\partial x_2^0)^2$$

and by using the above identity we find that

$$\begin{aligned}(\partial y)^0 &= (2\partial x_1^0 - 2\partial x_2^0 + \partial x_2^0)(2\partial x_1^0 - 2\partial x_2^0 - \partial x_2^0) = 0\\ &= (2\partial x_1^0 - \partial x_2^0)(2\partial x_1^0 - 3\partial x_2^0)\end{aligned}$$

Then,

$$2(x_1)^0 = (\partial x_2)^0 \quad \text{or} \quad (2\partial x_1)^0 \equiv 2(\partial x_2)^0$$

are contour directions. Directions of decrease in y are those for which ∂x_1 and $2\partial x_1 - 3\partial x_2$ have opposite signs.

Although it may be disappointing to find a stationary point, located perhaps after considerable effort, to be a saddle rather than a minimum, there is comfort in being able to generate directions for further improvement. In the example, such favorable directions would not have been detected simply by holding one variable constant and perturbing the other.

Now consider a quartic (fourth-degree) function for which effects of higher than second order will be important.

$$y_5 = -x_2 + 2x_1x_2 + x_1^2 + x_2^2 - 3x_1^2x_2 - 2x_1^3 + 2x_1^4$$

Since this is only quadratic in x_2, the stationary points can be found algebraically by solving the gradient nullification equations.

$$y_{5x} = (2x_1 + 2x_2 - 6x_1x_2 - 6x_1^2 + 8x_1^3, -1 + 2x_1 + 2x_2 - 3x_1^2)$$

Setting $(\partial y/\partial x_2)^0$ to zero gives $(x_2)^0 = \frac{1}{2} - (x_1)^0 + \frac{3}{2}(x_1^0)^2$. Substitution of this into $(\partial y/\partial x_1)^0 = 0$ gives, after simplification, $(1 - x_1^0)^3 = 0$. Hence there is a triple root at the base case $(1, 1)^T$, which is consequently the only stationary point. The Hessian there is

$$(y_{5xx})^0 = \begin{pmatrix} 2 - 12x_1 - 6x_2 + 24x_1^2 & 2 - 6x_1 \\ 2 - 6x_1 & 2 \end{pmatrix}^0 = \begin{pmatrix} 8 & -4 \\ -4 & 2 \end{pmatrix}$$

which is singular and positive semidefinite, with a valley direction $\mathbf{v} = (1, 2)^T$ orthogonal to $(y_{5xx})^0$. Hence there appears to be no way to decrease y_5 along straight lines in the neighborhood of $\mathbf{x}^0$.

Consider, however, a differential step in the valley direction; let $\partial\mathbf{x} =$

$k\mathbf{v}$ with $k \neq 0$. Since the first- and second-order parts of the Taylor expansion in this direction are zero, ∂y_5 is approximated by the third-order term.

$$\partial y_5 \approx \frac{1}{3!}\sum_{i=1}^{2}\sum_{j=1}^{2}\sum_{k=1}^{2}\left(\frac{\partial^3 y_5}{\partial x_i\,\partial x_j\,\partial x_k}\right)^0 \partial x_i\,\partial x_j\,\partial x_k$$

$$= \frac{1}{6}\left[3\left(\frac{\partial^3 y_5}{\partial x_1^2\,\partial x_2}\right)^0(\partial x_1)^2(\partial x_2) + \left(\frac{\partial^3 y_5}{\partial x_1^3}\right)^0(\partial x_1)^3\right]$$

$$= \frac{1}{6}[18k^2(2k) + (-12 + 48x_1)^0 k^3] = \frac{1}{6}(36k^3 - 36k^3) = 0$$

Hence no improvement can be obtained to third order. If it could, then of course $\mathbf{x}^0$ would not be a local minimum. The contrapositive of this statement can be generalized into a necessary condition known to Lagrange.

Higher-Order Unconstrained Minimum Necessity Theorem: *If* $\mathbf{x}^*$ *is a local minimum, then in every direction z the lowest-order nonvanishing term in the expansion* $y(\mathbf{x}^* + k\mathbf{z}) - y(\mathbf{x}^*) \equiv y(k\mathbf{z})$ *must be of even degree in the scalar k.*

Proof: If, contrary to the theorem, there exists $\mathbf{z}$ such that $\partial y(k\mathbf{z}) = ck^{2p+1} + \cdots = ck(k^{2p} + \cdots)$, where $c \neq 0$ and p is an integer, then $\partial y(k\mathbf{z})$ can be made negative by letting ck be negative, contradicting the hypothesis that $\mathbf{x}^*$ is a local minimum.

Next consider the fourth-order Taylor expansion in the valley direction. All terms lower than fourth degree vanish, leaving

$$\partial y_5(\sim)^4 \frac{1}{4!}\sum_h\sum_i\sum_j\sum_k\left(\frac{\partial^4 y_5}{\partial x_h\,\partial x_i\,\partial x_j\,\partial x_k}\right)^0 \partial x_h\,\partial x_i\,\partial x_j\,\partial x_k$$

$$= \frac{1}{24}[48(\partial x_1)^4] = 2k^4 > 0 \qquad \text{for } k \neq 0$$

No direction of improvement is available, even to fourth order, along any straight line through $\mathbf{x}^0$. In this example, where y_5 is of fourth degree, the Taylor expansion to fourth order is exact, so no such improvement is possible to any order.

By now the evidence that $\mathbf{x}^0$ is a minimum is so great that even Lagrange, who believed in the converse of the preceding theorem, would be satisfied. Yet the point $\mathbf{x}^0$ is *not* a minimum, and the converse is false. This is proven by exhibiting another point where y has the same value as at $\mathbf{x}^0$ but is not stationary. The origin will do: $y_5(\mathbf{0}) = y_5(\mathbf{x}^0) = 0$, and since $\mathbf{x}^0$ is the only stationary point, $y_{5x}(\mathbf{0})$ cannot vanish; in fact, $y_{5x}(\mathbf{0}) = (0, -1)$. Hence y_5

can be decreased near 0 by making x_2 positive. As a check, $y_5(0, \frac{1}{4}) = -(\frac{1}{4}) + (\frac{1}{16}) < 0$. So be very careful with high-degree functions.

The way in which this annoying function was constructed furnishes a lesson in poor modeling. The function is a product of two parabolic functions $(x_1^2 - x_2)$ and $[2(x_1 - \frac{1}{2})^2 - (x_2 - \frac{1}{2})]$, each of which passes through the base case with the same slope $\partial x_2/\partial x_1 = 2$. Figure 2-9 shows that these parabolic functions have the same sign above as well as below both parabolas where the functions vanish. Hence their product is positive in these regions. Between them, however, the signs are different, making the product negative. But this negative region is hard to detect because it lies entirely above the common tangent, along which the product vanishes to second order but increases to fourth. If $\bar{x}_1 \equiv x_1^2 - x_2$ and $\bar{x}_2 \equiv 2x_1^2 - 2x_1 - x_2 + 1$ were the independent variables, the character of $y_5(\bar{x}) = \bar{x}_1\bar{x}_2$ would have been immediately apparent. It was the representation in **x**-space that obscured the nature of y_5.

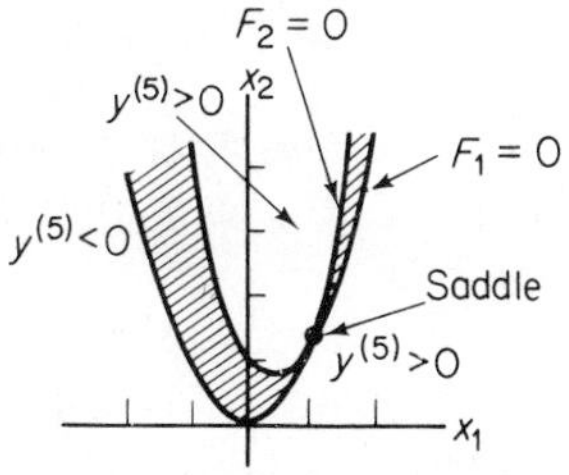

Figure 2-9. Hidden saddle.

This section, together with Section 2.12 on singular valleys, intends to alert the reader, not only to the paradoxes of high-order objective functions, but also to the opportunities for piercing their mysteries. "From this nettle, danger, we pluck this flower, safety" (Shakespeare's *Henry IV*).

2-15 EQUALITY CONSTRAINTS—ELIMINATION

When the objective function must satisfy side conditions given as equations relating the independent variables, the optimum necessarily lies on a boundary of the feasible region $\mathfrak{F}$. Indirect methods are valid only for interior optima, but equality constrained problems can often be transformed into new ones having the optimum inside the feasible region. The next five sections give two classical ways of accomplishing this transformation and a new method based on classical ideas. At the same time the important distinction between "decision" variables and "solution" variables will be introduced. In this section we consider the *elimination* method for handling equality constraints, best described by example.

Suppose in the hypothetical design problem that instead of allowing any positive pressure and recycle ratio, we must for technical reasons consider only those satisfying the following equality constraint.

$$x_1 x_2 = 9000 \tag{2-83}$$

The points satisfying this equation lie on the dashed line of Fig. 2-2. Let this constraint be solved for one of the variables, say x_2, as a function of the other.

$$x_2 = 9000 x_1^{-1} \equiv s(x_1)$$

The right member is an example of a *solution function*, and x_2 is called a *solution variable*. Now use the solution function to eliminate the solution variable from the objective function.

$$\begin{aligned} y(x_1, x_2) = y(x_1, s(x_1)) &= 1000x_1 + 4 \times \frac{10^9}{9000} + (2.5 \times 10^5)\left(\frac{9000}{x_1}\right) \\ &= 1000x_1 + 2.25 \times 10^9 x_1^{-1} + 4.44 \times 10^5 \equiv z(x_1) \end{aligned} \tag{2-84}$$

This new *substituted objective function* $z(x_1)$ depends only on x_1, called a *decision variable* because it is free to be manipulated without restriction. The minimum of this function of x_1 alone is found by the classical technique described already. The result is $x_1^* = 1500$ psia, which, when substituted into Eq. (2-1), gives the constrained minimum cost as $y' = \$3.44 \times 10^6$. Substitution into the constraint gives the optimal recycle ratio as $x_2' = 6$. This constrained optimum is shown in Fig. 2-2 on the dashed line. The minimum cost has gone up because the unconstrained minimum has been forbidden by the constraints.

Elimination, when it can be used, is effective because it reduces the number of independent variables and, consequently, the number of derivative equations that must be solved. It does require, however, the initial solution of the constraint equations in closed form, often an impossible task.

2-16 DECISION VARIABLES

The remaining indirect ways to solve equality constrained optimization problems do not require solving the constraints. They involve working with the linear terms of Taylor expansions, not only of the objective function, but also of each constraint. This approach leads us to partition the independent variables into two groups, called *decision variables* and *solution variables*[1] for reasons to be made clear in this section. Then the linear equations are solved for the solution differentials as linear functions of the decision differentials.

[1]Solution variables may also be referred to as *state* variables.

The coefficients of these linear expansions, which will be known as *decision derivatives*, are important in developing the generalizations in optimization theory that fill the rest of the book. Of comparable future value will be the concepts of solution and decision. Thus the ideas of this section are powerful enough to help solve systems much more complicated than the simple equality constrained case. Although the derivation here involves neglecting second- and higher-order differentials, a rigorous proof based on the implicit function theorem has been given by Wilde (September 1965).

Let there be M differentiable constraints

$$f_m(\mathbf{x}) = 0; \qquad m = 1, \ldots, M \tag{2-85}$$

The feasible region $\mathfrak{F}$ now consists of all points $\mathbf{x}$ satisfying Eq. (2-85). Since every point of this region is a boundary point, there are no interior points at all, much less an interior optimum. In every neighborhood of a feasible point $\mathbf{x}$ there are both infeasible points where some of the differentials $\partial f_m \neq 0$, and feasible points where all $\partial f_m = 0$. For points in the feasible neighborhood $\mathfrak{N}$, the differentials $\partial\mathbf{x}$ must satisfy, to a first-order approximation, the M linear equations

$$0 = \partial f_k = \sum_{n=1}^{N} \left(\frac{\partial f_k}{\partial x_n}\right) \partial x_n = \nabla f_k \cdot \partial\mathbf{x} \tag{2-86}$$

Recall that a similar expansion for the objective function gives, with ∂y not necessarily zero,

$$\partial y = \sum_{n=1}^{N} \left(\frac{\partial y}{\partial x_n}\right) \partial x_n = \nabla y \cdot \partial\mathbf{x} \tag{2-87}$$

Since at $\mathbf{x}$ the partial derivatives are presumed to be known constants, Eqs. (2-86) and (2-87) constitute $M + 1$ linear equations in $N + 1$ unknown differentials: ∂y and the N components of the feasible differential vector $\partial\mathbf{x}$. Let us assume that the equations are linearly independent (Birkhoff and Mac Lane, p. 167); if they are not, then take the largest number that form an independent set and discard the rest as redundant. This automatically rules out the case where there are more equations than unknowns ($M > N$), and the case where $M = N$ is of no interest because then the only solution to Eq. (2-86) would be $\partial\mathbf{x} = 0$, meaning there is no feasible neighborhood at all. Therefore, assume from now on that $M < N$.

Rearrange Eqs. (2-86) and (2-87), placing on the left terms involving ∂y and the first M components of $\partial\mathbf{x}$.

$$-\partial y + \sum_{n=1}^{M} \left(\frac{\partial y}{\partial x_n}\right) \partial x_n = -\sum_{n=M+1}^{N} \left(\frac{\partial y}{\partial x_n}\right) \partial x_n \tag{2-88}$$

$$\sum_{n=1}^{M} \left(\frac{\partial f_k}{\partial x_n}\right) \partial x_n = -\sum_{n=M+1}^{N} \left(\frac{\partial f_k}{\partial x_n}\right) \partial x_n \tag{2-89}$$

Actually, any M among the N independent variables could have been chosen; the first M were selected for convenience and without sacrificing generality. There is, however, one restriction; the left members of the M equations (2-89) must be linearly independent. That Eqs. (2-86) are independent assures us of the existence of such an independent subset.

To emphasize the distinction between the differentials on the right and those on the left, we introduce new notation and terminology. Let

$$s_m \equiv x_n; \qquad m, n = 1, \ldots, M \tag{2-90}$$

be called the *solution variables* of the problem. The difference

$$P \equiv N - M \tag{2-91}$$

is known as the *number of degrees of freedom*, and a new set of indices $p = 1, \ldots, P$ is given by

$$p \equiv n - M; \qquad n = M + 1, \ldots, N \tag{2-92}$$

The variables with these indices are called *decision variables* d_p.

$$d_p = x_p \tag{2-93}$$

In this terminology Eqs. (2-88) and (2-89) become

$$-\partial y + \sum_{m=1}^{M} \left(\frac{\partial y}{\partial s_m}\right) \partial s_m = -\sum_{p=1}^{P} \left(\frac{\partial y}{\partial d_p}\right) \partial d_p \tag{2-94}$$

$$\sum_{m=1}^{M} \left(\frac{\partial f_k}{\partial s_m}\right) \partial s_m = -\sum_{p=1}^{P} \left(\frac{\partial f_k}{\partial d_p}\right) \partial d_p; \qquad k = 1, \ldots, M \tag{2-95}$$

There must be exactly M *solution* and P decision variables, but in the problem at hand, it does not matter which of the original $M + P$ independent variables are placed in these categories as long as linear independence is preserved. In more complicated situations, it may be clear which are to be the decision variables. However they happen to be chosen, any arbitrary specification of the decision differentials permits calculation of numerical values for the right members of Eqs. (2-94) and (2-95). Then Eqs. (2-95) can be solved for the unique values of the solution differentials $\partial \mathbf{s}$ which keep the new point $\mathbf{x} + \partial \mathbf{x}$ inside the feasible region. The resulting change ∂y in the objective function, calculated from Eq. (2-94), can then be used to see if the perturbation was an improvement.

Decision variables can be manipulated freely, while solution variables adjust to keep the new point feasible. Arbitrary adjustment of more than P variables would move $\mathbf{x} + \partial \mathbf{x}$ out of the feasible region; specification of fewer than P variables would leave too many unknown quantities, making it impossible to locate the new point. Since P is the exact number of decisions that can be made with no regard for feasibility, it measures the number of "degrees of freedom" in the system. Each additional constraint cuts down the number of degrees of freedom, and by reducing the number of decision variables, actually makes the optimization problem easier. The dimension of the feasible region $\mathfrak{F}$ is not the total number of "independent" variables, but the number of degrees of freedom. An optimization problem in a million and one variables with a million equality constraints can in principle be reduced to an unconstrained problem in a single variable.

2-17 CONSTRAINED DERIVATIVES

The dependence of the solution (state) variables on the decisions is made even clearer by solving Eqs. (2-90) and (2-95). This will lead to the useful concept of *constrained derivative*. Some notational preliminaries are needed before the solutions can be written down explicitly. The M by M matrix of the coefficients of the solution variables is

$$\begin{pmatrix} \partial f_1/\partial s_1 & \cdots & \partial f_1/\partial s_M \\ \vdots & & \vdots \\ \partial f_M/\partial s_1 & \cdots & \partial f_M/\partial s_M \end{pmatrix} \equiv \frac{\partial \mathbf{f}}{\partial \mathbf{s}} \tag{2-96}$$

Also defined are

$$\left(\frac{\partial y}{\partial s_1}, \ldots, \frac{\partial y}{\partial s_M}\right) \equiv \frac{\partial y}{\partial \mathbf{s}} \tag{2-97}$$

$$\left(\frac{\partial y}{\partial d_1}, \ldots, \frac{\partial y}{\partial d_P}\right) \equiv \frac{\partial y}{\partial \mathbf{d}} \tag{2-98}$$

$$\begin{pmatrix} \partial f_1/\partial d_1 & \cdots & \partial f_1/\partial d_P \\ \vdots & & \vdots \\ \partial f_M/\partial d_1 & \cdots & \partial f_M/\partial d_P \end{pmatrix} \equiv \frac{\partial \mathbf{f}}{\partial \mathbf{d}} \tag{2-99}$$

Then Eqs. (2-94) and (2-95) become, respectively,

$$-\partial y + \frac{\partial y}{\partial \mathbf{s}}\,\partial \mathbf{s} = \frac{\partial y}{\partial \mathbf{d}}\,\partial \mathbf{d} \tag{2-100}$$

$$\frac{\partial \mathbf{f}}{\partial \mathbf{s}}\,\partial \mathbf{s} = -\frac{\partial \mathbf{f}}{\partial \mathbf{d}}\,\partial \mathbf{d} \tag{2-101}$$

Since $\partial \mathbf{f}/\partial \mathbf{s}$ is nonsingular by the way it was constructed, it has an inverse, whence, by Eq. (2-101),

$$\partial \mathbf{s} = -\left(\frac{\partial \mathbf{f}}{\partial \mathbf{s}}\right)^{-1} \frac{\partial \mathbf{f}}{\partial \mathbf{d}} \partial \mathbf{d} \equiv \frac{\partial \mathbf{s}}{\partial \mathbf{d}} \partial \mathbf{d} \tag{2-102}$$

which, when substituted into Eq. (2-100) gives

$$\partial y = \left[\frac{\partial y}{\partial \mathbf{d}} - \frac{\partial y}{\partial \mathbf{s}} \left(\frac{\partial \mathbf{f}}{\partial \mathbf{s}}\right)^{-1} \frac{\partial \mathbf{f}}{\partial \mathbf{d}}\right] \partial \mathbf{d} \equiv \frac{\partial z}{\partial \mathbf{d}} \partial \mathbf{d} \tag{2-103}$$

Here, $z \equiv z(\mathbf{d})$ is the *reduced objective function*, from which the solution variables have been eliminated. Since this function is not constrained and depends only on the decision variables, the more descriptive name, *unconstrained objective function*, is used in this book.

Equation (2-103) gives the rate of change of the objective resulting from *feasible* (not arbitrary) perturbations in the pth decision variable, which will cause the solution variables to readjust. This partial derivative is given the special symbol $\partial z/\partial \mathbf{d}$ to distinguish it from $\partial y/\partial \mathbf{d}$, which is obtained for nonfeasible perturbations with all the solution variables held constant. The quantity $\partial \mathbf{z}/\partial \mathbf{d}_p$ will be called "the *constrained* derivative of y with respect to d_p" or, more simply, the pth *decision derivative.*

In the example, $N = 2$ and $M = P = 1$. Let x_1 be the *solution* (i.e., state) variable s_1, and x_2 the decision variable d_1. The single constraint equation (2-83) can be written in the form of Eq. (2-85) as

$$f_1(s_1, d_1) \equiv s_1 d_1 - 9000 = 0 \tag{2-104}$$

The matrices corresponding to Eqs. (2-96), (2-97), (2-98), and (2-99) are

$$\frac{\partial f_1}{\partial s_1} = d_1 \tag{2-105}$$

$$\frac{\partial y}{\partial s_1} = 1000 - 4 \times 10^9 (s_1)^{-2} (d_1)^{-1} \tag{2-106}$$

$$\frac{\partial y}{\partial d_1} = -4 \times 10^9 (s_1)^{-1} (d_1)^{-2} + 2.5 \times 10^5$$

$$\frac{\partial f_1}{\partial d_1} = s_1$$

In the example, Eq. (2-103) gives

$$\frac{\partial z}{\partial d_1} = 2.5 \times 10^5 - 1000 \frac{s_1}{d_1} \tag{2-107}$$

Since the decision differentials can be arbitrary without causing $\mathbf{x} + \partial \mathbf{x}$ to leave the P-dimensional feasible neighborhood $\mathcal{P}$, $\mathbf{x}$ is now *interior* to $\mathcal{P}$, even though it is on the boundary of the feasible region $\mathfrak{F}$ embedded in the N-dimensional space of the original "independent" variables $x_1, \ldots, x_N$. Figure 2-2 shows this, for any point on the dashed constraint line $\mathfrak{F}$ is a boundary point when both x_1 and x_2 can change arbitrarily in two dimensions, but it is interior when only movement along the unidimensional feasible line is permitted. Hence when decision derivatives are used, the classical theory for *interior* optima applies. Therefore, at a feasible optimum it is necessary that the decision derivatives vanish.

$$\left(\frac{\partial z}{\partial d_p}\right)^* = 0; \qquad p = 1, \ldots, P \tag{2-108}$$

In the example this gives, in view of Eq. (2-107),

$$2.5 \times 10^5 - 1000\frac{s_1^*}{d_1^*} = 0$$

which, together with the constraint equation (2-83),

$$s_1^* d_1^* = 9000$$

permits the constrained solution to be identified as $s_1^*(\equiv x_1^*) = 1500$; $d_1^*(\equiv x_2^*) = 6$, confirming the result obtained by direct elimination.

This technique expands the decision derivatives $\partial z/\partial dp$ as functions of the independent variables and sets them to zero. Ultimately, N nonlinear simultaneous equations must be solved—M from the constraints and $N - M$ from the decision derivatives. Since unconstrained problems also require solving N equations, the constraints do not really increase the computational difficulty when this method is used. This illustrates the power of the differential approach, coupled with the concepts of solution, decision, degrees of freedom, and constrained derivative. These ideas will reappear in different roles throughout the book.

2-18 SENSITIVITY ANALYSIS

Rarely does one know the constraint functions f_m with absolute precision. As in the example, the constraint often has a term which, although nominally a constant because of its independence of the problem variables, may really change with fluctuating sales, raw material quality, or other uncontrollable factors. Design problems have many constant terms which, being related to actual performance, cannot be predicted before operations begin. Before going ahead with an enterprise, a decision maker wants to estimate the impact of all these uncertainties upon the optimum he has calculated. Determining the rate of change of the optimum value with respect to perturbations in the constraint function is called *sensitivity analysis.*

To perform a sensitivity analysis, we make the constraint functions infinitesimally different from zero and find the new optimum. Let $\bar{f}_k$ be the numerical value of the constraint function $f_k(\mathbf{x})$. Keep clearly in mind the distinction between $\bar{f}_k$, a scalar, and $f_k(\mathbf{x})$, a function of $\mathbf{x}$. In the example, $f_1(\mathbf{x})$ was always $(x_1x_2 - 9000)$, whereas $\bar{f}_1$ was the number zero. From now on $f_k(\mathbf{x})$ will retain its original character, but the $\bar{f}_k$ should be considered as M new variables which can assume arbitrary values. Let the $f_k(\mathbf{x})$ be assembled into a column M-vector $\mathbf{f}$ and let it be understood that $\partial f_k/\partial x_n$ is an abbreviation for $\partial f_k(\mathbf{x})/\partial x_n$.

Let $\mathbf{x}^*$ be a feasible optimum where $\mathbf{f} = 0$, and let $\partial\mathbf{f}$ be a vector of perturbations of $\mathbf{f}$ resulting from perturbations $\partial\mathbf{s}$ and $\partial\mathbf{d}$. Equation (2-86) must be replaced by

$$\partial\mathbf{f} = \frac{\partial\mathbf{f}}{\partial\mathbf{s}}\,\partial\mathbf{s} + \frac{\partial\mathbf{f}}{\partial\mathbf{d}}\,\partial\mathbf{d} \tag{2-109}$$

The left member was always zero in the previous section, and the matrices have been defined in Eqs. (2-96) and (2-99). Multiplication of all vectors on the left by $(\partial\mathbf{f}/\partial\mathbf{s})^{-1}$ and rearrangement gives

$$\partial\mathbf{s} = \left(\frac{\partial\mathbf{f}}{\partial\mathbf{s}}\right)^{-1}\partial\mathbf{f} - \left(\frac{\partial\mathbf{f}}{\partial\mathbf{s}}\right)^{-1}\frac{\partial\mathbf{f}}{\partial d}\,\partial\mathbf{d} \tag{2-110}$$

which can be substituted into Eq. (2-100) to give

$$\partial z = \frac{\partial y}{\partial\mathbf{s}}\left(\frac{\partial\mathbf{f}}{\partial\mathbf{s}}\right)^{-1}\partial\mathbf{f} + \frac{\partial z}{\partial\mathbf{d}}\,\partial\mathbf{d} \tag{2-111}$$

But since $\mathbf{x}^*$ is a local optimum, the *gradient projection* (or *constrained derivative* or *reduced gradient*) $\partial z/\partial\mathbf{d}$ must vanish there [Eq. (2-108)], so

$$\partial y^* = \frac{\partial y}{\partial\mathbf{s}}\left(\frac{\partial\mathbf{f}}{\partial\mathbf{s}}\right)^{-1}\partial\mathbf{f} \equiv \left(\frac{\partial z}{\partial\mathbf{f}}\right)^*\partial\mathbf{f} \tag{2-112}$$

Therefore,

$$\left(\frac{\partial z}{\partial\mathbf{f}}\right)^* = \left(\frac{\partial y}{\partial\mathbf{s}}\right)^*\left(\frac{\partial\mathbf{f}}{\partial\mathbf{s}}\right)^{-1}_* \tag{2-113}$$

The elements of this row M-vector are partial derivatives of y^*, the *optimum* value, with respect to perturbations $\mathbf{f}$ in the constraint functions. Because of this they will be called *sensitivity coefficients*. In the example,

$$\left(\frac{\partial z}{\partial f_1}\right)^* = \frac{\partial y/\partial s_1}{\partial f_1/\partial s_1} = \frac{1000 - 4 \times 10^9(s_1^*)^{-2}(d_1^*)^{-1}}{d_1^*} = \$117.30/\text{atm} \tag{2-114}$$

Here $s_1^* = x_1^* = 1500$ and $d_1^* = x_2^* = 6$, since $\mathbf{x}^*$ is the *constrained* minimum. Equation (2-114) means that if $\bar{f}_1$ were made 1 instead of 0, the minimum cost at the new optimum satisfying the new constraint $x_1x_2 = 9001$ would be approximately \$117.30 higher than at the old minimum where $x_1x_2 = 9000$. Keep in mind that this is an approximation, good only near the point $\mathbf{x}^*$ where the derivatives were evaluated.

The values of the matrices in Eq. (2-113) depend on which variables were chosen as decisions. One would at first suppose that there could therefore be a different value of the sensitivity coefficient $(\partial z/\partial f_k)^*$ for every possible set of decisions. In Section 2-19 it will be shown that each sensitivity coefficient is in fact unique, and one set of decisions is as good as any other for calculating $(\partial z/\partial f_k)^*$.

2-19 LAGRANGE'S UNDETERMINED MULTIPLIERS

The final indirect method covered can be considered "classical" since it goes back to Lagrange (1760–1761). It is, in fact, often the only method mentioned when optimization theory is under discussion. Briefly, it introduces the M sensitivity coefficients $(\partial z/\partial f_k)^*$ into the solution, treating them as unknowns, designated λ_k for abbreviation.

$$\left(\frac{\partial z}{\partial f_k}\right)^* \equiv \lambda_k \tag{2-115}$$

This ultimately reduces the problem to solving $M + N$ simultaneous equations in the $M + N$ unknowns (the N x_n and M λ_k). Although this may seem harder than merely solving the N equations required by the method of Section 2-17 and it often is, there are times when the $M + N$ equations resulting are easier to solve. Moreover, Lagrange's method gives the sensitivity coefficients as by-products of the computations, and is a very useful tool in the development of computational algorithms.

Although the results derived here are the same as Lagrange's, the development is entirely different. Lagrange did not view the λ_k as derivatives, but merely as "undetermined multipliers"—arbitrary constants introduced for no other purpose than mathematical convenience. Let us proceed now with the demonstration.

In view of how the sensitivity coefficients were defined in the preceding section, Eq. (2-111) can be written

$$\partial y^* = \left(\frac{\partial z}{\partial \mathbf{f}}\right)^* \partial \mathbf{f} = \boldsymbol{\lambda}\,\partial \mathbf{f} \tag{2-116}$$

where $\boldsymbol{\lambda}$ is the row M-vector of sensitivity coefficients, for which the conventional terminology is *Lagrange multipliers.* Differentiation with respect to any of the N independent variables x_n and subsequent rearrangement gives

$$\begin{aligned} \frac{\partial y^*}{\partial x_n} - \boldsymbol{\lambda}\left(\frac{\partial \mathbf{f}}{\partial x_n}\right) &= 0 \\ &= \frac{\partial}{\partial x_n}(y^* - \boldsymbol{\lambda}\mathbf{f}) \end{aligned} \tag{2-117}$$

The quantity inside the parentheses is called the *Lagrangian* (function) L.

$$L(\mathbf{x}, \boldsymbol{\lambda}) \equiv y(\mathbf{x}) - \boldsymbol{\lambda}\mathbf{f}(\mathbf{x}) \tag{2-118}$$

Its derivatives with respect to the $\boldsymbol{\lambda}$ are the constraint functions, which by Eq. (2-85) must be zero at a feasible point

$$\frac{\partial L}{\partial \boldsymbol{\lambda}} = -\mathbf{f}(\mathbf{x}) = 0 \tag{2-119}$$

Equations (2-117) and (2-119) establish that any feasible local optimum $\mathbf{x}^*$ is a stationary point for the Lagrangian function. Jointly they furnish $M + N$ equations in the M unknown λ_k, and N unknown x_n, which in principle can be solved for $\mathbf{x}^*$ and $\boldsymbol{\lambda}^*$.

There being exactly as many equations as unknowns, any local optimum $\mathbf{x}^*$ has a unique set of sensitivity coefficients $\boldsymbol{\lambda}^*$ associated with it, provided that Eqs. (2-117) and (2-119) are independent in the neighborhood of $\mathbf{x}^*$. Therefore, no matter what set of variables are designated decisions in defining the $\boldsymbol{\lambda}$ of Eq. (2-116), $\boldsymbol{\lambda}$ will be the same. Basically, this comes about because the constrained derivative $(\partial z/\partial d)^*$ vanishes at the optimum, making the term $(\partial z/\partial d)^* \, \partial \mathbf{d}$ vanish in Eq. (2-111). With it disappears any dependence of the sensitivity coefficients on the choice of the set of states. This settles the point raised in the preceding section.

The lack of distinction between decision and solution variables in the Lagrangian formulation makes it useful in theoretical work where numerical computations need not be performed and where, as in the example, there is no basis for singling out particular variables as "decisions." And even when the $M + N$ equations must be solved, the symmetric Lagrange formulation sometimes exposes an advantageous order of solution which may not show up in the other method.

In the example the Lagrangian is

$$L \equiv 1000x_1 + 4 \times 10^9(x_1x_2)^{-1} + 2.5 \times 10^5 x_2 - \lambda_1(x_1x_2 - 9000)$$

Setting the partial derivatives with respect to x_1, x_2, and λ_1 to zero gives

$$\begin{aligned} 1000 - 4 \times 10^9(x_1^*)^{-2}(x_2^*)^{-1} - \lambda_1^* x_2^* &= 0 \\ 2.5 \times 10^5 - 4 \times 10^9(x_1^*)^{-1}(x_2^*)^{-2} - \lambda_1^* x_1^* &= 0 \\ -x_1^* x_2^* + 9000 &= 0 \end{aligned}$$

The reader can verify that the values of x_1^*, x_2^*, and λ_1^* found by the other methods also satisfy these equations. In this case the more widely known Lagrange method has no advantages over direct elimination or the method of Section 2-17, taking more effort than either of them. Lagrange's method will prove its worth later in more complicated situations.

2-20 SUMMARY

Despite this chapter's opening allegorical allusion to "paths to the top of the mountain" the mathematical treatment confined itself entirely to describing peaks, not paths. Knowledge of the mathematical topography of summits leads to indirect methods which, when applicable, find peaks by a procedure resembling a parachute jump straight to the peak more than a mountaineer's climbing ascent. The theory depends heavily on a differential approach, which will, in fact, be the key unifying idea throughout the book.

After a warning about careful employment of differentials, they were manipulated to give necessary and sufficient conditions for a local optimum. Stationary points other than optima were described, and it was shown how to identify them by completing the square on second-order terms of a Taylor expansion. See Wilde (1978) for global optimization tests.

In adapting the classical methods for finding interior optima to handle equality constraints, the concepts of decision, solution, and degrees of freedom were defined. They in turn generated the decision derivatives needed to transform boundary optimization problems into interior ones capable of solution by easy extensions of classical techniques. Similar developments lead to sensitivity coefficients for analyzing the effects on the value of the optimum of changing constraints After completing the development of the classical indirect method for finding interior optima, we derived the different approach of Lagrange multipliers.

This is the present state of knowledge about indirect methods that proceed straight to the peak without touching the mountainsides. In future chapters, obscured visibility will force us to proceed more cautiously, taking shorter steps and keeping to the trails.

BIBLIOGRAPHY

BEIGHTLER, C. S., and D. J. WILDE, "Diagonalization of quadratic forms by Gauss elimination," *Man. Sci.*, **12** (January 1966), 371–379.

BELL, E. T., *The Development of Mathematics* (McGraw-Hill, New York, 1940).

BERKELEY, G., BISHOP, *The Analyst; or, a Discourse Addressed to an Infidel Mathematician.*

BERNOULLI, JOHANN, *Essai d'une nouvelle théorie de la manoeuvre des vaisseaux* (J. G. König, Basle, 1714), pp. 32, 36.

BIRKHOFF, G., and S. M. MAC LANE, *A Survey of Modern Algebra*, 3rd ed. (Macmillan, New York, 1965).

BROOKS, R., and A. GEOFFRION, "Finding Everett's multipliers by linear programming," *Operations Res.*, **14** (1966), pp. 1149–1153.

CAJORI, F., *A History of Mathematics* (Macmillan, New York, 1919).

CAUCHY, AUGUSTIN L., *Cours d'analyse* (Paris, 1821).

COURANT, R., and D. HILBERT, *Methods of Mathematical Physics*, Vol. I (Interscience, New York, 1953).

CRAMER, GABRIEL (for Cramer's rule, see Rosenbach and Whitman, Chap. 18).

DESCARTES, RÉNÉ, cited in Bell, p. 127.

DUFFIN, R. J., "Cost minimization problems treated by geometric means," *Operations Res.*, **10** (1962), 668–675.

———, "Dual programs and minimum cost," *J. Soc. Ind. Appl. Math.*, **10** (1962), 119–123.

EDELBAUM, T. N., "Theory of maxima and minima," in Leitman, p. 16.

EULER, L., *Calc. diff.* (1755), cited in Hancock, p. 18.

EVERETT, H., "Generalized Lagrange multiplier method for solving problems of optimum allocation of resources," *Operations Res.*, **11** (1963), pp. 399–471.

FERMAT, PIERRE DE, "Methodus disquirendum maximam et minimam" (1638), *Oeuvres*, **1** (Gauthier-Villars, Paris, 1841), pp. 133–136, cited in Bell, p. 128.

———, "De maximus et minimus," *Oeuvres*, **1** (Gauthier-Villars, Paris, 1841), pp. 147–150, cited in Cajori, p. 164.

GAUSS, CARL FRIEDRICH, *Disquisitiones Arithmeticae* (Leipzig, 1801), p. 1278, cited in Hancock, p. 86.

GENOCCHI, A., and G. PEANO, *Calcolo Differenziale e Principii di Calcolo Integrale* (1884), prob. 133–136, cited in Hancock, p. 33 *et seq.*

HANCOCK, HARRIS, *Theory of Maxima and Minima* (1917) (Dover, New York, 1960).

HUYGENS, CHRISTIAN, *Bibliothèque universelle et historique* (September 1693).

JACOBI, C. G. J., cited in Beil, p. 400.

KEPLER, JOHANNES, cited in Cajori, p. 163.

KUNZ, K. S., *Numerical Analysis* (McGraw-Hill, New York, 1957).

LAGRANGE, COMTE JOSEPH LOUIS, "Recherches sur la méthode de maximis et minimis," *Miscellanea Taurinensia* (1759), *Oeuvres*, **1**, pp. 3–20, cited in Hancock, p. 86.

———, "Essai d'une nouvelle méthode pour détérminer les maxima et les minima," *Miscellanea Taurinensia*, **2** (1760–1761) *Oeuvres*, **1**, pp. 356–357, 360.

———, *Théorie des fonctions analytiques* (1797), p. 290, cited in Hancock, p. 33.

———, *Calcul des fonctions* (1799), cited in Bell, p. 267.

———, *Oeuvres de Lagrange* (Gauthier-Villars, Paris, 1867).

LEIBNIZ, GOTTFRIED WILHELM, *The Early Mathematical Manuscripts of Leibniz*, translated by J. M. Child from Latin texts published by Carl Immanuel Gerhardt (The Open Court Publishing Co., London, 1920).

MACLAURIN, COLIN, *A Treatise on Fluxions* (1742), pp. 238, 857, cited in Cajori, p. 229.

MENCKEN, H. L., *A New Dictionary of Quotations* (Alfred A. Knopf, New York, 1942).

NEWTON, SIR ISAAC, cited in Bell, p. 203.

PHILLIPS, D. T., A. RAVINDRAN, and J. J. SOLBERG. *Operations Research: Principles and Practice* (Wiley, New York, 1976).

RAPHSON, JOSEPH, *Analysis aequationem universalis*, cited in Cajori, p. 203.

RENAU, CHEVALIER, *Théorie de la manoeuvre des vaisseaux* (Paris, 1689).

ROSENBACH, J. B., and E. A. WHITMAN, *College Algebra* (Ginn and Company, Boston, 1939).

SCHEEFER, LUDWIG, "Über die Bedeutung der Begriffe Maximum und Minimum in der Variationsrechnung," *Math. Ann.*, **26** (1886), 197–208, cited in Hancock, Chap. 4.

———, "Theorie der Maxima und Minima einer Function von zwei Variabeln," *Math. Ann.*, **35** (1890), 541–576, cited in Hancock, Chap. 4.

STOLZ O., cited in Hancock, Chap. 4, pp. 39–42.

TAYLOR, BROOK, *Methodus incrementorum directa et inversa* (1715–1717), cited in Cajori, p. 226.

WEIERSTRASS, K., cited in Courant and Hilbert, p. 164.

WILDE, D. J., "A unified approach to multivariable optimization theory," *Ind. Eng. Chem.*, **57**, 8 (August 1965), 18–30.

———, *Globally Optimal Design* (Wiley-Interscience, New York, 1978).

EXERCISES

2-1. Prove Eqs. (2-17) and (2-18).

2-2. Perform two iterations of the Newton–Raphson method to find the positive root of Eqs. (2-39) and (2-40), starting at (a) $\mathbf{x} = (800, 3)$; (b) $\mathbf{x} = (1000, 5)$; (c) $\mathbf{x} = (800, 4)$.

2-3. Prove Maclaurin's principle that a stationary point is a minimum if and only if the lowest-order nonvanishing derivative present is positive and of even order.

2-4. (Genocchi and Peano's counterexample) Let $y = (x_1 - a_1^2 x_2^2)(x_1 - a_2^2 x_2^2)$, with a_1 and a_2 constants.

(a) Show that Lagrange's criterion indicates a maximum for y at the origin.

(b) Show that y is minimum at the origin for all points on the curve $x_1 = (a_1^2 + a_2^2)x_2^2/2$, if $a_1 \neq a_2$.

2-5. Use the Newton–Raphson method to solve the following system of equations, starting at $\mathbf{x} = (2, 3, 4)$:

$$x_1 e^{-x_2^2} + x_3 = 4$$
$$x_1^2 + x_2^2 + x_3^2 = 45$$
$$x_1 + x_2 + x_3 = 11$$

2-6. Solve the following nonlinear programming problem using the method of Lagrange multipliers: Find real numbers $x_1, x_2, \ldots, x_N$ that

$$\text{minimize } y = \sum_{i=1}^{N} i x_i^2$$

and that satisfy

$$\sum_{i=1}^{N} x_i = R$$

where N is a given positive integer and R is a given real number.

2-7. Use the technique of Lagrange multipliers and Newton's method to solve the following nonlinear programming problem: Find values x_1, x_2 that

$$\text{minimize } y = x_1^2 - 10x_1 - x_2$$

and that satisfy

$$6x_1 + 16x_2 - x_1^2 - 4x_2^2 \geq 9$$

2-8. A function $y(\mathbf{x})$ is *strictly convex* if it is always overestimated by a linear interpolation between any two points, $\bar{\mathbf{x}}$, $\hat{\mathbf{x}}$; that is, for every number $0 < \alpha < 1$, $y[\alpha\bar{\mathbf{x}} + (1 - \alpha)\hat{\mathbf{x}}] < \alpha y(\bar{\mathbf{x}}) + (1 - \alpha)y(\hat{\mathbf{x}})$. Prove that a positive-definite quadratic function is strictly convex.

2-9. Use the Lagrangian method to find the maximum value of the function

$$y = x_1^2 + x_2^2 - 3x_1x_2$$

where x_1 and x_2 must lie on the circle

$$x_1^2 + x_2^2 = 6$$

Ans. $(\sqrt{3}, -\sqrt{3})$ and $(-\sqrt{3}, \sqrt{3})$.

Linear Programming, Sensitivity Analysis, and Integer Programming 3

When the Well's dry,
they know the Worth of Water.

BENJAMIN FRANKLIN, *Poor Richard's Almanack* (1758)

Poor Richard's maxim raises a subtle question: What is the value of water? To a man dying of thirst, water is worth a great deal, but to a drowning man it has no value at all. The answer often depends on how much water is available, how much is needed, and for what it is to be used.

The availability of a commodity is expressed mathematically by an inequality or equality constraint. If at the optimum this constraint is tight, then its value to the optimum solution is important and restrictive. In this case, incremental changes in the scarce commodity will normally cause improvements in the objective function. Finding the enhancement of a material's worth due to its scarcity is therefore closely related to optimization theory. The quantitative study of such questions is called *sensitivity analysis* because it involves computing the effects of changes in availability upon the optimum value of the objective.

The impact of constraints on an optimum is especially marked when the objective and constraints are all linear functions of the independent variables. Since in this case it is impossible to have an interior optimum, the optimum is profoundly affected by the constraints and, consequently, by availabilities of scarce commodities. Moreover, the linearity implies that incremental changes in the objective function relevant to a particular solution variable will remain constant over finite ranges of availability changes, which is not the case in general. This makes possible precise predictions of the economic effects of availability shifts and the point at which a scarce commodity no longer affects the objective function.

Since linear programming involves optimization over both a linear objective function and linear constraints, it is possible to develop an algorithm that will predict exactly how changes in decision variables affect constraints and program objectives. The simplex algorithm of George Dentzig accomplishes this goal and its computational use will be explained in detail.

A major topic of this chapter, linear sensitivity analysis, is illustrated in full detail by an example due to Symonds involving a hypothetical petroleum refinery. Changes in availability of raw material or in the demands for products force alterations in the production plan which may at first contradict what one's intuition might suggest. The imputed values of certain materials fluctuate remarkably as the availabilities of other commodities change. When other availabilities are held constant, a commodity's imputed value decreases in jumps as more of it becomes available, until ultimately its worth to the system drops to zero. This illustrates the well-known economic "law of diminishing returns," even though the problem is completely linear. Detailed study of this idealized example gives great insight into the behavior of complicated multivariable systems.

Aside from the ease with which its sensitivity analysis can be performed, the fully linear case has many interesting characteristics described in this chapter. Historically, the linear case was the first inequality constrained optimization problem to be solved, and Dantzig's successful *simplex* algorithm stimulated the current revived interest in optimization theory. By now, there are hundreds of applications of linear programming to subjects ranging from agriculture to zinc smelting [see Dantzig (1963)] and existing computer codes can handle several thousand independent variables and a thousand constraints.

The linearity also allows exploitation of the structure of a large problem to permit its decomposition into small problems of manageable size. This *decomposition principle* of Dantzig and Wolfe, illustrated here with a numerical example having intriguing economic interpretations, is a foretaste of the dynamic programming conditional optimization approach of Chapter 7.

The *transportation problem*, in which one wishes to distribute goods from depots to customers in an optimal fashion, is a special case of the linear programming problem. Its elegant solution algorithm and simple sensitivity analysis justify a brief discussion here. The chapter ends by describing the difficult *integer programming* problem in which the independent variables must be whole numbers. The practical significance and abundance of integer programming problems warrants a separate investigation of these solution procedures.

3-01 LINEAR PROGRAMMING

The linear programming problem is a special case of the general nonlinear programming problem in which there are no terms of second degree or higher in the objective function or constraints. Thus the problem is to find nonnegative x_n, maximizing the *linear* form

$$y = \sum_{n=1}^{N} c_n x_n \tag{3-1}$$

subject to the linear constraints,[1]

$$\sum_{n=1}^{N} a_{mn} x_n \leq b_m; \qquad m = 1, \ldots, M \tag{3-2}$$

In order to illustrate the formulation of a linear programming problem, consider the following (highly simplified) example.

The Oilcan Drilling Corporation is a leading manufacturer of two major drilling bits, types A and B. Owing to increased exploration efforts, Oilcan, Inc., can sell all they can produce. Type A sells for $3000 and type B sells for $2000. Each bit is manufactured in two phases: phase I and phase II. Drill bit type A requires 2 hours of phase I and 1 hour of phase II. Drill bit type B requires 1 hour of phase I and 3 hours of phase II. If phase I and phase II availability is only 8 and 15 hours, respectively, over a 24-hour period, how many drill bits of each type can be produced per day?

The objective is to maximize daily production, subject to availability constraints. Define the following:

$$x_1 = \text{units of type A produced daily}$$

$$x_2 = \text{units of type B produced daily}$$

Hence the following linear programming problem represents the manufacturing process.

$$\text{Maximize } Z = 3x_1 + 2x_2$$

[1]The problems of more general formulations containing equality constraints and reversed ($\geq$) constraints will be considered shortly.

subject to

$$2x_1 + x_2 \leq 8$$
$$x_1 + 3x_2 \leq 15$$
$$x_1, \quad x_2 \geq 0$$

This problem can be solved graphically in two dimensions, and we will now do so in an effort to reveal several important properties of linear programming problems which hold in the more general cases. First, the linear inequalities in this formulation are known as *half-lines* and serve to bound the region of our solution. At points along these half-lines where the constraint functions hold as equalities, the constraints form hyperplanes. Intersections of hyperplanes define extreme points in the solution space, and it is this set of extreme points that is of interest to us.

Wehl first proved that the optimal solution must lie on at least one of the extreme points, which is intuitively clear in two and three dimensions. This two-dimensional problem is solved graphically, as shown in Fig. 3-1. The objective function is a family of parallel straight lines that move farther away from the origin as y is increased through positive values. We want to maximize y and at the same time satisfy the constraints. These constraints consist of the nonnegativity conditions plus the inequalities, and serve to define the convex set bounded by the lines $x_1 = 0$, $x_2 = 0$, $x_1 + 3x_2 = 15$, and $2x_1 + x_2 = 8$. In the general linear programming problem defined by Eqs. (3-1) and (3-2), the convex set will be bounded by $M + N$ hyperplanes, each of which is N-dimensional, and the objective function will consist of a family of N-dimensional hyperplanes. When $N = 2$, the hyperplanes are lines; when $N = 3$, they are planes. In either case, the extreme value (maximum or minimum) of the objective function will be reached by that member of the family of lines (or planes) which touches the convex set on the boundary, and this must include at least one extreme point. In a two-dimensional problem, the objective hyperplane (line) can touch the convex set at no more than two extreme points. This can occur when the slope of the line is the same as the slope of a constraint; then the optimal solution is achieved at all points on the line joining these extreme points, producing alternative optimal solutions. In a three-dimensional problem, the objective plane may be optimized when it strikes a point (intersection of three constraint planes), a line (intersection of two constraint planes), or a plane (the direction cosines of the objective function coincide with those of a constraint). In an N-dimensional problem the optimal solution may be achieved at up to N different points. In high-dimensional problems, therefore, alternative optima are the rule rather than the exception. For the present problem, the optimum is achieved at the unique point $x_1 = \frac{9}{5}$, $x_2 = \frac{22}{5}$, determined by the intersection

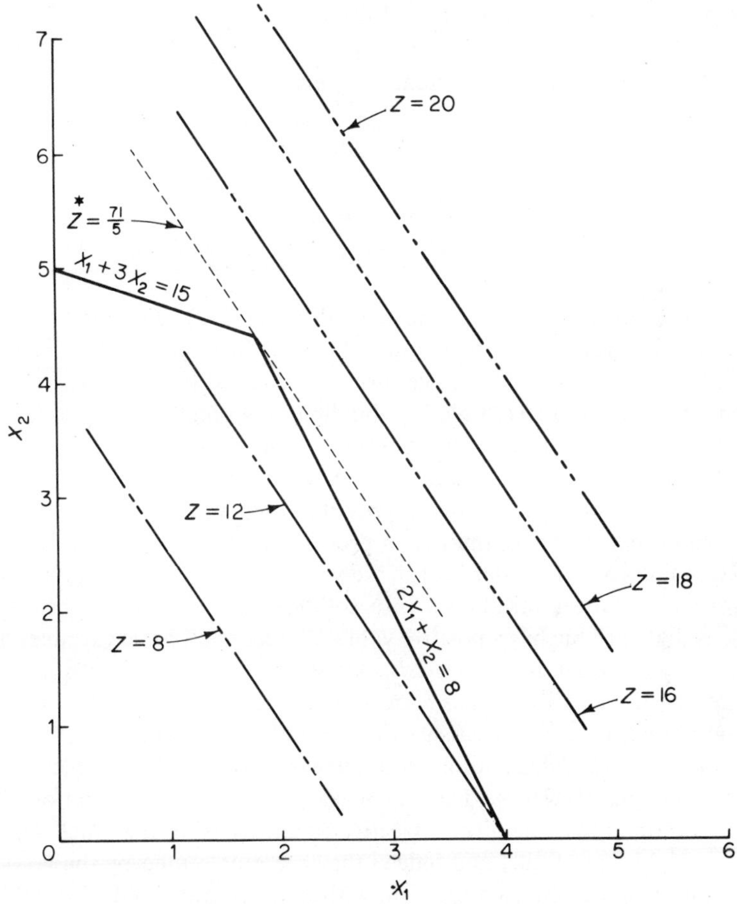

Figure 3-1. Graphical solution of two-dimensional example problem.

of the constraint lines $x_1 + 3x_2 \doteq 15$, and $2x_1 + x_2 = 8$, as shown in Fig. 3-1.

3-02 THE SIMPLEX ALGORITHM

From the previous example, it should now be intuitively clear that we seek the intersection of half-lines in N-dimensional space. These intersections form the *vertices* of the solution space, and we are assured that the optimal solution will lie at one of these points. Simply speaking, the central mathematical problem of linear programming is to find the solution to a system of linear equations which maximizes or minimizes a give objective function. Before proceeding, we define the *standard form* of a linear programming problem with m constraints and n variables as follows:

$$\text{maximize/minimize} \sum_{j=1}^{n} c_j x_j \tag{3-3}$$

subject to

$$\begin{aligned} \sum_{j=1}^{n} a_{kj} x_j &= b_k & k &= 1, 2, \ldots, M \\ x_j &\geq 0 & j &= 1, 2, \ldots, N \\ b_k &\geq 0 & k &= 1, 2, \ldots, N \end{aligned} \tag{3-4}$$

Since the standard form requires the inequalities of (3-2) to become equality constraints, *slack* variables $\bar{S}_k$ will be added to each inequality restriction ($\leq$). Note that these new "slack" variables are also required to be nonnegative. If the inequality constraint is "tight" or binding at optimality, the associated slack variable will be zero. If the constraint is "loose" or nonbinding at optimality, the slack variable will be positive. Slack variables play a central role in the algorithmic processes of linear programming, and they will be treated equally with the x_j in the optimization process.

Returning to the standard form, we will define a variable x_j to be *basic* (or a *state* or solution variable) if in a solution to the m equations and n unknowns that variable has a positive value. The set of all basic variables in a particular solution set forms a *basis.* Those variables that are not basic (hence zero) will be called *nonbasic*, or *decision* variables. If the entire set of basic and nonbasic variables satisfies the original constraint set, the solution vector is called a *feasible* solution to the linear programming problem. The simplex technique of George Dantzig seeks a basic, feasible solution that maximizes (minimizes) the given objective function. Note that such sets of solution variables are finite, and indeed they are represented by the vertices of the solution space. Given a problem with M constraints and N variables, the *maximum* number of basic solutions is given by

$$\binom{N}{M} = \frac{N!}{M!(N-M)!}$$

By construction, all basic solutions are also basic feasible solutions. A solution procedure that readily presents itself is to examine all basic solutions and choose the one that maximizes (minimizes) the objective function. Unfortunately, this approach would involve a great deal of effort for even moderate-sized formulations. The simplex procedure allows us to determine the optimal solution (or verify its nonexistence) with a fraction of the effort.

The simplex technique requires that one obtain a basic feasible solution to begin the procedure. By construction, a choice that is always available is to initially choose each slack variable as basic, which yields a feasible solution of $x_n = 0$; $n = 1, 2, \ldots, N$ and $\bar{S}_k = b_k$; $k = 1, 2, \ldots, M$. Since the slack variables contribute nothing to the original objective function, the optimal solution at this point is zero. The simplex procedure is best presented through the use of a working tableau, which is illustrated in the following general form.

THE SIMPLEX TECHNIQUE

The Simplex tableau   Column I

		x_1	x_2	x_3	$\cdots$	x_N	s_1	s_2	$\cdots$	s_M	b	θ
$c_j \rightarrow$												
$\downarrow P_i \downarrow$	Basic											
$\vdots$	$\vdots$	$\vdots$	$\vdots$	$\vdots$	$\vdots$	$\vdots$	$\vdots$	$\vdots$	$\vdots$	$\vdots$	$\vdots$	$\vdots$
	z_j											
v_j	$c_j - z_j$											

m rows

Row I

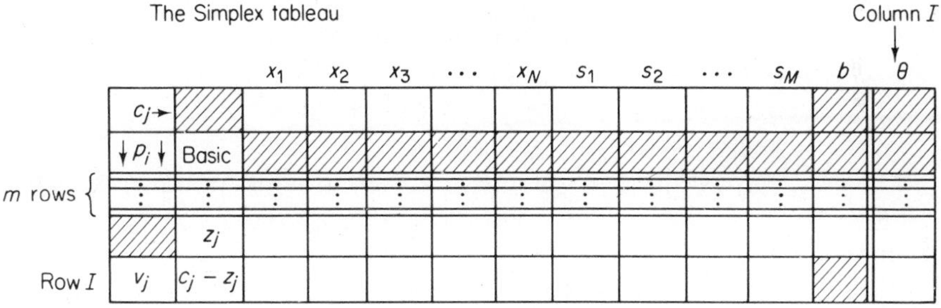

The general steps in the simplex procedure are as follows:

The simplex procedure

Step 1:   Convert all inequalities to equalities by adding slack variables.

Step 2:   Enter the coefficients of the objective function in the first row; original solution variables first, then slack variables.

Step 3:   Form a *basis* by setting $(N - M)$ variables to zero—the slack variables are a logical choice for the original basis since this corresponds to the *origin* (a basic, feasible solution). This puts the slack variables into the first state variable set.

Step 4:   Enter the coefficients of the equality constraints by rows in the body of the table.

Step 5:   Enter the solution variables in the basis column corresponding to the equality constraints in each row. (These are initially the slack variables.)

Step 6:   Enter the objective function coefficient corresponding to the solution variables in the P_i *column*. Note slack variable coefficients will be zero.

For the example that we previously solved graphically, we obtain the following simplex tableau.

		x_1	x_2	$\bar{s}_1$	$\bar{s}_2$	b	θ
$c_j \rightarrow$		3	2	0			
$\downarrow p_i \downarrow$	Basis						
0	$\bar{s}_1$	2	1	1	0	8	
0	$\bar{s}_2$	1	3	0	1	15	
	z_j	0	0	0	0		↑ Column I
	v_j	3	2	0	0	← Row I	

where

$$Z_j = \sum_{i=1}^{M} P_i a_{ij} \qquad j = 1, 2, \ldots, n \quad (n = N + M)$$
$$v_j = c_j - Z_j \tag{3-5}$$

Note that the "θ" column is blank. This column will be utilized in determining basis change strategies and will be discussed shortly.

We shall now attempt to explain the steps of the simplex procedure by way of algebraic example. First, we should note that in the above tableau, the v_j elements of row I indicate the contribution of a unit change in the variables $x_1, x_2, \bar{S}_1$, and $\bar{S}_2$ to objective function maximization. The v_j variables which appear in the last row of the simplex tableau will play an important role in the developments which fill this chapter. Since they measure the rate of change of program objective relative to increases in each decision variable, they will be referred to as *sensitivity coefficients.*[2] Since x_1 contributes the most, we will try to bring it into the basis. This represents an attempt to change x_1 from a nonbasic variable to a basic variable. The column corresponding to the nonbasic variable just selected will be called the *pivotal column.* We must now decide which basic variable ($\bar{S}_1$ or $\bar{S}_2$) to remove from the basis. This decision is made by calculating the elements of the θ column (column I) according to the formula

$$\theta = \frac{b_i}{a_{ij}} \qquad \begin{array}{l} i = 1, 2, \ldots, M \\ j = \text{index of the pivotal column} \end{array} \tag{3-6}$$

For this example,

θ
4
15

$$\theta_1 = \tfrac{8}{2} \qquad \theta_2 = \tfrac{15}{1}$$

[2]These are sometimes called "shadow prices" or the "pricing vector."

The basic variable which will be removed is the one that corresponds to the smallest positive entry. Hence S_1 will be removed from the basis. The row corresponding to the basic variable just chosen will be called the *pivotal row*. An algebraic interpretation of these two steps is as follows. From the initial tableau, we can write the two constraint equations.

$$2x_1 + x_2 + \bar{S}_1 = 8$$
$$x_1 + 3x_2 + \bar{S}_2 = 15$$

Expressing the original basic variables in terms of the original nonbasic variables we obtain

$$\bar{S}_1 = 8 - 2x_1 - x_2$$
$$\bar{S}_2 + 15 - x_1 = 3x_2$$

Now, recall that any basic variable cannot be driven negative lest we obtain a nonfeasible solution vector. How large can we increase the two nonbasic variables x_1 and x_2 before a basic variable is driven negative? By inspection, x_1 can be raised to 4 and x_2 to 5 (individually, of course). Which change will contribute the most to objective function maximization? From the original objective function coefficients (c_j entries) we obtain an increase of 12 through a change in x_1 and an increase of 10 through a change in x_2. Hence we will place x_1 in the basis as our new basic variable and remove $\bar{S}_1$ from the current basis and designate it as a decision (nonbasic) variable. Note that this is *precisely* the same decision obtained through our simplex procedure. The current tableau will now be transformed through application of the following rules.

The intersection of the pivotal row and the pivotal column will be called the *pivotal element*.

1. Divide the Pivotal Element by itself.
2. Change all other elements in the *pivotal column* to zero (except the Z_j and v_j elements).
3. Divide all elements in the *pivotal row* by the pivot (except the P_i and θ_i elements).
4. Replace the old basic variable with the new entering basic variable in the basis column, and place the objective function coefficient associated with the new basic variable in the P_i column.
5. Transform the rest of the tableau according to the following formula.

$$\hat{E} = E - \frac{(R)(C)}{PT} \tag{3-7}$$

where $\hat{E}$ = new value of tableau element
E = old value of tableau element
R = number in the pivotal row corresponding to the same column as E
C = number in the pivotal column corresponding to the same column as E
PT = pivotal element

For the current example, these rules generate the following tableau.

		x_1	x_2	$\bar{s}_1$	$\bar{s}_2$	b	θ
c_j		3	2	0	0		
p_i	SOL						
3	x_1	1	1/2	1/2	0	4	8
0	$\bar{s}_2$	0	(2 1/2)	−1/2	1	11	22/5
	z_j	3	3/2	3/2	0		
	v_j	0	1/2	−3/2	0		

Proceeding as before, we will attempt to explain this transformation algebraically. In order to generate a new basis change, we must obtain an expression relating the new basic variables (basis) to the new decision (that is, nonbasic) variables. Since the new basis is x_1 and $\bar{S}_2$, we wish to solve our constraint set for these two variables. We obtain the following.

$$x_1 = 4 - \tfrac{1}{2}x_2 - \tfrac{1}{2}\bar{S}_1$$
$$\bar{S}_2 = 11 - \tfrac{5}{2}x_2 - \tfrac{1}{2}\bar{S}_1$$

Since $\bar{S}_1$ and x_2 are now our new decision (nonbasic) variables, we must assess the impact of a change in these variables on the current objective function. Expressing the original objective function in terms of these new nonbasic variables, we obtain

$$Z = 3x_1 + 2x_2 = 3(4 - \tfrac{1}{2}x_2 - \tfrac{1}{2}\bar{S}_1) + 2x_2$$

or

$$Z = 12 + \tfrac{1}{2}x_2 - \tfrac{3}{2}\bar{S}_1$$

From the decision variable/basic variable relationships, it is clear that $\bar{S}_1$ can only be increased to 8, and x_2 can only be increased to $\frac{22}{5}$. From the objective function equation, we see that a unit increase in x_1 causes a gain of $\frac{1}{2}$ toward objective maximization while any increase in $\bar{S}_1$ will cause a decrease. Hence we choose to move x_2 to $\frac{22}{5}$. This will make x_2 a new basic variable and $\bar{S}_2$ a new decision variable ($\bar{S}_2 \equiv 0$). Referring to the transformed simplex tableau, it is clear that the new constraint relationships are given in the body of the second tableau, while the new objective function is given in the last row. Since the column corresponding to x_2 has the largest positive element in this row, it becomes the pivotal column. By calculating the elements of the θ column; $\theta_1 = 8$ and $\theta_2 = \frac{22}{5}$, the row corresponding to the state variable $\bar{S}_2$ becomes the pivotal row. Again, using the simplex tableau transformation rules, one obtains the following. (The pivotal element is circled in the second tableau.) Since there are no positive elements in the last row, no further

		x_1	x_2	$\bar{s}_1$	$\bar{s}_2$	b	θ
c_j		3	2	0	0		
p_i	SOL						
3	x_1	1	0	3/5	−1/5	9/5	
2	x_2	0	1	−1/5	2/5	22/5	
	z_j	3	2	7/5	1/5		
	v_j	0	0	−7/5	−1/5		

improvement is possible in the current objective function. Hence the optimal solution is given by $x_1^* = \frac{9}{5}$; $x_2^* = \frac{22}{5}$; $\bar{S}_1^* = \bar{S}_2^* = 0$. The optimal (maximum) value of the objective function is given by $Z^* = \frac{71}{5}$.

Note that the simplex method, at each iteration, has moved the objective function line parallel to itself and away from the origin ($y = 0$) until a further move in this direction would yield an infeasible solution, as shown in Fig. 3-1. Values such as $Z = 16$ correspond to solutions that are optimal (since all the sensitivity coefficients would be negative) but not feasible, whereas solution points anywhere along the lines $Z = 8$ or $Z = 12$ (and within the convex set) are feasible but not optimal. The line $Z = 8$ does not pass through an extreme point of the convex set, and therefore no basic feasible solution would produce this value of the objective function.

The simplex method moves from one basic feasible solution to another (vertex to vertex) and usually never returns to a previous solution point, because it improves the value of the objective function at each iteration. There being only a finite number of extreme points, each corresponding to a unique basic feasible solution, the simplex method will find the optimal solution to any linear programming problem in a finite number of steps. This algorithm does not usually investigate every extreme point of the convex set, of course, but example problems have been constructed (Goldman and Kleinman) for which the method *does* encounter every extreme point. Consequently, upper bounds on the maximum number of iterations required to solve a general problem containing N variables and M inequality constraints are determined solely by the total number of such points, or vertices (Saaty). This number, of course, is considerably smaller than the number of all intersections of the constraint and coordinate hyperplanes, given by the combinatorial $(M + N)!/M!\,N!$, which includes such infeasible points, for example, as (0, 8) and (15, 0) in Fig. 3-1. The *expected* number of iterations required by the simplex method has been estimated empirically as $2M$, that is, twice the number of inequality constraints (3-2). Both the average and the maximum number of iterations, however, can be changed by modifying the rule for selecting the order in which decision variables are brought into the basis (Quandt and Kuhn).

Linear programming has been used extensively in business and industry, and its applicability to a wide variety of problems has resulted in numerous computer codes and techniques for handling special formulations of the problem. Although few practical problems are actually linear, the simplex method is so efficient that the approximate solution given by the linear programming model is sometimes the most practical one for very large problems. In a later section, we shall describe briefly some of the algorithms for solving certain special classes of linear programming problems. These are all modifications of the general simplex method previously described, and we shall not attempt a complete listing of them, nor of the numerous applications of linear programming, as these are well documented elsewhere (Dantzig, 1963). Finally, note that since the first two rows of the simplex tableau never change, and the Z_j row is only used to calculate the v_j decision row, further applications of the simplex technique will normally use only the body of the simplex tableau and the v_j decision row. We will follow this convention in subsequent examples.

3-03 DEGENERACY

Since the components of a basic feasible solution generated at each simplex tableau will always be positive (in a maximization problem), the objective function will increase with each iteration. The sole exception occurs when a θ-element is generated which exactly equals zero; when this happens, the minimal pivotal row will certainly be chosen by that element and the next simplex iteration will produce no change in the value of the objective function. Note that this will generate a new basic variable with value of zero. If at any time during the solution procedure one or more of the basic variables take on a value of zero, this condition is referred to as *degeneracy*. Although degeneracy occurs often in linear programming problems, it has never caused any difficulties in the solution of problems arising in practice. If, however, the value of the objective function does not change for several iterations, it is possible for the simplex method to return to a previous solution point, forming an endless *cycle*, and thus never reach the optimal solution (Hoffman). Cycling can always be prevented by recourse to a perturbation method due to Charnes, which resolves degeneracy from the theoretical point of view. Computationally, however, this method is rarely used; even if degenerate solutions *are* encountered in a problem, the objective function will fail to improve only during those iterations in which a basic variable having a zero value is removed from the basis. This has never caused any real-world problem to cycle back to a previous solution point, however, so that degeneracy is nothing to worry about. The following example shows some of the effects produced by degenerate basic feasible solutions.

$$\text{Maximize } y = x_1 + 2x_2 \tag{3-8}$$

subject to the constraints

$$\begin{aligned} x_1 + 3x_2 &\leq 105 \\ -x_1 + x_2 &\leq 15 \\ 2x_1 + 3x_2 &\leq 135 \\ -3x_1 + 2x_2 &\leq 15 \\ x_1, \quad x_2 &\geq 0 \end{aligned}$$

The solution to this problem requires four iterations, as shown in the tableaux of Fig. 3-2; again we have circled the pivot numbers to indicate which variables will enter and leave the basis at the next iteration. The value of the objective function is also given at each iteration in the lower right-hand corner. The objective function value can also be determined at each

	x_1	x_2	$\bar{s}_1$	$\bar{s}_2$	$\bar{s}_3$	$\bar{s}_4$			
	1	2	0	0	0	0	b	θ	
$\bar{s}_1$	1	3	1	0	0	0	105	35	
$\bar{s}_2$	−1	1	0	1	0	0	15	15	
$\bar{s}_3$	2	3	0	0	1	0	135	45	
$\bar{s}_4$	−3	(2)	0	0	0	1	15	15/2	
	+1	+2	0	0	0	0	0	← PROFIT	

(a) Initial tableau

Figure 3-2. Tableaux for problem illustrating degeneracy.

								θ
$\bar{s}_1$	11/2	0	1	0	0	−3/2	165/2	15
$\bar{s}_2$	1/2	0	0	1	0	−1/2	15/2	15
$\bar{s}_3$	13/2	0	0	0	1	−3/2	225/2	225/13
x_2	−3/2	1	0	0	0	1/2	15/2	−
	+4	0	0	0	0	−1	15	← PROFIT

(b) Tableau resulting from x_2 replacing f_4 in state set

								θ
$\bar{s}_1$	0	0	1	−11	0	4	0	0
x_1	1	0	0	2	0	−1	15	−
$\bar{s}_3$	0	0	0	−13	1	5	15	3
x_2	0	1	0	3	0	−1	30	−
	0	0	0	−8	0	+3	75	← PROFIT

(c) Tableau resulting from x_1 replacing $\bar{s}_2$ in state set

								θ
$\bar{s}_4$	0	0	1/4	−11/4	0	1	0	−
x_1	1	0	1/4	−3/4	0	0	15	−
$\bar{s}_3$	0	0	−5/4	3/4	1	0	15	20
x_2	0	1	1/4	1/4	0	0	30	120
	0	0	−3/4	+1/4	0	0	75	← PROFIT

(d) Tableau resulting from $\bar{s}_4$ replacing $\bar{s}_1$ in state set

$\bar{s}_4$	0	0	−13/3	0	11/3	1	55	
x_1	1	0	−1	0	1	0	30	
$\bar{s}_2$	0	0	−5/3	1	4/3	0	20	
x_2	0	1	2/3	0	−1/3	0	25	
	0	0	−1/3	0	−1/3	0	80	← OPTIMAL PROFIT

(e) Final tableau

Figure 3-2. (continued)

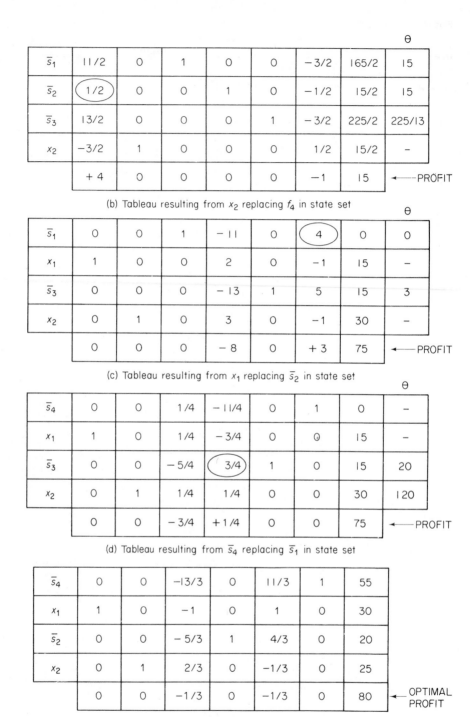

iteration through the application of standard tableau transformation rules. Notice that there is no increase in the objective function between the tableaux (c) and (d) in this figure, since $\bar{S}_4$ enters the basis with a zero value. This does not prevent the process from reaching the optimal solution on the next step, and indeed, in many problems the value of the objective function may not change for several iterations before once again beginning to improve.

3-04 FINDING A FIRST FEASIBLE SOLUTION

When all constraints are of the form of equations (3-2), and every b_m on the right-hand side of each inequality is positive, then a first basic feasible solution is immediately at hand. Simply set all the x_n; $n = 1, 2, \ldots, N$ variables to zero; then the slack variables $\bar{S}_m$; $m = 1, 2, \ldots, M$ consitute the basis set and the first feasible solution is given by $\bar{S}_m = b_m$; $m = 1, 2, \ldots, M$. However, if there are any greater-than or equal-to constraints, the procedure fails. Note the following.

If

$$\sum_{n=1}^{N} a_{kn}x_n \geq b_k \qquad k = 1, 2, \ldots, l \tag{3-9}$$

then

$$\sum_{n=1}^{N} a_{kn}x_n - \bar{S}_k = b_k \qquad k = 1, 2, \ldots, l \tag{3-10}$$

In this case $\bar{S}_k = -b_k$ in the initial solution, and this is infeasible. To take care of this situation, we add an *artificial* (or dummy) variable, u_k, to each such constraint, which thus becomes

$$\sum_{n=1}^{N} a_{kn}x_n - \bar{S}_k + u_k = b_k; \qquad k = 1, \ldots, l \tag{3-11}$$

where we have renumbered these constraints from 1 to l with no loss in generality. If any of the constraints (3-2) are *equalities*, rather than inequalities, we add a dummy variable to these constraints also, after first multiplying through by -1 if necessary to make the right-hand side positive. These will then be of the form of Eqs. (3-11) except that, of course, they will not contain any slack variables. Thus we have generated the *artificial* basic feasible solution: $u_k = b_k$, $k = 1, \ldots, l$; $\bar{S}_m = b_m$, $m = l + 1, \ldots, M$. This does not satisfy the original constraints, but we may now use the simplex method to move to another basic feasible solution which does satisfy the constraints. This is accomplished by *minimizing*

$$K = \sum_{k=1}^{l} u_k \tag{3-12}$$

subject to Eqs. (3-11), the constraints

$$\sum_{n=1}^{N} a_{mn}x_n + \bar{S}_m = b_m; \qquad m = l + 1, \ldots, M \tag{3-13}$$

and nonnegativity conditions on all the variables. It can be shown (Hadley) that if *any* feasible solution exists to Eqs. (3-12), then a basic feasible solution also exists. Therefore, unless there is no solution to the original constraints, the optimal solution to this auxiliary linear programming problem will be $K = 0$. Since the u_k were required to be nonnegative, their sum can be zero only when each variable is itself zero. We have driven the dummy variables to zero while retaining a basic feasible solution that may now be used as the first feasible solution for the original problem. The dummy variables will be dropped, of course, since they were introduced only in order to obtain a basic feasible solution in terms of the x_n and $\bar{S}_m$.

The final tableau of the auxiliary problem can now be used as the initial tableau for the original problem, except for the bottom row, which must now express the *original* objective function in terms of the current decision (nonbasic) variables. This is accomplished as before by replacing all basic variables in the original objective function by their expressions in terms of the decision variables, as given by the simplex tableau. The current basic variables can always be expressed in terms of the remaining decision variables. In general, one might obtain the following equations.

$$S_m = B_m - \sum_{n=M+1}^{M+N} \alpha_{mn}x_n$$

Note that if each state variable happened by chance to equal the original slack variables, then $S_m \equiv \bar{S}_m$, $B_m \equiv b_m$, and $\alpha_{mn} \equiv a_{mn}$. Thus, if the original function is given by Eq. (3-1), substitution of the expressions for the basic variables x_m yields the substituted objective function:

$$\begin{aligned} Z &= \sum_{m=1}^{M} p_m\left(\beta_m - \sum_{n=M+1}^{M+N} \alpha_{mn}x_n\right) + \sum_{n=M+1}^{M+N} c_n x_n \\ &= \sum_{m=1}^{M} p_m\beta_m + \sum_{n=M+1}^{M+N}\left(c_n - \sum_{m=1}^{M} p_m\alpha_{mn}\right)x_n \end{aligned} \tag{3-14}$$

where the basic variables have been renumbered from 1 to M, as before, and the associated cost coefficients are designated as P_m. The sensitivity coefficients can thus be found from the expressions

$$v_n = c_n - \sum_{m=1}^{M} p_m\alpha_{mn}; \qquad n = M + 1, \ldots, M + N \tag{3-15}$$

Equation (3-15) can always be used to find the v_n elements in the bottom row for any linear programming problem having numbers α_{mn} in the body of the tableau and known coefficients from the objective function. At this point, the reader should recognize Eq. (3-15) as the one previously used (without proof) in generating row I of each simplex tableau. It will also be of fundamental importance in the following section on sensitivity analysis.

In practice, the problem of linear programming minimization is no more difficult than that of maximization. There are two basic ways to treat a maximization problem. The first is to simply multiply the objective function through by a minus sign and proceed with a standard simplex maximization procedure. The second way is more direct and retains the basic integrity of the original formulation. Suppose that the objective function coefficients are entered directly into the simplex tableau and the sensitivity coefficients (v_j elements) are calculated in the usual manner. Since the objective is to minimize, one would prefer to choose the v_j entry with the smallest (most negative) v_j rather than the largest v_j. Following this rule to select the pivotal column (entering basic variable), and proceeding exactly as before, optimality will be attained. The procedure terminates when all v_j elements are nonnegative. We will use the latter procedure for linear programming minimization problems.

As a numerical example of determining a first (basic) feasible solution, consider the problem of finding nonnegative x_1, x_2, x_3, which *minimize*

$$y = 6x_1 + 2x_2 + 3x_3 \tag{3-16}$$

and which satisfy the constraints

$$30x_1 + 20x_2 + 40x_3 \geq 34 \tag{3-17}$$

$$x_1 + x_2 + x_3 = 1 \tag{3-18}$$

$$10x_1 + 70x_2 \leq 11 \tag{3-19}$$

This problem is so simple that a first feasible solution is immediately evident. It will, however, serve to illustrate the systematic method for finding such a solution in much larger problems, where this task is almost as difficult as finding the *optimal* solution.

The third constraint (3-19) requires no artificial variable, since it is of the form of Eq. (3-13), and the slack variable $\bar{S}_3$ may serve as the basic variable. We shall add artificial variables to the other two constraints, of course, but Eq. (3-18) will not have a slack variable. The first tableau for the auxiliary problem, shown in Fig. 3-3(a), corresponds to the artificial basic feasible solution $u_1 = 34$, $u_2 = 1$, and $\bar{S}_3 = 11$. The bottom row may be read as

$$K = u_1 + u_2 = 35 - 31x_1 - 21x_2 - 41x_3 + \bar{S}_1 \tag{3-20}$$

Since this is a minimization problem, x_3 is the first decision variable to be

	x_1	x_2	x_3	$\bar{s}_1$	$\bar{s}_3$	u_1	u_2	
	6	2	3	0	0	0	0	
u_1	30	20	40	−1	0	1	0	34
u_2	1	1	1	0	0	0	1	1
$\bar{s}_3$	10	70	0	0	1	0	0	11
	−31	−21	−41	−1	0	0	0	35

(a) Initial tableau for auxiliary problem

	x_1	x_2	x_3	$\bar{s}_1$	$\bar{s}_3$	u_1	u_2	
x_3	1	1	1	0	0	0	1	1
$\bar{s}_1$	10	20	0	1	0	−1	40	6
$\bar{s}_3$	10	70	0	0	1	0	0	11
	0	0	0	0	0	+1	+1	0

(b) Final tableau for auxiliary problem

	x_1	x_2	x_3	$\bar{s}_1$	$\bar{s}_3$	
x_3	1	1	1	0	0	1
$\bar{s}_1$	10	20	0	1	0	6
$\bar{s}_3$	10	70	0	0	1	11
	+3	−1	0	0	0	3

(c) Initial tableau for original problem

	x_1	x_2	x_3	$\bar{s}_1$	$\bar{s}_3$	
x_3	0.857	0	1	0	−0.0143	0.843
$\bar{s}_1$	7.14	0	0	1	−0.286	2.86
x_2	0.143	1	0	0	0.0143	0.157
	−3.143	0	0	0	+0.0143	2.843

(d) Final tableau for original problem

Figure 3-3. Finding a first feasible solution.

adjusted, and it will replace u_1 in the state set, as indicated by the circled pivot number. Continuing with the simplex method, we arrive (after one more iteration) at the optimal solution to the auxiliary problem given by the tableau of Fig. 3-3(b). The artificial variables have been driven to zero, so we now have a first basic feasible solution to the original problem; namely, $x_3 = 1$, $\bar{S}_1 = 6$, and $\bar{S}_3 = 11$.

This basis produces the initial tableau of Fig. 3-3(c), where the bottom row has been obtained by replacing the state (basic) variable x_3 in the objective function (3-16) by the expression $1 - x_1 - x_2$, as given by the first row of the tableau. Alternatively, one may simply use Eq. (3-15) to obtain the numbers in the bottom row. For example, we find

$$v_1 = 6 - [3(1) + 0(10) + 0(10)] = 3$$
$$v_2 = 2 - [3(1) + 0(20) + 0(70)] = -1$$

so that x_2 is increased, driving $\bar{S}_3$ out of the state (solution) set. This produces the tableau of Fig. 3-3(d), which is the optimal solution to the original problem.

Before leaving the problem of finding a first feasible solution, we think that the reader should be aware of certain dangers in handling equality constraints. One is tempted to solve each such constraint for one of the variables and then use the resulting expression to eliminate this variable from the remaining constraints and the objective function, thereby reducing the dimension of the problem. That this is an unjustifiable practice is well illustrated by the present example. If the equality constraint (3-18) is solved for x_1 and then substituted for x_1 in (3-16), (3-17), and (3-19), the reduced problem becomes one of minimizing

$$y = 6 - 4x_2 - 3x_3 \tag{3-21}$$

subject to the usual nonnegativity conditions and the constraints

$$-5x_2 + 5x_3 \geq 2 \tag{3-22}$$
$$60x_2 - 10x_3 \leq 1 \tag{3-23}$$

which clearly has an unbounded solution. We may, of course, obtain a reduced problem having an optimal solution which *is* the correct solution to the original problem by adding the constraint, $x_2 + x_3 \leq 1$, to (3-21), (3-22), and (3-23). All this work, however, has not reduced the number of constraints, which is the determining factor in the number of iterations needed to solve a problem, and in addition, we still must find a basic feasible solution to the reduced problem. Furthermore, in a large problem, eliminating variables using the equality constraints is not a trivial process, nor is the necessary resubstitution of the solution values of the reduced problem into the equality constraints in order to obtain the optimal values of the deleted variables. In general, then, this procedure should be avoided.

3-05 DUAL LINEAR PROBLEMS

Duality relations arise in many mathematical systems, and these relations are subject to various interpretations, depending upon the particular context represented by the mathematical model. Here we are concerned with duality only as it relates to linear programming problems, although dual problems have also been devised for nonlinear programming problems (Chapter 5). Indeed, the fundamental ideas can be obtained directly from Legendre's dual transformation and form the basis for the geometric programming algorithms of Chapter 6.

We consider two linear programming problems:

(I)
$$\text{Maximize } p = \sum_{n=1}^{N} c_n x_n \tag{3-24}$$

subject to

$$\sum_{n=1}^{N} a_{mn} x_n \leq b_m; \qquad m = 1, \ldots, M \tag{3-25}$$

$$x_n \geq 0; \qquad n = 1, \ldots, N \tag{3-26}$$

(II)
$$\text{Minimize } y = \sum_{m=1}^{M} b_m z_m \tag{3-27}$$

subject to

$$\sum_{m=1}^{M} a_{mn} z_m \geq c_n; \qquad n = 1, \ldots, N \tag{3-28}$$

$$z_m \geq 0; \qquad m = 1, \ldots, M \tag{3-29}$$

where the c_n, b_m, and a_{mn} are given constants. Problems I and II are said to be *dual* linear programming problems; each is the dual of the other.

The nature of this duality may be derived by writing out the Lagrangian functions for problems I and II and computing the appropriate partial derivatives. Designating the Lagrange multipliers for problem I as λ_m, we find the conditions for optimality in that problem to be

$$x_n\Big(c_n - \sum_{m=1}^{M} \lambda_m a_{mn}\Big) = 0; \qquad n = 1, \ldots, N \tag{3-30}$$

$$c_n - \sum_{m=1}^{M} \lambda_m a_{mn} \leq 0; \qquad n = 1, \ldots, N \tag{3-31}$$

$$\bar{S}_m \lambda_m = 0; \qquad m = 1, \ldots, M \tag{3-32}$$

$$\lambda_m \geq 0; \qquad m = 1, \ldots, M \tag{3-33}$$

where the $\bar{S}_m$ are the slack variables for Eqs. (3-25). It can now be seen that Eqs. (3-31) and (3-33) are the constraint equations (3-28) and (3-29) of *dual* problem II, and that *the Lagrange multipliers λ_m are the decision variables*, z_m, of that problem. Equations (3-30) and (3-32) are complementary slackness conditions, and the former of these suggest that the decision variables, x_n of *primal* problem I are the Lagrange multipliers for the constraints of problem II. It is left as an exercise to show that this is indeed the case, and that Eqs. (3-25) and (3-26) are optimality conditions for problem II. This leads to the important result that the conditions for *feasibility* of the dual solution are also the conditions for *optimality* of the primal solution, and vice versa. Thus, for any linear programming problem, the (absolute) values of the dual variables are given by the *sensitivity coefficients* corresponding to the slack variables of the primal. (These values may be *opposite in sign*, depending upon the direction of the constraint inequalities and whether the objective function is to be minimized or maximized.)

Introducing the slack variables $\bar{S}_m$ and w_n into constraint equations (3-35) and (3-38), respectively, and then multiplying the former through by z_m, and the latter through by x_n, we may sum the resulting equations over m and n, respectively, to obtain

$$\sum_{m=1}^{M}\sum_{n=1}^{N} a_{mn}x_n z_m + \sum_{m=1}^{M} \bar{S}_m z_m = \sum_{m=1}^{M} b_m z_m \tag{3-34a}$$

$$\sum_{n=1}^{N}\sum_{m=1}^{M} a_{mn}z_m x_n - \sum_{n=1}^{N} w_n x_n = \sum_{n=1}^{N} c_n x_n \tag{3-34b}$$

If the double-summation terms in these equations are equated, there results

$$\sum_{m=1}^{M} b_m z_m - \sum_{m=1}^{M} \bar{S}_m z_m = \sum_{n=1}^{N} c_n x_n + \sum_{n=1}^{N} w_n x_n \tag{3-35}$$

We now observe that since z_m is the sensitivity coefficient for the slack variable $\bar{S}_m$, the product $\bar{S}_m z_m$, for $m = 1, \ldots, M$, must at all times be zero. To see this, note that if $\bar{S}_m$ is nonzero, it must necessarily be a solution variable, and its sensitivity coefficient z_m will be zero, whereas, if z_m is nonzero, then $\bar{S}_m$ must be a decision variable, and hence have a value of zero. An analogous argument shows that the products $w_n x_n$ must likewise be zero, since x_n is the sensitivity coefficient for w_n. Accordingly, we have shown that at all times

$$\sum_{n=1}^{N} c_n x_n = \sum_{m=1}^{M} b_m z_m \tag{3-36}$$

Corresponding to an optimal-feasible solution to either problem I or II, all the x_n and z_m in Eq. (3-36) will be nonnegative. A solution that is feasible but not optimal for problem I will have dual variables z_m, which provide an optimal but not feasible (since some of them will be negative) solution to problem II, and vice versa. When a linear programming problem is solved using the simplex method, the final tableau contains the optimal (and feasible) solutions to both the primal and the dual problem.

The following statements concerning dual linear problems are of particular importance.

1. The primal problem has a finite optimal-feasible solution if and only if the dual has a finite optimal-feasible solution, and in such a case, the optimal values are equal.
2. When one problem has no feasible solution, then the other either has no feasible solution or has an unbounded optimal-feasible solution.
3. If one problem has an unbounded optimal-feasible solution, then the other has no feasible solution.

As an illustration of dual linear problems, consider the following:

$$\text{Maximize } z = 35x_1 + 60x_2 + 30x_3$$

subject to

$$\begin{aligned} 3x_1 + 5x_2 + 2x_3 &\leq 50 \\ 2x_1 + 6x_2 + 3x_3 &\leq 40 \\ x_1, \quad x_2, \quad x_3 &\geq 0 \end{aligned}$$

The dual to the preceding problem is then:

$$\text{minimize } y = 50z_1 + 40z_2$$

subject to

$$\begin{aligned} 3z_1 + 2z_2 &\geq 35 \\ 5z_1 + 6z_2 &\geq 60 \\ 2z_1 + 3z_2 &\geq 30 \\ z_1, \quad z_2 &\geq 0 \end{aligned}$$

The optimal simplex tableaux for these problems are given in Fig. 3-4, and one can see that the complete solution to both problems can be obtained from either tableau.

The reader may be interested in solving this two-dimensional dual problem graphically to see that the second constraint is dominated by the other two, and thus to infer that x_2, the dual variable for this constraint, must be zero.

	35	60	30			
	x_1	x_2	x_3	$\bar{s}_1$	$\bar{s}_2$	
x_1	1	3/5	0	3/5	-2/5	14
x_3	0	8/5	1	-2/5	3/5	4
	0	-9	0	-9	-4	610

(a) Primal problem

	50	40				
	z_1	z_2	w_1	w_2	w_3	
z_1	1	0	-3/5	0	2/5	9
z_2	0	1	2/5	0	-3/5	4
w_2	0	0	-3/5	1	-8/5	9
	0	0	+14	0	+4	610

(b) Dual problem

Figure 3-4. Optimal tableaux for illustrative dual problems.

3-06 THE DUAL-SIMPLEX TECHNIQUE

The previous discussion established a unique relationship between the primal linear programming problem and the dual linear programming problem. From this discussion, one might suspect that a linear programming problem may be solved either by the simplex method, which maintains a series of feasible solutions that approach optimality, or by an alternative method, which maintains a series of optimal solutions that approach feasibility. This latter method was developed by Lemke, and it is usually referred to as the dual-simplex method. The importance of this method will surface in the treatment of sensitivity analysis, and it also plays a significant role in duality and integer programming, topics to be discussed later in this chapter. The technique is easily mastered once the principles of the primal simplex method and the role of the decision row (v_j variables) are understood. Consider a general linear programming tableau such as that of Fig. 3-5. At optimality the decision coefficients $v_1, \ldots, v_N$ will all be nonpositive at an optimal (maximizing) solu-

c_n	...	c_1	c_2	...	c_j	...	c_c	...	c_N	
P_n	SOL	x_1	x_2		x_j		x_c		x_N	
	S_1	α_{11}	α_{12}		α_{1j}				α_{1N}	b_1
	S_2	α_{21}	α_{22}		α_{2j}				α_{2N}	b_2
	⋮								⋮	
	S_i	α_{i1}			α_{ij}				α_{iN}	b_i
	S_r						α_{rc}		⋮	b_r
	⋮								⋮	
	S_M	α_{M1}			α_{Mj}				α_{MN}	
	...	v_1	v_2	...	v_j	...	v_c	...	v_N	

Maximization Optimality $v_k \leq 0$ → (last row) ← Minimization Optimality $v_k \geq 0$

Figure 3-5. General linear programming tableau.

tion. Further, the final basis will form a feasible, positive solution. Indeed, the primal simplex method starts feasible and remains feasible until optimality is achieved. The dual-simplex procedure starts with optimality criteria satisfied and a nonfeasible solution vector. Provided that optimality criteria ($v_j \leq 0$) are maintained, an optimal/feasible solution is found once a feasible solution vector is generated. Note that since the decision coefficients are always nonpositive, the only components of interest in the solution (b_i; $i = 1, 2, \ldots, M$) vector are those which are *negative*.

In general, if a solution variable S_r is to be removed from the basis, it is necessary that a decision variable be chosen in such a way that optimality is preserved and feasibility is approached. Feasibility can only be generated if the final pivot selection is negative, since the current value of S_r chosen is already negative or zero. Further, the necessary condition for each v_j to remain positive in the next tableau for a particular pivot selection α_{rc} is

$$v_j - \frac{\alpha_{rj}}{\alpha_{rc}} v_c \leq 0; \qquad j = 1, 2, \ldots, N; \quad j \neq r \tag{3-37}$$

or

$$v_j \leq \frac{\alpha_{rj}}{\alpha_{rc}} v_c \tag{3-38}$$

It is clear that if α_{rj} is nonnegative, then the inequality above is satisfied for *any* negative α_{rc}. If α_{rj} is negative, however, then optimality can only be preserved by selecting that decision variable x_c to enter the state set whose v_c and α_{rc} values satisfy the relation

$$\frac{v_c}{\alpha_{rc}} \leq \frac{v_j}{\alpha_{rj}} \qquad \text{all for } \alpha_{rj} \leq 0 \tag{3-39}$$

Hence, to select the variable to come into the basis compute for each column having a negative entry in the chosen pivot row, the ratio of the bottom row number (v_j) in this column to that entry in the pivot row, and choose the ratio smallest in value.

The following rules can now be stated for the dual simplex procedure for maximization problems. Starting with an optimal ($v_j \leq 0$) solution and a nonfeasible solution vector:

Step 1: Choose the most negative b_r value to establish the pivotal row:

$$b_r = \min_j (b_j) \qquad b_j \leq 0$$

Step 2: Search the rth row for negative elements:

$$\alpha_{rn} \leq 0$$

Step 3: Select the pivotal column from the columns corresponding to non-basic variables according to the following rule:

$$x_c = \min_{\alpha_{rn} \leq 0} \left(\frac{v_n}{\alpha_{rn}}\right) = \alpha_{rc}$$

Step 4: Exchange basic variable x_r with nonbasic variable x_c, use α_{rc} as the pivot, and transform the matrix according to normal simplex procedures.

Step 5: Repeat steps 1–4 until an optimal, feasible solution is found.

The other problem of interest is to minimize a linear program. For objective function minimization, steps 1, 2, and 4 remain the same. Step 2 is modified to that of choosing the *maximum* ratio.

$$x_c = \max_{\alpha_{rn} < 0} \left(\frac{v_n}{\alpha_{rn}}\right) = \alpha_{rc} \tag{3-40}$$

Of course, if minimization is the goal, all v_j will be nonnegative at the optimal solution. To illustrate the dual-simplex procedure, consider the following dual linear program previously introduced.

$$\text{Minimize } Y = 50Z_1 + 40Z_2$$

subject to

$$3Z_1 + 2Z_2 \geq 35$$
$$5Z_1 + 6Z_2 \geq 60$$
$$2Z_1 + 3Z_2 \geq 30$$
$$Z_1, \quad Z_2 \geq 0$$

Suppose that a cannonical form is generated through the introduction of slack variables $\bar{S}_1$, $\bar{S}_2$, and $\bar{S}_3$ such that

$$-3Z_1 - 2Z_2 + \bar{S}_1 = -35$$
$$-5Z_1 - 6Z_2 + \bar{S}_2 = -60$$
$$-2Z_1 - 3Z_2 + \bar{S}_3 = -30$$

so that an initial simplex tableau is as follows:

	50	40	0	0	0	
	Z_1	Z_2	$\bar{S}_1$	$\bar{S}_2$	$\bar{S}_3$	
$\bar{S}_1$	-3	-2	1	0	0	-35
$\bar{S}_2$	-5	-6	0	1	0	-60
$\bar{S}_3$	-2	-3	0	0	1	-30
V_j	50	40				

Note that the optimality criteria ($v_1, v_2 \geq 0$) is satisfied but the solution is infeasible. We will choose $\bar{S}_2$ to leave the basis. Note that Z_1 and Z_2 are both candidates for entering the basis. Since $-\frac{50}{5} < -\frac{40}{6}$, choose Z_2 to enter the basis.

	50	40				
	Z_1	Z_2	$\bar{S}_1$	$\bar{S}_2$	$\bar{S}_3$	
$\bar{S}_1$	-4/3	0	1	-1/3	0	-15
Z_2	5/6	1	0	-1/6	0	10
$\bar{S}_3$	1/2	0	0	-1/2	1	0
V_j	100/6	0	0	40/6	0	400

	50	40				
	Z_1	Z_2	$\bar{S}_1$	$\bar{S}_2$	$\bar{S}_3$	
Z_1	1	0	-3/4	1/4	0	45/4
Z_2	0	1	5/8	-9/24	0	5/8
$\bar{S}_3$	0	0	3/8	-5/8	1	-45/8
V_j	0	0	25/2	5/2	0	1175/2

	50	40				
	Z_1	Z_2	$\bar{S}_1$	$\bar{S}_2$	$\bar{S}_3$	
Z_1	1	0	-3/5	0	2/5	9
Z_2	0	1	2/5	0	-3/5	4
$\bar{S}_2$	0	0	-3/5	1	-8/5	9
V_j	0	0	14	0	4	610

The final solution is given by $Z_1 = 9$, $Z_2 = 4$, and $Y = 610$. This corresponds to the solution previously given. In general applications, it is not always an easy task to find a starting solution. Hence the primal method is generally preferred. The real value of the dual-simplex technique will arise in subsequent applications to linear programming sensitivity analysis and integer linear programming.

3-07 SENSITIVITY ANALYSIS

One particularly important feature of linear problems is the ease with which a *sensitivity analysis* can be carried out. This technique analyzes the sensitivity of the optimal solution to changes or uncertainty in the input data, without re-solving the problem for each new value. Such analyses provide insight into the structure of the problem which cannot be gained by examining only the optimal solution. Sensitivity analysis can also be performed on certain nonlinear problems, but the calculations become exceedingly difficult. For even large linear programming problems, however, the necessary computations for carrying out the analysis are simple enough to be done manually, or occasionally, even by inspection. All the information needed for sensitivity analysis is contained in the final (optimal) tableau, and thus is generated automatically by the simplex method in the course of finding the optimal solution.

Symonds has shown how to use linear programming to find the most profitable plan for processing crude petroleum in a hypothetical refinery. We base our discussion of sensitivity analysis on this example, which is the prototype of certain scheduling problems common to the entire chemical industry. No knowledge of petroleum technology is needed to understand this simple example.

Four different kinds of crude petroleum are available for purchase by an oil company: 100,000 barrels per week each of crudes 1, 2, and 3, and 200,000 barrels per week of crude 4. We shall designate by x_1, x_2, x_3, the amounts of crudes 1, 2, and 3, respectively, which are purchased (and processed), expressed in thousands of barrels per week. Accordingly, we may write the following constraint equations for these crudes:

$$x_i + \bar{S}_i = 100; \qquad i = 1, 2, 3 \tag{3-41}$$

where the $\bar{S}_i$ are the unpurchased quantities of the respective crudes.

Slightly different handling is required for crude 4 because it can be processed two ways. Let x_4 by the amount of crude 4 processed primarily to make heating oil, and let x_5 be the amount processed mainly to make lubricating oil. Then

$$x_4 + x_5 + \bar{S}_4 = 200 \tag{3-42}$$

where $\bar{S}_4$ is the unpurchased crude 4.

Four products—gasoline, heating oil, lubricating oil, and jet fuel—are made from these crudes, as shown schematically in Fig. 3-6. Table 3-1 gives the amount of each product which can be sold. The bottom row of Table 3-1 shows the profit gained per 1000 barrels (bbl) of crude processed. These numbers are obtained by adding the market value of the products coming

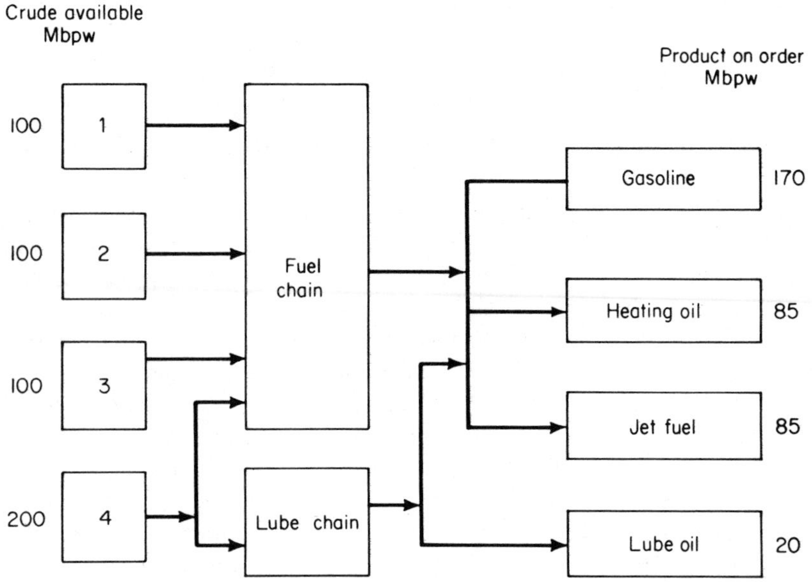

Figure 3-6. Schematic diagram of refinery processing operations.

TABLE 3-1

Profits, Yields, and Availabilities for Refinery Problem

		Crude					Product
		1	2	3	4		Product on order barrels/wk
					Fuel process	Lube process	
Yield, bbl product per bbl crude	Gasoline	0.6	0.5	0.3	0.4	0.4	170,000
	Heating oil	0.2	0.2	0.3	0.3	0.1	85,000
	Lube oil	0	0	0	0	0.2	20,000
	Jet fuel	0.1	0.2	0.3	0.2	0.2	85,000
	Loss	0.1	0.1	0.1	0.1	0.1	
Crude available, bbl/wk		100,000	100,000	100,000	200,000		
Profit, $/1000 bbl crude processed		100	200	70	150	250	

from 1000 bbl of the crude in question and then deducting the costs of production, sales, and the crude itself. The yields (inside the double lines) are fixed by process technology and remain constant throughout the week. On the other hand, the other data—availabilities, orders, and profits—are only estimates and may change between the receipt of the computer solution and the actual crude run.

The gasoline yields in the top row can be used to show that the weekly gasoline production (in thousands of barrels) in terms of the crude consumption is $0.6x_1 + 0.5x_2 + 0.3x_3 + 0.4x_4 + 0.4x_5$. It will be assumed that we are permitted to make *less* product than is ordered, but not more. Thus, letting $\bar{S}_g$ be the amount by which the gasoline demand is unsatisfied, we may write the following constraint on gasoline production.

$$0.6x_1 + 0.5x_2 + 0.3x_3 + 0.4x_4 + 0.4x_5 + \bar{S}_g = 170 \tag{3-43}$$

Similarly, we obtain production constraints on heating oil:

$$0.2x_1 + 0.2x_2 + 0.3x_3 + 0.3x_4 + 0.1x_5 + \bar{S}_h = 85 \tag{3-44}$$

lube oil:

$$0.2x_5 + \bar{S}_l = 20 \tag{3-45}$$

and jet fuel:

$$0.1x_1 + 0.2x_2 + 0.3x_3 + 0.2x_4 + 0.2x_5 + \bar{S}_j = 85 \tag{3-46}$$

The bottom line of Table 3-1 enables us to calculate the profit Z (in dollars/week) in terms of the quantities of each crude processed:

$$Z = 100x_1 + 200x_2 + 70x_3 + 150x_4 + 250x_5 \tag{3-47}$$

Naturally, the crude consumptions x_1, x_2, x_3, x_4, and x_5 cannot be negative, and from the way the slack variables $\bar{S}_1$, $\bar{S}_2$, $\bar{S}_3$, $\bar{S}_4$, $\bar{S}_g$, $\bar{S}_h$, $\bar{S}_l$, and $\bar{S}_j$ are defined, they cannot be negative either. Thus, this scheduling problem takes the form of a linear programming problem concerned with maximizing the objective function (3-47), subject to the foregoing nonnegativity conditions and the constraints (3-41) through (3-46). The initial tableau is shown in Fig. 3-7, where the slack variables form the state set for this first basic feasible solution. Since every x_n is a decision variable, no raw materials are processed, and all orders go unfilled, so the profit at this beginning point is zero. Five simplex iterations are required to find the optimal solution to this scheduling problem as shown in Figs. 3-7(a) through (e) and 3-8. The lines in each of these tableaux define the variables that will enter and leave the state set at the next iteration, and thus they intersect at the pivot number.

	x_1	x_2	x_3	x_4	x_5	$\bar{s}_1$	$\bar{s}_2$	$\bar{s}_3$	$\bar{s}_4$	$\bar{s}_g$	$\bar{s}_h$	$\bar{s}_\ell$	$\bar{s}_j$	
$\bar{s}_1$	1					1								100
$\bar{s}_2$		1					1							100
$\bar{s}_3$			1					1						100
$\bar{s}_4$				1					1			−5		100
$\bar{s}_g$	0.6	0.5	0.3	0.4						1		−2		130
$\bar{s}_h$	0.2	0.2	0.3	0.3							1	$-\frac{1}{2}$		75
x_5					1							5		100
$\bar{s}_j$	0.1	0.2	0.3	0.2								−1	1	65
	+100	+200	+70	+150	0	0	0	0	0	0	0	−1250	0	25,000

(a)

	100	200	70	150	250									
	x_1	x_2	x_3	x_4	x_5	$\bar{s}_1$	$\bar{s}_2$	$\bar{s}_3$	$\bar{s}_4$	$\bar{s}_g$	$\bar{s}_h$	$\bar{s}_\ell$	$\bar{s}_j$	
$\bar{s}_1$	1					1								100
$\bar{s}_2$		1					1							100
$\bar{s}_3$			1					1						100
$\bar{s}_4$				1	1				1					200
$\bar{s}_g$	0.6	0.5	0.3	0.4	0.4					1				170
$\bar{s}_h$	0.2	0.2	0.3	0.3	0.1						1			85
$\bar{s}_i$					0.2							1		20
$\bar{s}_j$	0.1	0.2	0.3	0.2	0.2								1	85
	+100	+200	+70	+150	+250	0	0	0	0	0	0	0	0	0

(b)

Figure 3-7. (a) Starting simplex tableau for refinery problem. (b) Results of first iteration. (c) Results of second iteration. (d) Results of third iteration. (e) Results of fourth iteration.

	x_1	x_2	x_3	x_4	x_5	$\bar{s}_1$	$\bar{s}_2$	$\bar{s}_3$	$\bar{s}_4$	$\bar{s}_g$	$\bar{s}_h$	$\bar{s}_\ell$	$\bar{s}_j$	
f_1	1					1								100
x_2		1					1							100
f_3			1					1						100
x_4				1					1			−5		100
f_g	0.6		0.3				−0.5		−0.4	1		0		40
f_h	0.2		0.3				−0.2		−0.3		1	1		25
x_5					1							5		100
f_j	0.1		0.3				−0.2		−0.2			0	1	25
	+100	0	+70	0	0	0	−200	0	−150	0	0	−500	0	60,000

(c)

	100	200	75	150	250									
	x_1	x_2	x_3	x_4	x_5	$\bar{s}_1$	$\bar{s}_2$	$\bar{s}_3$	$\bar{s}_4$	$\bar{s}_g$	$\bar{s}_h$	$\bar{s}_\ell$	$\bar{s}_j$	
f_1	1					1								100
x_2		1					1							100
f_3			1					1						100
f_4				1					1			−5		100
f_g	0.6		0.3	0.4			−0.5			1		−2		80
f_h	0.2		0.3	0.3			−0.2				1	−0.5		55
x_5					1							5		100
f_j	0.1		0.3	0.2			−0.2					−1	1	45
	+100	0	+70	+150	0	0	−200	0	0	0	0	−1250	0	45,000

(d)

Figure 3-7. (continued)

	100	200	75	150	250									
	x_1	x_2	x_3	x_4	x_5	$\bar{s}_1$	$\bar{s}_2$	$\bar{s}_3$	x_4	$\bar{s}_g$	$\bar{s}_h$	$\bar{s}_\ell$	$\bar{s}_j$	
$\bar{s}_g$			−0.5			1	$-\frac{5}{6}$		$-\frac{2}{3}$	$-\frac{5}{3}$				$\frac{100}{3}$
x_2		1					1							100
x_3			1					1						100
x_4				1					1			−5		100
x_1	1		0.5				$-\frac{5}{6}$		$-\frac{2}{3}$	$-\frac{5}{3}$				$\frac{200}{3}$
$\bar{s}_h$			0.2				$-\frac{1}{30}$		$-\frac{1}{6}$	$-\frac{1}{3}$	1	1		$\frac{35}{3}$
x_5					1							5		100
$\bar{s}_j$			0.25				$-\frac{7}{60}$		$-\frac{2}{15}$	$-\frac{1}{6}$			1	$\frac{55}{3}$
	0	0	+20	0	0	0	$-\frac{350}{3}$	0	$-\frac{250}{3}$	$-\frac{500}{3}$	0	−500	0	$\frac{200{,}000}{3}$

(e)

Figure 3-7. (continued)

	x_1	x_2	x_3	x_4	x_5	$\bar{s}_1$	$\bar{s}_2$	$\bar{s}_3$	$\bar{s}_4$	$\bar{s}_g$	$\bar{s}_h$	$\bar{s}_\ell$	$\bar{s}_j$	
$\bar{s}_1$						1	$\frac{3}{4}$		$\frac{1}{4}$	$-\frac{5}{2}$	$\frac{5}{2}$	$\frac{5}{2}$		62.5
x_2		1					1							100
$\bar{s}_3$							$\frac{1}{6}$	1	$\frac{5}{6}$	$\frac{5}{3}$	−5	−5		41.67
x_4				1					1			−5		100
x_1	1						$-\frac{3}{4}$		$-\frac{1}{4}$	$\frac{5}{2}$	$-\frac{5}{2}$	$-\frac{5}{2}$		37.5
x_3			1				$-\frac{3}{6}$		$-\frac{1}{6}$	$-\frac{5}{3}$	5	5		58.33
x_5					1							5		100
$\bar{s}_j$							$-\frac{3}{40}$		$\frac{3}{40}$	$\frac{1}{4}$	$-\frac{5}{4}$	$-\frac{5}{4}$	1	3.75
	0	0	0	0	0	0	$-\frac{340}{3}$	0	$-\frac{200}{3}$	$-\frac{400}{3}$	−100	−600	0	\$ 67,833

Figure 3-8. Final tableau for refinery problem.

The optimal tableau given in Fig. 3-8 shows that for the given inputs, the maximum weekly profit will be \$67,833. This profit is achieved by running all available quantities of crudes 2 and 4, but only 37.5 and 58.3 Mbpw (thousand barrels per week), respectively, of crudes 1 and 3; the weekly demands for all products are exactly met except for jet fuel ($\bar{S}_j = 3.75$). The reader may verify that this optimal state set is made up of x_1 through x_5, supplemented by $\bar{S}_1$, $\bar{S}_3$, and $\bar{S}_j$.

3-08 AVAILABILITY CHARTS

It is now of interest to show how the production schedule and the associated maximum profit would change with variations in the availability of the crudes (Beightler and Wilde). With the methods to be described, we shall be able to construct availability charts similar to Figs. 3-9 and 3-10, which trace the behavior of raw material consumption, product sales, and net profit as a function of the amount of crude 2 available. Before showing how to do this, we shall discuss several interesting properties of the system, which, although shown quite clearly on the graphs, would possibly not be evident without sensitivity analysis. Some of these results would even seem to contradict our

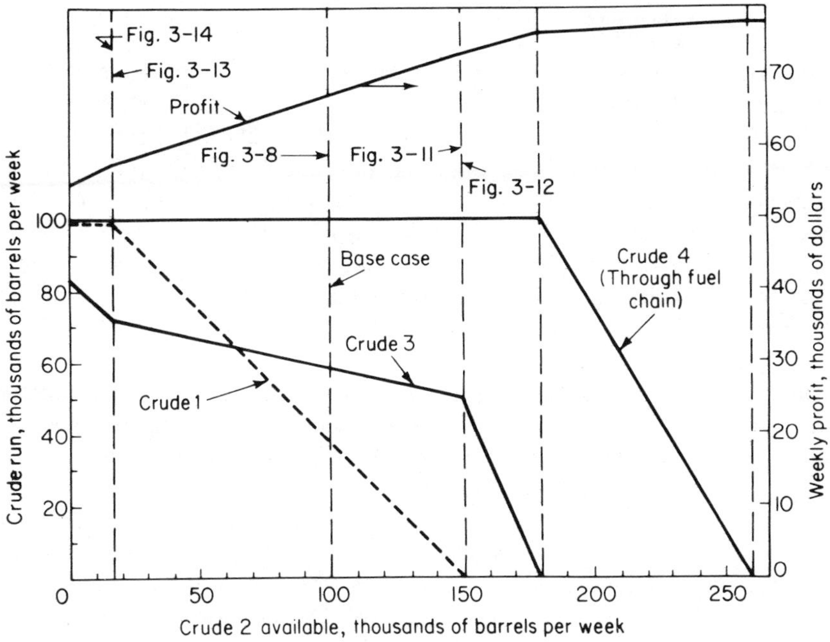

Figure 3-9. Maximum profit and optimal raw-material consumption as functions of crude 2 availability.

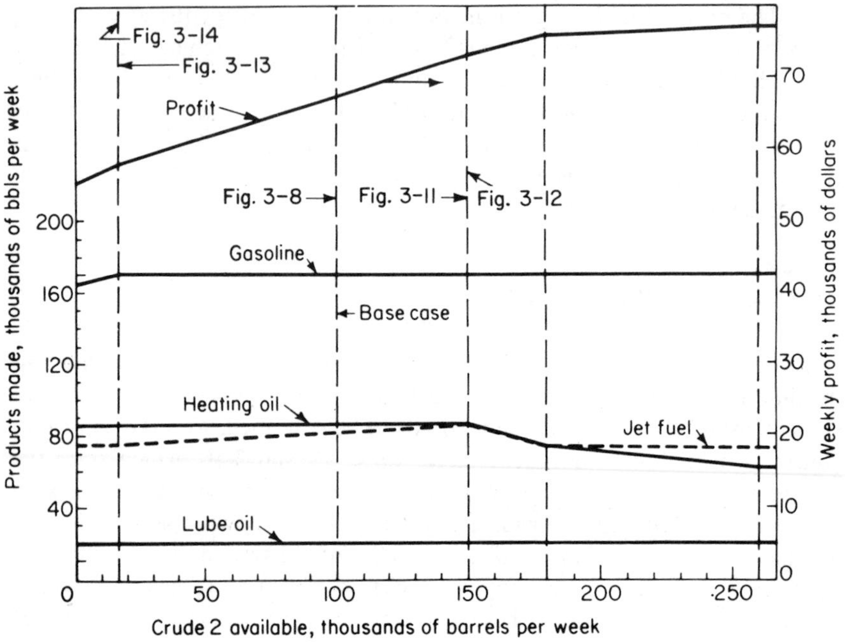

Figure 3-10. Maximum profit and optimal product sales as functions of crude 2 availability.

intuition, which is not always able to weigh properly the effects of complicated interactions.

Notice, for example, that increased availability of crude 2 cuts down the consumption of crude 1 faster than that of crude 3, even though the profit per barrel for crude 3 is 30% less than that for crude 1. This paradox occurs because both crudes 1 and 2 are rich in gasoline, whereas crude 3 contains relatively little gasoline. Thus it is the limit on gasoline sales which forces crude 1 out of the state set (when 150,000 bbl/week of crude 2 is available) before crude 3 (at 180,000 bbl/week of 2).

Another result that may seem surprising is shown in Fig. 3-10. As more crude 2 becomes available, more jet fuel is produced at first. But, when more than 150,000 bbl/week of crude 2 can be run, jet fuel production steadily drops off. Similarly, heating oil production no longer meets the sales demands when large amounts of crude 2 are available.

It is not surprising that the profit climbs steadily as more crude 2 becomes available, for one would expect any loosening of restrictions to increase the profit possibilities. Nevertheless, it may be of some interest to notice that the slope of the profit curve never increases as the restrictions on crude 2 are loosened. This occurs because the slope changes only when a new restriction on some other material is encountered, and these additional limitations always eliminate the most profitable alternatives. To see how this works, let us trace the effects of increasing availability of crude 2. When there is very little crude 2, it is not economical to meet the gasoline requirements, and any additional crude 2 can be run without affecting the consumption of the rather profitable crude 1. The first change in the state set occurs when $x_2 = 16{,}667$ bbl/week. Since at this point no more gasoline can be produced, it is necessary to reduce the amount of crude 1 run, and the profit does not climb as steeply as before. The profit rate stays constant until $x_2 = 150{,}000$, when it is no longer feasible to run crude 1. This additional constraint further slows down the rate of profit increase. When $x_2 = 180{,}000$, the amount of crude 3 run drops to zero, and it becomes necessary to cut down on the crude 4 run through the fuel chain. From this point on, the rate of climb of the profit curve is very small. Since it is not economical to process more than 260,000 bbl/week of crude 2, the profit line becomes horizontal for greater values of x_2.

That the profit line is concave illustrates what economists call the *law of diminishing returns*; that is, increases in availability of crude 2 do not always bring proportional profits. Thus, if one already has 100,000 bbl per week of crude 2 (the base case) one can afford to pay up to \$5667 for 50,000 additional bbl and still make a profit. But Fig. 3-9 shows that the next 50,000 bbl will bring in only an additional \$2750, and the premium should be set accordingly.

A final fact of general interest is that all the curves are piecewise linear. The slopes change whenever the state set (basis) changes. Sensitivity analysis shows exactly at what availabilities these changes take place.

3-09 SMALL AVAILABILITY CHANGES

Having seen some of the advantages of sensitivity analysis, let us now consider how to perform one. First we shall study the effects of changes of availability so small that a change in the basis is not called for. Suppose that the availability of crude 2 changes from the present value of 100 Mbpw by a small amount Δb_2. *This is equivalent to changing the corresponding slack variable $\bar{S}_2$ by an amount $-\Delta b_2$.* Thus the equations of the final tableau in Fig. 3-8 affected by the change become

$$
\begin{aligned}
\bar{S}_1 + \tfrac{3}{4}\bar{S}_2 - \tfrac{3}{4}\Delta b_2 \quad &+ \tfrac{1}{4}\bar{S}_4 - \tfrac{5}{2}\bar{S}_g + \tfrac{5}{2}\bar{S}_h + \tfrac{5}{2}\bar{S}_l &&= 62.5 \\
x_2 + \bar{S}_2 - \Delta b_2 \quad & &&= 100 \\
\tfrac{1}{6}\bar{S}_2 - \tfrac{1}{6}\Delta b_2 + \bar{S}_3 &+ \tfrac{5}{6}\bar{S}_4 + \tfrac{5}{3}\bar{S}_g - 5\bar{S}_h - 5\bar{S}_l &&= 41.7 \\
x_1 - \tfrac{3}{4}\bar{S}_2 + \tfrac{3}{4}\Delta b_2 \quad &- \tfrac{1}{4}\bar{S}_4 + \tfrac{5}{2}\bar{S}_g - \tfrac{5}{2}\bar{S}_h - \tfrac{5}{2}\bar{S}_l &&= 37.5 \\
x_3 - \tfrac{1}{6}\bar{S}_2 + \tfrac{1}{6}\Delta b_2 \quad &- \tfrac{5}{6}\bar{S}_4 - \tfrac{5}{3}\bar{S}_g + 5\bar{S}_h - 5\bar{S}_l &&= 58.3 \\
-\tfrac{3}{40}\bar{S}_2 + \tfrac{3}{40}\Delta b_2 \quad &+ \tfrac{3}{40}\bar{S}_4 + \tfrac{1}{4}\bar{S}_g - \tfrac{5}{4}\bar{S}_h - \tfrac{5}{4}\bar{S}_l + \bar{S}_j &&= 3.75
\end{aligned}
$$

Transposing the terms in Δb_2 to the right-hand side of each equation and noting that the decision variables are zero, we see that none of the state variables becomes negative so long as all the following quantities remain nonnegative:

$$62.5 + \tfrac{3}{4}\Delta b_2, \qquad 100 + \Delta b_2, \qquad 41.7 + \tfrac{1}{6}\Delta b_2,$$
$$37.5 - \tfrac{3}{4}\Delta b_2, \qquad 58.3 - \tfrac{1}{6}\Delta b_2, \qquad 3.8 - \tfrac{3}{40}\Delta b_2$$

Therefore, the present optimal running plan will remain feasible as long as $-83.3 \leq \Delta b_2 \leq 50$, or equivalently, when b_2 lies between 16.7 and 150 Mbpw. Notice that this range can be found directly from the final tableau by dividing each value in the right-hand column by the value on the same line in the $\bar{S}_2$ column (unless it is zero), and then selecting the positive and negative quotients that are smallest in absolute value.

Within the foregoing range, the optimal values for the present basic variables will change as follows:

$$\Delta\bar{S}_1 = \tfrac{3}{4}\Delta b_2 \qquad (3\text{-}48)$$

$$\Delta x_2 = \Delta b_2 \qquad (3\text{-}49)$$

$$\Delta\bar{S}_3 = \tfrac{1}{6}\Delta b_2 \qquad (3\text{-}50)$$

$$\Delta x_1 = -\tfrac{3}{4}\Delta b_2 \qquad (3\text{-}51)$$

$$\Delta x_3 = -\tfrac{1}{6}\Delta b_2 \qquad (3\text{-}52)$$

$$\Delta\bar{S}_j = -\tfrac{3}{40}\Delta b_2 \qquad (3\text{-}53)$$

and the maximum profit will then change by

$$\Delta z = 100(-\tfrac{3}{4}\Delta b_2) + 200(\Delta b_2) + 70(-\tfrac{1}{6}\Delta b_2) = \tfrac{340}{3}\Delta b_2 \qquad (3\text{-}54)$$

It is instructive to note that this result could have been read directly from Fig. 3-8 as the value of v_2 in the $\bar{S}_2$ column. Thus, if additional crude 2 were available at the present price plus any premium cost less than \$113 (more precisely, \$340/3) per thousand barrels, it would pay to buy it, run it, and decrease the amounts of crudes 1 and 3 purchased according to Eqs. (3-51) and (3-52). In an optimal, feasible tableau, the values of v_k are *sensitivity coefficients* for the associated decision variable.

Notice that this change in one of the b_m of Eqs. (3-2) has not affected any of the sensitivity coefficients, and this is how we know that the final tableau will remain *optimal* in the Δb_2 range previously found. In fact, from Eq. (3-15) we see that the sensitivity coefficients are altered only by changes either in the c_n of the original objective function, or in the α_{mn} of the final tableau. Thus perturbations in the b_m can affect the *feasibility* of the optimal solution to a linear programming problem, but they never destroy the *optimality*.

3-10 LARGE AVAILABILITY CHANGE

If the availability of crude 2 were to increase by more than 50,000 bbl/wk (or decrease by more than 83,300 bbl/week), then some of the variables would become negative, which, of course, would produce an infeasible solution. Figure 3-11 shows that when Δb_2 reaches 50, both x_1 and $\bar{S}_j$ are zero. To keep these variables from becoming negative, we must change the variables in the basis set in such a way that the associated production schedule will be both feasible and optimal in spite of the altered availability of crude 2. Fortunately, it is not necessary to reshuffle all the variables in the set; we need only replace one of them at a time—the one on the verge of becoming negative. In this case, there are two such variables (x_1 and $\bar{S}_j$), but instead of replacing them both simultaneously, we shall find it easier to remove only one and formally keep the other as a basic variable with its value equal to zero. The choice being arbitrary so far as the ultimate consequences are concerned, we shall remove x_1.

	100	200	70	150	250									
	x_1	x_2	x_3	x_4	x_5	$\bar{s}_1$	$\bar{s}_2$	$\bar{s}_3$	$\bar{s}_4$	$\bar{s}_g$	$\bar{s}_h$	$\bar{s}_\ell$	$\bar{s}_j$	
$\bar{s}_1$						1	$\frac{3}{4}$		$\frac{1}{4}$	$-\frac{5}{2}$	$\frac{5}{2}$	$\frac{5}{2}$		100
x_2		1					1							150
$\bar{s}_3$							$\frac{1}{6}$	1	$\frac{5}{6}$	$\frac{5}{3}$	-5	-5		50
x_4				1			1					-5		100
x_1	1						$-\frac{3}{4}$		$-\frac{1}{4}$	$\frac{5}{2}$	$-\frac{5}{2}$	$-\frac{5}{2}$		0
x_3			1				$-\frac{1}{6}$		$-\frac{5}{6}$	$-\frac{5}{3}$	5	5		50
x_5					1							5		100
$\bar{s}_j$							$-\frac{3}{40}$		$\frac{3}{40}$	$\frac{1}{4}$	$-\frac{5}{4}$	$-\frac{5}{4}$	1	0
	0	0	0	0	0	0	$-\frac{340}{3}$	0	$-\frac{200}{3}$	$-\frac{400}{3}$	-100	-600	0	$ 73,500

Figure 3-11. Optimal tableau for $b_2 = 150$.

Next, we must decide which decision variable should replace x_1 in the basis. The new basis must be feasible and optimal for the altered availability of crude 2. Finding the change that will do this is straightforward, since it actually amounts to one dual simplex iteration. One should again realize the difference between this operation and the usual simplex procedure, since here we first decide which variable is to be removed from the basis, and then select the variable that will replace it, rather than the other way around. Also, only a negative value can be chosen for the pivot number, since a positive pivot would make the $\bar{S}_2$ column entry in the x_1 row negative for any further increase in Δb_2, whereas a negative pivot will allow the new variable to increase with Δb_2. Further, the pivot must be chosen in such a way that the optimality of the solution is preserved. These goals are all achieved through the dual simplex procedure.

Since we are interested in objective function maximization, we compute, for each column having a negative entry in the pivot row, the ratio of the bottom row number in this column to the entry in the pivot row and choose the ratio smallest in *value*. In Fig. 3-11 this smallest ratio will be $100 : \frac{5}{2}$, so that $\bar{S}_h$ is the variable that replaces x_1, with the results as given in Fig. 3-12.

	100	200	70	150	250									
	x_1	x_2	x_3	x_4	x_5	$\bar{S}_1$	$\bar{S}_2$	$\bar{S}_3$	$\bar{S}_4$	$\bar{S}_g$	$\bar{S}_h$	$\bar{S}_\ell$	$\bar{S}_j$	
$\bar{S}_1$	1					1				6				100
x_2		1					1							150
$\bar{S}_3$	−2						$\frac{5}{3}$	1	$\frac{4}{3}$	$\frac{-10}{3}$				50
x_4				1					1			−5		100
$\bar{S}_h$	$-\frac{2}{5}$						$\frac{3}{10}$		$\frac{1}{10}$	−1	1	1		0
x_3	2	1					$-\frac{5}{3}$		$-\frac{4}{3}$	$\frac{10}{3}$				50
x_5					1							5		100
$\bar{S}_j$	$-\frac{1}{2}$						$\frac{3}{10}$		$\frac{1}{5}$	−1			1	0
	−40	0	0	0	0	0	$-\frac{250}{3}$		$-\frac{170}{3}$	$-\frac{700}{3}$	0	−500	0	\$ 73,500

Figure 3-12. Results of $\bar{S}_h$ replacing x_1 in dependent set of Fig. 3-11.

3-11 FURTHER INCREASES IN AVAILABILITY

With this new basis, increases in the amount of crude 2 run will no longer tend to decrease $\bar{S}_j$ (which measures the amount by which the weekly production of jet fuel falls short of maximum weekly sales), but will actually cause it to increase. That is, the maximum total profit z can be increased by producing *less* total products. (At $\Delta b_2 = 50$, all maximum weekly sales of products were exactly met.) Proceeding as before, it can be seen that further increases in the availability of crude 2 will not cause another change in the state set until the term $50 - \frac{5}{3}\Delta b_2'$ becomes negative, where $\Delta b_2' = \Delta b_2 - 50$.

Therefore, in the range $50 \leq \Delta b_2 \leq 80$, the optimal amount of crude 3 to run decreases at $\frac{5}{3}$ the rate of increase of crude 2 available, so that the best amount to run will be $50 - \frac{5}{3}\Delta b_2'$, or $\frac{400}{3} - \frac{5}{3}\Delta b_2$. In this same range, the amounts of jet fuel and heating oil produced will each continue to decrease at $\frac{3}{10}$ the rate of increase in the amount of crude 2 available, and the profit will increase at a rate of $\frac{250}{3}$ of this same value. Again, all these figures can be read directly from Fig. 3-12.

When $\Delta b_2 = 80$ (which means that 180 Mbpw of crude 2 is available), x_3 is driven to zero, and further increases in Δb_2 would give this variable a negative value were it to remain in the basis set. The reader may verify that when x_3 is removed from this set, $\bar{S}_4$ must be brought in to replace it, and this new basis set will be optimal in the range $80 \leq \Delta b_2 \leq 160$.

At the upper limit of this range ($b_2 = 260$), x_4 reaches zero and must be removed from the basis, to be replaced by $\bar{S}_2$. The only effect of further increases in Δb_2 will be to increase the slack variable $\bar{S}_2$ (which measures the amount of the available crude 2 which is *unpurchased*), causing no change in the maximum profit, nor in the values of the other basic variables. The meaning of this result is now clear; it does not pay to run more than 260 Mbpw of crude 2 as long as there is no change in the availabilities of the other crudes nor in the maximum weekly sales of the products. The reader may verify that when b_2 is greater than 260, the maximum profit will be \$77,000, achieved by running 260 Mbpw of crude 2 and 100 Mbpw of crude 4 (through the lube chain). At these availabilities of crude 2, it does not pay to purchase any of crude 1 or 3, or to run any of crude 4 through the fuel chain. Thus we have reached the upper limit for Δb_2, and it now remains only to investigate the effects on the problem structure of diminished supplies of crude 2.

3-12 DECREASED AVAILABILITY

When the availability of crude 2 decreases, this is equivalent to making Δb_2 negative, so that, for example, the change in the amount of crude 1 run will increase $\frac{3}{4}$ for each unit decrease in crude 2, whereas the amount of jet fuel produced will decrease by $\frac{3}{40}$. These values can be read directly from Fig. 3-8, and are valid until Δb_2 reaches -83.3, below which the basis set in that tableau is no longer feasible.

Proceeding as before, we find that $\bar{S}_1$ is the first variable to be driven negative (Fig. 3-13), being replaced by $\bar{S}_g$ as shown in Fig. 3-14. If the availability of crude 2 continues to decrease below the value of 16.7 Mbpw, we

	x_1	x_2	x_3	x_4	x_5	$\bar{s}_1$	$\bar{s}_2$	$\bar{s}_3$	$\bar{s}_4$	$\bar{s}_g$	$\bar{s}_h$	$\bar{s}_\ell$	$\bar{s}_j$	
	100	200	70	150	250									
$\bar{s}_1$						1	$\frac{3}{4}$		$-\frac{1}{4}$	$-\frac{5}{2}$	$-\frac{5}{2}$	$\frac{5}{2}$		0
x_2		1					1							16.7
$\bar{s}_g$							$\frac{1}{6}$	1	$\frac{5}{6}$	$\frac{5}{3}$	−5	−5		27.8
x_4				1					1			−5		100
x_2	1						$-\frac{3}{4}$		$-\frac{1}{4}$	$\frac{5}{2}$	$-\frac{5}{2}$	$-\frac{5}{2}$		100
x_3			1				$-\frac{1}{6}$		$-\frac{5}{6}$	$-\frac{5}{3}$	5	5		72.2
x_5					1							5		100
$\bar{s}_j$							$-\frac{3}{40}$		$\frac{3}{40}$	$\frac{1}{4}$	$-\frac{5}{4}$	$-\frac{5}{4}$	1	10
	0	0	0	0	0	0	$-\frac{340}{3}$	0	$-\frac{200}{3}$	$-\frac{400}{3}$	−100	−600	0	\$ 58,389

Figure 3-13. Optimal tableau for $b_2 = 16.7$.

	x_1	x_2	x_3	x_4	x_5	$\bar{s}_1$	$\bar{s}_2$	$\bar{s}_3$	$\bar{s}_4$	$\bar{s}_g$	$\bar{s}_h$	$\bar{s}_\ell$	$\bar{s}_j$	
$\bar{s}_g$						$-\frac{2}{5}$	$-\frac{3}{10}$		$-\frac{1}{10}$	1	−1	1		0
x_2		1					1							16.7
$\bar{s}_g$						$\frac{2}{3}$	$\frac{2}{3}$	1	1		$-\frac{10}{3}$	$-\frac{10}{3}$		27.8
x_4				1					1			−5		100
x_2	1					1								100
x_3			1			$-\frac{2}{3}$	$-\frac{2}{3}$		−1		$\frac{10}{3}$	$\frac{10}{3}$		72.2
x_5					1							5		100
$\bar{s}_j$						$\frac{1}{10}$			$\frac{1}{10}$		−1	−1	1	10
	0	0	0	0	0	$-\frac{160}{3}$	$-\frac{460}{3}$	0	−80	0	$-\frac{700}{3}$	$-\frac{2200}{3}$	0	\$58,389

Figure 3-14. Results of f_g replacing f_1 in dependent set of Fig. 3-13.

see from Fig. 3-14 that the basis for this tableau will remain feasible as long as $\Delta b_2 \geq -100$.

Since $x_2 - \Delta b_2 = 100$, Δb_2 cannot be less than -100. Thus the present running plan will be feasible for the remaining range of Δb_2: $-100 \leq \Delta b_2 \leq -83.3$. Within this range, there will be no further changes in the amount of jet fuel produced or in the amount of crude 1 run. The amount of crude 3 run, however, now increases (for *decreasing* availability of crude 2) by a factor of 4 (from $\frac{1}{6}$ to $\frac{2}{3}$), and the maximum total profit decreases more rapidly, to a rate of $\frac{460}{3}$, as compared to the previous rate of $\frac{340}{3}$, which held throughout the range $-83.3 \leq \Delta b_2 \leq 50$. These results are all shown graphically in Figs. 3-9 and 3-10.

Similar analyses, of course, can be made concerning the effects on the optimal solution of changes in the availability of the other crudes and also changes in the maximum amounts of the various products which can be sold (that is, perturbations in $\bar{S}_g$, $\bar{S}_h$, $\bar{S}_l$, and $\bar{S}_j$). These analyses are carried out in exactly the same manner as that described for crude 2 availability, and are left as an exercise for the reader.

3-13 SIMULTANEOUS AVAILABILITY CHANGES

The situation in which the availabilities of several of the crudes change during the same time period can be analyzed by a method which is a direct extension of that just described. As a simple example, consider the effects of simultaneous availability changes in crudes 2 and 4, by amounts Δb_2 and Δb_4, respectively. Proceeding as before, we see from Fig. 3-8 that the basic variables are affected by these changes in the following manner:

$$\bar{S}_1 = 62.50 + \tfrac{3}{4}\Delta b_2 + \tfrac{1}{4}\Delta b_4 \tag{3-55}$$

$$x_2 = 100 + \Delta b_2 \tag{3-56}$$

$$\bar{S}_3 = 41.67 + \tfrac{1}{6}\Delta b_2 + \tfrac{5}{6}\Delta b_4 \tag{3-57}$$

$$x_4 = 100 + \Delta b_4 \tag{3-58}$$

$$x_1 = 37.50 - \tfrac{3}{4}\Delta b_2 - \tfrac{1}{4}\Delta b_4 \tag{3-59}$$

$$x_3 = 58.33 + \tfrac{1}{6}\Delta b_2 - \tfrac{5}{6}\Delta b_4 \tag{3-60}$$

$$\bar{S}_j = 3.75 - \tfrac{3}{40}\Delta b_2 + \tfrac{3}{40}\Delta b_4 \tag{3-61}$$

The optimal solution of Fig. 3-8 will remain feasible as long as the right-hand sides of Eqs. (3-55)–(3-61) remain nonnegative. This nonnegativity restriction changes the foregoing equations into inequalities that define the shaded area of Fig. 3-15. Within this area, the simultaneous changes Δb_2 and Δb_4 result in alteration of the objective function by the amount

$$\Delta z = 100\Delta x_1 + 200\Delta x_2 + 70\Delta x_3 + 150\Delta x_4$$

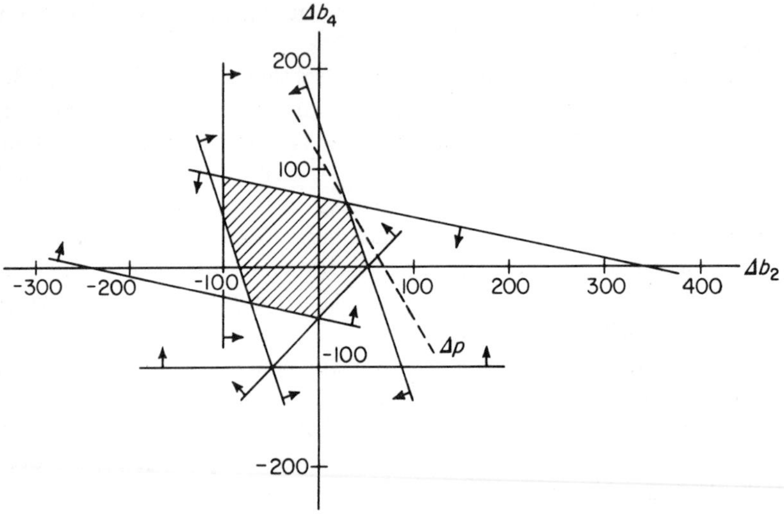

Figure 3-15. Region of feasibility for optimal solution under simultaneous availability changes in crudes 2 and 4.

or, by Eqs. (3-55)–(3-61),

$$\Delta z = 100(-\tfrac{3}{4}\Delta b_2 - \tfrac{1}{4}\Delta b_4) + 200(\Delta b_2) + 70(-\tfrac{1}{6}\Delta b_2 - \tfrac{5}{6}\Delta b_4) + 150\Delta b_4$$

which may be written as

$$\Delta z = \tfrac{340}{3}\,\Delta b_2 + \tfrac{200}{3}\,\Delta b_4 \tag{3-62}$$

Notice that the coefficients in Eq. (3-62) are the negatives of the sensitivity coefficients v_2 and v_4 corresponding to $\bar{S}_2$ and $\bar{S}_4$ in Fig. 3-8, so that this expression for Δz could have been obtained directly from the bottom row of the optimal tableau. A similar statement holds for the case in which more than two crude availabilities are changed simultaneously, so that in general we have

$$\Delta z = \sum_i (-v_i)\,\Delta b_i \tag{3-63}$$

where the v_i are the sensitivity coefficients corresponding to the slack variables $\bar{S}_i$. If we are able to purchase unlimited quantities of crudes 2 and 4, but do not wish to change the optimal running plan of Fig. 3-8 (that is, do not want to change the state set), it is important to know what quantities of these crudes to purchase so as to maximize the total profit. Clearly, this is equivalent to maximizing (3-62) subject to the inequalities resulting from the nonnegativity restrictions on Eqs. (3-55)–(3-61). This new linear programming problem is solved graphically in Fig. 3-15. The solution is easily found to be

$\Delta b_2 = 28.6$, $\Delta b_4 = 64.2$, for an increase in profit of \$7520. From Eqs. (3-55)–(3-61) it can be seen that this additional profit is achieved by running 128,600 bbl/week of crude 2; 164,200 bbl/week of crude 4, and none of crudes 1 and 3.

Notice that in this new linear programming problem, we should not limit all the Δb_i to nonnegative values. For example, if the sensitivity coefficients corresponding to $\bar{S}_2$ and $\bar{S}_4$ had had the values -50 and -300, respectively, then the point $\Delta b_2 = -100$, $\Delta b_4 = 90$, would have resulted in an increase in profit of \$22,000, and for these values of the sensitivity coefficients, this would indeed have been the maximum profit attainable.

Thus the general problem of maximizing (3-63) subject to constraints of the form

$$\kappa + \sum \alpha_{ik} \Delta b_i \geq 0 \tag{3-64}$$

cannot be solved by the simplex method (which forces all $\Delta b_i \geq 0$) without some modification. The modification required, however, is quite simple; we merely express each variable Δb_i as the difference of two nonnegative variables as follows:

$$\Delta b_i = \Delta'' b_i - \Delta' b_i \tag{3-65}$$

Each Δb_i is then replaced in (3-63) and (3-64) by the value given in (3-65), and the standard simplex method is used to find the optimal solution to this equivalent problem, where the $\Delta'' b_i$ and $\Delta' b_i$ are all required to be nonnegative.

3-14 SENSITIVITY ANALYSIS ON THE OBJECTIVE COEFFICIENTS

Analyses of the effects on the optimal solution of perturbations in the coefficients c_n in the objective function (3-1) are performed in much the same manner as were those concerned with the constants b_m in the constraint equations. From Eq. (3-15) it is clear that a change in any of the c_n will affect the value of some of th sensitivity coefficients, v_n, but will leave the state variables unchanged. Thus, such changes can affect the optimality of the final solution to a linear programming p:oblem, but never the feasibility.

For a maximization problem, a solution is optimal as long as the sensitivity coefficients are all nonpositive. Thus, if an objective coefficient, c_n, corresponding to a decision variable, d_n, is changed by an amount Δc_n, then from Eq. (3-15), the final solution will remain optimal if

$$\Delta c_n \leq -v_n \tag{3-66}$$

In other words, the objective coefficient for a *decision* variable may *decrease* by any amount without affecting the optimality of the solution, but must *not increase* by more than the value of the bottom row entry for that variable. Thus such a change affects only the sensitivity coefficient, and its effect can be evaluated by inspection.

Changes in some objective coefficient p_m corresponding to a basic variable can, however, affect every sensitivity coefficient. Indeed, we see from Eq. (3-15) that

$$\Delta v_n = -\alpha_{mn}\,\Delta p_m; \qquad n = M+1, \ldots, M+N \tag{3-67}$$

and so the solution remains optimal within the range

$$\frac{v_L}{\alpha_{mL}} \leq \Delta c_m \leq \frac{v_U}{\alpha_{mU}} \tag{3-68}$$

where

$$\frac{v_U}{\alpha_{mU}} = \min_{\alpha_{mn}<0}\left(\frac{v_n}{\alpha_{mn}}\right) \tag{3-69}$$

and

$$\frac{v_L}{\alpha_{mL}} = \max_{\alpha_{mn}>0}\left(\frac{v_n}{\alpha_{mn}}\right) \tag{3-70}$$

since in both Eqs. (3-69) and (3-70), we want to choose the ratio of v_n to α_{mn} which is smallest in absolute value. For *minimization* problems, of course, the sensitivity coefficients must remain *nonnegative*, and Eqs. (3-66) and (3-68) must be adjusted accordingly.

Let us now suppose that in the refinery scheduling problem the *cost* of crude 1 changes by an amount $-\Delta c_1$; this is equivalent to changing the objective coefficient of x_1 from the present value of 100 to a value of 100 $+ \Delta c_1$. Changes in the selling price of the products could also affect the net *profit*, c_1, as could changes in the production and sales costs. For brevity, we shall speak of changes Δc_m without reference to the changes in costs of market values that produced them.

From Fig. 3-8 we see that the *bottom-row entries* affected by the change Δc_1 in the profit of crude 1 become

$$-v_2 = \tfrac{340}{3} - \tfrac{3}{4}\Delta c_1; \qquad -v_4 = \tfrac{200}{3} - \tfrac{1}{4}\Delta c_1$$

$$-v_g = \tfrac{400}{3} + \tfrac{5}{2}\Delta c_1$$

$$-v_h = 100 - \tfrac{5}{2}\Delta c_1 \quad \text{and} \quad -v_l = 600 - \tfrac{5}{2}\Delta c_1$$

Since the present basis set will remain optimal only as long as all of these numbers are nonnegative, no change in this set will be required in the range $-53.33 \leq \Delta c_1 \leq 40$. The values -53.33 and 40, of course, are obtained from Eqs. (3-70) and (3-69), respectively. In this range, the values taken on by the basic variables do not change, and therefore, the maximum profit will be given by $z = 67{,}833 + 37.5\Delta c_1$, since $x_1 = 37.5$, and no other terms in the objective function will change.

If the profit from running crude 1 were to decrease by more than \$53.33 per 1000 bbl, then some of the sensitivity coefficients would become positive, indicating that the basis which generated them was no longer optimal. Figure 3-16 shows that when Δc_1 reaches -53.33, only the sensitivity coefficient associated with $\bar{S}_g$ has been driven to zero. All the other coefficients have decreased, because of negative entries for these columns in the x_1 row (that is, the fifth row, in which x_1 is the state variable). Any further decrease in Δc_1 will cause the sensitivity coefficient v_g to become positive, meaning that the total profit can be increased by bringing $\bar{S}_g$ into the state set.

	c_n	$\frac{140}{3}$	200	70	150	250									
p	$\bar{S}_m$	x_1	x_2	x_3	x_4	x_5	$\bar{S}_1$	$\bar{S}_2$	$\bar{S}_3$	$\bar{S}_4$	$\bar{S}_g$	$\bar{S}_h$	$\bar{S}_\ell$	$\bar{S}_j$	
$\bar{S}_1$							1	$\frac{3}{4}$		$\frac{1}{4}$	$-\frac{5}{2}$	$\frac{5}{2}$	$\frac{5}{2}$		62.5
x_2	200		1					1							100
$\bar{S}_g$								$\frac{1}{6}$	1	$\frac{5}{6}$	$\frac{5}{3}$	-5	-5		41.67
x_4	150				1					1			-5		100
x_1	$\frac{140}{3}$	1						$-\frac{3}{4}$		$-\frac{1}{4}$	$\frac{5}{2}$	$-\frac{5}{2}$	$-\frac{5}{2}$		37.5
x_3	70			1				$-\frac{1}{6}$		$-\frac{5}{6}$	$-\frac{5}{3}$	5	5		58.3
x_5	250					1							5		100
$\bar{S}_j$								$-\frac{3}{40}$		$\frac{3}{40}$	$\frac{1}{4}$	$-\frac{5}{4}$	$-\frac{5}{4}$	1	3.75
		0	0	0	0	0	0	$-\frac{460}{3}$	0	-80	0	$-\frac{700}{3}$	$-\frac{2200}{3}$	0	\$ 65,833

Figure 3-16. Optimal tableau for $c_1 = 46.67$ ($\Delta c_1 = -53.33$).

Using the simplex method, we replace x_1 in the basis with $\bar{S}_g$, obtaining the tableau of Fig. 3-17. Here, x_1 and $\bar{S}_j$ tied as the variable to be removed, so that our arbitrary choice of x_1 has produced a degenerate solution with the basic variable $\bar{S}_j$ having a zero value.

Unlike the case of variation in the availability of crudes, here the change in production schedule is discontinuous and accompanied by substantial changes in the values of many of the basic variables. The solutions in Figs. 3-16 and 3-17, however, produce exactly the same profit of \$65,833, even though they represent considerable differences in the running plans employed.

	c_n	$\frac{140}{3}$	200	70	150	250									
p	$\bar{s}_m$	x_1	x_2	x_3	x_4	x_5	$\bar{s}_1$	$\bar{s}_2$	$\bar{s}_3$	$\bar{s}_4$	$\bar{s}_g$	$\bar{s}_h$	$\bar{s}_\ell$	$\bar{s}_j$	
$\bar{s}_1$		1					1								100
x_2	200		1					1							100
$\bar{s}_g$		$-\frac{2}{3}$						$-\frac{2}{3}$	1	1		$-\frac{10}{3}$	$-\frac{10}{3}$		16.67
x_4	150				1					1			-5		100
$\bar{s}_g$		$\frac{2}{5}$						$-\frac{3}{10}$		$-\frac{1}{10}$	1	-1	-1		15
x_3	70	$\frac{2}{3}$		1				$-\frac{2}{3}$		-1		$\frac{10}{3}$	$\frac{10}{3}$		83.3
x_5	250					1								5	100
$\bar{s}_j$		$-\frac{1}{10}$								$\frac{1}{10}$		-1	-1	1	0
		0	0	0	0	0	0	$-\frac{460}{3}$	0	-80	0	$-\frac{700}{3}$	$-\frac{2200}{3}$	0	\$ 65.833

Figure 3-17. Results of replacing x_1 with S_g in the basis of Fig. 3-16.

Since this is a linear programming problem, the decrease in the profit, z, has taken place at a constant rate of \$37.50 for each dollar decrease in the net profit of crude 1, for a total decrease between Figs. 3-8 and 3-16 of \$37.50 $\times$ \$53.33 $=$ \$2000.

Figure 3-17, however, shows that this decrease is now at an end, for x_1 has been removed from the basis. Thus we see that when the net profit from running 1000 bbl of crude 1 falls below \$46.67, it is no longer profitable to run any of this crude. Although the net profit has been uniformly decreased from the original value of \$100 per 1000 bbl to \$46.67, no changes in the running plan have been desirable, so that the maximum profit has been achieved by continually processing 37,500 bbl/week of crude 1. At the \$46.67 figure, however, crude 1 is suddenly dropped out of the running plan altogether, with compensating increases in the amounts of crude 3 processed. Notice that this change in running plan has also reduced the gasoline production, since crude 3 is not as rich in gasoline as is crude 1. (See Table 3-1; this also explains the increase in production of jet fuel.)

The effects of *increased* net profit from running crude 1 are handled in the same manner as were the decreases. Thus, until the unit profit from this crude has increased by more than \$40 per 1000 bbl, no change in the basic set will occur since all the sensitivity coefficients will remain nonpositive. Unlike the values of the basic variables, these coefficients do change continuously with uniform changes in the unit profit for crude 1. For example, the sensitivity coefficient for $\bar{S}_h$ will be $v_h = \frac{5}{2}\Delta c_1 - 100$, so that it increases linearly to zero as Δc_1 increases to 40. This is the first coefficient to be driven to zero, and the reader may verify that $\bar{S}_h$ replaces x_3 as a result of the required simplex iteration.

Figures 3-18 and 3-19 summarize the effects of changes in the unit profit

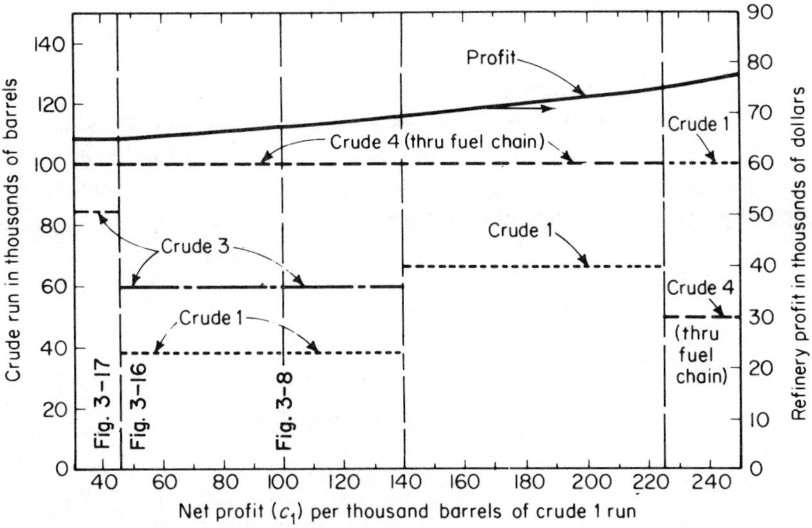

Figure 3-18. Maximum profit and optimal raw material consumption as functions of crude 1 unit profit.

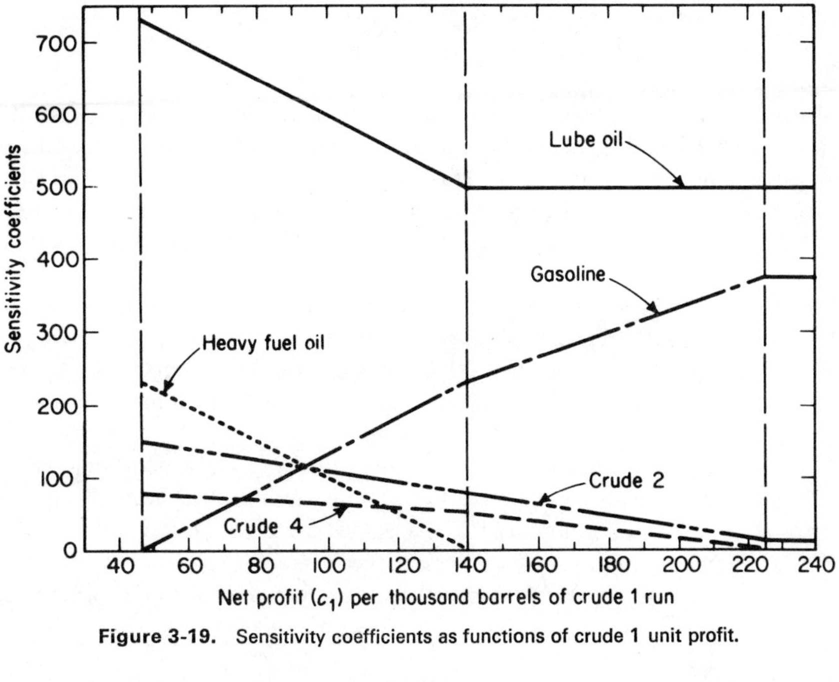

Figure 3-19. Sensitivity coefficients as functions of crude 1 unit profit.

c_1; notice that the curves are piecewise linear, and that the three products are affected strongly by the profit of crude 1. Crude 2, not shown in Fig. 3-18, has a constant value of 100 Mbpw throughout the entire range of c_1, even though the per barrel profit on the crude is exceeded by that of crude 1 for $c_1 > 200$. From Fig. 3-19, however, we see that the penalty that would be incurred by reducing the amount of crude 2 steadily declines as crude 1 contributes more and more profit per barrel run.

Simultaneous changes in the objective coefficients are handled in the same manner as were such changes in the constraint constants b_m. For example, if the unit profit from both crudes 1 and 3 were to change by the respective amounts Δc_1 and Δc_3, then the solution of Fig. 3-8 would remain optimal as long as the following inequalities were satisfied:

$$v_2 = \tfrac{3}{4}\Delta c_1 + \tfrac{1}{6}\Delta c_3 - \tfrac{340}{3} \leq 0$$

$$v_4 = \tfrac{1}{4}\Delta c_1 + \tfrac{5}{6}\Delta c_3 - \tfrac{200}{3} \leq 0$$

$$v_g = -\tfrac{5}{2}\Delta c_1 + \tfrac{5}{3}\Delta c_3 - \tfrac{400}{3} \leq 0$$

$$v_h = \tfrac{5}{2}\Delta c_1 - 5\Delta c_3 - 100 \leq 0$$

$$v_l = \tfrac{5}{2}\Delta c_1 - 5\Delta c_3 - 100 \leq 0$$

Within the region defined by the foregoing inequalities, the total refinery profit would then be given by the expression $z = 37.5\,\Delta c_1 + 58.33\,\Delta c_3$.

3-15 COMBINED EFFECTS OF PROFIT AND AVAILABILITY CHANGES

We now turn to a consideration of the effects of simultaneous changes in the unit profit and availability of a crude. It may happen that as larger quantities of a crude become available, the market price will decrease, or alternatively, larger amounts of the crude may be made available by paying a premium to get them. Figure 3-20 summarizes the results obtained from an investigation of all possible combinations of changes in the profits and availabilities of crude 2. In this figure, changes in availability are plotted horizontally; profit changes are plotted vertically. The origin may be taken as the point $\Delta b_2 = 0$, $\Delta c_2 = 0$, which represents the tableau of Fig. 3-8. At this point, $b_2 = 100$ Mbpw, $c_2 = \$200$ per 1000 bbl processed, and the total refinery profit is \$67,833.

The vertical dashed lines represent the values of b_2 at which a change in the basis set is required (because some basic variable has been driven to zero). The horizontal boundaries of the shaded areas define the values of c_2 at which a change in the basis is necessary (because some sensitivity coefficient has been driven to zero). All these values are computed exactly as described

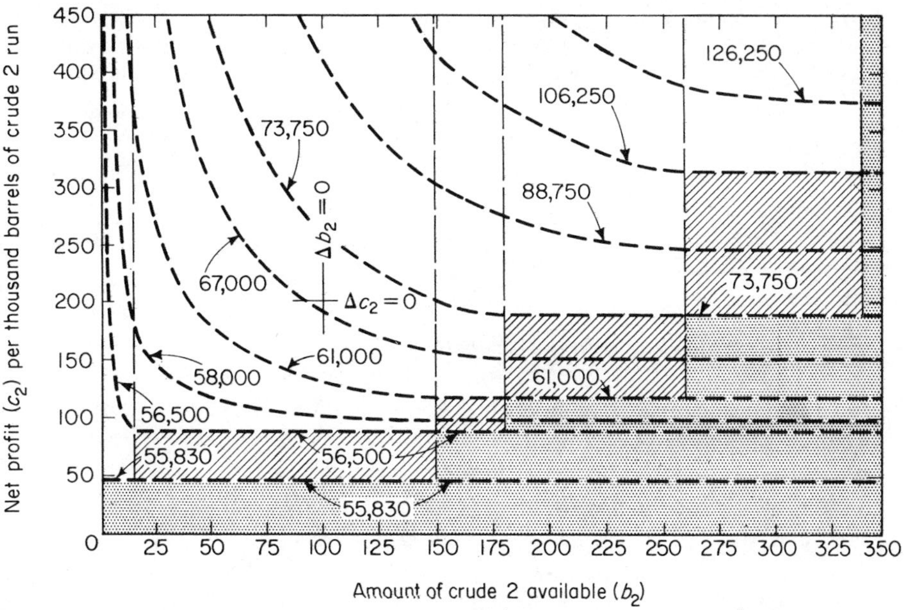

Figure 3-20. Summary of simultaneous changes in unit profit and availability of crude 2.

earlier, where changes were made in only one parameter. For example, having obtained the tableau resulting from the change in the basis set required when $\Delta b_2 = 160$, one can now examine changes in c_2, and compute the range $-12.5 \leq \Delta c_2 \leq 112.5$ for which the basis of this tableau is optimal. Thus, this tableau would give the optimal solution to the refinery problem for the combined range $260 \leq b_2 \leq 340$, $187.5 \leq c_2 \leq 312.5$, as shown in Fig. 3-20. The remaining areas in the figure are calculated in a similar manner, and clearly, the results obtained are independent of the order in which the changes Δb_2 and Δc_2 are computed.

An interesting situation occurs at the point $\Delta b_2 = 240$, $\Delta c_2 = \frac{225}{2}$; for these changes, crude 2 has become so profitable, and so much of it is available, that no other crude is being processed. Thus, when the profit from each 1000 bbl of this crude exceeds \$312.50, and at least 340 Mbpw of it is available, the entire refinery should be run on this one crude alone. If more than 340 Mbpw is available, it cannot be run, since we have already reached the sales limit on gasoline.

The dashed lines on Fig. 3-20 are constant profit lines; for example, a maximum total refinery profit of \$88,750 will be achieved at $\Delta b_2 = 0$, $\Delta c_2 = 210$; at $\Delta b_2 = 80$, $\Delta c_2 = 75$; and at $\Delta b_2 = 160$, $\Delta c_2 = 45$. The line becomes horizontal to the right of the latter point, indicating that increased availabilities of crude 2 beyond that point cannot be processed. The constant profit lines define areas of profitable change in the quantities of crude 2 purchased. For example, if 50 Mbpw of the crude were now being run at a net profit of $300 per 1000 bbl, Fig. 3-20 shows that the total refinery profit would be $67,000. Any change in b_2 and c_2 that resulted in a point above this line would be profitable. Thus, if unlimited amounts were available at a price increase of less than $150 per 1000 bbl, it would pay to purchase an additional 130 Mbpw, for a total of 180 Mbpw. Beyond this amount, of course, all available quantities are not processed, so further purchases could not be justified.

3-16 PARAMETRIC PROGRAMMING

For some linear programming problems, it is important to know the optimal solution for all values of a particular parameter, usually one of the b_m or c_n. As we have seen in the case of b_2 in the refinery problem, this can be accomplished by solving the problem first for some numerical value of the parameter, and then investigating the effects of changes Δb_2 on this solution. Alternatively, one may proceed by finding the set of optimal solutions corresponding to various ranges of the parameter without introducing a specific numerical value. This procedure is called *parametric programming* (Gass and Saaty), and we illustrate it by parameterizing the objective coefficient c_1 in the following problem:

Find nonnegative x_1, x_2, which maximize

$$Z = c_1x_1 + x_2 \tag{3-71}$$

and which satisfy the constraints:

$$2x_1 + x_2 \leq 10 \tag{3-72}$$

$$x_1 + x_2 \leq 8 \tag{3-73}$$

$$x_1 \leq 3; \qquad x_2 \leq 7 \tag{3-74}$$

An initial basic feasible solution for this problem is immediately at hand since the basic set can be selected to consist of the slack variables $\bar{S}_1$–$\bar{S}_4$, corresponding to the constraints (3-72)–(3-74), respectively. This produces the tableau of Fig. 3-21(a), which clearly cannot be an optimal solution for any value of c_1, since the sensitivity coefficient for x_2 is positive. Therefore, we select x_2 to enter the basis, replacing $\bar{S}_4$, with the results shown in Fig. 3-21(b). The basis for this tableau will be optimal as long as c_1 is negative, for then all the sensitivity coefficients will be negative. When c_1 becomes positive, the sensitivity coefficient for x_1 also becomes positive, and the solution can be improved by replacing $\bar{S}_2$ with x_1 in the basic set, producing the tableau of Fig. 3-21(c). Since now the sensitivity coefficients for $\bar{S}_2$ and $\bar{S}_4$ will remain nonpositive if c_1 is respectively nonnegative and not greater than 1, the basis for this tableau is optimal in the range $0 \leq c_1 \leq 1$. When c_1 exceeds a value of $+1$, $\bar{S}_4$ is made a basic variable with the results shown in Fig. 3-21(d). Continuing in this manner, we compute the tableau of Fig. 3-21(e), which remains the optimal solution for all values of c_1 greater than 2.

	c_1	1					
	x_1	x_2	$\bar{s}_1$	$\bar{s}_2$	$\bar{s}_3$	$\bar{s}_4$	
$\bar{s}_1$	2	1	1	0	0	0	10
$\bar{s}_2$	1	1	0	1	0	0	8
$\bar{s}_3$	1	0	0	0	1	0	3
$\bar{s}_4$	0	1	0	0	0	1	7
	$+c_1$	$+1$	0	0	0	0	0

(a) Initial tableau, nonoptimal for all values of c_1

$\bar{s}_1$	2	0	1	0	0	-1	3
$\bar{s}_2$	1	0	0	1	0	-1	1
$\bar{s}_3$	1	0	0	0	1	0	3
x_2	0	1	0	0	0	1	7
	$+c_1$	0	0	0	0	-1	7

(b) Optimal solution for $c_1 < 0$

$\bar{s}_1$	0	0	1	-2	0	1	1
x_1	1	0	0	1	0	-1	1
$\bar{s}_3$	0	0	0	-1	1	1	2
$\bar{s}_2$	0	1	0	0	0	1	7
	0	0	0	$-c_1$	0	$c_1 - 1$	$7 + c_1$

(d) Optimal solution in the range $0 \leq c_1 \leq 1$

Figure 3-21. Tableaux for parametric programming problem.

$\bar{s}_4$	0	0	−1	−2	0	1	1
x_1	1	0	1	−1	0	0	2
$\bar{s}_1$	0	0	−1	1	1	0	1
$\bar{s}_2$	0	1	−1	2	0	0	6
	0	0	$1 - c_1$	$c_1 - 2$	0	−1	$6 + 2c_1$

(d) Optimal solution in the range $1 \leq c_1 \leq 2$

$\bar{s}_4$	0	0	−1	0	2	1	3
x_1	1	0	0	0	1	0	3
$\bar{s}_2$	0	0	−1	1	1	0	1
$\bar{s}_3$	0	1	1	0	−2	0	4
	0	0	−1	0	$2 - c_1$	0	$4 + 3c_1$

(e) Optimal solution for $c_1 \geq 2$

Figure 3-21. (continued)

Although this simple problem can be worked by hand, most parametric programming must be performed on a digital computer, since it requires that numerous simplex iterations be performed on the full tableau.

3-17 OTHER SENSITIVITY ANALYSES

The sensitivity analyses most frequently required are those concerned with variations either in the constraint constants, b_m, or in the objective coefficients, c_n, as previously described. Changes in the constraint coefficients, a_{mn}, occur less often and are more difficult to analyze, especially when several of them change simultaneously. We shall describe the procedure for analyzing those cases in which one such coefficient changes, after first discussing the important topics of deleted variables and constraints.

For very large problems, computer storage and running time become critical considerations, and the original problem may have to be decreased in size by removing constraints and/or variables. After the reduced problem is solved, the effect of having eliminated a constraint or variable may be computed without re-solving the problem. The procedure for eliminated variables consists in testing the hypothesis that these variables would have been decision variables in the optimal solution and thus would not have affected the solution had they been retained. The analysis for the eliminated constraints consists in checking the implicit assumption that the slack variables for these constraints would have been basic variables, indicating that the constraints would not have been binding.

For example, suppose that, in the refinery scheduling problem, it was assumed that a constraint on total refinery capacity would not be exceeded by the optimal solution obtained when this constraint is deleted. Specifically, if the refinery were not able to process more than, say, 350,000 bbl of crude per week, then the eliminated constraint would be

$$x_1 + x_2 + x_3 + x_4 + x_5 + \bar{S}_r = 350 \tag{3-75}$$

where $\bar{S}_r$ is the slack variable that measures the unused refinery capacity. Substitution of the optimal solution from Fig. 3-8 into this constraint results in a value of $\bar{S}_r = -45.83$, assuming $\bar{S}_r$ to be the basic variable in this constraint row in the final tableau. This means that the refinery capacity has been exceeded by 45.83 Mbpw, so that the optimal solution obtained is no longer feasible when this constraint is included.

Since the optimal tableau of Fig. 3-8 expresses the basic variables in terms of the decision variables, we may use these expressions to eliminate all basic variables in Eq. (3-75) and produce the new constraint row in the optimal tableau:

$$-\tfrac{1}{12}\bar{S}_2 + \tfrac{1}{12}\bar{S}_4 - \tfrac{5}{6}\bar{S}_g - \tfrac{5}{2}\bar{S}_h - \tfrac{5}{2}\bar{S}_l + \bar{S}_r = -45.83 \tag{3-76}$$

Then the new optimal (but no longer feasible) tableau is as shown in Fig. 3-22, where $\bar{S}_h$ will replace $\bar{S}_r$ in the basic set on the next iteration, as indicated by the lines on the figure. Notice that the addition of an eliminated constraint never changes the sensitivity coefficients, since it affects only feasibility, not optimality. The reader may verify that two iterations of the dual-simplex method are required to obtain the new optimal feasible solution: $x_1 = 50$, $x_2 = 100$, $x_3 = 100$, $x_4 = 100$, $x_5 = 100$, which produces a total refinery profit of \$65,000.

Again, suppose that a variable, x_e, were eliminated from the original problem, and it was desired to check whether inclusion of this variable would have affected the optimal solution as obtained by ignoring it. If x_e would have been a decision variable in the final tableau, then its coefficient, α_{ie}, in the ith constraint row (the row in which x_i is the basic variable) could be calculated from

$$\alpha_{ie} \equiv \frac{\delta x_i}{\delta x_e} = \sum_k \frac{\delta x_i}{\delta x_k}\frac{\partial x_k}{\partial x_e} = \sum \alpha_{ik} a_{ke} \tag{3-77: i, e}$$

c_n	100	200	70	150	250										
c_m	x_1	x_2	x_3	x_4	x_5	g_1	g_2	g_3	g_4	g_g	g_h	g_l	g_j	g_r	
							$-\frac{1}{12}$		$\frac{1}{12}$	$-\frac{5}{6}$	$-\frac{5}{2}$	$-\frac{5}{2}$		1	−45.83
						1	$\frac{3}{4}$		$\frac{1}{4}$	$-\frac{5}{2}$	$\frac{5}{2}$	$\frac{5}{2}$			62.5
200		1					1								100
							$\frac{1}{6}$	1	$\frac{5}{6}$	$\frac{5}{3}$	−5	−5			41.67
150				1					1			−5			100
100	1						$-\frac{3}{4}$		$-\frac{1}{4}$	$\frac{5}{2}$	$-\frac{5}{2}$	$-\frac{5}{2}$			37.5
70			1				$-\frac{1}{6}$		$-\frac{5}{6}$	$-\frac{5}{3}$	5	5			58.33
250					1							5			100
							$-\frac{3}{40}$		$\frac{3}{40}$	$\frac{1}{4}$	$-\frac{5}{4}$	$-\frac{5}{4}$	1		3.75
	0	0	0	0	0	0	$\frac{340}{3}$	0	$\frac{200}{3}$	$\frac{400}{3}$	100	600	0	0	$67,833

Figure 3-22. Optimal tableau of Fig. 3-7 with additional constraint, $x_1 + x_2 + x_3 + x_4 + x_5 \leq 350$ included.

where the summation is taken over those variables, x_k, which were basis variables in the initial tableau and where the notation δ indicates differentiation in which only decision variables are held constant. [The reader familiar with matrix algebra will recognize that for this set of indices, k, the α_{ik} are the elements of a matrix B^{-1} ("inverse of the basis") which when postmultiplied by the initial simplex tableau matrix, yields the final (optimal) tableau matrix.] Thus the first term in the summation is taken from the final tableau, whereas the second term is obtained from the initial tableau. If any x_k is also a basic variable in the *final* tableau, then of course the first term vanishes for $i \neq k$, and is equal to unity for $i = k$.

From Eqs. (3-15) and (3-77), the sensitivity coefficient for x_e in the final tableau would be given by

$$v_e = c_e - \sum_i c_i \sum_k \alpha_{ik} a_{ke}$$

$$= c_e - \sum_k a_{ke}(c_k - v_k) \qquad (3\text{-}78: e)$$

where again the summation on k is defined as it was for Eq. (3-77). If v_e is negative (for a maximization problem), then the optimal solution would not have been changed by introducing the variable x_e into the problem. If v_e is positive, however, then the solution can be improved by bringing x_e into the basis. This is accomplished by performing a simplex iteration on the final tableau to which the column of elements, α_{ie} and v_e, has been added. (More than one iteration may be required to find the new optimal solution if this first one makes any of the other sensitivity coefficients positive.)

The methods for analyzing the effects of changes in the constraint coefficients, a_{ij}, consist of two separate procedures, depending upon whether the variable x_j is a basic or a decision variable in the final tableau (x_i, in our notation, if always a *basic* variable in this tableau).

First consider the case in which some x_h is a decision variable in the final tableau. In such a case, changes in an a_{ih} will not affect the elements α_{ik} in Eq. (3-77) which map the initial tableau entries a_{ij} into the final tableau entries α_{ij} (where j is indexed over *all* the variables in both tableaux). Therefore, from Eq. (3-77) we see that a change Δa_{lh} in the constraint coefficient a_{lh} changes the final tableau entries α_{ih} by an amount

$$\Delta\alpha_{ih} = \alpha_{il}\,\Delta a_{lh}; \qquad i = 1, \ldots, M \qquad \text{(3-79: } i, h\text{)}$$

where l must necessarily be one of the indices over which k was defined in Eqs. (3-77) and (3-78). From Eq. (3-15), this change in each α_{ih} will change the sensitivity coefficient v_h by an amount

$$\Delta v_h = -\sum_i c_i\,\Delta\alpha_{ih} \qquad \text{(3-80: } h\text{)}$$

so that from Eq. (3-79), the new sensitivity coefficient becomes

$$\begin{aligned}\hat{v}_h &= v_h - \sum_{i=1}^{M} c_i\alpha_{il}\,\Delta a_{lh} \\ &= v_h - \Delta a_{lh}(c_l - v_l)\end{aligned} \qquad \text{(3-81: } h, l\text{)}$$

and (for a maximization problem), the change in a_{lh} will not affect the optimal solution already found as long as

$$\Delta a_{lh}(c_l - v_l) \geq v_h \qquad \text{(3-82: } h, l\text{)}$$

If the magnitude of the change Δa_{lh} is large enough to cause v_h to exceed the left side of inequality (3-82), then one simply changes v_h and each of the α_{ih} in the final tableau by the amounts calculated from Eqs. (3-80) and (3-79) and proceeds with the simplex method (first bringing in x_h) as in the case of eliminated variables.

A change $\Delta a_{l\lambda}$, where x_λ is a basic variable in the final tableau, is best analyzed by introducing a new variable, $\hat{x}_\lambda$, having constraint coefficients $\hat{a}_{k\lambda}$ identical with the $a_{k\lambda}$ corresponding to x_λ, except for the one coefficient $\hat{a}_{l\lambda}$, which is set equal to $a_{l\lambda} + \Delta a_{l\lambda}$. Then, using Eq. (3-77: l, λ), the coefficients $\hat{\alpha}_{l\lambda}$ in the x_λ column of the final tableau can be calculated, as can the sensitivity coefficient $\hat{v}_\lambda$. One then introduces the new objective coefficient, $\hat{c}_\lambda$, making its value identical with that of the original c_λ, and replaces the original c_λ with a very large negative number (for a maximization problem). This latter operation has the effect of changing the sensitivity coefficients in the final tableau—since c_λ is one of the c_i in Eq. (3-15)—in such a way that subsequent application of the simplex method to this tableau will drive x_λ out of the basis, thus effectively removing it (and therefore $a_{i\lambda}$) from the problem.

3-18 THE DANTZIG–WOLFE DECOMPOSITION PRINCIPLE

One of the difficulties in certain practical linear programming problems is that the simplex tableau is often so large that it exceeds the storage capacity of the available computer. This is especially true when one is attempting to optimize the operation of a company consisting of several manufacturing plants, for here the number of variables and constraints may run into the tens of thousands. Furthermore, even when storage space is not at a premium, the optimization of an entire supersystem, such as a company, may not be efficient, since the time required for the solution of a linear programming problem on a computer is approximately proportional to the cube of the number of constraints. For these reasons, a method has been developed for decomposing the large problem, representing the supersystem, down into smaller problems, representing the component systems (say, plants), and then adjusting the optimal solutions of the smaller problems in such a way that they produce an optimal solution for the supersystem.

Dantzig and Wolfe developed a decomposition principle which imposes on each local system additional constraints that allow the system problems to be optimized individually, ignoring the effects of the solution on the other systems. These artificial restrictions, which reflect the interactions between the systems, are selected by the decomposition principle in such a way that the solutions found for each small problem will also be optimal for the large supersystem problem. This principle was developed mainly as a computational device for solving large linear programming problems. This aspect, although certainly still of practical value, is not discussed here because of its specialized character. As Dantzig and Wolfe pointed out, the calculation process itself is interesting because it suggests a rational method for reconciling the conflicting requirements of the various systems in a way that is best for the supersystem. This feature of the decomposition principle is discussed here.

The numerical example itself is something of a mathematical parable. It concerns two manufacturing plants that must share raw materials and markets. One plant is less effective than the other, say, because it is older and not designed for the materials being processed. If a coordinating board were to allocate raw material based on the profit each plant could make under the most favorable circumstances, it would award all material to the newer plant. This would leave none for the older one, which would presumably be shut down—a situation not uncommon when a strong company absorbs a weaker one in a merger. It happens, though, that it is really much more profitable to the entire company if the better plant uses an inferior material under less advantageous circumstances, for this releases other material to put the older plant into production. The increased profits from operating the older plant outweigh, in the example, the decreased effectiveness of the newer one, to the benefit of the entire company. This demonstrates not only the perils of sub-optimization, but also the fruits of cooperation.

The highly simplified numerical example (Wilde) involves a hypothetical manufacturing firm converting three raw materials (identified by the numbers 1, 2, and 3) into two products. The company has two plants (designated A and B), each located near the source of one of the raw materials—plant A near raw material 1 and plant B near raw material 3. Because of high transportation costs it is uneconomical for plant A to use material 3 or for plant B to use material 1, but material 2 is available to either plant and must be shared by them. Production is limited by the market for the two products. Both plants can sell their output of the principal product anywhere in the company's marketing area, but each secondary product must be sold locally, near the plant producing it. Thus the plants share the market for the primary product but not for the by-product.

In any given month, each plant manager knows two things: the local demand for his by-product and the amount available of his exclusive raw material. Both may know the total demand for the main product as well as the availability of the shared raw material 2, but neither has the authority to determine his own share of the market or of material 2. The president of the company has appointed a coordinator to decide this question. Raw material availability and product demands are given in Table 3-2.

TABLE 3-2

Monthly Supplies and Demands

Supply of raw material (tons)			*Product demands (tons)*		
				By-products	
1	2	3	*Main product*	*Plant A*	*Plant B*
80	160	180	180	100	200

Each local manager is, of course, familiar with the economic and technological performance of his plant. Given a production plan telling how much of each raw material to process, each manager can predict the amount of each product made and the corresponding profit. The specific technological and economic information for this problem is summarized in Table 3-3.

TABLE 3-3

Local Yields and Profits

Yield of main product	0.40	0.80	0.60	0.20
Yield of by-product	0.60	0.20	0.40	0.80
Profit ($/ton processed)	15	22	18	8

From the information in Tables 3-2 and 3-3, we may now write out the linear programming model representing the supersystem (company) maximization problem. Let x_{1a} and x_{2a} be, respectively, the tons of raw material 1 and 2 consumed per month in plant A only; then the limitations on the availability of material 1 and the sale of by-product from plant A lead to the constraints

$$x_{1a} + \bar{S}_1 = 80 \tag{3-83}$$

and

$$0.60x_{1a} + 0.20x_{2a} + \bar{S}_2 = 100 \tag{3-84}$$

Similarly, if x_{2b} and x_{3b} are, respectively, the monthly consumption of raw materials 2 and 3 by plant B, then

$$x_{3b} + \bar{S}_3 = 180 \tag{3-85}$$

and

$$0.40x_{2b} + 0.80x_{3b} + \bar{S}_4 = 200 \tag{3-86}$$

Equations (3-83) and (3-84) represent the constraints affecting only plant A; Eqs. (3-85) and (3-86), those affecting plant B alone. The limitation on material 2, which is shared by both plants, is

$$x_{2a} + x_{2b} + \bar{S}_5 = 160 \tag{3-87}$$

Demand for the main product, to be filled by production from both plants, is represented by

$$0.40x_{1a} + 0.80x_{2a} + 0.60x_{2b} + 0.20x_{3b} + \bar{S}_6 = 180 \qquad (3\text{-}88)$$

The $\bar{S}_m$ in Eqs. (3-83)–(3-88) are, of course, the slack variables for these constraints.

The supersystem problem then is to find the production plan that maximizes the monthly profit, Z, given by

$$Z = 15x_{1a} + 22x_{2a} + 18x_{2b} + 8x_{3b} \qquad (3\text{-}89)$$

and which satisfies the feasibility constraints (3-83)–(3-88). It would be very easy to find the optimal plan by solving this small problem directly, but in order to demonstrate the workings of the decomposition principle, let us assume that the two plants and the coordinating group each has a computer capable of solving linear programming problems having no more than four constraints. We shall, in fact, allow each plant manager to ignore the presence not only of the common constraints (3-87) and (3-88), but even of the other plant. Thus, manager A is responsible for constraints (3-83) and (3-84); manager B, for (3-85) and (3-86); while the coordinating group is custodian of the shared constraints (3-87) and (3-88). The objective function is split into two parts, one for each plant. The monthly profit Z_a from plant A is

$$Z_a = 15x_{1a} + 22x_{2a} \qquad (3\text{-}90)$$

and that from plant B is

$$Z_b = 18x_{2b} + 8x_{3b} \qquad (3\text{-}91)$$

3-19 LOCALLY OPTIMAL PLANS

The computations begin with each manager finding the plant that is the most profitable for his plant. In either case this is a linear programming problem involving only two constraints and solvable by direct inspection, for it is clear that each manager would like to use as much of the profitable raw material 2 as he can. Thus the optimal plan for plant A (labeled solution A1) is $x_{1\alpha1} = 0$ and $x_{2\alpha1} = 500$, with a profit $Z_{\alpha1} = \$11,000$ per month. Notice that the last number of the subscript identifies the number of the plan. The optimal plan for plant B is $x_{2b1} = 500$ and $x_{3b1} = 0$, which would give a profit of $Z_{b1} = \$9000$ per month.

When these locally optimal plans are transmitted to the coordinating group, the application of the decomposition principle begins. It is clear that the local proposals cannot be applied, for although each plant wants 500 tons of raw material 2, only 160 are available for both plants. Furthermore, the production plans would produce 700 tons of the main product, more than three times as much as the market will bear. Now if each plan is multiplied by a nonnegative weighting factor less than 1, the new plans generated will certainly be feasible locally. And if the factors are made small enough, the common constraints (3-87) and (3-88) can be satisfied.

3-20 WEIGHTING FACTORS

Let λ_{a1} and λ_{b1} be the weighting factors for solutions A1 and B1, respectively. The variables of the problem can be expressed in terms of these factors as follows:

$$\begin{aligned} x_{1a} &= x_{1a1}\lambda_{a1} = 0 \\ x_{2a} &= x_{2a1}\lambda_{a1} = 500\lambda_{a1} \\ x_{2b} &= x_{2b1}\lambda_{b1} = 500\lambda_{b1} \\ x_{3b} &= x_{3b1}\lambda_{b1} = 0 \\ Z_a &= Z_{a1}\lambda_{a1} = 11{,}000\lambda_{a1} \\ Z_b &= Z_{b1}\lambda_{b1} = 9000\lambda_{b1} \end{aligned}$$

In terms of the factors λ_{a1} and λ_{b1}, the shared constraints can be written

$$500\lambda_{a1} + 500\lambda_{b1} + \bar{S}_5 = 160 \tag{3-92}$$

for the availability of raw material 2, and

$$400\lambda_{a1} + 300\lambda_{b1} + \bar{S}_6 = 180 \tag{3-93}$$

for the main product demand. The requirement that the factors be between zero and unity imposes two more constraints,

$$\lambda_{a1} + \bar{S}_a = 1 \tag{3-94}$$

for plant A, and

$$\lambda_{b1} + \bar{S}_b = 1 \tag{3-95}$$

for plant B, where $\bar{S}_a$ and $\bar{S}_b$ are, of course, nonnegative slack variables.

3-21 FIRST MASTER PLAN

The coordinators now have the problem of maximizing the total profit

$$Z = 11{,}000\lambda_{a1} + 9000\lambda_{b1} \tag{3-96}$$

subject to the four constraints (3-92)–(3-95). Since the variables λ_{a1} and λ_{b1} cannot be negative, this is a linear programming problem that can be solved by the computation facilities assumed available. The optimal solution, obtainable in this case by inspection, is $\lambda_{a1} = 0.320$, $\lambda_{b1} = 0$, and $Z = \$3520$. Thus if only locally optimal solutions are considered, plant A will operate at 32% of its optimal rate and plant B will be shut down. This all-or-nothing aspect is often the way that capital improvement funds are allocated, various plants competing for funds that are given ultimately to the manager who can make the best use of them.

One would suspect, however, that it would be more profitable for the company as a whole to divert some of the raw material and market to plant B, even though this would mean compelling plant A to operate under conditions which, locally at least, would appear not to be optimal. Moreover, it might be possible to use some of the locally available raw materials 1 and 3 to add to the profit without using any of the scarce raw material 2, especially since Eq. (3-93) shows that only 400(0.32) = 128 tons of the main product are produced—52 tons short of the total demand. In reality, the allocation generated is "optimal" only relative to the two production plans submitted. Possibly, a more profitable overall plan could be found if more local plans were available.

3-22 GENERATING PROFITABLE ALTERNATIVES

Since the number of possible feasible local production plans is literally infinite, a means of picking out the more promising ones is needed. One intriguing feature of the decomposition principle is that the coordinating group actually can guide the local managers in their search for profitable alternatives. It accomplishes this by using the sensitivity coefficients generated by the solution to the coordinating problem. The bottom row of the optimal tableau for this first coordinating problem is found to be

$$Z = 3520 - 2000\lambda_{b1} - 22\bar{S}_5 \tag{3-97}$$

Now consider the effect of adding a new production plan A2 from plant A and giving it a weighting factor of λ_{a2}. Since at the moment the value of λ_{a2} is zero, we can treat it as a decision variable that would appear in Eq. (3-97) as follows:

$$Z = 3520 - 2000\lambda_{b1} - 22\bar{S}_5 + v_{a2}\lambda_{a2} \tag{3-98}$$

where v_{a2} is the sensitivity coefficient, as yet unknown, of the weight factor λ_{a2}. If we knew the value of v_{a2}, we could decide immediately whether to bring plan A2 into consideration, for the profit can be increased only if v_{a2} is positive. Let p_{a2} be the profit (not yet known) associated with plan A2. We know that

$$Z = 11{,}000\lambda_{a1} + 9000\lambda_{b1} + Z_{a2}\lambda_{a2} \tag{3-99}$$

An expression for v_{a2} is now obtained by equating (3-98) to (3-99) and differentiating partially with respect to λ_{a2}, holding λ_{a1} and λ_{b1}, but not $\bar{S}_5$, constant. Upon rearrangement, we get

$$v_{a2} = Z_{a2} + 22\frac{\partial \bar{S}_5}{\partial \lambda_{a2}} \tag{3-100}$$

It remains to evaluate the partial derivative. To do this, we first write Eq. (3-87) in terms of the weight factors and the still unknown variables x_{1a2} and x_{2a2} of plan A2. Since

$$x_{2a} = 500\lambda_{a1} + x_{2a2}\lambda_{a2}$$

we have

$$500\lambda_{a1} + x_{2a2}\lambda_{a2} + 500\lambda_{b1} + \bar{S}_5 = 160$$

whence

$$\frac{\partial \bar{S}_5}{\partial \lambda_{a2}} = -x_{2a2} \tag{3-101}$$

The profit Z_{a2} may be written in terms of plan A2 as

$$Z_{a2} = 15x_{1a2} + 22x_{2a2} \tag{3-102}$$

Combining Eqs. (3-100) and (3-101), we obtain

$$v_{a2} = 15x_{1a2} \tag{3-103}$$

Thus any plan using raw material 1 will be worthy of consideration.

It would seem attractive to make the coefficient v_{a2} as large as possible, since it is the rate of profit increase with respect to changes in the weight λ_{a2} placed on plan A2. To do this, plant manager A solves a new linear programming problem using the same constraints (3-83) and (3-84) as before, but with the altered profit function

$$Z'_{a2} \equiv v_{a2} = 15x_{1a2} \tag{3-104}$$

In comparing this with the original profit function $Z_a = 15x_{1a} + 22x_{2a}$, we see that the coordinating committee has effectively reduced the profit rate for raw material 2 to zero in plant A. Manager A is asked to generate the best plan he can, taking this handicap into account. The solution, designated solution A2, is evidently to use as much raw material 1 as possible:

$$x_{1a2} = 80, \qquad x_{2a2} = 0, \qquad Z_{a2} = 15(80) = \$1200$$

In a similar way the coordinating group instructs manager B to find, if he can, a production plant that is locally profitable even when the return from processing raw material 2 is artificially reduced to zero. Manager B simply maximizes

$$Z'_{b2} = 8x_{3b2} \tag{3-105}$$

subject to constraints (3-85) and (3-86), obtaining solution B2: $x_{2b2} = 0$, $x_{3b2} = 180$, $Z_{b2} = 8(180) = \$1440$.

3-23 SECOND MASTER PLAN

The coordinating group now has four local solutions to work with. As before, weighting coefficients λ_{a2} and λ_{b2} are assigned to the new solutions. The two solutions from plant A are combined, so that

$$x_{1a} = x_{1a1}\lambda_{a1} + x_{1a2}\lambda_{a2} = 80\lambda_{a2}$$

$$x_{2a} = x_{2a1}\lambda_{a1} + x_{2a2}\lambda_{a2} = 500\lambda_{a1}$$

and

$$Z_a = Z_{a1}\lambda_{a1} + Z_{a2}\lambda_{a2} = 11{,}000\lambda_{a1} + 1200\lambda_{a2}$$

These expressions for the plant A production plan will be feasible locally because they represent weighted averages of solutions which are themselves locally feasible. The weighting coefficients for plant A must, of course, be nonnegative and add up to unity or less.

Similar expressions can be derived involving the two plans for plant B. When these are combined with those for plant A and substituted into the shared constraints (3-87) and (3-88), the following equations result:

$$500\lambda_{a1} + 500\lambda_{b1} + \bar{S}_5 = 160 \tag{3-106}$$

$$400\lambda_{a1} + 32\lambda_{a2} + 300\lambda_{b1} + 36\lambda_{b2} + \bar{S}_6 = 180 \tag{3-107}$$

In addition, the definition of the weights requires that they satisfy

$$\lambda_{a1} + \lambda_{a2} + \bar{S}_a = 1 \tag{3-108}$$

and

$$\lambda_{b1} + \lambda_{b2} + \bar{S}_b = 1 \tag{3-109}$$

with $\bar{S}_a$ and $\bar{S}_b$ nonnegative. The coordinators wish to find a set of weights which satisfy these four constraints and which also maximize the profit

$$z = 11{,}000\lambda_{a1} + 1200\lambda_{a2} + 9000\lambda_{b1} + 1440\lambda_{b2} \tag{3-110}$$

The solution to this linear programming problem is $\lambda_{a1} = 0.2645$, $\lambda_{a2} = 0.7355$, $\lambda_{b1} = 0.0555$, and $\lambda_{b2} = 0.9445$, giving a monthly profit of \$5652—an increase of over \$2000.

This new coordinating plan diverts some of the scarce raw material 2 from plant A to plant B, since manager A can now make money by processing his local raw material 1. Although this increases his monthly profit only slightly (from \$3520 to \$3792), the released material 2 can now go to work in plant B, which previously was not even operating. Drawing also on some of its local material 3, plant B now makes \$1860 a month.

3-24 PENALTIES

The coordinating group must now check to see whether further improvement is possible. As before, they consider the effects of any new plan A3 submitted by manager A. The new profit equation, whose decision derivatives were generated during the solution of the most recent coordinating linear program, is more complicated than before because all four slack variables are now in it.

$$Z = 5652 - 3.75\bar{S}_5 - 21.55\bar{S}_6 - 510.8\bar{S}_a - 664.4\bar{S}_b + v_{a3}\lambda_{a3} \tag{3-111}$$

The result is differentiated partially with respect to λ_{a3}, holding the other weights constant but allowing the four slack variables to change. Rearranging this, we obtain

$$v_{a3} = Z_{a3} + 3.75\frac{\partial \bar{S}_5}{\partial \lambda_{a3}} + 21.55\frac{\partial \bar{S}_6}{\partial \lambda_{a3}} + 510.8\frac{\partial \bar{S}_a}{\partial \lambda_{a3}} + 664.4\frac{\partial \bar{S}_b}{\partial \lambda_{a3}} \tag{3-112}$$

To evaluate $\partial \bar{S}_5/\partial \lambda_{a3}$ we write Eq. (3-87) in terms of the weight factors and the still unknown variables x_{1a3} and x_{2a3} of plan A3. Since

$$x_{2a} = 500\lambda_{a1} + x_{2a3}\lambda_{a3}$$

we have

$$500\lambda_{a1} + x_{2a3}\lambda_{a3} + 500\lambda_{b1} + \bar{S}_5 = 160$$

whence

$$\frac{\partial \bar{S}_5}{\partial \lambda_{a3}} = -x_{2a3} \tag{3-113}$$

Similarly, Eq. (3-88) may be written

$$400\lambda_{a1} + 32\lambda_{a2} + (0.40x_{1a3} + 0.80x_{2a3})\lambda_{a3} + 300\lambda_{b1} + 36\lambda_{b2} + \bar{S}_6 = 180$$

This gives

$$\frac{\partial \bar{S}_6}{\partial \lambda_{a3}} = -0.40x_{1a3} - 0.80x_{2a3} \tag{3-114}$$

The definition of the weights requires that

$$\lambda_{a1} + \lambda_{a2} + \lambda_{a3} + \bar{S}_a = 1$$

from which it follows that

$$\frac{\partial \bar{S}_a}{\partial \lambda_{a3}} = -1 \tag{3-115}$$

Since $\bar{S}_b$ is the slack variable for the plant B weights only, it must be true that

$$\frac{\partial \bar{S}_b}{\partial \lambda_{a3}} = 0 \tag{3-116}$$

The remaining variable Z_{a3}, the profit associated with plan A3, is given by

$$Z_{a3} = 15x_{1a3} + 22x_{2a3} \tag{3-117}$$

Equations (3-112)–(3-118) give

$$v_{a3} = 6.38x_{1a3} + x_{2a3} - 510.8 \tag{3-118}$$

If manager A can find a plan that makes this expression positive, then the plan will be effective in increasing the system profit. He cannot affect the constant -510.8, but he is able to maximize the pseudo profit

$$Z'_{a3} \equiv 6.38x_{1a3} + x_{2a3} \tag{3-119}$$

subject to the plant A constraints. To do this he simply solves a linear programming problem as before, treating the factors 6.38 and 1 as profit decision derivatives. Notice that the availability of plans A1, A2, B1, and B2 has lowered the effective profit on material 1 from \$15 to \$6.38 per ton. On the other hand, the effective profit for the shared raw material 2, which went from its original value of \$22 per ton down to zero after solution A1 was available, is now back up to \$1.00. The coordinating group communicates this information to manager A by asking him if he can find a solution that would make at least \$510.80 a month if his profit were lowered \$8.62 per ton of material 1 and \$21.00 per ton of material 2. Manager A solves the corresponding linear programming problem and reports that if $x_{1a3} = 80$ and $x_{2a3} = 260$, he can make \$770.80 a month in spite of the penalties. Thus solution A3, for which the actual profit Z_{a3} is \$6920 per month, is submitted to the coordinating committee as capable of improving the company-wide profit. It is perhaps interesting that this solution is the first to use all of material 1 and to satisfy the demand for by-product from plant A.

3-25 GENERAL EQUATION

Until now we have used direct analysis to obtain the penalties to be imposed on the raw materials. We shall now show how to obtain a simple formula for the penalties. Let the decision derivatives of the slack variables $\bar{S}_k$ generated by the coordinating linear program be denoted by v_k so that, taking into account a potential new solution B3, we would have from the bottom row of the optimal tableau:

$$Z = Z_0 + \sum_k v_k \bar{S}_k + v_a \bar{S}_a + v_b \bar{S}_b + v_{b3}\lambda_{b3} \tag{3-120}$$

where Z_0 is the numerical value of the profit when all the $\bar{Z}_j$ are zero. Equations (3-98) and (3-111) are special cases of this equation written for other solutions. Let the equations for the shared constraints be written

$$\sum_j a_{kj}x_j + \bar{S}_k = b_k \tag{3-121}$$

Using the methods of analysis already described in the numerical cases, one can show that

$$v_{b3} = \sum_j (Z_j - \sum_k v_k a_{kj})x_{jb3} - v_b \tag{3-122}$$

For abbreviation, let the penalty $\hat{Z}_j$ to be assessed against material j in solution B3 be defined by

$$\hat{Z}_j = \sum_k v_k a_{kj} \tag{3-123:j}$$

Then

$$v_{b3} = \sum_k (Z_j - \hat{Z}_j)x_{jb3} - v_b \tag{3-124}$$

In this case the penalties to be transmitted to manager B are

$$\hat{Z}_{2b} = 3.75(1) + 21.55(0.6) = 16.68$$

and

$$\hat{Z}_{3b} = 3.75(0) + 21.55(0.2) = 4.31$$

Thus manager B will try to find a plan whose penalized profit, calculated from

$$Z'_{b3} = 1.32x_{2b3} + 3.69x_{3b3}$$

must exceed v_b (= \$664.40/month) to be worth considering by the coordinators. Solving the proper linear program, he generates solution B3, with $x_{2b3} = 140$, $x_{3b3} = 180$, and a penalized profit of \$795.00 per month which qualifies it as potentially profitable. Its true profit is $Z_{a3} = 140(18) + 180(8) = \$3960$. This solution uses all of material 3 and satisfies the local demand for by-product.

3-26 OPTIMAL MASTER PLAN

The reader may easily verify that the new coordinating constraints, using all six local solutions, are

$$\begin{aligned}
500\lambda_{a1} \qquad\qquad + 260\lambda_{a3} + 500\lambda_{b1} \qquad\qquad + 140\lambda_{b3} + \bar{S}_5 &= 160 \\
400\lambda_{a1} + 32\lambda_{a2} + 240\lambda_{a3} + 300\lambda_{b1} + 36\lambda_{b2} + 120\lambda_{b3} + \bar{S}_6 &= 180 \\
\lambda_{a1} + \lambda_{a2} + \lambda_{a3} \qquad\qquad\qquad\qquad + \bar{S}_a &= 1 \\
\lambda_{b1} + \lambda_{b2} + \lambda_{b3} + \bar{S}_b &= 1
\end{aligned}$$

The coordinating group is to find weights satisfying these equations and maximizing the total profit

$$Z = 11{,}000\lambda_{a1} + 1200\lambda_{a2} + 6920\lambda_{a3} + 9000\lambda_{b1} + 1440\lambda_{b2} + 3960\lambda_{b3}$$

The solution is $\lambda_{a1} = 0$, $\lambda_{a2} = 0.693$, $\lambda_{a3} = 0.307$, $\lambda_{b1} = 0$, $\lambda_{b2} = 0.427$, and $\lambda_{b3} = 0.573$. This gives a total profit of \$5840 per month, an increase of about 4 per cent over the previous case.

3-27 TERMINATION

We shall see that this allocation is optimal and that we need no longer search for new local solutions. To find this out, we generate penalties as before to guide the local managers in selecting production plans. Given the coefficients $v_5 = 6.22$, $v_6 = 19.6$, $v_a = 572$, and $v_b = 733$, obtained from the coordinating linear program, the reader may use Eq. (3-123) to verify that the penalties in this case are $\hat{Z}_{1a} = 7.85$, $\hat{Z}_{2a} = 22.00$, $\hat{Z}_{2b} = 18.00$, and $\hat{Z}_{3b} = 4.07$.

Manager A then tries to generate a solution that will give a penalized profit of more than \$572 a month with only a unit profit of \$7.15 allowed on raw material 1 and no profit at all permitted on material 2. He finds he cannot do this, the "new" proposal turning out to be identical to solution A2 already under consideration. With the present penalties, this proposal will bring in only \$572 per month, exactly the threshold specified. Thus no better solution from plant A is available. It turns out that plant B cannot improve things either, so the coordinator is able to conclude that the allocation problem is at last solved.

The six local plans and the final coordinated plan are shown in Table 3-4. All the restrictions are satisfied except the demands for by-products. It is

TABLE 3-4

Optimal Allocation

Proposal	*Solution* *A1*	*A2*	*A3*	*Solution* *B1*	*B2*	*B3*
Consumed						
1	0	80	80	—	—	—
2	500	0	260	500	0	140
3	—	—	—	0	180	180
Produced						
Main	400	32	240	300	36	120
By-product	100	48	100	200	144	200
Profit	11,000	1200	6920	9000	1440	3960
Weight	0	0.693	0.307	0	0.427	0.573

Allocation				*Plant A total*				*Plant B total*	*Grand total*
Consumed									
1	0	55.4	24.6	80	—	—	—	—	80
2	0	0	79.8	79.8	0	0	80.2	80.2	160
3	—	—	—	—	0	76.9	103.1	180	180
Produced									
Main	0	22.2	73.7	95.9	0	15.4	68.8	84.2	180
By-product	0	33.3	30.7	64.0	0	61.5	114.6	176.1	—
Profit	0	832	2124	2956	0	615	2269	2884	5840

interesting that the scarce raw material 2 is divided almost equally between the two plants, for if only locally optimal plans were considered, plant A would get it all. Notice also that neither of the locally optimal plans is used; both plans require too much of the scarce material and do not permit any money to be made processing local materials. It may seem strange that plant A actually makes \$564 a month *less* with this optimal allocation than it would by operating with its locally optimal plan. This sacrifice is advantageous because it permits plant B to bring in \$2884 a month more. Plant B is, in fact, almost as profitable as plant A in the optimum company-wide plan. Recall that when locally optimal plans were considered alone, it did not appear economical to run plant B at all.

Here at least there are no problems of unemployment caused by technological change and corporate merger. The decomposition principle, with its rational rewards and penalties, has shown each manager how to run his plant for the maximum profit of the entire company. Other methods for optimizing interacting systems are developed in Chapter 7.

3-28 TRANSPORTATION PROBLEMS

An important special case of linear programming is called the *transportation* or *distribution* problem. Such problems are important both because they occur often in practice, and because they can be solved by algorithms which are more efficient for this class of problem than is the simplex method. Many different situations may produce a model having this simple structure, but in general these problems concern the distribution of limited resources to satisfy known demands so as to optimize the distribution cost (or profit). The distribution may, for example, involve assignment of facilities to operations, or it may require physical transportation of a homogeneous product from given sources to known destinations. For clarity in the following mathematical presentation, we use the latter concept, although the derivation is general and applies to all distribution problems.

In linear programming terms, let the amount of product available at origin m be given by a_m, the amount required at destination n by b_n, and the cost of shipping one unit from m to n by $c_{(m-1)N+n}$, where $m = 1, \ldots, M$, and $n = 1, \ldots, N$. Hence we wish to determine the amounts $x_{(m-1)N+n}$ to be shipped from each origin m to each destination n so as to minimize the total cost of transportation, while satisfying constraints on availabilities at the origins and requirements at the destinations.

Setting $k \equiv (m-1)N + n$, the problem is to minimize

$$y = \sum_{k=1}^{MN} c_k x_k \tag{3-125}$$

subject to the constraints on availabilities at the M origins:

$$\sum_{k=(m-1)N+1}^{mN} x_k = a_m; \qquad m = 1, \ldots, M \tag{3-126}$$

and the demands at the N destinations:

$$\sum_{j=0}^{M-1} x_{n+jN} = b_n; \qquad n = 1, \ldots, N \tag{3-127}$$

as well as the nonnegativity conditions:

$$x_k \geq 0; \qquad k = 1, \ldots, MN \tag{3-128}$$

which state that shipments may not be made from a destination to an origin. From Eqs. (3-125)–(3-128), it can be seen that the transportation problem is a special case of the general linear programming problem defined at the beginning of this chapter. Here, all the a_{mn} are either 0 or 1; furthermore, these constraint coefficients appear in the particular pattern shown in the simplex tableau of Fig. 3-23. Notice the echeloned structure of the first M equations, associated with the origin availabilites, and the diagonal structure of the last N equations, associated with the destination requirements. Each variable appears in exactly two equations (rows), and we would need to add M artificial variables to obtain a first basic feasible solution if this problem were solved by the simplex method. A transportation problem having M origins and N destinations then requires a simplex tableau with $M + N$ rows and $(M + 1)(N + 1)$ columns (including the column for the a_m and b_n constants). Owing to the special structure of this problem, we may modify the simplex method in such a way that we need work with only an M by N table.

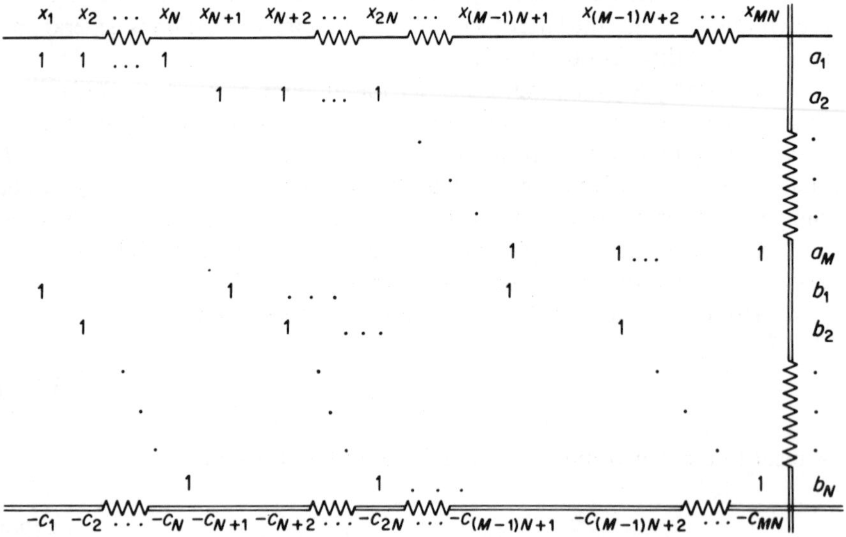

Figure 3-23. Simplex tableau for a transportation problem.

3-29 FINDING A FIRST FEASIBLE SOLUTION

Each variable appears in exactly one of the first M equations, and also in exactly one of the last N equations, so by summing these two sets of equations, we obtain

$$\sum_{k=1}^{MN} x_k = \sum_{m=1}^{M} a_m = \sum_{n=1}^{N} b_n \equiv T \tag{3-129}$$

Since the sum of the first M rows is equal to the sum of the last N rows, the $M + N$ equations are not linearly independent. In fact, we can demonstrate that any $M + N - 1$ rows are linearly independent and that therefore we need work with only $M + N - 1$ basic variables. First, we delete the $M +$

Nth (last) row, and choose $x_1, x_2, \ldots, x_N, x_{2N}, x_{3N}, \ldots, x_{MN}$, as the basic variables, setting the remaining variables to zero. Rearranging these first $M + N - 1$ equations, we may write them in terms of the basic variables, as shown in Fig. 3-24. If each of the last $N - 1$ equations ($x_n = b_n$; $n = 1, \ldots, N - 1$) is now subtracted from the first equation, each basic variable will appear in one and only one equation, showing that the $M + N - 1$ equations are linearly independent. Considering that any one of the $M + N$ equations may be written as a linear combination of the remaining $M + N - 1$ equations, it is clear that we have lost no generality in selecting the last equation as the one to be deleted; any other equation could just as well have been selected. Thus we have shown that a basic feasible solution to the constraints (3-126) and (3-127) will contain exactly $M + N - 1$ basic variables. Now we shall show that a feasible solution to these constraints can always be found; in fact, an obvious solution may be obtained simply by setting

$$x_{(m-1)N+n} = \frac{a_m b_n}{T}; \qquad m = 1, \ldots, M; \quad n = 1, \ldots, N \tag{3-130}$$

$$\begin{array}{lr}
x_N + \qquad\qquad + x_1 + x_2 + \cdots + x_{N-1} & = a_1 \\
\quad x_{2N} & = a_2 \\
\qquad x_{3N} & = a_3 \\
\qquad\quad \ddots & \vdots \\
\qquad\qquad x_{MN} & = a_M \\
\qquad\qquad\quad x_1 & = b_1 \\
\qquad\qquad\qquad x_2 & = b_2 \\
\qquad\qquad\qquad\quad \ddots & \vdots \\
\qquad\qquad\qquad\qquad\qquad x_{N-1} & = b_{N-1}
\end{array}$$

Figure 3-24. Finding a feasible solution.

This solution, although feasible, makes each of the MN variables positive; to obtain no more than $M + N - 1$ positive variables, one may use the *north-west corner* rule (Charnes and Cooper, 1954): set $x_1 = \min(a_1, b_1)$; if $a_1 > b_1$, set $x_2 = \min(a_1 - b_1, b_2)$; if $a_1 < b_1$, set $x_{n+1} = \min(b_1 - a_1, a_2)$, and so on. In other words, at each step, either an origin or a destination constraint is satisfied, until at the last step, the value of x_{MN} satisfies simultaneously the Mth origin and the Nth destination constraint. As long as none of the other variables selected satisfies both an origin and a destination constraint simultaneously, we will have chosen exactly $M + N - 1$ positive variables; otherwise, we will have less than $M + N - 1$ positive variables.

Hence we can find a basic feasible solution from among the original variables. It will always be possible then to drive the artificial variables to zero; in fact, we shall find that the artificial variables need never be introduced at all.

From the manner in which we solved for the basic variables in Fig. 3-24 it is clear that all the coefficients are either 0, +1, or −1. Furthermore, one can verify that in calculating the ratios involved in determining the basic variable/decision variable interchange, this ratio will always be positive. Therefore, in transportation problems, it is not necessary to perform the division (simplex) operations associated with basis changes, since the value of any basic variable can be found by adding and subtracting the values of some subset of the decision variables. If some decision variable, say d_r, is found in terms of the basic variables (holding all other decision variables at zero), then in the resulting expression the coefficients of the basic variables will also be 0, +1, or −1, a factor of considerable importance in the special tableau for solving transportation problems.

We find it convenient to rewrite Eqs. (3-125)–(3-128), using a *double subscript* notation on both the variables and the objective coefficients. Let

$$c_{mn} \equiv c_{(m-1)N+n} \quad \text{and} \quad x_{mn} \equiv x_{(m-1)N+n} \qquad \text{for all } m \text{ and } n$$

With this notation change, the transportation problem now appears as:

$$\text{minimize } y = \sum_{m=1}^{M} \sum_{n=1}^{N} c_{mn} x_{mn} \tag{3-131}$$

subject to

$$\sum_{n=1}^{N} x_{mn} = a_m; \qquad m = 1, \ldots, M \tag{3-132}$$

$$\sum_{m=1}^{M} x_{mn} = b_n; \qquad n = 1, \ldots, N \tag{3-133}$$

and

$$x_{mn} \geq 0; \qquad m = 1, \ldots, M; \qquad n = 1, \ldots, N \tag{3-134}$$

Double subscripts are natural in the formulation of a transportation problem; c_{mn} and x_{mn} are, respectively, the unit shipping cost and amount shipped from origin m to destination n. Transportation problems are usually solved using the special tableau format shown in Fig. 3-25. Since the decision variables in a linear programming problem are always zero, their values (x_{mn}) would not actually be entered in this form of tableau when working a problem, but would be left blank. Thus, any zeros appearing in the x_{mn} portion of a cell would be those belonging to basic variables in a degenerate basis.

From the appearance of the transportation tableau of Fig. 3-25, it is clear why the name "northwest corner rule" is appropriate to the rule previously described for finding a first feasible solution containing no more than $M + N - 1$ positive variables. [Allocations are begun in the northwest

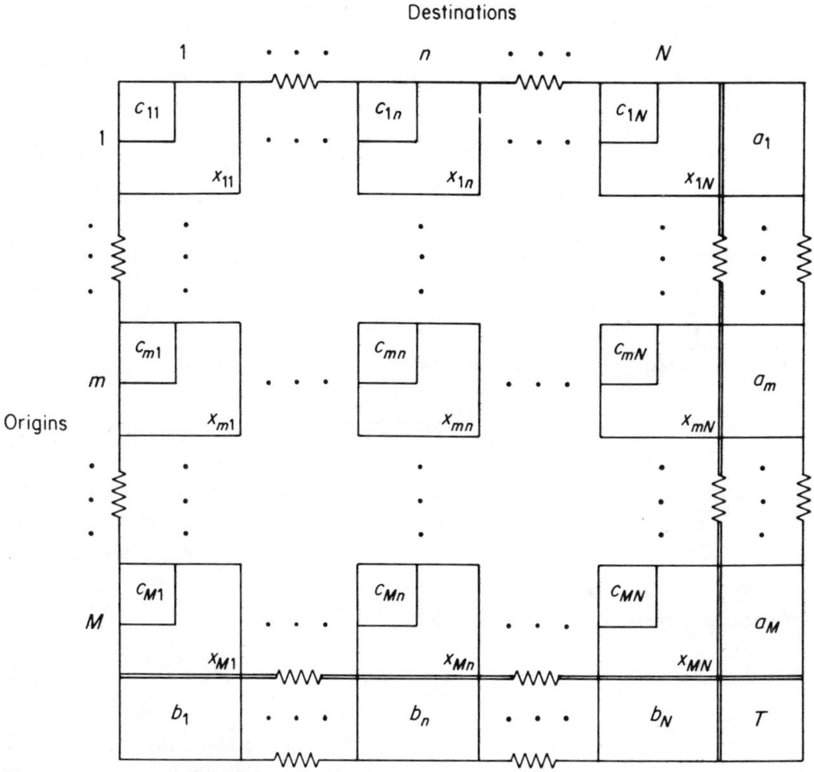

Figure 3-25. Transportation problem tableau format.

(upper left) corner of the tableau, and continued in a path moving south and east.] Modifying this rule to select exactly $M + N - 1$ variables, we may use these as the basic variables, even though some of them may occasionally have a zero value. Since all the coefficients are either 0, +1, or −1, a great deal of work can be saved by not eliminating each basic variable from all but one constraint equation, as required in the simplex method. Thus, in the transportation problem, each constraint equation will contain one *or more* basic variables. To handle the problem of degeneracy, the northwest corner rule is modified as follows. When a row (origin) and a column (destination) constraint are both satisfied by the assignment of a value to some variable, increase the requirement for the row by a small amount ϵ, and continue with the northwest corner rule. In general, this means that we perturb the values of the original constants a_m and b_n to produce new constants a'_m and b'_n defined by

$$
\begin{aligned}
a'_m &= a_m + \epsilon; \qquad & m &= 1, \ldots, M \\
b'_n &= b_n; \qquad & n &= 1, \ldots, N - 1 \\
b'_N &= b_N + M\epsilon
\end{aligned}
\tag{3-135}
$$

Notice that these new constants still satisfy Eq. (3-129):

$$\sum_{m=1}^{M}\sum_{n=1}^{N} x_{mn} = \sum_{m=1}^{M} d'_m = \sum_{n=1}^{N} b'_n \equiv T' = T + M\epsilon \tag{3-136}$$

Upon completion of the steps of the rule, ϵ is set to zero to provide a degenerate basic feasible solution containing exactly $M + N - 1$ basic variables. In practice, of course, it is not actually necessary to introduce the ϵ's, since it is clear where zeros must be added in order to carry out the steps of the modification.

3-30 EVALUATING A SOLUTION

By the northwest corner rule, there must be at least one basic variable in each equation. Thus, when it is necessary to express some decision variable x_{mn} in terms of the basic variables, there will be some basic variable x_{ms} in this same (mth) row, and from the pattern produced by the northwest corner rule, there must also then be at least one basic variable, x_{ps}, in the column s constraint equation. In fact, it can be shown that any decision variable x_{mn} may be expressed uniquely in terms of the basic variables as

$$x_{mn} = x_{ms} - x_{ps} + x_{pt} - \cdots - x_{wr} + x_{wn} \tag{3-137}$$

Equation (3-137) shows that the expression of a decision variable in terms of the basic variables always contains an odd number of basic variables, and also that the plus and minus coefficients of the basic variables alternate, beginning and ending with a plus sign. In the transportation tableau of Fig. 3-25, Eq. (3-137) corresponds to the formation of a *path* which begins and ends at the cell containing the decision variable x_{mn}. All other cells in the path are occupied by basic variables which form the expression in Eq. (3-137).

This path must be traced out by alternating horizontal and vertical moves, which change direction at the cells containing basic variables ("stepping stones") and must also be such that any two adjacent cells in the path lie in the same row or same column. Each horizontal move must be followed by a vertical move, and vice versa; thus no three adjacent cells in the path will lie in the same row or same column. Also, since the representation of a decision variable in terms of the basic variables is unique, the required path is also unique. Note that if one could trace out a path of the type just described which, however, contained only basic cells (beginning and ending with a basic cell), then this would express one of the basic variables in terms of some of the others, indicating that the basic variables were not linearly independent. Hence such a "loop" among the basic variables should never be possible.

The illustrative problem shown in Fig. 3-26(a) will clarify the foregoing discussion. Here, $c_{11} = 1$, $c_{12} = 4$, $c_{13} = 6$, $c_{21} = 0$, $c_{22} = 2$, $c_{23} = 5$, $c_{31} = 2$, $c_{32} = 7$, $c_{33} = 5$, $a_1 = 5$, $a_2 = 4$, $a_3 = 6$, $b_1 = 3$, $b_2 = 5$, $b_3 = 7$, and $T = 15$. Beginning in the northwest corner [cell (1, 1)], we set $x_{11} = \min(a_1, b_1) = \min(5, 3) = 3$, which satisfies the first column constraint; then $x_{12} = \min(a_1 - b_1, b_2) = \min(2, 5) = 2$, and so on, producing the first feasible solution of Fig. 3-26(b). The evaluation path for cell (3, 1) is also

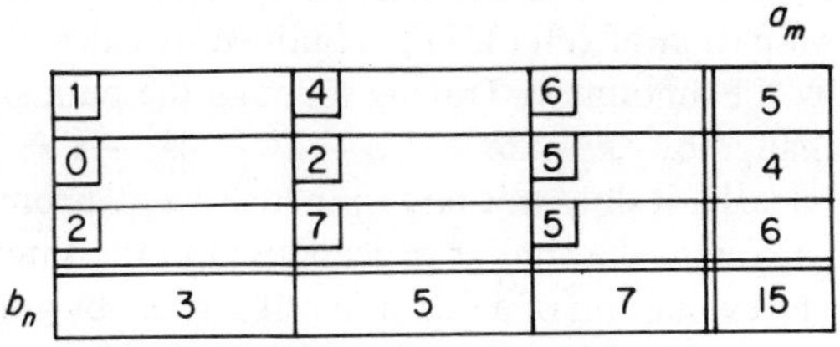

(a) Tableau containing problem statement

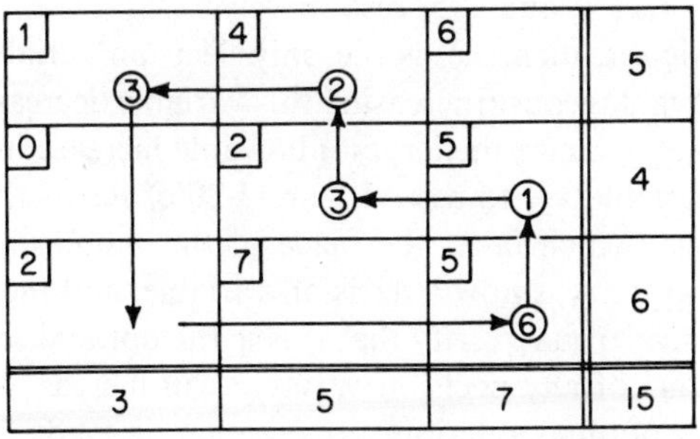

(b) Results of northwest corner rule, showing evaluation path for cell (3, 1)

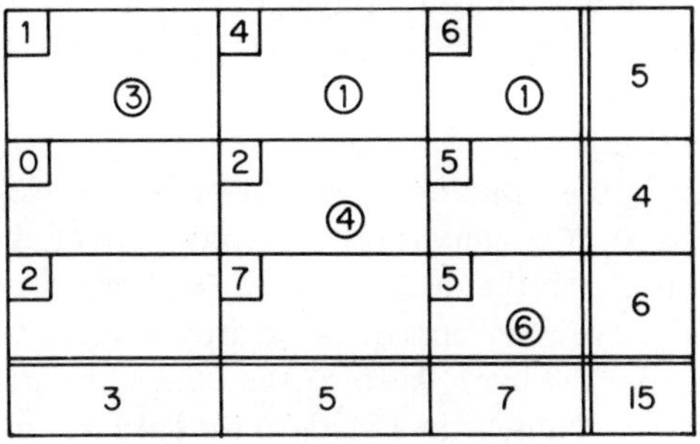

(c) Optimal tableau; $y^* = 51$

Figure 3-26. Tableaux for first transportation problem.

shown in this figure, and it happens to contain all the basic cells. This is not usually the case, of course, and in fact all the other decision cells in this tableau can be written in terms of exactly three of the basic variables.

If now x_{31} is increased (from zero) by an amount k, then we see that x_{33} must be *decreased* by k in order that the sum of the variables in the third row will remain equal to a_3 ($=6$). Decreasing x_{33} then requires that x_{23} be increased by k so that the sum of the elements in column 3 will still add up to 7; this in turn requires that x_{22} be decreased by k, and so on. Now it can be seen that this evaluation path, or closed loop, will always be such as to leave the row and column totals unchanged; a feasible solution will be maintained at all times. The evaluation of cell (3, 1) is obtained by calculating the effect of increasing x_{31} by the amount k. Tracing through the path shows that the total cost would change by $2k - 5k + 5k - 2k + 4k - k = 3k$, or a unit *increase* in y of 3; clearly, it does not pay to increase x_{31}. From this analysis (and from the analogy with the similar process used in the simplex method), it can be seen that the evaluation of a decision cell is given by a number which is the *decision derivative* for that cell. Proceeding with the evaluation of the other decision cells, the decision derivatives for cells (1, 3), (2, 1), and (3, 2), are found to be -1, $+1$, and $+5$, respectively.

For this basic set, then, x_{13} is the only decision variable for which an increase will be advantageous. Increasing this variable decreases both x_{12} and x_{23}, and increases x_{22}; hence the largest allowable increase in x_{13} is 1, as this drives the basic variable x_{23} to zero. Figure 3-26(c) shows the results of this change in the basis (which takes the place of the simplexing operations of linear programming); x_{23} is now a decision variable, and therefore cell (2, 3) is left blank. The reader may verify that this is the optimal solution, since an increase in the value of any decision variable will increase the total cost y. Notice that in this optimal solution, no shipment is made from origin 2 to destination 1, even though zero cost would be incurred for such a shipment. In this simple problem, one can see that any shipments made through cell (2, 1) will decrease the shipments through cells (1, 1) and (2, 2) while increasing the amount shipped through cell (1, 2), causing a net increase in the total cost. In larger problems, the analysis is far more complex, and consequently less intuitive.

If an increase in the value of a decision variable drives more than one basic variable to zero, it is important that only one of these variables be dropped from the basis, so that at all times we will retain a total of $M + N - 1$ basic variables. We may choose arbitrarily which of the zero-valued variables to drop from the basis, keeping the others as part of a degenerate basic feasible solution. Thus, if in Fig. 3-26(b), both x_{12} and x_{23} had been equal to 2, the decision variable x_{13} would have been increased to 2, and we would have retained either x_{12} or x_{23} in the basis at a zero level, and made the other a decision variable (left its cell blank).

3-31 SIMPLIFIED METHOD FOR EVALUATING THE DECISION DERIVATIVES

The process just described for evaluating the decision cells can become tedious in even moderate-sized transportation problems. We now present a simplified method (Dantzig, 1951) for making these evaluations. By writing the Lagrangian function for the transportation problem, designating as u_m the Lagrange multiplier for the mth row constraint, and as v_n the Lagrange multiplier for the nth column constraint, and taking partial derivatives, we obtain as necessary conditions for an optimal solution:

$$x_{mn}(c_{mn} - u_m - v_n) = 0; \qquad m = 1, \ldots, M; \quad n = 1, \ldots, N \tag{3-138}$$

$$c_{mn} - u_m - v_n \geq 0; \qquad m = 1, \ldots, M; \quad n = 1, \ldots, N \tag{3-139}$$

Notice that the u_m and v_n are *not* required to be nonnegative, and that this relaxation of the nonnegativity restriction is due to the absence of slack variables in the constraints (3-132) and (3-133); all constraints in the transportation problem are equations rather than inequalities.

Equation (5-125) states that if any x_{pq} is positive, then $c_{pq} - u_p - v_q = 0$. Thus, for all basic variables, a necessary condition for optimality is

$$u_p + v_q = c_{pq}; \qquad \text{where } x_{pq} \text{ is a basic variable} \tag{3-140}$$

From Eq. (3-137), the unit change in the objective function as a consequence of increasing a decision variable x_{mn} is given by

$$\frac{\delta y}{\delta x_{mn}} = c_{mn} - c_{ms} + c_{ps} - c_{pt} + \cdots + c_{wr} - c_{wn} \tag{3-141}$$

which, by Eq. (3-140), becomes

$$\frac{\delta y}{\delta x_{mn}} = c_{mn} - (u_m + v_s) + (u_p + v_s) - (u_p + v_t) + \cdots + (u_w + v_r) - (u_w + v_n)$$

$$= c_{mn} - u_m - v_n \tag{3-142}$$

since each of the "stepping stones" in the path is in a basic cell. Comparison of Eqs. (3-141) and (3-142) with Eq. (3-15) shows that the number $c_{mn} - u_m - v_n$ is the *decision derivative* associated with cell (m, n). (Recall that all nonzero α_{mn} are ± 1 in the transportation problem.)

Equation (3-139) thus provides a simple means for checking the optimality of a solution; simply compute the sum $u_m + v_n$ for each decision cell, and if this sum exceeds the unit cost c_{mn} of that cell, then the total cost y can be decreased by increasing x_{mn}. Therefore, at each iteration, we need find only the closed loop path for the one decision variable that is to be increased. The numbers u_m and v_n, of course, are found from Eq. (3-140). Since there will be a total of $M + N$ such numbers, and each basic solution will contain exactly $M + N - 1$ state variables, Eqs. (3-140) will always consist of $M + N - 1$ equations in $M + N$ unknowns. Hence these equations have one degree of freedom, and any one of the u_p or v_q may be chosen arbitrarily.

For example, in the problem of Fig. 3-26(b), we might choose $u_1 = 0$; because cells (1, 1) and (1, 2) are basic cells, v_1 and v_2 must then be 1 and 4, respectively. Cell (2, 2) is also a basic cell, and since $v_2 = 4$, we must set $u_2 = -2$; this, in turn, forces $v_3 = 7$, as (1, 3) is a basic cell, and finally, $u_3 = -2$, so that $u_3 + v_3 = c_{33} \equiv 5$. The results of these calculations are shown in Fig. 3-27(a).

Using these numbers, we may now evaluate each decision cell without needing to trace out the entire closed-loop path for the cell. For example, the

u_m \ v_n	1	4	7	a_m
0	1 ③	4 ②	6	5
−2	0	2 ③	5 ①	4
−2	2	7	5 ⑥	6
b_n	3	5	7	15

(a) First feasible solution as given by northwest corner rule

u_m \ v_n	1	4	6	a_m
0	1 ③	4 ①	6 ①	5
−2	0	2 ④	5	4
−1	2	7	5 ⑥	6
b_n	3	5	7	15

(b) Optimal solution

Figure 3-27. Evaluation of independent cells using u_m and v_n numbers.

decision derivative for cell (3, 1) is $c_{31} - (u_3 + v_1) = 2 - (-2 + 1) = 3$, and the decision derivative for cell (1, 3) is $6 - (0 + 7) = -1$. Since among all the blank cells, only (1, 3) has a negative decision derivative, x_{13} is the only decision variable that can be increased to improve the solution. The procedure for bringing this variable into the basis is exactly the same as that described previously, and it produces the tableau of Fig. 3-27(b). With this change in the basis, we must recalculate at least some of the u_m and v_n numbers; here it was necessary to change only u_3 and v_3. From Fig. 3-27(b), we see that all decision derivatives are positive, indicating that the optimal solution has been reached.

The decision derivatives in a transportation problem may, of course, be used to perform sensitivity analyses just as in the more general linear programming problem. For example, the reader may easily verify that the solution given in Fig. 3-26(c) will remain optimal as long as c_{32} is no less than 3, or c_{13} is any nonnegative number, and so on. Again, if some unit cost is changed by an amount large enough to render the solution nonoptimal, then the new optimal solution can be obtained from the final tableau without re-solving the problem. As an illustration, suppose that c_{23} were changed from 5 to 3 in this problem, causing the decision derivative for cell (2, 3) to become negative. Beginning with the tableau of Fig. 3-26(c), this cell is brought into the basis, and a new optimal solution obtained: $x_{11}^* = 3$, $x_{12}^* = 2$, $x_{22}^* = 3$, $x_{23}^* = 1$, $x_{33}^* = 6$, and $y^* = 50$.

3-32 VOGEL'S METHOD

We now describe what would seem to be a more effective method for obtaining a first feasible solution than is the northwest corner rule, since it takes the unit cell costs into consideration. We shall refer to this method as *Vogel's method* (Reinfeld and Vogel). In practice, the method has often been more successful in determining an initial solution that requires fewer iterations to reach the optimum than do other methods. It is not possible, however, to prove any general theorem to this effect as problems can be devised (by placing the smallest c_{mn} numbers along the main diagonal of the transportation tableau) for which the northwest corner rule yields the best beginning solution.

We shall use the problem described by the tableau of Fig. 3-28(a) to illustrate Vogel's method as well as some computational simplifications which can result from making scale changes in the problem parameters. For example, if in Eq. (3-131) we replace each c_{mn} with $c'_{mn} \equiv kc_{mn} + K$, there results the modified objective function:

$$y' = \sum_{m=1}^{M} \sum_{n=1}^{N} (kc_{mn} + K)x_{mn} = ky + KT \qquad (3\text{-}143)$$

						a_m
	1.5	2.0	1.4	1.2	1.9	78
	1.1	1.8	1.2	1.0	1.6	30
	0.6	0.7	0.6	0.5	0.7	48
b_n	24	54	6	48	18	150

(a) Original transportation tableau

	5	11	6	5	9	a'_m
2	10	15	9	7	14	13
1	6	13	7	5	11	5
1	1	2	1	0	2	7
b'_n	4	9	1	8	3	25

(b) Modified problem including first Vogel numbers

						a'_m
	10	15 (2)	9 (1)	7 (8)	14 (2)	13
	6 (4)	13	7	5	11 (1)	5
	1	2 (7)	1	0	2	7
b'_n	4	9	1	8	3	25

(c) First feasible solution as given by Vogel's method

Figure 3-28. Scale effects and Vogel's method.

where y is the original objective function, k and K are arbitrary constants, and T is the sum of the x_{mn} as defined by Eq. (3-129). Clearly, minimizing y' is equivalent to minimizing y as long as k is a positive number (K may be any real number). In addition, both sides of the constraint equations (3-132) and (3-133) may be multiplied by the same positive constant $\mathfrak{K}$ with the only effect being that of multiplying the optimal solution $\{x^*_{mn}\}$ by $\mathfrak{K}$. Hence we may make use of these scale effects to simplify the calculations required to solve transportation problems by hand. For this illustrative problem, let us choose $k = 10$, $K = -5$, and $\mathfrak{K} = \frac{1}{6}$, to produce the modified problem shown in Fig. 3-28(b).

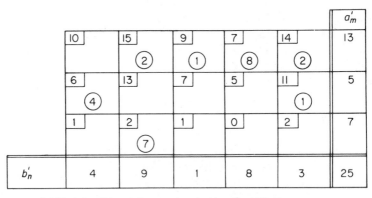

To obtain a first feasible solution by Vogel's method, we proceed as follows. For each row, compute the absolute value of the difference between the lowest unit cell cost and the next lowest unit cell cost in the row, and write this number to the left of the row. For example, in Fig. 3-28(b), this difference for the first row is $9 - 7 = 2$. Repeat this procedure for the columns, writing the differences just above the corresponding columns, thus obtaining a total of $M + N$ nonnegative Vogel numbers for all the rows and columns, as shown in Fig. 3-28(b).

Put as much allocation as possible through the cell having the *smallest* unit cost in the row or column which has the *largest* Vogel number. This allocation will then satisfy either a row or a column constraint (not necessarily the row or column corresponding to the largest Vogel number). In the present problem, this process selects column *two*, and we would allocate 7 units to cell (3, 2), since it has the lowest c_{mn} cost in that column, and min $(a_3, b_2) = 7$. None of the remaining cells in row three can be included in this first feasible solution, so these cells should be marked in some way as to exclude them from further consideration in carrying out the remaining steps in the method. This means that at each step, either a row or a column is eliminated from the computational procedure, and new Vogel numbers are computed. The process is repeated until all constraints are satisfied. The Vogel numbers may be looked upon as penalty costs which would be incurred if an allocation were not made to the cell having the lowest unit cost in a particular row or column. By analogy with the argument advanced for the northwest corner rule, this process of adding basic cells one at a time will produce (in the absence of degeneracy) a total of exactly $M + N - 1$ such cells. As stated earlier, degeneracy occurs only if the addition of a basic cell satisfies simultaneously both a row and a column constraint. Under such circumstances, one again perturbs the row constraint, a_m, by an amount ϵ, and continues with Vogel's method. At the conclusion of this procedure, setting the ϵ's to zero will always yield an initial feasible solution containing exactly $M + N - 1$ cells (Eisemann).

Continuing with Vogel's method on the foregoing problem, we obtain the initial solution shown in Fig. 3-28(c), which also happens to be the optimal solution, as the reader may verify by computing the u_m, v_n, numbers. (It is important not to confuse these evaluation numbers with the Vogel numbers used to obtain a feasible solution.) The resulting total minimum cost for this derived problem is readily found to be $y' = 172$. Since the original objective function y is related to this modified function through the expression $y' = \mathfrak{R}(ky + KT)$, and the optimal solution $\{x^*_{mn}\}$ to the original problem is given in terms of the derived optimum by $\{x^*_{mn}\} \equiv \mathfrak{R}\{x'^*_{mn}\}$, we have at once

$$y^* = \frac{1}{10}\left[\frac{172}{1/6} - (-5)(150)\right] = 178.2; \qquad x^*_{12} = 12, \quad x^*_{13} = 6, \quad x^*_{14} = 48,$$

$$x^*_{15} = 12, \quad x^*_{21} = 24, \quad x^*_{25} = 6,$$

$$\text{and} \quad x^*_{32} = 42$$

In some transportation or distribution problems, it is desired to *maximize* the objective function (3-131) rather than to minimize it (that is, the c_{mn} then represent unit *profits*). For such problems, one may replace each c_{mn} with $c'_{mn} \equiv c_{rs} - c_{mn}$, where c_{rs} is the largest of the c_{mn}, and then employ the foregoing solution algorithm to minimize $\sum_{m=1}^{M} \sum_{n=1}^{N} c'_{mn} x_{mn}$. Alternatively, one may work with the original c_{mn} and merely reverse the direction of the inequality signs in Eq. (3-139), which prescribes the conditions for optimality. In this case, the Vogel numbers would be computed as the absolute value of the difference between the *largest* and the *next largest* c_{mn} in each row and column. Vogel's method would then consist of allocating as much as possible to the cell having the largest unit profit in the row or column with the largest Vogel number. This latter procedure is better for hand computation; the former may be simpler when a computer program for minimizing the objective function is available.

3-33 INEQUALITY CONSTRAINTS

The description of the transportation problem constraints as given by Eq. (3-132) and Eq. (3-133) may appear somewhat artificial in that the total shipping requirements are always exactly equal to the total amount available for shipment, as shown in Eq. (3-128). This is not generally the case in practical problems, as when, for example, more units are available at the origins than are required at the destinations. In such a case, Eq. (3-133) remains unchanged, but Eq. (3-132) becomes

$$\sum_{n=1}^{N} x_{mn} \leq a_m; \qquad m = 1, \ldots, M \tag{3-144}$$

In order to obtain equality constraints, and thus satisfy the requirements of the solution algorithm, one merely adds slack variables, $x_{m,N+1}$ to Eq. (3-144), much as in the simplex method. Equation (3-129) then becomes

$$\sum_{m=1}^{M} \sum_{n=1}^{N+1} x_{mn} = \sum_{m=1}^{M} a_m = \sum_{n=1}^{N+1} b_n = T \tag{3-145}$$

from which follows the expected result

$$\sum_{m=1}^{M} x_{m,N+1} = \sum_{m=1}^{M} a_m - \sum_{n=1}^{N} b_n = b_{N+1} \tag{3-146}$$

where b_{N+1} is the total of the slack variables (that is, the total available quantity that is *not* shipped). By adding an $N + 1$st column to the M by N transportation tableau, the $a_1, \ldots, a_M, b_1, \ldots, b_{N+1}$ are once again the row and column totals required for the new tableau, and the standard solution algorithm may be used on this modified problem. The following numerical example will clarify these concepts.

Orders for 15, 20, and 10 automobile engines have come in from three assembly plants, A, B, and C, respectively. These orders are to be met from three factories, 1, 2, and 3, which have respective availabilities of 35, 30, and 10 engines. The shipping distances in miles from the various factories to each assembly plant are shown in Fig. 3-29(a). The cost for shipping each engine is 10 cents per mile, and it is desired to find which factories should supply which plants so as to minimize the total cost of shipping the 45 engines on order.

		Plants A	Plants B	Plants C
Factories	1	700	800	400
	2	100	200	600
	3	300	400	700

(a) Shipping distances from factories to assembly plants

		Plants A	Plants B	Plants C	Dummy	a_m
Factories	1	70	80	40 (10)	0 (25)	35
	2	10 (10)	20 (20)	60	0	30
	3	30 (5)	40	70	0 (5)	10
b_n		15	20	10	30	75

(b) Optimal solution; $y^* = \$1,050$

Figure 3-29. Transportation problem illustrating inequality constraints.

In this problem, 30 more engines are available than are required to be shipped, so a column must be added to the transportation tableau to represent a dummy destination to which no engines will be sent. Since no cost is associated with such fictitious shipments, the c_{mn} in the dummy column will all be zero, as shown in Fig. 3-29(b). Application of Vogel's method and the stepping-stone algorithm to this modified tableau produces the optimal solution given by the circled numbers in Fig. 3-29(b), which shows that 25 of the 35 engines available at factory 1, and 5 of the 10 engines available at factory 3, will not be shipped.

An interesting situation has arisen in this problem; the decision derivative for cell (3, 2) is *zero*. The reader will recall that this indicates the existence of an *alternative optimum*, for it means that the value of y^* would not change if the decision variable x_{32} were brought into the basis. Indeed, one can easily see that all optimal solutions to this problem may be written as $x_{21}^* = 10 + Z$, $x_{22}^* = 20 - Z$, $x_{31}^* = 5 - Z$, $x_{32}^* = Z$, $x_{13}^* = 10$, $x_{14}^* = 25$, and $x_{34}^* = 5$, where $Z = 0, 1, \ldots, 5$. As we stated earlier in this chapter, alternative optima are especially common in practical problems of high dimensionality.

Other inequalities in the constraints of a transportation problem are handled in an analogous manner. For example, if more units are demanded at the destinations than are available at the origins, a dummy *row* is added (to take care of shipments from a nonexistent source), which converts the constraints to equalities, as required by the solution algorithm. For a further discussion of inequalities in transportation problems as well as a generalized version of this type of problem, see the thorough treatment by Hadley.

3-34 INTEGER PROGRAMMING

Some linear programming problems require that some or all of the variables take on only *integral* values, and this additional restriction actually makes the programming problem nonlinear. In spite of this nonlinearity, the name *linear* is still attached to such problems, since both the objective function (3-1) and the constraints (3-2) are linear; the only change in the mathematical description of the problem is that the nonnegativity conditions are replaced by the requirements

$$x_n = 0, 1, 2, \ldots; \qquad n = 1, \ldots, N \tag{3-147}$$

Integer programming is a valuable tool in operations research, having tremendous potential for applications. Although there has been considerable theoretical research in the last two decades, progress in the computational aspects of large-scale integer programming problems are not yet impressive. In this section we will briefly review techniques that have been reported, and numerically investigate several approaches.

The general integer linear programming problem is directed toward finding $\mathbf{x} = (x_1, x_2, \ldots, x_n)$ such that:

$$\text{maximize } x_0 = \sum_{j=1}^{n} c_j x_j \tag{3-148}$$

subject to

$$\sum_{j=1}^{n} a_{ij} x_j \le b_i \qquad i = 1, 2, \ldots, m \tag{3-149}$$

$$x_j \ge 0, \quad \text{integer} \qquad j = 1, 2, \ldots, n \tag{3-150}$$

The zero–one integer programming problem is one in which each solution variable is restricted to only binary values. Any general integer programming problem can be converted to a zero–one integer programming problem using the following transformation on each solution variable.

$$x = \sum_{l=0}^{k} 2^l Y_l = Y_0 + 2Y + 4Y_2 + \cdots + 2^k Y_k \tag{3-151}$$

where k is the smallest integer such that $2^{k+1} \geq U + 1$ and U is the smallest upper bound on x. This procedure obviously increases the size of the original problem, but transforms the general problem into the zero–one problem for which available solution techniques are more efficient.

In order to illustrate the transformation above, consider the following integer programming problem.

$$\text{Maximize } f(\mathbf{x}) = 3x_1 + 2x_2 \tag{3-152}$$

subject to

$$5x_1 + 4x_2 \leq 23.7$$

$$x_1, x_2 \text{ integer}$$

$$\mathbf{x} \geq 0$$

It is clear that due to the integrality requirements on the solution variable, the following constraints are implied:

$$x_1 \leq 4 \quad \text{and} \quad x_2 \leq 5$$

Since the upper bounds on x_1 and x_2 are $U_1 = 4$ and $U_2 = 5$, we can reformulate the problem as follows. Consider replacing x_1 with two binary variables; that is, $k = 1$. From the above,

$$2^{k+1} \geq U_1 + 1 \geq 5 \tag{3-153}$$

or

$$2^2 = 4 \geq 5$$

Since the inequality is not satisfied, we will require $k = 2$; that is,

$$2^3 = 8 \geq 5$$

Hence

$$x_1 = \sum_{l=0}^{2} 2^l Y_l = Y_0 + 2Y_1 + 4Y_2 \tag{3-154}$$

In a similar manner, the transformation for x_2 in terms of a binary vector Z is given by

$$x_2 = Z_0 + 2Z_1$$

The new problem is formulated as follows:

$$\text{maximize } f(\mathbf{Y}, \mathbf{Z}) = 3Y_0 + 6Y_1 + 12Y_2 + 2Z_0 + 4Z_1$$

subject to

$$5Y_0 + 10Y_1 + 20Y_2 + 4Z_0 + 8Z_1 \leq 23.7$$
$$\mathbf{Y}, \mathbf{Z} \geq 0$$
$$\mathbf{Y}, \mathbf{Z}, \text{ binary}$$

The solution is given by

$$Y_0 = 1, \quad Y_1 = 1, \quad Y_2 = 0$$
$$Z_0 = 0, \quad Z_1 = 1, \quad f^*(\mathbf{Y}, \mathbf{Z}) = 13$$

Transforming back to the original solution space, we obtain the following:

$$x_1^* = 3 \; x_2^* = 2 \; f^*(\mathbf{x}) = 13$$

Note the striking difference between the optimal solution to this integer programming problem and its continuous counterpart. In particular, the constraint is *loose* at optimality, but it is necessary for problem solution. This constraint is guaranteed to be *tight* in any continuous solution.

History and progress

Several algorithms have been proposed for a solution to the integer programming problem. A systematic method for solving the general integer linear programming problem has been given by Gomory. In this method, the simplex algorithm is used to obtain the solution that is optimal when the integrality requirements are ignored, and then a new constraint is added to the original formulation which eliminates a portion of the feasible region near this solution point. The result of this procedure is to render this optimal point nonfeasible, and the dual-simplex method of Lemke is then employed to move from this point to an extreme point of the modified feasible set. If this is a lattice point, the process terminates; if not, another constraint ("cut") is added, and the foregoing procedure is repeated. We will demonstrate this procedure numerically in the following section.

This algorithm yields dual feasible solutions, so that a primal feasible integer solution is not available until the optimal integer solution is reached. This is a major disadvantage in the Gomory schemes. However, primal algorithms which continually maintain a feasible integer solution have recently been developed by Young and Glover. Other recent algorithms use branch-and-bound methods for implicitly (or explicitly) enumerating the space of all feasible integer solutions. These include the algorithms developed by Land and Doig, Balas, Cook and Cooper, Krolak, Hillier, and Geoffrion. We will also explore the numerical aspects of the branch-and-bound procedure in a following section. A short literature review will be conducted at this time to aid the interested reader.

Geoffrion describes a general framework for solving integer programming problems. He bases his general framework on three key notions, "separation," "relaxation," and "fathoming." *Separation* is breaking down the original problem into subproblems. *Relaxation* is enlarging the set of feasible solutions by modifying the constraints in some judicious fashion. *Fathoming* is checking for feasibility, improving the current solution, and establishing optimality. He also describes different integer programming algorithms and explains their solution procedure within this general framework.

Egon Balas has developed an algorithm for solving problems with strictly zero–one variables. The general approach of the algorithm is to start the solution procedure with a superoptimal but infeasible solution. A procedure which uses additive implicit enumeration techniques is developed that forces the solution toward feasibility.

Glover developed an algorithm following the approximate line of approach of Balas with the following modifications. The algorithm is based upon an underlying tree-search structure upon which a series of tests is superimposed to exclude large portions of the tree of all possible zero–one solutions from examination. Special types of constraints called *surrogate* constraints were used for this purpose. A surrogate constraint is a single constraint that enforces restrictions upon the optimal solution which cannot be determined from any of the individual constraints. A surrogate constraint is formed from multiple (original) problem constraints. The function of this constraint is to serve as a substitute for some of the original problem constraints so that fathoming can be done more efficiently.

The Land and Doig algorithm is based on branch-and-bound procedures which create successive parallel shifts in the objective hyperplane toward the interior of the solution space, such that each new shift will coincide with an integer value of at least one integer variable. These successive shifts are made in an orderly manner, so that it is never possible to bypass a superior integer point in the solution space. Writing a computer code accounting for all possible branches of the problem usually creates a severe demand on storage, which is a major limitation of the algorithm. Dakin overcame the limitation in Land and Doig's algorithm to a certain extent. He modified the algorithm so that it would guarantee exactly two branches at each node. Rather than forcing the variables at each node to take exact integral values, Dakin suggested that bounds can be used to cover the entire range of each variable. By considering these two branches at each node, one is actually accounting for all possible branches that may be generated for a particular variable in the original Land–Doig algorithm. The search terminates either when all the branches in the tree have integer solutions, a feasible solution does not exist, or a solution worse than the current best solution is found. The best solution is then selected from among all feasible solutions.

Through Dakin's algorithm, the essence of an important development was incorporated, commonly known as penalties. *Penalties* are changes in the objective function value for assigning an integer value to a particular variable. Davis, Kendrick, and Weitzman gave one of the first systematic developments of penalties in the context of a powerful enumerative scheme. The DKW paper addresses mixed integer programs (ones for which only some of the variables need be integers) in which all integer variables are represented as binary variables (zero–one). During "separation" it dichotomizes the "candidate problem" (candidate problem is a subproblem from the list of subproblems, which is considered currently for solution) in the manner of Dakin. The variable selected is the one with the greatest "up penalty" (the penalty for assigning integer value of 1). The associated bounds of the two new problems are derived from the relaxed problem plus the appropriate penalties. Once bounds are established, the next candidate problem is selected.

Tomlin has proposed two ways of improving DKW penalties. The first method of obtaining improved bounds is based on the following observations. Forcing a fractional variable in a linear programming solution to an integer value requires increasing at least one nonbasic variable above zero, but any such increase must be by at least one if this nonbasic variable is specified as integer in the original problem statement. This later condition was ignored in the earlier derivation of simple penalties by Davis, Kendrick, and Weitzman. Tomlin showed how this requirement could be used to yield stronger up and down penalty (penalty for assigning next lower integer value, that is, zero if the variables are of zero–one type) bounds.

Toyoda Yoshiaki and Senju Shizuo solved a large-scale knapsack type zero–one integer programming problem by setting all the variables equal to 1 and then calculating the effective gradient (a measure of effectiveness based upon the ratio of objective function coefficients to constraint coefficients) of each variable. The effective gradient was then sorted to find out which variables "should" be 1 and still satisfy the constraints. The remaining set of variables are then assigned a value of zero. Toyoda developed a method that is similar to the previous one. In the first method the problem is initially infeasible, and the feasibility is attained by assigning zero values to most nonpromising variables (variables with least effective gradient) successively. In this method the problem is started as feasible, and optimality is attained by successive assignment of 1-values to the most promising variables (variables with most effective gradient).

Bradley developed a technique by which a bounded integer linear programming problem can be transformed to an equivalent integer linear programming problem with a single constraint and the same number of solution variables. The solution to the original problem may be obtained by solving the single constraint problem.

Brocklehurst developed an algorithm to find an approximate solution to general integer linear programming problems. His algorithm starts from an infeasible continuous LP solution, then tries to reach a feasible integer solution from that point by a combination of search techniques.

Geoffrion introduced the idea of Lagrangian relaxation, which seems promising for integer programming problems with special structures. Lagrangian relaxation means including a set of constraints into the objective function to form an unconstrained or reduced constraint set problem with an expanded set of solution variables. Any problem that has a large number of multiple-choice constraints may be amenable to solution by this technique if the optimum multipliers could be obtained by an efficient iterative procedure.

Integer programming problems in which the objective function or the constraints may be quadratic have also been studied (Witzgall). The integrality requirements usually result from combinatorial considerations, such as those found in replacement and inventory problems, which will be discussed in Chapter 7, using a dynamic programming formulation. Many other practical problems require integer solutions; for example, scheduling and sequencing problems involve ordering restrictions which are often formulated by the introduction of a variable (Kronecker delta) that may take on only the values zero or 1. Although the dynamic programming solution algorithm works quite efficiently for problems of this type, it is computationally infeasible if more than three state (basic) variables are present. Since each constraint corresponds to one state variable, only a limited (but important) class of integer programming problems can be solved with this algorithm. However, dynamic programming can be used to solve *nonlinear* integer programs as well.

It is worth noting in conclusion that a general approach developed by Benders shows great promise for mixed integer programming problems. His approach, called Benders decomposition, essentially forms the dual equivalence of a new problem projected from the space of the old one. The procedure alternates between a pure integer problem in the dual space and a continuous problem in a primal space. The details can be found in the original paper by Benders.

3-35 THE CUTTING-PLANE ALGORITHM

One of the earliest attempts to solve the integer programming problem was presented by Gomory. In addition to being historically significant, the basic procedure is still very popular for solving small-to-moderate all-integer, integer linear programs. We will illustrate the basic procedure at this time.

In order to introduce the algorithmic procedures, we must define what is known as *congruence relationships* between two real numbers. A number $\bar{\delta}$ is said to be *congruent* to a second number ϕ if their difference is zero or an integer. For example:

$$\tfrac{1}{3} \text{ is congruent to } \tfrac{10}{3}$$
$$\tfrac{2}{3} \text{ is congruent to } -\tfrac{1}{3}$$
$$\tfrac{1}{4} \text{ is congruent to } \tfrac{5}{4}$$
$$0 \text{ is congruent to } 3$$

There is a simple way to determine the number $\bar{\delta}$ which is congruent to a number ϕ. Define $[\phi]$ as the largest integer less than or equal to the number ϕ. Then $\bar{\delta}$ is defined as

$$\bar{\delta} = \phi - [\phi]$$

For example,

$\bar{\delta}$	Ø	[Ø]
2/3	- 4/3	- 2
1/2	5/2	2
0	- 5	- 5
0	1	1
5/6	- 7/6	- 2
3/4	3/4	0

Note that $\bar{\delta}$ is always fractional or zero; $1 > \bar{\delta} \geq 0$. This definition will be used in a subsequent discussion of Gomory's technique.

Before proceeding, consider the following integer linear programming formulation.

$$\text{Maximize } Y(\mathbf{x}) = \sum_{j=1}^{N} c_j x_j \tag{3-155}$$

subject to

$$\sum_{j=1}^{N} a_j x_j = b \tag{3-156}$$

$$x_j \geq 0; \quad j = 1, 2, \ldots, N \tag{3-157}$$
an integer

By definition, the x_j^* must be integer-valued. Consider for the moment *any* set of solution variables that satisfy the equality constraint. If these variables are all integers, they are a candidate for the optimal solution. For any set of variables, the following relation is true.

$$\sum_{j=1}^{N} \{[a_j] + \bar{\delta}_j\}x_j = [b] + \bar{\delta} \qquad (3\text{-}158)$$

Since $[a_j]$ and $[b]$ are by previous definition integers, then we can assert that

$$\sum_{j=1}^{N} [a_j]x_j \leq [b] \qquad (3\text{-}159)$$

or

$$\sum_{j=1}^{N} [a_j]x_j + S = [b] \qquad (3\text{-}160)$$

where S is an *integer-valued* slack variable. If we subtract Eq. (3-158) from Eq. (3-160), we obtain the following constraint:

$$\sum_{j=1}^{N} (-\bar{\delta}_j)x_j + S = -\bar{\delta} \qquad (3\text{-}161)$$

In order to illustrate the significance of the discussion, let us consider how it might be applied to the solution of integer linear programming problems. Suppose that we solve the integer programming problem as a continuous linear program. If the (optimal) continuous solution is all integer, we have the optimal solution. If one or more of the solution (basic) variables have noninteger values, examine that row in the final simplex tableau which defines the value of a noninteger basic variable. Assume that x_B is that basic variable and the row is as follows:

$$x_B + \sum_K B_K x_K = \bar{b}$$

Note that by construction, the x_K variables are all *nonbasic*. Writing Eq. (3-161) for this relationship, we obtain

$$\sum_K (-\bar{\delta}_K)x_K + S = -\bar{\delta} \qquad (3\text{-}162)$$

where

$$\bar{\delta}_K = \phi_K - [B_K]$$
$$\bar{\delta} = \phi - [b]$$

As before, the x_K variables remain nonbasic, and the (new) solution is augmented by $S = -\bar{\delta}$. It is clear that if we append Eq. (3-162) to the current tableau, then we have an infeasible solution, violation occurring in Eq. (3-162). However, feasibility is easily obtained by means of a single dual-simplex operation, since the current solution is superoptimal. Equation (3-162) is called a *cut*, because it effectively "cuts off" a portion of the previous solution space. Indeed, one can show that by construction this cut will exclude the previous solution from feasibility, but will not exclude any all-integer candidate solutions not already considered. The following example will help to clarify the computational details.

$$\text{Maximize } f(\mathbf{x}) = \quad x_1 + x_2$$

subject to

$$-2x_1 + 5x_2 \leq 8$$
$$6x_1 + x_2 \leq 30$$
$$x_1, \quad x_2 \geq 0, \quad \text{integer}$$

Step 1: Solve the preceding problem as a continuous linear program.

	x_0	x_1	x_2	s_3	s_4	RHS
	1	-1	-1	0	0	0
s_3	0	-2	5	1	0	8
s_4	0	6	1	0	1	30

	x_0	x_1	x_2	s_3	s_4	RHS
	1	0	- 5/6	0	1/6	5
s_3	0	0	16/3	1	1/3	18
x_1	0	1	1/6	0	1/6	5

	x_0	x_1	x_2	s_3	s_4	RHS
	1	0	0	5/32	7/32	125/16
x_2	0	0	1	3/16	1/16	27/8
x_1	0	1	0	- 1/32	5/32	71/16

The optimal solution is given by

$$x_1^* = 4\tfrac{7}{16}$$
$$x_2^* = 3\tfrac{3}{8}$$

Step 2: Since both solution variables are noninteger, we choose one to form a cut. Choosing x_1 at random and applying Eq. (3-162), we obtain

$$-\tfrac{3}{16}S_3 - \tfrac{1}{16}S_4 + S_1 = -\tfrac{3}{8}$$

Appending this to the second simplex tableau, we obtain the following.

	x_0	x_1	x_2	s_3	s_4	s_1	RHS
	1	0	0	5/32	7/32	0	125/16
x_2	0	0	1	3/16	1/16	0	27/8
x_1	0	1	0	-1/32	5/32	0	71/16
s_1	0	0	0	-3/16	-1/16	1	-3/8

Using the dual-simplex procedure, we generate the following optimal tableau.

	x_0	x_1	x_2	s_3	s_4	s_1	RHS
	1	0	0	0	1/6	5/6	15/2
x_2	0	0	1	0	0	1	3
x_1	0	1	0	0	1/6	-1/6	9/2
s_3	0	0	0	1	1/3	-16/3	2

The current solution is given by

$$x_2^* = 3, \qquad x_1^* = 4\tfrac{1}{2}, \qquad S_3^* = 2$$

Since x_1^* is noninteger, we repeat step 2 using that variable.

Step 2 (repeat):

$$x_1 + \tfrac{1}{6}S_4 - \tfrac{1}{6}S_1 = \tfrac{9}{2}$$

Using Eq. (3-162) we obtain the cut:

$$-\tfrac{1}{6}S_4 - \tfrac{5}{6}S_1 + S_2 = -\tfrac{1}{2}$$

Appending this to the previous tableau:

	x_0	x_1	x_2	s_3	s_4	s_1	s_2	RHS
	1	0	0	0	1/6	5/6	0	15/2
x_2	0	0	1	0	0	1	0	3
x_1	0	1	0	0	1/6	−1/6	0	9/2
s_3	0	0	0	1	1/3	−16/3	0	2
s_2	0	0	0	0	−1/6	−5/6	1	−1/2

Using dual-simplex we again reach optimality.

	x_0	x_1	x_2	s_3	s_4	s_1	s_2	RHS
	1	0	0	0	0	0	1	7
x_2	0	0	1	0	0	1	0	3
x_1	0	1	0	0	0	−1	1	4
s_3	0	0	0	1	0	−7	2	1
s_4	0	0	0	0	1	5	−6	3

The answer is

$$x_1^* = 4, \quad x_2^* = 3, \quad x_0^* = 7, \quad S_3^* = 1, \quad S_4^* = 3$$

Since this solution is all-integer, the procedure is terminated.

Before recapitulation, let us examine the solution procedure. Recall at step 2 we first appended the following equation.

$$-\tfrac{1}{6}S_4 - \tfrac{5}{6}S_1 + S_2 = -\tfrac{1}{2}$$

or

$$\tfrac{1}{6}S_4 + \tfrac{5}{6}S_1 \geq \tfrac{1}{2}$$

but

$$S_3 = 8 + 2x_1 - 5x_2$$
$$S_4 = 30 - 6x_1 + x_2$$

Hence

$$x_2 \leq 3$$

The second added constraint was

$$S_2 - \tfrac{1}{6}S_4 - \tfrac{5}{6}S_1 = -\tfrac{1}{2}$$

Hence

$$x_1 + x_2 \leq 7$$

These restrictions are plotted along with the original constraints in Fig. 3-30. It is clear that these "cuts" force our solution to an all-integer vertex.

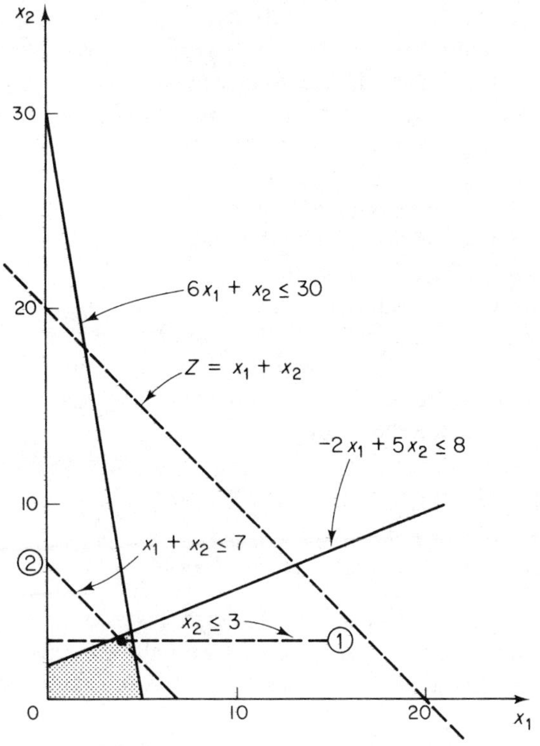

Figure 3-30. Generation of Gomery cuts.

Another approach to this type of problem is simply to round off each nonintegral value in the optimal solution obtained by the simplex method. Care must be taken to round off to integral values that will satisfy the constraints, of course, and if the noninteger values are large, the resulting solution may not be far from the true integer optimum. This method cannot, however, be used in the important class of combinatorial problems that require some of the variables to be restricted to the values zero and 1 only. Furthermore, rounding to the nearest integer can be misleading, especially in higher dimensions. Consider the difficulties encountered in the following *two*-dimensional problem.

$$\text{Maximize } Z = -7x_1 + 106x_2$$

subject to

$$\begin{aligned} -x_1 + 15x_2 &\leq 90 \\ x_1 + 2x_2 &\leq 35 \\ -3x_1 + 4x_2 &\leq 12 \end{aligned}$$

$$x_1, x_2, \text{ to be nonnegative integers}$$

Using the simplex method, three iterations are required to reach the nonintegral optimal solution, $x_1 = \frac{345}{17}$, $x_2 = \frac{125}{17}$. If these values are now rounded to the nearest feasible lattice point, one finds $\hat{x}_1 = 20$, $\hat{x}_2 = 7$, giving the objective function a value of $\hat{Z} = 602$. By drawing a graph of this problem, the reader will be interested to see that the optimal integer solution is actually $x_1^* = 15$, $x_2^* = 7$, for a value of $Z^* = 637$.

In conclusion, several comments seem in order. First, in describing the Gomory procedure, we assumed that all variables terminated at integer solutions. To guarantee this property, the original problem formulation must involve only integer parameters in the constraint formulations. This might require rescaling of constraint functions. Second, by adding "cuts" at each tableau, the size of the problem grows considerably. One modification is to drop constraints out of active participation if they are loose for two consecutive iterations. Third, for the serious integer programmer there are several significant improvements to this basic approach in the technical literature; a version suited to mixed integer programs is also available.

3-36 BRANCH-AND-BOUND ALGORITHMS

A technique particularly useful in solving large classes of mathematical programming problems is branch-and-bound. Originally conceived as *backtrack programming*, the general procedure has found extensive applications in integer and mixed integer problem solutions.

Consider the following mixed integer programming problem.

$$\text{Maximize } f(\mathbf{x}) = \sum_{j=1}^{N} c_j x_j$$

subject to

$$\begin{aligned} &\sum_{j=1}^{N} a_{ij} x_j \leq b_i && i = 1, 2, \ldots, M \\ &U_j \geq x_j \geq L_j && j = 1, 2, \ldots, N \\ &x_k \geq 0 && k = 1, 2, \ldots, \bar{N} \\ &x_l \text{ integer} && l = \bar{N} + 1, \bar{N} + 2, \ldots, N \end{aligned}$$

Note that we require upper and lower bounds on all variables. Without loss of generality, we can assume that all lower bounds are zero since a simple transformation will always translate all variables to the origin. If $\bar{N}$ is zero, we have an all integer linear program; if all variables are restricted to zero–one, we have a binary integer program. All sets of problems can be attacked using branch-and-bound procedures.

The underlying concept of branch-and-bound strategies is the observation that at any value for x_j, one can always state that

$$[x_j] + 1 \geq x_j \geq [x_j]$$

where $[x_j]$ is the largest integer less than (or equal to) the value x_j.

For example, suppose that we are seeking an integer value for x_j and by the normal simplex procedure we obtain $x_j^* = 4.3$. Since we are seeking an integer solution, we can dichotomize our present problem to two mutually exclusive and exhaustive sets; those which satisfy $U_j \geq x_j \geq 5.0$ and those which satsify $4.0 \geq x_j \geq L_j$. Using this observation, we create from our original problem two new problems, one that adds $U_j \geq x_j \geq 5.0$ to the original problem constraints and one that adds $4.0 \geq x_j \geq L_j$. That is, we *branch* to two new problems. Note that the original solution provides an available *upper bound* on the optimal solution. This is true since by adding additional constraints at each branch of the problem one can only decrease the value of the original solution. Note further that this process of creating new problems rapidly creates a *tree* of possible alternatives.

As one proceeds to dichotomize the noninteger variables at each branch of the tree, one of two events will occur. First, a solution might be obtained which does not satisfy the integrality requirements of the specified variables, and has an objective function value equal to or less than another branch whose solution is all-integer. In this case we may *abandon* this branch. Second, we may generate a specified integer-valued solution whose value is equal to or less than that of another solution, which also satisfies the integrality restrictions. In this case, that branch can also be abandoned. Each feasible solution obtained represents a *lower bound* on the optimal solution, or the optimal solution itself. It should be clear that in the process of branching and possibly creating bounds on the optimal solution, feasible solutions satisfying all constraints might be found. Of course, one cannot be sure that such a solution is optimal until the entire process is complete. This can only occur when one (feasible) solution dominates every other branch of the tree. In the event that no such solution can be found, there exists no feasible solution. The entire procedure is best illustrated by way of an example.

$$\text{Maximize } f(\mathbf{x}) = 9x_1 + 6x_2 + 5x_3$$

subject to

$$2x_1 + 3x_2 + 7x_3 \leq \tfrac{35}{2}$$

$$4x_1 \qquad + 9x_3 \leq 15$$

$$x_1 \text{ nonnegative integer}$$

Step 1: Solve the problem as an ordinary linear program.

x_0	x_1	x_2	x_3	x_4	x_4	RHS
1	−9	−6	−5	0	0	0
0	2	3	7	1	0	35/2
0	4	0	9	0	1	15

x_0	x_1	x_2	x_3	x_4	x_5	RHS
1	0	15/2	53/2	9/2	0	35/4 x 9
0	1	3/2	7/2	1/2	0	35/4
0	0	−6	−5	−2	1	−20

x_0	x_1	x_2	x_3	x_4	x_5	RHS
1	0	0	81/4	2	5/4	35/4 x 9 − 10/3 x 15/2
0	1	0	23/4	0	1/4	15/4
0	0	1	5/6	1/3	−1/6	10/3

$x_1 = 15/4,\ x_2 = 10/3,\ x_3 = 0,\ f(\mathbf{x}) = 215/4$

Step 2: This solution does not satisfy the integer restrictions, so we divide the original problem into two mutually exclusive and exhaustive subproblems.

(a) Maximize

$$f(\mathbf{x}) = 9x_1 + 6x_2 + 5x_3$$

subject to

$$\begin{aligned} 2x_1 + 3x_2 + 7x_3 &\leq \tfrac{35}{2} \\ 4x_1 \qquad\quad + 9x_3 &\leq 15 \\ x_1 \qquad\qquad\qquad &\geq 4 \quad \text{(integer)} \\ x_2, x_3 &\geq 0 \end{aligned}$$

(b) Maximize

$$f(\mathbf{x}) = 9x_1 + 6x_2 + 5x_3$$

subject to

$$
\begin{aligned}
2x_1 + 3x_2 + 7x_3 &\le \tfrac{35}{2} \\
4x_1 \qquad\quad + 9x_3 &\le 15 \\
0 \le x_1 \qquad\qquad &\le 3 \quad \text{(integer)} \\
x_2, x_3 &\ge 0
\end{aligned}
$$

It is clear that problem (a) has no feasible solution. The solution to problem (b) is given by: $x_1 = 3$, $x_2 = \frac{23}{6}$, $x_3 = 0$:

$$f(\mathbf{x}) = 50$$

Since this satisfies all problem restrictions, it is a feasible and optimal solution.

To further illustrate the branch-and-bound procedure, consider the following similar problem.

$$\text{Maximize } f(\mathbf{x}) = 9x_1 + 6x_2 + 5x_3$$

subject to

$$
\begin{aligned}
2x_1 + 3x_2 + 7x_3 &\le \tfrac{35}{2} \\
4x_1 \qquad\quad + 9x_3 &\le 15 \\
x_1, x_2, x_3 &\ge 0 \text{ and are integers}
\end{aligned}
$$

Step 1: First we obtain the following linear programming solution.

$$x_1 = \tfrac{15}{4}, \quad x_2 = \tfrac{10}{3}, \quad x_3 = 0, \quad f(\mathbf{x}) = \tfrac{215}{4} \text{ (not all integers)}$$

Step 2: Choosing the variable x_1, we create two new problems, (a) and (b). One has no feasible solution and one generates the following solution.

$$x_1 = 3, \quad x_2 = \tfrac{23}{6}, \quad x_3 = 0, \quad f(\mathbf{x}) = 50$$

Step 3: The problem in (b) of step 2 can be divided into two more problems.

(c) Maximize

$$f(\mathbf{x}) = 9x_1 + 6x_2 + 5x_3$$

subject to

$$
\begin{aligned}
2x_1 + 3x_2 + 7x_3 &\le \tfrac{35}{2} \\
4x_1 \qquad\quad + 9x_3 &\le 15 \\
0 \le x_1 &\le 3 \\
x_2 &\ge 4 \\
x_3 &\ge 0
\end{aligned}
$$

(d) Maximize

$$f(\mathbf{x}) = 9x_1 + 6x_2 + 5x_3$$

subject to

$$\begin{aligned} 2x_1 + 3x_2 + 7x_3 &\le \tfrac{35}{2} \\ 4x_1 \qquad\quad + 9x_3 &\le 15 \\ 0 \le x_1 &\le 3 \\ 0 \le x_2 &\le 3 \\ x_3 &\ge 0 \end{aligned}$$

Problem (c) has no feasible solution, while the solution to problem (d) is given by $x_1 = 3$, $x_2 = 3$, $x_3 = 0$, $f(\mathbf{x}^*) = 45$.

Since this solution is all-integer, the procedure terminates. For a summary of the branch-and-bound procedures, see Fig. 3-31.

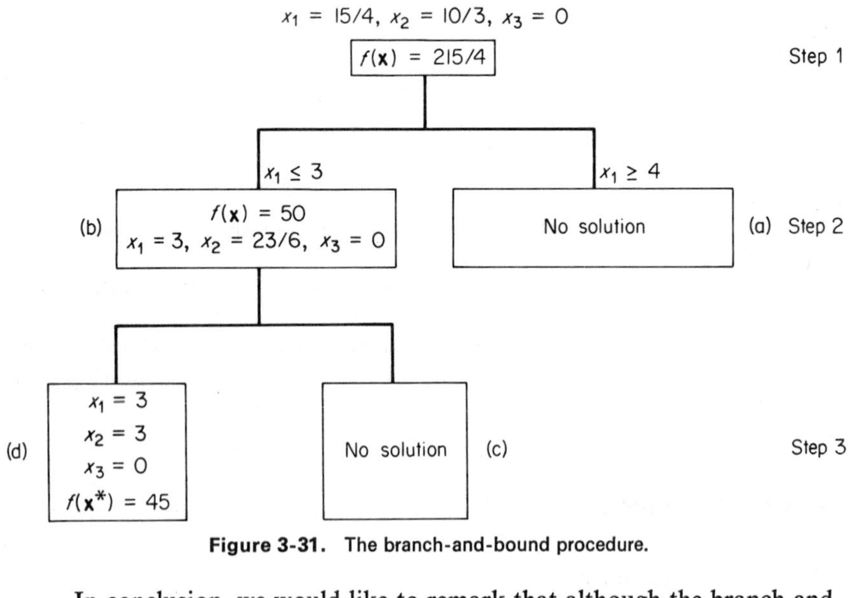

Figure 3-31. The branch-and-bound procedure.

In conclusion, we would like to remark that although the branch-and-bound procedures and the cutting-plane techniques represent important classes of solution procedures for general integer programs, there are other methods which can be more effectively used for zero–one integer programming problems. In particular, the algorithm of Balas using implicit enumeration procedures is widely used. Geoffrion has modified Balas's original algorithm and implemented surrogate (composite) constraints and stronger branch abandoning procedures. Each should be carefully studied by the serious integer programmer. Finally, specialized algorithms dealing with structured problems, such as the zero–one knapsack problem, have experienced success in solving relatively large problems. The interested reader is referred to Taha or Garfinkle and Nemhauser for further details.

3-37 COMPUTATIONAL EXPERIENCE

In general, the earlier cutting-plane algorithms of Gomory exhibit erratic behavior and can fail to converge on even small problems. However, Gomory codes remain well established and popular probably due to their historical significance. Certainly they are not generally recommended for problems over 50–75 variables except in problems with special structure. Perhaps the next-most-popular form of integer programming solution procedures is that of branch and bound. This popularity is attributed to the ease of programming and the availability of computer codes. Like the cutting-plane algorithms, branch-and-bound codes use linear programming subproblems at each step. Computational improvements have centered around better branching procedures and improving the bounds. Like cutting-plane algorithms, the success story is better than the practical results. The approach is clearly applicable to most general problems, although problems on the order of several hundred variables require large amounts of computer time in general. Spectacular results can probably be achieved in specific cases by using an interactive approach and placing man "in the loop." Such approaches remain largely unexplored.

For the general 0–1 problems, Balas algorithms have proved most reliable. Modifications such as surrogate constraints have produced efficient algorithms. Special structures such as those reported by Toyoda and later improved by Phillips have produced solutions to problems with over 5000 variables. However, the general Balas algorithm enjoys few reported applications with more than 200 variables in the problem statement. For the mixed integer programming problem, a branch-and-bound methodology appears to be the most attractive. We would like to suggest, however, that one should not fail to investigate a Benders decomposition approach. The primary disadvantage in a Benders solution is lack of a readily available commercial code.

3-38 COMPUTER SOFTWARE

Reliable computer software for general integer programming solution is available, but the generality and limits of existing commercial codes may preclude usage on many real-world problems. Of those available, the branch-and-bound codes of Balas variants seem to offer the most durable options. RIP30C, a Balas variant with surrogate constraints, appears to be a good general code with access through Geoffrion. The IBM SHARE package offers a general branch-and-bound program, BBMIP. Other algorithms currently referenced in the literature include OPHELIA MIXED, MPSX/MIP, UMPIRE, and MARISABETH. BENDERS codes have been tried in MIDAS-2 and FMPS/MIP. Finally, we would like to cite some particularly spectacular results:

Author	*Machine*	*Constraints*	*Continuous variables*	*Integer variables*	*Time (min)*
Tomlin (1971)	UNIVAC 1108	5	—	50	1.5
Benichou (1970)	IBM 360	721	39	1117	18
Tomlin (1970)	UNIVAC 1108	1000	—	200	45
Shetty[a] (1971)	UNIVAC 1108	73	—	450	$\frac{1}{12}$
Toyoda[b] (1975)	—	100	—	1000	3
Phillips[c] (1976)	IBM 370	200	—	6000	4
Geoffrion[d] (1972)	CO6 3600	150	—	7000	50
Speilberg[e] (1972)	IBM 360	184	8464	92	46

[a]Assembly line balance problem.
[b]Knapsack problem.
[c]Allocation problem.
[d]Set-covering problem.
[e]Plant location.

BIBLIOGRAPHY

Balas, E., "An additive algorithm for solving linear programs with zero-one variables," *Operations Res.*, **13**, 4 (July–August 1965), 517–546.

Balinski, M. L., "Integer programming: methods, uses, computations," *Man. Sci.*, **12**, 3 (November 1965), 253–313.

Beale, E. M. L., "Survey of integer programming," *Operational Res. Quart.*, **16**, 2 (June 1965), 219–228.

Beightler, C. S., and D. J. Wilde, "Sensitivity analysis gives better insight into linear programming," *Hydrocarbon Proc. Petrol. Ref.*, **44** (February 1965), 111–126.

Benders, J. F., "Portioning procedures for solving mixed-variables programming problems," *Numerische Math.*, **4** (1962), 238–251.

Boot, J. C. G., "On sensitivity analysis in convex quadratic programming problems," *Operations Res.*, **11** (September 1963), 771–786.

Bradley, G. H., "Transformation of integer programs to knapsack problems," Yale Univ. Administrative Sciences Report 37, November 16, 1970.

———, "Heuristic solution methods and transformed integer linear programming problems," Yale Univ. Administrative Sciences Report 43, March 1971.

BROCKLEHURST, E. R., "A heuristic algorithm for integer linear programming problems," National Physical Laboratory Report 18, June 1972.

BROOKS, R., and A. GEOFFRION, "Finding Everett's multipliers by linear programming," *Operations Res.*, **14** (1966), 1149–1153.

CHARNES, A., "Optimality and degeneracy in linear programming," *Econometrica*, **20** (1952), 160–170.

———, and W. W. Cooper, "The stepping stone method of explaining linear programming calculations in transportation problems," *Man. Sci.*, **1** (January 1954), 49–69.

———, and W. W. Cooper, *Management Models and Industrial Applications of Linear Programming*, Vol. 2 (Wiley, New York, 1961).

DAKIN, R. J., "A tree search algorithm for mixed integer programming problems," *Com. J.*, **8**, 3 (1965), 250–255.

DANTZIG, G. B., "Maximization of a linear function of variables subject to linear inequalities," in *Activity Analysis of Production and Allocation*, T. C. Koopmans, ed. (Wiley, New York, 1951), pp. 339–347.

———, *Linear Programming and Extensions* (Princeton Univ. Press, Princeton, N.J., 1963).

———, and P. WOLFE, "Decomposition principle for linear programming," *Operations Res.*, **8** (January 1960), 101–111.

DAVIS, R. E., D. A. KENDRICK, and M. WEITZMAN, "A branch and bound algorithm for zero-one mixed integer programs," Development Economic Report 69, Center for International Affairs, Harvard Univ., 1967.

DENNIS, J. B., *Mathematical Programming and Electrical Networks* [Wiley (Technology Press), New York, 1959].

DORN, W. S., "Duality in quadratic programming," *Quart. Appl. Math.*, **18**, 2 (July 1960), 155–162.

EISEMANN, K., "Simplified treatment of degeneracy in transportation problems," *Quart. Appl. Math.*, **14**, 4 (1957).

EVERETT, H., "Generalized Lagrange multiplier method for solving problems of optimum allocation of resources," *Operations Res.*, **11** (1963), 399–471.

FIACCO, A. V., and G. P. MCCORMICK, *Nonlinear Programming: Sequential Unconstrained Minimization Techniques* (Wiley, New York, 1968).

FRANKLIN, BENJAMIN, *Poor Richard's Almanack* (*The Sayings of Poor Richard* (1733–1758), T. H. Russell, ed.

GASS, S. I., and T. L. SAATY, "The computational algorithm for the parametric objective function," *Naval Res. Logistics Quart.*, **2** (June 1955), 39–45.

GEOFFRION, A. M., "An improved implicit enumeration approach for integer programming," *Operations Res.*, **13**, 6 (November–December 1965), 879–919.

———, "Integer programming by implicit enumeration and Balas' method," *Siam Rev.*, **9**, 2 (April 1967), 178–190.

———, "Lagrangean relaxation for integer programming," Working Paper 195, Western Management Science Institute, Univ. of California, Los Angeles, December 1973.

———, and R. E. MARTSEN, "Integer programming algorithms: a framework and state-of-the-art survey," in A. M. Geoffrion, ed., *Perspectives on Optimization: A Collection of Expository Articles* (Addison-Wesley, Reading, Mass., 1972).

GLOVER, F., "A multiphase-dual algorithm for the zero-one integer programming problem," *Operations Res.*, **13**, 6 (November–December 1965), 879–919.

———, "Surrogate constraints," *Operations Res.*, **16** (1968), 741–749.

GOLDMAN, A. J., and D. KLEINMAN, "Examples relating to the simplex method," *Operations Res.*, **12** (January 1964), 159–161.

GOMORY, R. E., "An algorithm for integer solutions to linear programs," *Princeton—IBM Mathematics Research Project*, Technical Report 1, November 1958.

———, "On the relation between integer and non-integer solutions to linear programs," *Proc. Natl. Acad. Sci.*, **53** (1965), 260–265.

GOTTFRIGO, B. S., and J. WEISMAN *Introduction to Optimization Theory* (Prentice-Hall, Englewood Cliffs, N.J., 1973).

GREENBERG, HAROLD, *Integer Programming* (Academic Press, New York, 1971).

———, and R. L. HEGGORICH, "A branch search algorithm for the knapsack problem," *Man. Sci.*, **16**, 5 (January 1970), 327–332.

HADLEY, G. H., *Linear Programming* (Addison-Wesley, Reading, Mass., 1962).

HILLIER, F. S., "A bound-and-scan algorithm for pure integer linear programming with general variables," *Operations Res.*, **17** (1969), 638–679.

HOFFMAN, A. J., "Cycling in the simplex algorithm," *National Bureau Standards Report 2974* (December 1953).

KROLAK, P., "Computational results of an integer programming algorithm," *Operations Res.*, **7** (1969), 743–749.

LANCZOS, C., *The Variational Principles of Mechanics*, 2nd ed. (Toronto, 1962), pp. 161–165.

LAND, A. H., and A. G. DOIG, "An automatic method of solving discrete programming problems," *Econometrica*, **28** (1960), 497–520.

LEMKE, C. E., "The dual method of solving the linear programming problem," *Naval Res. Logistics Quart.*, **1** (March 1954), 48–54.

NEMHAUSER, GEORGE L., and R. S. GARFINKEL, *Integer Programming* (Wiley, New York, 1972).

PETERSON, C. C., "Computational experience with variants of the Balas algorithm applied to the selection of RSD projects," *Man. Sci.*, **13**, 9 (May 1967), 736–750.

———, "A capital budgeting heuristic algorithm using exchange operations," *AIIE Transactions*, **6**, 2 (June 1974), 143–150.

PHILLIPS, D. T., and R. L. LYTTON, "Optimal highway rehabilitation and maintenance using integer programming, *Proc. AIIE Natl. Conf.*, Dallas, Tex., May 1977.

PHILLIPS, D. T., A. RAVINDRAN, and J. J. SOLBERG, *Operations Research: Principles and Practice* (Wiley, New York, 1976).

QUANDT, R. E., and H. W. KUHN, "On upper bounds for the number of iterations in solving linear programs," *Operations Res.*, **12** (January 1964), 161–165.

REINFELD, N. V., and W. R. VOGEL, *Mathematical Programming* (Prentice-Hall, Englewood Cliffs, N.J., 1958).

SAATY, T. L., "The number of vertices of a polyhedron," *Amer. Math. Monthly*, **62** (May 1955), 326–331.

SALKIN, H. M., *Integer Programming* (Addison-Wesley, Reading, Mass., 1975).

SHAPIRO, J. F., "Dynamic programming algorithms for the integer programming problem—I: the integer programming problem viewed as a knapsack type problem," *Operations Res.*, **16**, 1 (January–February 1968), 103–121.

SINDEN, F. W., "Duality in convex programming and in projective space," *J. Soc. Ind. Appl. Math.*, **11** (1963), 535–552.

SPIELBERG, KURT, "Minimal preferred variable reduction for zero-one programming," *IBM Philadelphia Scientific Center, Technical Report 320–3013*, July 1972.

SYMONDS, G. H., *Linear Programming: The Solution of Refinery Problems* (Esso, New York, 1955).

TAHA, H. A., "On the solution of zero-one linear programs by ranking the extreme points," *Department of Industrial Engineering, Univ. of Arkansas, Technical Report 71-2*, February 1971.

———, *Integer Programming Theory, Applications, and Computations* (Academic Press, New York, 1975).

TOMLIN, J. A., "An improved branch and bound method for integer programming," *Operations Res.*, **19**, 4 (July–August 1971), 1070–1075.

TOYODA, YOSHIAKI, "A simplified algorithm for obtaining approximate solutions to zero-one programming problems," *Man. Sci.*, **21**, 12 (August 1975), 1417–1426.

———, and SHIZUO SENJU, "An approach to linear programming with 0-1 variables," *Man. Sci.*, **15**, 4 (December 1968), B196–207.

WEHL, H., "Elementare Theorie der konvexen Polyeder," *Comm. Math. Helv.*, **7** (1934), 290–306.

WILDE, D. J., "Production planning of large systems," *Chem. Eng. Progr.*, **59**, 1 (January 1963), 46–51.

WITZGALL, C., "An all-integer programming algorithm with parabolic constraints," *J. Soc. Ind. Appl. Math.*, **11** (1963), 855–871.

WOLFE, P., "A duality theorem for non-linear programming," *Quart. Appl. Math.*, **19** (1961), 239–244.

YOUNG, R. D., "A primal (all integer) integer programming algorithm," *J. Res. Nat. Br. Stds.*, (1965), 213–250.

EXERCISES

3-1. (a) Solve the following problem, using the simplex method.

$$\text{Maximize } p = 2x_1 + x_2$$

subject to

$$x_1 + x_2 \leq 14$$
$$x_1 - 2x_2 \leq 4$$
$$x_1, \ x_2 \geq 0$$

(b) Re-solve the problem when the first constraint is changed to:

$$x_1 + x_2 \leq 15$$

What is the change in the optimal value of the objective function p? Compare this change with the value of the sensitivity coefficient for the slack variable in the first constraint, as found in the optimal tableau in part (a).

3-2. The Relthgieb Company produces four products, I through IV. The raw-material requirements, space needed for storage, production rates, and profits are given in the accompanying table. The total amount of raw material available per day for all four products is 360 lb, the total storage space for all products is 475 ft., and a maximum of 7 hr/day can be used for production. Assume that all products manufactured each day are shipped out of the storage area at the end of the day. Thus, the four products must share the total available storage space, production time, and raw material.

	Product			
	I	*II*	*III*	*IV*
Raw material (pounds per piece)	4	4	3	5
Storage space (square feet per piece)	4	5	4	3
Production rate (pieces per hour)	30	60	20	30
Profit (dollars per piece)	10	13	10	11

(a) Find the number of pieces of each of the products that the company should produce per day in order to maximize the total profit.
(b) What hourly rate could the company afford to pay for more production time?
(c) How much should the company be willing to pay for additional storage space?
(d) What premium could the company pay for more raw material?

3-3. (a) Solve the following linear programming problem, using the simplex method.
Find nonnegative numbers x_1, x_2, and x_3 that *maximize* the function

$$p = 2x_1 + x_2 + x_3$$

and which satisfy the constraints:

$$x_1 + x_2 + x_3 \leq 10$$
$$x_1 + 5x_2 + x_3 \geq 20$$

(b) Find the optimal solution to part (a) if the first constraint is changed to $x_1 + x_2 + x_3 \leq 5$, without re-solving the entire problem. (Assume that the second constraint is left unchanged.)
(c) What would be the effect on the optimal solution to part (a) of a change Δc_1 in the coefficient of x_1 ($c_1 = 2$) in the objective function?

3-4. A company has five warehouses, numbered 1, . . . , 5, containing 40, 50, 90, 30, and 60 units of its product, respectively. In the next month 20, 30, 40, 80, 60, 25, and 15 units, respectively, must be shipped to seven retail outlets numbered 1, . . . , 7.

The unit cost of shipment from any warehouse to any retail outlet is contained in the following matrix:

		Outlet						
		1	2	3	4	5	6	7
Warehouse	1	8	6	10	12	9	11	5
	2	3	7	6	9	8	7	8
	3	5	4	2	6	3	9	3
	4	17	12	11	13	9	10	12
	5	7	11	4	5	6	5	7

Figure E3-4

Find the minimum-cost shipping schedule.
What is the minimum cost?

3-5. The Bergwerk Corporation operates three coal mines, A, B, and C, which provide 400, 500, and 700 tons, respectively, per week. Orders for 500, 400, 300, 300, and 600 tons per week have been received from customers I through V, respectively. A schedule of transportation costs in dollars per ton from each mine to each customer is given in the following table:

Customer

Mine	I	II	III	IV	V
A	4	16	1	16	14
B	18	10	8	12	12
C	6	1	4	13	2

Figure E3-5

The company will ship all available tonnage each week, but will obviously not be able to satisfy the demands of all five customers, resulting in a loss of future business which is estimated to cost 3 dollars for each ton demanded but not supplied, to customers II, III and V, and 5 dollars for each ton demanded but not supplied to customers I and IV.

Find the weekly shipping schedule that minimizes the total cost.

3-6. A machine-tool manufacturer has received orders for 190, 110, 150, 50, and 70 lathes of types I–V, respectively. These orders must be filled by the end of the month, and the manufacturer has four production lines, A, B, C, and D, each of which is capable of producing all five different types of lathe. The monthly production capacities of these lines are 110, 150, 230, and 190 lathes, respectively, regardless of the type(s) made. The dollar profit resulting from the manufacture of a given type of lathe on a given production line is shown in the following table:

Lathe type

Production line	I	II	III	IV	V
A	340	410	260	590	470
B	375	445	300	600	450
C	345	400	270	545	500
D	330	420	250	570	445

Figure E3-6

Find how many lathes of each type each production line should produce so as to maximize the total profit.

3-7. Write out the dual of the refinery problem described in Fig. 3-7, and give an economic interpretation of this dual problem. Show that the dual variable associated with a primal *equality* constraint is unrestricted in sign (a "free" variable). Do this by writing the constraint as two equivalent *inequality* constraints, and thus explain the absence of nonnegativity requirements on the dual variables u_m, v_n of the transportation problem.

Show that the following problem is the dual of the transportation problem: Find the values of u_m and v_n that maximize

$$p = \sum_{m=1}^{M} a_m u_m + \sum_{n=1}^{N} b_n v_n$$

and satisfy the constraints

$$u_m + v_n \leq c_{mn}; \qquad m = 1, \ldots, M; \quad n = 1, \ldots, N$$

3-8. (a) Attempt to solve the following problem using the simplex method. Then, make a change of variable to show why the solution is unbounded.

$$\text{Maximize } p = 4x_1 + 3x_2 + 2x_3 - 2x_4$$

subject to

$$\begin{aligned} -7x_1 + 2x_2 + 5x_3 + 4x_4 &\leq 7 \\ -9x_1 - x_2 + 4x_3 + 7x_4 &\leq 4 \\ x_1 + 3x_2 + 2x_3 - x_4 &\leq 8 \\ x_1, \ x_2, \ x_3, \ x_4 &\geq 0 \end{aligned}$$

(b) Show that the constraints of the problem which is dual to the one in part (a) have no solution.

3-9. Solve the following problem.

$$\text{Maximize } p = 3x_1 + 6x_2 + 2x_3$$

subject to

$$\begin{aligned} 3x_1 + 4x_2 + x_3 &\leq b_1 \\ x_1 + 3x_2 + 2x_3 &\leq b_2 \end{aligned}$$

for $b_1 = 2$ and $b_2 = 1$.

(a) For what simultaneous changes Δb_1, Δb_2, in the values of b_1 and b_2, respectively, will the final state set be optimal? Draw a graph of this region in b_1, b_2 space. How will the optimal value of p change within this region?

(b) If only b_1 is changed, show the optimal simplex tableaux for all values of b_1 and plot the optimal values of x_1, x_2, x_3, and p as functions of b_1.

(c) Write out the dual of the given problem, keeping b_1 as a parameter (with $b_2 = 1$). Solve this dual both graphically and by the simplex method. Compare the results with those obtained in part (b).

3-10. (a) Solve the following linear programming problem, choosing x_2 as the first variable to enter the state set (to simplify the calculations).

$$\text{Maximize } p = 12x_1 + 9x_2 + 7x_3$$

subject to

$$\begin{aligned} 3x_1 + 2x_2 + x_3 &\leq 20 \ (= b_1) \\ x_1 + x_2 + x_3 &\leq 11 \ (= b_2) \\ 12x_1 + 4x_2 + x_3 &\leq 48 \ (= b_3) \\ x_j \geq 0; \ j &= 1, 2, 3 \end{aligned}$$

(b) Assume that the optimal state set found in part (a) cannot be changed, but that *any two* of b_1, b_2, b_3 may be changed by any amounts as long as none of the x_j are driven out of the state set. For what values of Δb_1, Δb_2, Δb_3 (not all nonzero) will p be maximized?

(c) Re-solve the problem given in (a) when the following constraints are added to the original three:

$$2x_1 + x_2 + x_3 \leq 12 \quad \text{and} \quad 4x_1 + 3x_2 + 2x_3 \leq 33$$

Do this by starting with the final optimal tableau from part (a).

(d) Suppose that c_1 and c_2 (the coefficients of x_1 and x_2 in p) are subject to market fluctuations. What must be the relation between their values if the original optimal solution in (a) is to remain optimal? In this range of fluctuation of c_1 and c_2, by how much does p^* vary?

(e) Starting with the optimal tableau from (a), write the equation for p as a function of Δb_1 and Δc_1 and draw the line for $p^* = 97$.

(f) In solving the problem given in (a), we ignored the variable x_4, where $c_4 = 10$, $a_{14} = 2$, $a_{24} = 1$, and $a_{34} = 5$. Re-solve the problem, taking into account this new variable, by starting with the optimal tableau from (a).

3-11. Prove Farkas's lemma: The statement "$\mathbf{c}'\mathbf{x} \leq 0$ for all $\mathbf{x}$ such that $\mathbf{A}\mathbf{x} \geq \mathbf{0}$" is equivalent to: "There exist $\boldsymbol{\lambda} \geq \mathbf{0}$ such that $\mathbf{c} + \mathbf{A}'\boldsymbol{\lambda} = \mathbf{0}$."
Hint: Consider the following linear programming problem: maximize $\mathbf{c}'\mathbf{x}$, subject to $\mathbf{A}\mathbf{x} \geq \mathbf{0}$, and also its dual program.

3-12. Using the branch-and-bound method, find nonnegative integers, x_1 and x_2, that

$$\text{maximize } y = x_1 + 2x_2$$

and satisfy

$$\begin{aligned} 2x_1 + 6x_2 &\leq 21 \\ x_1 + x_2 &\leq 5 \\ x_1 - x_2 &\leq 0 \\ x_1 + 2x_2 &\geq 2 \end{aligned}$$

3-13. Find nonnegative integers x_1, x_2, x_3, and x_4 that

$$\text{maximize } y = 4x_1 + 2x_2 + 5x_3 + x_4$$

and satisfy

$$\begin{aligned} x_1 + 2x_2 + x_3 + 16x_4 &\geq 4 \\ -5x_1 + 3x_2 + 10x_3 - x_4 &\leq -4 \\ 2x_1 + x_2 + 5x_3 + 3x_4 &\leq 10 \end{aligned}$$

Use the branch-and-bound method to solve this problem.

3-14. Solve the following problems using (a) the branch-and-bound method; (b) Gomory's cutting-plane procedure.

(a) Maximize $y = 2x_1 + 3x_2 + 10x_3$

subject to

$$-x_1 + 4x_2 + 7x_3 \leq 10$$
$$5x_1 - 2x_2 + 6x_3 \leq 10$$
$$0 \leq x_1, x_2, x_3 \leq 4$$
$$\mathbf{x} \text{ an integer}$$

(b) Maximize $y = x_1 + 2x_2 + 4x_3 + 3x_4$

subject to

$$x_1 + x_2 + 3x_3 + 2x_4 \leq 6$$
$$\mathbf{x} = 0, 1$$

(c) Minimize $y = x_1 + 3x_2 + 5x_3$

subject to

$$x_1 + 2x_2 \geq 7$$
$$3x_1 + x_2 + 2x_3 \geq 8$$
$$\mathbf{x} \text{ a nonnegative integer}$$

Unconstrained Nonlinear Optimization 4

The superior man, when he sees what is good, moves toward it; and when he sees his errors, he turns from them.

The Book of Changes, Appendix II, Hexagram 42
(China, ca. 1200 B.C., Translated by J. LEGGE)

PART I: Direct Elimination Procedures

A fundamental problem that arises naturally in general optimization theory is that of finding the optimal solution to unconstrained nonlinear functions, especially where movement along gradient lines is required. These important problems fall naturally into two major classes: single-variable and multi-variable problems. These may be further subdivided into two categories; convex and nonconvex functions. The strategies developed tend to follow the 3000-year-old advice from the *I Ching* quoted at the beginning of this chapter. As information is accumulated, one moves into regions where the optimum may lie and eliminates areas where it cannot be.

Imagine, then, an optimization problem where an analytic expression for the objective function is either unavailable or too complicated to manipulate by indirect methods. Assume, however, that given a specific set of values of the independent variables, one can compute the corresponding value of the objective function. The computation device may be a table of numbers, a set of graphs in a handbook, a complicated computer program, a manufacturing process, or the estimating department of an engineering design firm. In such circumstances there are two approaches to optimization. One may evaluate the objective at many points in order to approximate it by an expression amenable to indirect methods. Or, one may eschew such attempts at complete description and drive directly toward the peak, using incomplete information generated along the way. Methods based on the former approach will be called *approximation techniques*; those embodying the latter viewpoint, *direct methods*.

Since approximation techniques are founded on curve-fitting procedures well known to engineers and numerical analysts (see the books of Lapidus et al. and of Southworth and Deleeuw), little space is given to them here. Most of this chapter is concerned with direct methods, which fall into two major categories: (1) the *elimination* techniques, which by bold moves continually strive to shrink the region in which the peak must lie; (2) the *climbing* procedures, which cautiously move in directions where, based on local measurements, the objective appears to be improving. The first part of this chapter deals with the elimination techniques; the second with climbing. Since elimination methods are extremely effective when there is but one independent variable, and since this case frequently arises as subproblems in constrained optimization algorithms, we shall develop efficient procedures to deal with one-dimensional case. Although few practical optimization problems involve only one decision variable, many multidimensional climbing procedures involve one-dimensional optimizations as subroutines. Hence, study of these techniques is essential for applying constrained optimization. But aside from their usefulness, unidimensional elimination methods are of great interest because their effectiveness as optimization procedures can itself be optimized. It is therefore possible to develop *optimum* elimination techniques, which can rarely be done for hill-climbing methods.

4-01 EXPLICIT AND IMPLICIT OBJECTIVES

When the objective function $y(\mathbf{x})$ can be computed directly from the N independent variables, it is said to depend on them *explicitly*. Most of the elimination methods described here will be appropriate for explicit objective functions, which occur often in practice. There are two major ways to calculate the gradient ∇y at a point $\mathbf{x}$. One can differentiate the objective directly and substitute the components of $\mathbf{x}$ into the resulting equations. Alternatively, one can perturb each coordinate x_i a small amount, say, ϵ_i, and estimate the partial derivatives as difference quotients. Let $\boldsymbol{\epsilon}_i$ represent an N vector having its ith component equal to ϵ_i, the others being zero. Then the quotient approximating the partial derivative $\partial y/\partial x_i$, evaluated at $\mathbf{x}$, is

$$\frac{y(\mathbf{x} + \epsilon_i) - y(\mathbf{x})}{\epsilon_i} \approx \left.\frac{\partial y}{\partial x_i}\right|_x \tag{4-1}$$

By either method, N additional computations are needed to find the N components of the gradient, which is needed whenever a climbing technique is employed. Most elimination techniques do not bother with the gradient, preferring to invest the computations saved in evaluating the objective at more points, spread widely over the feasible region.

Horn has pointed out that a different situation arises when the objective function is implicit rather than explicit. This means that the objective appears together with the independent variables in a function that cannot be solved directly for y, as in the following equations:

$$f(y, \mathbf{x}) = 0 \tag{4-2}$$

To find the value of y at a given point $\mathbf{x}$, one must substitute the numerical value of x into the expression and determine the value of y making $f(y, \mathbf{x})$ zero. This problem in solving a nonlinear equation numerically is handled most conveniently by the Newton–Raphson method, described in Chapter 2. As a by-product of the Newton–Raphson procedure, one obtains the derivative $\partial f/\partial y$. If the derivatives $\partial f/\partial x_i$ are easy to compute, one can therefore calculate $\partial y/\partial x_i$ from the relation

$$\frac{\partial y}{\partial x_i} = -\frac{\partial f/\partial x_i}{\partial f/\partial y} \tag{4-3}$$

since the numerical value of $\partial f/\partial y$ is known. It is therefore easier to compute the gradient at a point where y has been evaluated than to try to find y at a new point, which would involve the relatively long numerical solution of Eq. (4-2) again. Hence elimination methods using gradients are attractive whenever the objective function is implicit. In each circumstance for which an explicit technique is indicated in this chapter, there is often a modified version appropriate for implicit objectives. Any such implicit methods will be described together with the corresponding explicit procedure.

4-02 POLYNOMIAL APPROXIMATION

Before developing the elimination techniques, let us describe the sort of approximation methods that would occur naturally to most people faced with an optimization problem and disciplined in mathematics. By *polynomial approximation* we mean fitting a polynomial of degree K to the objective function, for simplicity considered here a function of a single variable x. The polynomial is

$$p(x) \equiv \alpha_0 + \alpha_1 x + \alpha_2 x^2 + \cdots + \alpha_K x^K$$

or

$$p(x) = \sum_{k=0}^{K} \alpha_k x^k \tag{4-4}$$

To construct this equation, one must measure $y(x)$ at $K + 1$ points, say x_0, $x_1, \ldots, x_K$, and solve the following $K + 1$ equations linear in the $K + 1$ coefficients α_k.

$$\sum_{k=0}^{K} \alpha_k x_j^k = y(x); \qquad j = 0, 1, \ldots, K \tag{4-5}$$

Where to place the measurements, a subject covered in most books on numerical analysis (see Lapidus et al.), need not concern us here, except to point out that the computations are simplified by spacing them equally. Suffice it to say that once the approximating polynomial $p(x)$ has been determined, one can apply the appropriate indirect method, such as setting its first derivative to zero. Since high-degree polynomials require many measurements, most of which will be far from the desired optimum, and since finding the roots of the $(K - 1)$-degree polynomial for the first derivative is a tedious task, approximations usually stick to low-degree polynomials. If for example four points are taken, $p(x)$ is cubic.

$$p(x) = \alpha_0 + \alpha_1 x + \alpha_2 x^2 + \alpha_3 x^3 \tag{4-6}$$

Since its first derivative is quadratic,

$$\frac{\partial p}{\partial x} = \alpha_1 + 2\alpha_2 x + 3\alpha_3 x^2 \tag{4-7}$$

the predicted optimum $\hat{x}$ is one of the roots given in closed form by

$$\hat{x} = \frac{-\alpha_2 \pm \sqrt{\alpha_2^2 - 3\alpha_1\alpha_3}}{3\alpha_3} \tag{4-8}$$

One could take a new measurement at $\hat{x}$, throwing away the point farthest from this prediction in order to keep the approximating polynomial cubic. The new values of α_1, α_2, and α_3 can then be substituted into Eq. (4-8) for an adjusted prediction.

When the objective is implicit, one need take measurements at only two points to find all four coefficients of a cubic approximation, because each point furnishes not only y, but $\partial y/\partial x$ as well—two independent pieces of data. Hence Eqs. (4-6) and (4-7) give four equations in the four unknown coefficients. Equation (4-8) can be used as before for predicting the location of the optimum once α_1, α_2, and α_3 have been estimated.

Many such procedures can be imagined. They would work very well on smooth functions which are approximated well by low-degree polynomials, but they could behave badly on arbitrary objective functions. Brooks has shown how much in error such methods can be whenever the prediction falls outside the range of the measurements, since in this case extrapolation is called for. Another objection is that there is no good way to estimate how far the prediction is from the true optimum. Approximation methods must be viewed as quick paths to rough guesses in which luck plays a large part. Nevertheless, these procedures are frequently used. Later in this chapter we shall review two optimum seeking methods that employ polynomial approximation schemes.

4-03 INTERVAL ELIMINATION

In this section and throughout most of Part I of this chapter, our attention focuses on unimodal objective functions of a single variable defined in a closed interval. Suppose to be definite that we seek the location x^* where $y(x)$ achieves its maximum value y^* in the unit interval $0 \leq x \leq 1$. The unimodality of $y(x)$ assures us that there is only one local maximum; in addition, we assume, for the sake of the explicit elimination methods to be developed, that there are no intervals of finite length in which the slope of y is zero, that is, where y is horizontal. More precisely, $y(x)$ is assumed to increase monotonically up to the maximum, after which it decreases monotonically. That is, if

$$x_1 < x_2 < x^* \tag{4-9a}$$

then

$$y_1 < y_2 < y^* \tag{4-9b}$$

whereas, if

$$x^* < x_1 < x_2 \tag{4-10a}$$

then

$$y^* > y_1 > y_2 \tag{4-10b}$$

Functions satisfying this definition, which is slightly more restrictive than the definition of unimodality employed so far in this book, will be called *strictly unimodal.* Notice that this definition does not require smoothness or even continuity; Fig. 4-1 shows three unimodal functions of varying character.

Strict unimodality makes it possible to say, after examining the results of any pair of evaluations of the objective function, that the maximum lies in

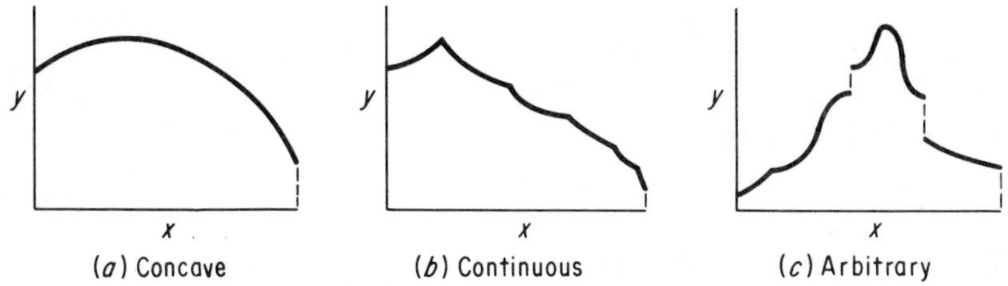

Figure 4-1. Unimodal functions.

some interval shorter than the original one. Consider two points $x_1 < x_2$. The three possible outcomes are shown in Fig. 4-2: $y_1 > y_2$, $y_1 < y_2$, or $y_1 = y_2$. When $y_1 \geq y_2$, the maximum cannot lie to the right of y_2 without contradicting the definition of unimodality, and so we can conclude that $x^* < x_2$ in this case. Similarly, $y_1 \leq y_2$ implies that $x^* > x_1$. When the two outcomes are exactly equal $(y_1 = y_2)$, the peak must lie between the points $(x_1 < x^* < x_2)$.

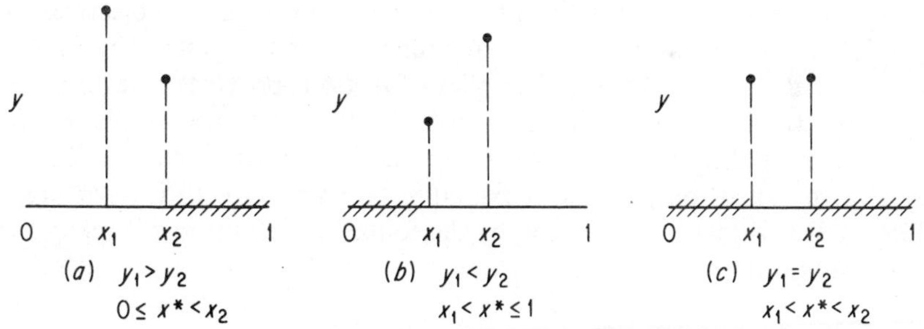

Figure 4-2. Possible outcomes of two experiments.

To express these results in a form extendable to k function evaluations, let the left and right ends of the original interval be denoted by x_0 and x_{k+1}, respectively. Let m_k be the location of the best point among them, and let l_k and r_k, respectively, be the points immediately to the left and right of m_k. Formally,

$$y(m_k) = \max_{1 \leq j \leq k} [y(x_j)] \tag{4-11}$$

$$l_k = \max_{x_j < m_k} (x_j) \tag{4-12}$$

$$r_k = \min_{x_j > m_k} (x_j) \tag{4-13}$$

Then, if y is strictly unimodal, the true maximum must lie between l_k and r_k.

$$l_k < x^* < r_k \tag{4-14}$$

If

$$y(m_k) = y(r_k) \qquad \text{(4-15)}$$

then the interval is smaller:

$$m_k < x^* < r_k \qquad \text{(4-16)}$$

Figure 4-3 shows all these relationships for $k = 7$.

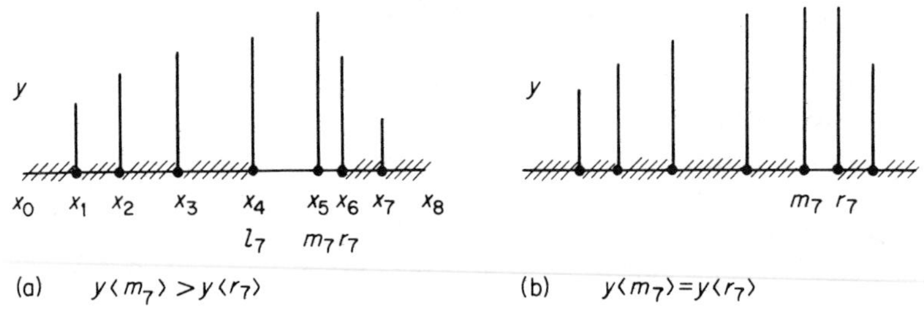

Figure 4-3. *k* (= 7) function evaluations.

The preceding development is relevant for explicit objectives; the implicit case requires only half as many function (and derivative) evaluations. Computation of y from an implicit function provides the derivative y' ($\equiv \partial y/\partial x$) with little additional effort. To take advantage of this, we must impose two additional restrictions on y: that it be differentiable and that the derivative vanish only at the maximum. To emphasize this departure, let Eqs. (4-9) and (4-10) be replaced by the following definition of *differentiable unimodality*. If

$$x < x^* \qquad \text{(4-17a)}$$

then

$$y'(x) > 0 \qquad \text{(4-17b)}$$

and if

$$x > x^* \qquad \text{(4-18a)}$$

then

$$y'(x) < 0 \qquad \text{(4-18b)}$$

From this it follows that $y'(x) > 0$ implies $x^* > x$, that $y'(x) < 0$ implies $x^* < x$, and that $y'(x) = 0$ implies $x^* = x$. When y and y' have been evaluated at k points, and with m_k, r_k, and l_k being defined as in Eqs. (4-11), (4-12), and (4-13), the true maximum x^* must lie in the interval immediately to the right or left of m_k according to whether $y'(m_k)$ is positive or negative. That is, if

$$y'(m_k) > 0 \qquad \text{(4-19a)}$$

then

$$m_k < x^* < r_k \tag{4-19b}$$

while if

$$y'(m_k) < 0 \tag{4-20a}$$

then

$$l_k < x^* < m_k \tag{4-20b}$$

The range defined by Eq. (4-14) for explicit objectives and by Eq. (4-19b) or (4-20b) for implicit objectives is called the *interval of uncertainty after k measurements* and denoted i_k. The ratio i_k/i_0 is a reasonable measure of the success of a particular deployment of k evaluations $x_1, \ldots, x_k$, which will be called a *k-measurement search plan*, abbreviated $\mathbf{x}_k$. With such a measure of effectiveness available, one can compare various schemes and look for a plan that is optimal in the sense of minimizing this ratio. Examination shows, however, that this ratio depends not only on the location of the measurements, but also upon the unknown location of the peak.

$$\frac{i_k}{i_0} = \frac{r_k - l_k}{x_{k+1} - x_0} \tag{4-21}$$

Specifically, the numerator $r_k - l_k$ is determined by where the sample maximum m_k happens to fall, which cannot be predicted in advance. In order to have a measure of *planning* effectiveness independent of such unknown factors, it is more reasonable to consider the worst (that is, longest) interval that might arise. This is given by

$$I^k(\mathbf{x}_k) \equiv \max_{1 \le j \le k} (x_{j+1} - x_{j-1}) \tag{4-22}$$

called the *maximum interval of uncertainty after k measurements*. Since I^k depends only on the search plan $\mathbf{x}_k$, it is an a priori measure of any plan's effectiveness. Therefore, any plan for which I^k attains its minimum I^{k*} with respect to all other search schemes is in a sense optimal.

$$\frac{I^{k*}}{I^0} \equiv \min_{\mathbf{x}_k} \left[\frac{I_k(\mathbf{x}_k)}{I^0}\right] = \min_{\mathbf{x}_k} \left[\max_{1 \le j \le k} \left(\frac{x_{j+1} - x_{j-1}}{x_{k+1} - x_0}\right)\right] \tag{4-23}$$

Because of the form of the right member of Eq. (4-23), a search plan giving such an interval is said to be *minimax*.

The minimax approach is completely conservative, chance determining only the *position* of the final interval, not its *length*. Despite this pessimistic point of view, the minimax concept leads to surprisingly effective search methods. Prudence of this sort is an ancient idea; in the *I Ching* (appendix III, sec. II, chap. V, hexagram 39) is written

> He who keeps danger in mind will rest safe in his seat;
> He who keeps ruin in mind will preserve his interests secure.

4-04 RESOLUTION AND DISTINGUISHABILITY

In deriving minimax simultaneous search plans, we frequently encounter a minimum separation δI^k ($= \epsilon I^0$) to be maintained between adjacent measurements. This quantity, called the *resolution*, appears as a constraint on the search plan.

$$x_{j+1} - x_j \geq \delta I^k = \epsilon I^0; \qquad j = 1, \ldots, k-1 \tag{4-24}$$

Note that Eq. (4-24) simply implies that a minimum separation must be maintained between adjacent experiments. This *resolution* is given as a known fraction (ϵ) of the initial interval, or a known factor (δ) of the final interval.

In some cases, one may know the smallest detectable difference between values of the objective function rather than the closest possible packing of measurements. This minimum difference in y will be called the *distinguishability* and denoted η. The distinguishability is such that for any pair of measurements x' and x'', one can say that

$$y(x') \neq y(x'') \tag{4-25a}$$

only if

$$|y(x') - y(x'')| \geq \eta \tag{4-25b}$$

When the difference is less than η, $y(x')$ and $y(x'')$ are said to be *indistinguishable*. Note that the resolution (as defined by either δ or ϵ) can be obtained from the distinguishability η by estimating the behavior of the function near the optimum, which can often be done even when the location of x^* is unknown. Specifically, ϵ is taken as the positive quantity satisfying the implicit relation

$$\min\,[|y(x^*) - y(x^* + \epsilon I^0)|, |y(x^*) - y(x^* - \epsilon I^0)|] = \eta \tag{4-26}$$

Figure 4-4 shows these quantities graphically.

The concept of distinguishability makes it possible to apply elimination methods when measurements of the objective are obscured by random experimental error. When the statistical distribution of this error can be predicted in advance, one can establish confidence limits about each estimate of y which correspond to the indistinguishability limits η. Doing this is a

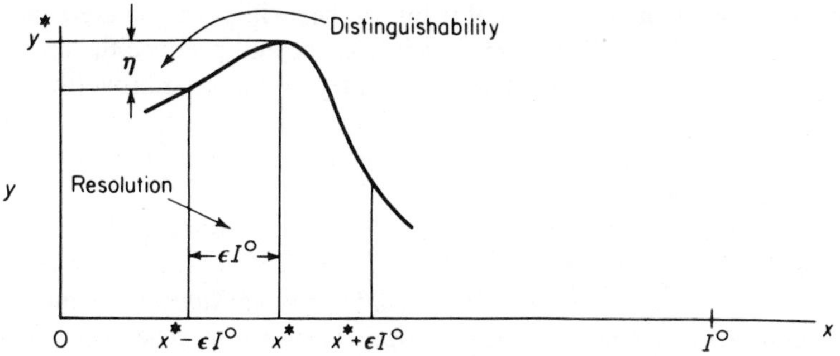

Figure 4-4. Resolution and distinguishability.

straightforward statistical task that will not be detailed here (see Mood or any other statistics text).

This discussion of resolution and distinguishability is justified because it enables us to rigorously define experimental procedures for the sequential methods that follow.

4-05 POPULATION EXPLOSIONS, RABBITS, AND OPTIMIZATION

There is an instructive analogy between rabbit breeding and a sequence of numbers which arise in optimal minimax variable optimization schemes. Let us digress for a moment to the theories of animal husbandry. In 1202 A.D. the famed Leonardo of Pisa concerned himself with the genetics of rabbit breedings. He assumed that each pair of mature adult rabbits give rise to exactly one couple per litter ($k = 1$), allowed a month for maturation, and computed the number of couples at the end of each month of the year. Table 4-1 gives the results. At time zero—midnight on New Year's Eve—there are no rabbits, but a newly born couple arrives on New Year's Day. It achieves maturity after one month, and at the beginning of March brings a new immature pair into the world—a total of 2 couples alive. In April the original

TABLE 4-1

Fibonacci's Numbers A^n

Months elapsed, n	0	1	2	3	4	5	6	7	8	9	10	11	12
Immature	0	1	0	1	1	2	3	5	8	13	21	34	55
Mature	0	0	1	1	2	3	5	8	13	21	34	55	89
Total, A^n	0	1	1	2	3	5	8	13	21	34	55	89	144

pair is blessed with another set of twins, as they will be during every month of the year; their first offspring reach adolescence and produce the first grandchildren in the fifth month. Hence each entry in the second row (the number of mature couples) appears again in the first row a month later as the young are born. Three couples alive in April mean 3 mature in May and 3 pairs of offspring in June.

Leonardo's nickname ("Fibonacci," or son of Bonacci) has been given to the sequence of numbers $A^n = 0, 1, 1, 2, 3, 5, 8, 13, \ldots$, in the third row, for they first appeared in his *Liber Abaci* (*Book of Numbers*), which helped introduce Arabic numerals into Europe. They are in general given by the recursion relation (Girard)

$$A^0 \equiv 0 \tag{4-27a}$$

$$A^1 \equiv 1 \tag{4-27b}$$

$$A^n = A^{n-1} + A^{n-2}; \qquad n = 2, 3, \ldots \tag{4-27c}$$

That is, each is the sum of the two numbers preceding. Having made ourselves experts in the genetics of rabbit reproduction, let us proceed to apply this knowledge to optimization theory.

4-06 UNIVARIATE OPTIMIZATION: FIBONACCI SEARCH

In this section we will derive a procedure that will find the optimum value of an unconstrained objective function of a single variable. In order to derive the Fibonacci search procedure, suppose that we are dealing with a unimodal function bounded over a fixed interval, I^0, known to contain the optimum (maximum/minimum) value of the decision variable. Suppose further that we are given exactly n experiments to find this value. Then, after n experiments, we will have bracketed our solution in an interval I^n.

Consider the next-to-last interval, I^{n-1}; it is optimal for the last two experiments to be placed ϵI^0 apart exactly in the middle of the interval, since we do not know which experiment will give the better value of the objective function. If this is done, then we will execute a minimax philosophy as discussed previously.

Rescale everything so that $I^0 = 1$ to simplify notation and let E_k^+ indicate the experiment that gave the best value of the first k experiments, and let x^k be the kth experiment. Then, Fig. 4-5(a) shows how the next-to-last interval will look. From this figure, it follows that $I^{n-1} = 2I^n - \epsilon$, and the preceding interval must have been as shown in Fig. 4-5(b), from which we see that

$$I^{n-2} = I^{n-1} + I^n = 3I^n - \epsilon$$

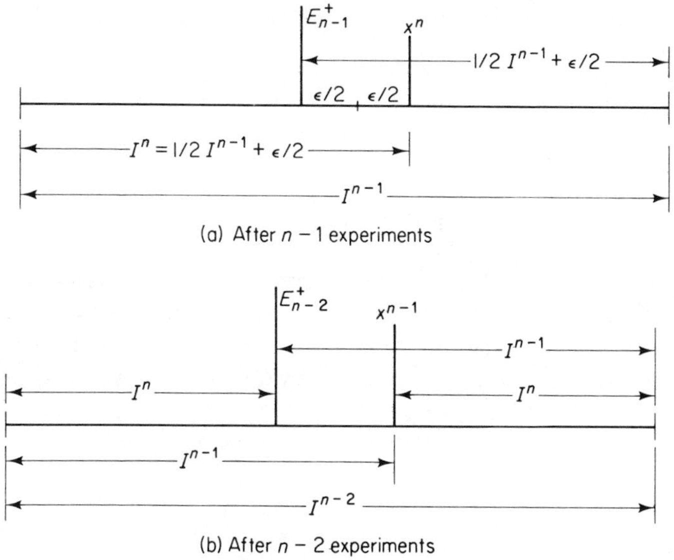

Figure 4-5. Maximum remaining intervals of uncertainty.

We can show that, in general,

$$I^{j-1} = I^j + I^{j+1} \qquad \text{for } j = 2, \ldots, n-1 \tag{4-28}$$

Thus: given $I^0 = I^1 = 1$, then $I^{n-1} = 2I^n - \epsilon$. Hence $I^{n-2} = 3I^n - \epsilon$, $I^{n-3} = 5I^h - 2\epsilon$, and $I^{n-4} = 8I^n - 3\epsilon$, $I^{n-5} = 13I^n - 4\epsilon$, and so on.

Next, a sequence of numbers is defined: $A^0 = 0$, $A^1 = 1$,

$$A^n = A^{n-1} + A^{n-2} \qquad \text{for } n = 2, 3, \ldots$$

Thus the first eight numbers of the sequence would be: 0, 1, 1, 2, 3, 5, 8, 13. Note that this sequence of numbers corresponds to that derived by Fibonacci in our study of rabbits!

Using this sequence of numbers we can write Eq. (4-28) as

$$I^{n-j} = A^{j+2}I^n - A^j\epsilon; \qquad j = 0, 1, 2, \ldots, n-1$$

Recall that, for simplicity, we defined the initial interval of uncertainty as $I^0 \equiv 1$. Since with only one experimental result we gain no additional insight, $I^1 = 1$.

Using Eq. (4-28), we obtain the following result by letting $I^1 = 1$, and defining $j = n - 1$:

$$I^n = \frac{1}{A^{n+1}} + \frac{A^{n-1}}{A^{n+1}}\epsilon \tag{4-29}$$

Proceeding further, with $j = n - 2$ we can arrive at a similar result:

$$I^2 = \frac{A^n}{A^{n+1}} + \frac{A^n A^{n-1}}{A^{n+1}} - A^{n-2}\epsilon$$

or

$$I^2 = \frac{A^n}{A^{n+1}} + \frac{(-1)^n}{A^{n+1}}\epsilon \tag{4-30}$$

Equation (4-30) is all that is necessary to begin a Fibonacci search. To perform a Fibonacci search, the first experiment must be placed a distance I^2 units from either end of the original interval I^0. The subsequent experiments are merely placed symmetrically in the interval that remains. The question is how to easily define I^2.

If $I^0 \neq 1$, then Eqs. (4-29) and (4-30) become

$$I^n = I^0\left(\frac{1}{A^{n+1}} + \frac{A^{n-1}\epsilon}{A^{n+1}}\right) \tag{4-31}$$

$$I^2 = I^0\left[\frac{A^n + (-1)^n\epsilon}{A^{n+1}}\right] \tag{4-32}$$

The *reduction ratio* is a measure of the search effectiveness. This is given by rearranging equation (4-29) as

$$\frac{I^0}{I^n} = \frac{A^{n+1}}{A^{n-1}\epsilon + 1} \tag{4-33}$$

Note that this result defines how closely we have bracketed the optimal solution in a sequence of n sequential experiments using a Fibonacci search procedure.

Neglecting resolution, it would take at least 11 observations to reduce the interval to less than 1% of its original length.

This scheme, called *Fibonacci search* for obvious reasons, was first devised and proved minimax for $\delta = 0$ by J. Kiefer. Oliver and Wilde showed informally how to include resolution, and the rigorous proof for this modification was given later by Avriel and Wilde. An example of Fibonacci search follows.

Example 4-1: Example of Fibonacci Search

Suppose that one wishes to reduce an interval $5.11 \leq x \leq 23.64$ to one only 10% as long, the resolution being 0.545 unit. Hence $I^0 = 18.53$, $I^n \leq 1.853$, and from Eq. (4-24), $\epsilon = 0.545/18.53 = 0.0294$. The number of observations n is the unique integer satisfying

$$A^n(A^{n-2}\epsilon + 1)^{-1} < \frac{I^0}{I^n} \leq A^{n+1}(A^{n-1}\epsilon + 1)^{-1} \tag{4-34}$$

which gives $n = 6$, since (see Table 4-1)

$$8[3(0.0294) + 1]^{-1} = 7.35 < 10.00 < 13[5(0.0294) + 1]^{-1} = 11.33$$

The final interval will be shorter than 1.853, so we shall take the full reduction ratio of 11.33 and have a final interval of length $18.53(11.33)^{-1} = 1.636$. In terms of this final interval, the resolution is

$$\delta = \frac{\epsilon I^1}{I^6} = (0.0294)(11.33) = 0.334$$

Since $\delta < \frac{1}{2}$, there is no danger of having the last two observations too close together.

Fibonacci search places the first observation at

$$\begin{aligned} 5.11 + x^1 &= 5.11 + (A^6 - A^{6-2}\delta)I^6 \\ &= 5.11 + [8 - 3(0.334)](1.636) \\ &= 5.11 + 11.45 = 16.56 \end{aligned}$$

The second is placed symmetrically at

$$23.64 - x^1 = 12.19$$

Suppose that $y(16.56) = 8.73$ and $y(12.19) = 9.07$. Then if we seek the maximum, there is no need to explore values greater than 16.56.

$$5.11 \leq x^* < 16.56$$

The third measurement is placed symmetrically with respect to the one still in the interval, namely at $16.56 - (12.19 - 5.11) = 9.48$. Let the result be $y(9.48) = 7.89$. This eliminates the left portion and implies that

$$9.48 < x^* < 16.56$$

The best result is still at 12.19.

The fourth is located symmetrically at

$$9.48 + 16.56 - 12.19 = 13.85$$

If $y(13.85) = 9.32 > y(12.19)$, then

$$12.19 < x^* < 16.56$$

and the fifth observation is made at

$$12.19 + 16.56 - 13.85 = 14.90$$

If $y(14.90) = 9.27 < y(13.85)$, then

$$12.19 < x^* < 14.90$$

and the objective is measured finally at

$$12.19 + 14.90 - 13.85 = 13.24$$

Notice that this is 0.61 unit from the nearest measurement at 13.85, a close approximation to the desired resolution, 0.545. The discrepancy is due to accumulated error caused by rounding off in the second decimal place.

Suppose that the sixth result is $y(13.24) = 9.36 > y(13.85)$. Then the final interval is

$$12.19 < x^* < 13.85$$

Its length is 1.66, slightly above the predicted 1.64 because of rounding error, but still well below the 1.85 required.

4-07 UNKNOWN RESOLUTION

All the search schemes developed so far require prediction of the resolution δ or ϵ before the beginning of the search. When this information is unavailable, a slightly less efficient procedure due to Kiefer must be employed.

Since the length of the final interval cannot be predicted until, at the last observation, the resolution is known, the search plan must be scaled in terms of the original interval I^1. The first experiment is located at

$$x^1 = \frac{I^1 A^n}{A^{n+1}}$$

and the second symmetrically at

$$x^2 = I^1 - x^1 = \frac{I^1 A^{n-1}}{A^{n+1}} \qquad (4\text{-}35)$$

thus giving an interval after two observations of

$$I^2 = \frac{I^1 A^n}{A^{n+1}}$$

This is repeated, giving in general after $n - 1$ observations an interval of length

$$I^{n-1} = \frac{I^1 A^3}{A^{n+1}} = \frac{2I^1}{A^{n+1}}$$

with an observation at

$$m^{n-1} = \frac{I^1 A^2}{A^{n+1}} = \frac{I^1}{A^{n+1}}$$

Since m^{n-1} is exactly in the middle of the interval, it would do no good to place the final observation x^n symmetrically because this would only duplicate effort without shortening the interval. Instead, one should put x^n as close as possible to m^{n-1}, either to the left or right. This involves estimating the resolution ϵI^1, presumable possible now after $n - 1$ measurements have given information about the objective and narrowed down the range containing the optimum.

$$x^n = m^{n-1} \pm \epsilon I^1$$

This will give two possible intervals—either $I^1[(A^{n+1})^{-1} + \epsilon]$ or $I^1[(A^{n+1})^{-1} - \epsilon]$. According to the cautious minimax approach, we must use the larger in measuring the effectiveness of the search plan, so

$$\frac{I^0}{I_n} = \frac{I^1}{I^n} = A^{n+1}(1 - \delta) \tag{4-36}$$

Kiefer proved that this plan is minimax for any given $\epsilon > 0$. The slightly decreased effectiveness compared to Oliver's method [see Eq. (4-33)] is the price of not knowing the resolution in advance.

If the resolution ϵ is estimated from the distinguishability η of the objective function as in Section 4-04, then the question of regularity must be reexamined.

A strictly unimodal function $y(x)$ is said to be *regular* for an n-point search with resolution δ if

$$|y(x') - y(x'')| \geq \eta$$

for all x' and x'' such that

(1) either $x^* \leq x' < x''$ or $x'' < x' \leq x^*$

(2) $|x' - x^*| \leq I^{n-j} = (A^{j+2} - A^j\delta)I^n$

(3) $|x'' - x'| \geq I^{n-j-1} = (A^{j+3} - A^{j+1}\delta)I^n$

for all $j = 2, 3, \ldots, n$. These conditions are satisfied by convex functions. One can prove that if the best result m^j after j observations cannot be distinguished from the observation on its right at $m^j + \delta I^n$, then the optimum x^* must lie between them. That is, if

$$|y(m^j) - y(m^j + \delta I^n)| < \eta$$

then

$$m^j < x^* < m^j + \delta I^n \tag{4-37}$$

(The proof is left as Exercise 4-3.)

4-08 THE GOLDEN SECTION

Often an experimenter begins searching for an optimum without knowing in advance exactly how many experiments to use. He simply keeps experimenting until the criterion of interest becomes good enough to satisfy him. But even when totally unconcerned about the interval of uncertainty itself, he would like to use a search plan that would rapidly close in on the optimum, since such a plan would reasonably give good values of the criterion as early as possible in the search.

Unfortunately, unless the number of experiments to be performed is known in advance, one cannot use the Fibonacci technique, because I^2, which must be known before the first experiment can be located, depends entirely on n, the number of trials [Eq. (4-30)]. There is, however, another technique which, although nearly as effective as the Fibonacci method, is completely independent of the number of experiments available.

As before, let j represent the number of experiments already run. We may use the same reasoning as before to deduce that the experimental plan should place successive experiments such that [Eq. (4-28)]

$$I^{n-j} = I^{n-(j-1)} + I^{n-(j-2)}$$

just as for the Fibonacci technique.

Since we do not know what n is, let us hold the ratio of successive lengths constant. Following Coxeter and calling this ratio τ (after τομη, the *section*), we have

$$\frac{I^{n-j}}{I^{n-(j-1)}} = \tau = \frac{I^{n-(j+1)}}{I^{n-j}} \tag{4-38}$$

By dividing Eq. (4-28) throughout by $I^{n-(j-2)}$ and noting that

$$\frac{I^{n-(j+1)}}{I^{n-(j-1)}} = \tau^2$$

we obtain $\tau^2 = \tau + 1$, as shown in Fig. 4-6. Only one root of this quadratic equation is positive, and so we see that

$$\tau = \frac{1+\sqrt{5}}{2} = 1.618033989\ldots \tag{4-39}$$

Notice that the negative root is $-1/\tau$. The results of the two experiments will

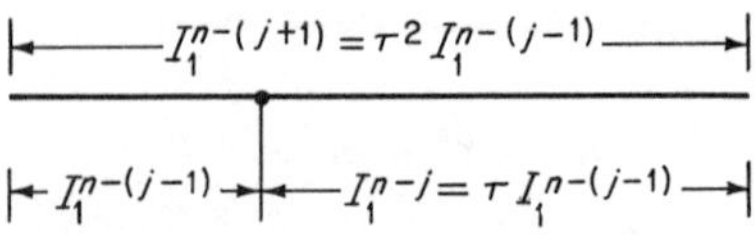

Figure 4-6. The golden section.

determine which segment is to be explored further. As usual, this remaining segment will contain one of the previous trials, and to continue the search one merely places the next experiments symmetrically in the interval. Once begun, this process may be continued as long as desired. After n experiments, the interval I_1^n remaining is given by

$$I^n = \frac{1}{\tau^{n-1}} \tag{4-40}$$

This procedure is called *golden section search.*

Such a name demands justification. Although both Kiefer and Johnson have suggested this search technique in modern times, the ancient geometer Euclid knew how to divide a line segment in this way with only a ruler and compass. This ratio was, in fact, discovered two centuries before Euclid by the Pythagorean Brotherhood, whose star-shaped badge has all its lines divided in these proportions (Fig. 4-7). Coxeter describes how the many surprising properties of this construction gave it great mystical significance to ancient and medieval scholars, who accorded it the name "golden section." Architects have used it to design structures ranging from the Parthenon to the apartment houses of Le Corbusier. Thus the venerable and respected "golden section" appears again in a modern application, and it is only right to combine old terms with new in the name "search by golden section."

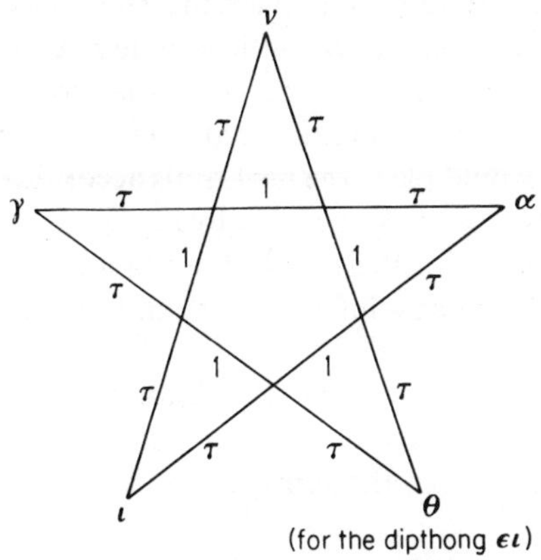

Figure 4-7. Badge of the Pythagoreans. Letters on the vertices spell ὑγίεια, meaning "health" (Ball).

In order to compare the performance of the search by golden section with the Fibonacci method, we make use of the following relation between the Fibonacci numbers and τ developed by Lucas:

$$A^n = \frac{\tau^n - (-\tau)^{-n}}{\sqrt{5}} \tag{4-41}$$

When n is very large, the second term becomes negligible, giving approximately

$$A^n \approx \frac{\tau^n}{\sqrt{5}} \tag{4-42}$$

If G^n is the interval left after n trials in a search by golden section, and if I^n is that remaining after n Fibonacci experiments, then, for large n,

$$\frac{G^n}{I^n} = \frac{\tau^n}{\sqrt{5}\,\tau^{n-2}} = \frac{\tau^2}{\sqrt{5}} = 1.1708$$

Thus a search by golden section will give a final interval only about 17% longer than that obtainable by Fibonacci search.

Lucas's formula also tells us that when n is large,

$$\frac{A^{n-1}}{A^n} \approx \frac{1}{\tau} \tag{4-43}$$

Combining Eqs. (4-35) and (4-43) we have that for large n, $I^2 \approx 1/\tau$. But by Eq. (4-40) $I^2 = 1/\tau$ for a search by golden section; thus, when n is large, both the Fibonacci technique and the search by golden section start at practically the same point. By examining Table 4-1 we see that the ratio of successive Fibonacci numbers approaches 0.618 very soon; even for $n = 5$, the ratio is already 0.600—only 3% low. We may therefore start a search using the golden section technique, switching to the Fibonacci methods when we are sufficiently close to the optimum to fix the remaining number of experiments.

For example, we could begin with an indefinite number of experiments, placing early trials according to the golden section. At some point in the search we might be able to predict that four more experiments should be sufficient. We then would place the next trials according to the Fibonacci plan for five experiments (counting the one already in the interval). In this way we would recover most of the efficiency of the Fibonacci technique, even though we had started out with absolutely no idea of how many experiments would be needed.

This section has shown how to seek an optimum without knowing in advance how many measurements are needed. Although the golden section scheme is less efficient than those presented earlier, it also requires the least prior information.

4-09 FIBONACCI LATTICE SEARCH SCHEME FOR INTEGER VALUES

When the independent variable x is *not* continuous, experiments can only be made at those points where x is defined.

For example, suppose that x can take on only the values 3.0, 3.2, 3.4, 4.0, 4.2, 4.4, and 5.0 (note unequal spacing). Associate the integers 1, 2, . . . , 7, respectively, with these points.

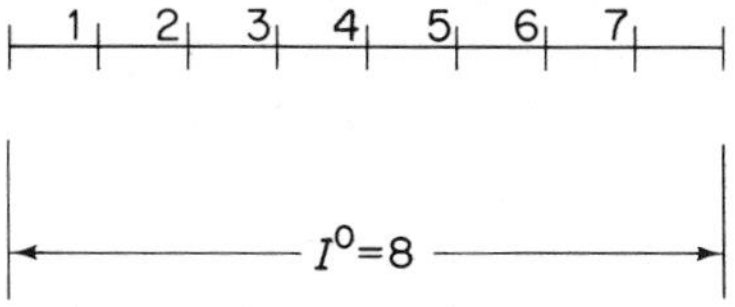

$A^0 = 0 \quad A^1 = 1$

$A^2 = 1 \quad A^3 = 2$

$A^4 = 3 \quad A^5 = 5$

$A^6 = 8 \quad A^7 = 13$

Note that we must extend I^0 one unit beyond the end points because Fibonacci never searches the end points. Taking $\epsilon = 0$,

$$I^2 = I^0 \frac{A^n}{A^{n+1}} \quad \text{and} \quad I^n = \frac{I^0}{A^{n+1}} = 1$$

Hence $I^0 = A^{n+1} = 8$; and since $A^6 = 8$, then five experiments will be required ($n = 5$). Then

$$I^2 = 8 \times \frac{A^5}{A^6} = 8 \times \frac{5}{8} = 5$$

The first two experiments are always placed symmetrically, so they fall on the lattice points 3 and 5.

Since we took $\epsilon = 0$, the fifth experiment (in general, the nth experiment) if run, would coincide with one already run.

Therefore, if the original interval was of length A^{n+1} (that is, the interval contained $A^{n+1} - 1$ lattice points), a Fibonacci search will find the best lattice point in $n - 1$ experiments. (In the example above, $n = 5$.)

When the number of lattice points does *not* happen to be exactly one less a Fibonacci number, *add* enough *fictitious points* to bring the total up to that value.

4-10 UNBRACKETED SEARCH PROCEDURES: AN INFINITE INTERVAL OF UNCERTAINTY

All of the preceeding discussion related to techniques which require that we begin with the optimum bracketed. In practice we may not have such information: we may only be given the direction of search. What do we do in those unfortunate (but common) circumstances? In the absence of further information, we can proceed to apply a simple bracketing scheme.

Step 1: Select an arbitrary point x^0 and evaluate the objective function at that point; $y(x^0)$.

Step 2: Let $x^1 = x^0 + \Delta x$, and evaluate the objective function at that point; $y(x^1)$.
If $y(x^1) < y(x^0)$, let $k = 2$ and go to step 3; otherwise, let $\Delta x = -\Delta x$ and repeat step 2.

Step 3: Let $x = 2\Delta x$ and $x^k = x^{k-1} + \Delta x$.

Step 4: If $y(x^k) < y(x^{k-1})$, $k = k + 1$ and go to step 3; otherwise, stop with x^k and x^{k-1}, forming a bracket on the optimal solution.

Once the interval of uncertainty has been bracketed, one can proceed to apply the Fibonacci or golden search schemes as previously described.

4-11 EXTRAPOLATION AND INTERPOLATION METHODS

Although the Fibonacci search procedures are optimum in a minimax sense, other methods exist which utilize simultaneous calculations at multiple points. These unidimensional search methods attempt to locate a point x near the optimum value x^* by extrapolation and interpolation rather than bracketing the minimum as Fibonacci does. Two such algorithms are described below (Himmelblau).

DSC search (Davies, Swann, and Campey)

In the DSC unidimensional search, steps of increasing size are taken until the minimum is overshot, and then a single quadratic interpolation is performed. Figure 4-8 illustrates the procedure graphically (x^m is the first value of x that overshoots the minimum and Δx is the step size). The search proceeds as follows (Himmelblau):

Step 1: Evaluate $y(x)$ at the initial point x^0. If $y(x^0 + \Delta x) < y(x^0)$, go to step 2. If $y(x^0 + \Delta x) > y(x^0)$, let $\Delta x = -\Delta x$ and go to step 2. If $y(x^0 + \Delta x) = y(x^0)$, go directly to step 6.

Step 2: Compute $x^{k+1} = x^k + \Delta x$.

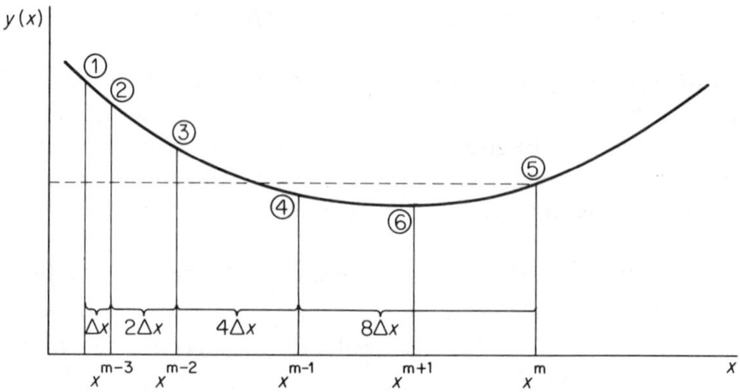

Figure 4-8. DSC unidimensional minimization.

Step 3: Compute $y(x^{k+1})$.

Step 4: If $y(x^{k+1}) \leq y(x^k)$, double Δx and return to step 2 with $k = k + 1$. If $y(x^{k+1}) > y(x^k)$, denote x^{k+1} by x^m, x^k, by x^{m-1}, and so on, reduce Δx by one-half, and return to steps 2 and 3 for one more (only) calculation.

Step 5: Of the four equally spaced values of x in the set $\{x^{m+1}, x^m, x^{m-1}, x^{m-2}\}$, discard either x^m or x^{m-2}, whichever is farthest from the x corresponding to the smallest value of $y(x)$ in the set. Let the remaining three values of x be denoted by x^a, x^b, and x^c, where x^b is the center point and $x^a = x^b - \Delta x$ and $x^c = x^b + \Delta x$.

Step 6: Carry out a quadratic interpolation to estimate x^*.

$$x^* \cong \hat{x}^* = x^b + \frac{\Delta x[y(x^a) - y(x^c)]}{2[y(x^a) - 2y(x^b) + y(x^c)]}$$

These steps complete the first stage of the DSC method. If it is continued, a new start is made from $\hat{x}^*$ [or x^c if $(x^c) < (\hat{x}^*)$] and Δx is reduced.

The procedure is illustrated in the following example.

Example 4-2: DSC Search

Minimize $y(x) = 3x^2 - 18x + 27$

	$x^{(k)}$	$y[x^{(k)}]$	Δx
$k = 0$	0	27	0.5
$k = 1$	0.5	18.75	1
$k = 2$	1.5	6.75	2
$k = 3$	3.5	0.75	4
$k = 4$	7.5	60.75	−2

Step 1: Initial $x^{(0)} = 0$, $\Delta x = 0.5$.

$y[x^{(0)}] = 27$, $y[x^{(0)} + \Delta x] = 18.75$

$y[x^{(0)} + \Delta x] < y[x^{(0)}]$

Step 2: $x^{(1)} = x^{(0)} + \Delta x = 0.5$

Step 3: $y[x^{(1)}] = 18.75$

Step 4: $y[x^{(1)}] < y[x^{(0)}]$

Let $\Delta x = 1$.

Step 2: $x^{(2)} = x^{(1)} + \Delta x = 1.5$

Step 3: $y[x^{(2)}] = 6.75$

Step 4: $y[x^{(2)}] < y[x^{(1)}]$

Let $\Delta x = 2$.

Step 2: $x^{(3)} = x^{(2)} + \Delta x = 3.5$

Step 3: $y[x^{(3)}] = 0.75$

Step 4: $y[x^{(3)}] < y[x^{(2)}]$
Let $\Delta x = 4$.

Step 2: $x^{(4)} = x^{(3)} + \Delta x = 7.5$

Step 3: $y[x^{(4)}] = 60.75$

Step 4: $y[x^{(4)}] > y[x^{(3)}]$
Denote $x^{(4)}$ by $x^{(m)}$
$x^{(3)}$ by $x^{(m-1)}$.
Reduce $\Delta x = -2$.

Step 2: $x^{(m+1)} = x^{(m)} + \Delta x = 5.5$

Step 3: $y[x^{(m+1)}] = 18.75$

Step 5: $y[x^{(4)}]$ is farthest from the x corresponding to the smallest value of $y[x]$, so discard $y[x^{(m)}]$.

$$\text{Let } x^{(b)} = x^{(m-1)} = x^{(3)}$$
$$x^{(a)} = x^{(m-2)} = x^{(2)} \qquad \text{where} \quad x^{(a)} = x^{(b)} - \Delta x$$
$$x^{(c)} = x^{(m+1)} \qquad x^{(c)} = x^{(b)} + \Delta x.$$

Step 6:
$$x^* \approx \tilde{x}^* = x^{(b)} + \frac{\Delta x\{y[x^{(a)}] - y[x^{(c)}]\}}{2\{y[x^{(a)}] - 2y[x^{(b)}] + y[x^{(c)}]\}}$$
$$= 3.5 + \frac{2(6.75 - 18.75)}{2(6.75 - 2 \times 0.75 + 18.75)}$$
$$= 3$$

At this point, no $y[x^{(c)}] < y[\tilde{x}^*] \Rightarrow \tilde{x}^* = 3$ is the final answer.

4-12 POWELL'S SEARCH

In Powell's algorithm a quadratic approximation is carried out using the first three points obtained in the direction of search, and these quadratic approximations are continued until the minimum of $y(x)$ is located to the required precision. We will assume that the minimum is bracketed. Powell's algorithm proceeds as follows (Himmelblau):

Step 1: From the base vector x^1, compute $x^2 = x^1 + \Delta x$.
Step 2: Compute $y(x^1)$ and $y(x^2)$.

Step 3: If $y(x^1) > y(x^2)$, let $x^3 = x^1 + 2\Delta x$.
If $y(x^1) < y(x^2)$, let $x^3 = x^1 - \Delta x$.

Step 4: Compute $y(x^3)$.

Step 5: Estimate the value of x at the minimum of $y(x)$, x^*, by

$$\hat{x}^* = \frac{1}{2}\frac{[(x^2)^2 - (x^3)^2]y(x^1) + [(x^3)^2 - (x^1)^2]y(x^2) + [(x^1)^2 - (x^2)^2]y(x^3)}{(x^2 - x^3)y(x^1) + (x^3 - x^1)y(x^2) + (x_1 - x^2)y(x^3)}$$

Step 6: If $\hat{x}^*$ and whichever of x^1, x^2, x^3 corresponding to the smallest $y(x)$ differ by less than the prescribed accuracy in x [or (x)], terminate the search. Otherwise, evaluate $y(x^*)$ and discard from the set x^1, x^2, x^3 the one that corresponds to the greatest current value of $y(x)$; unless the bracket on the minimum of $y(x)$ will be lost by so doing, in which case discard the x so as to maintain the bracket.

Step 7: Let $x^1 = x^*$, and go back to step 1. Eventually, it will be necessary to successively reduce Δx and the algorithm continues until the desired precision listed in step 6 is obtained. The Powell search is illustrated in the following example.

Example 4-3: Powell's Search

Minimize: $y(x) = 3x^2 - 18x + 27$

	$x^{(k)}$	$y[x^{(k)}]$
$k = 1$	0	27
$k = 2$	0.5	18.75
$k = 3$	1	12
$\tilde{x}_1^*$	3	0
$\tilde{x}_2^*$	3	0

Step 1: Let the accuracy in $x = 0.01$

accuracy in $y[x] = 0.01$.

$x^{(1)} = 0,\ \Delta x = 0.5$

$x^{(2)} = x^{(1)} + \Delta x = 0.5$

Step 2: $y[x^{(1)}] = 27$

$y[x^{(2)}] = 18.75$

Step 3: $y[x^{(1)}] > y[x^{(2)}]$ Let $x^{(3)} = x^{(1)} + 2\,\Delta x$

$$= 0 + 2x - 5 = 3.$$

Step 4: $y[x^{(3)}] = 12$

Step 5:

$$x^* = \frac{1\{[x^{(2)}]^2 - [x^{(3)}]^2\}y[x^{(1)}] + \{[x^{(3)}]^2 - [x^{(1)}]^2\}y[x^{(2)}] + \{[x^{(1)}]^2 - [x^{(2)}]^2\}y[x^{(3)}]}{2[x^{(2)} - x^{(3)}]y[x^{(1)}] + [x^{(3)} - x^{(1)}]y[x^{(2)}] + [x^{(1)} - x^{(2)}]y[x^{(3)}]}$$

$$= \frac{1(0.5^2 - 1^2)\cdot 27 + (1 - 0)\cdot 18.75 + (0 - 0.5^2)\cdot 12}{2(0.5 - 1)\cdot 27 + (1 - 0)\cdot 18.75 + (0 - 0.5)\cdot 12} = 3$$

Step 6: $y[x^*] = 0$ accuracy $x > 0.01$

discard $x^{(1)}$ accuracy in $y[x] > 0.01$

$$\text{Step 7:}\quad x^* = \frac{1}{2} \cdot \frac{(0.5^2 - 1^2) \cdot 0 + (1 - 3^2) \cdot 18.75 + (3^2 - 0.5^2) \cdot 12}{(0.5 - 1) \cdot 0 + (1 - 3) \cdot 18.75 + (3 - 0.5) \cdot 12}$$

$$= 3 \Longrightarrow x^* = 3$$

4-13 A COMBINATION ALGORITHM

Notice that DSC search does not require the optimum to be bracketed; however, it moves rapidly (Δx increases) to bracket the optimum. Powell's method is superior, but Powell's method requires the minimum to be bracketed. G. E. P. Box, Davies, and Swann recommended that these algorithms be combined. Initially, we perform a single stage of the DSC method to obtain a bracket on x^*; then we switch to Powell's algorithm. Thus, we benefit from the advantages of both techniques. The reader will be asked to perform this technique in Exercise 4-15.

4-14 SIMULTANEOUS SEARCHES

In prior discussions, we were examining *sequential search* methods; that is, only one experiment is performed at a time and we are able to benefit from the information gained from this experiment before proceeding. It should be clear that this is the best condition. However, the situation may easily arise where this is not possible. We may have to conduct *all* experiments *simultaneously*. For example, the occurrence may be a "one-time occurrence" such as the "eclipse of the sun in the Okefenokee Swamp," or it may be that the setup cost of the experiments is so high that we can only afford it once. We refer to the procedure for this type of optimization as *simultaneous search*. As before, define:

$$\text{resolution} = \epsilon I^0 = \delta I^K \qquad (I^K \equiv \text{final interval})$$

and let P be the number of *pairs* of experiments in K.

Even plan (*uniform pair search*)

The minimax approach suggests that we locate the experiments such that *all* possible final intervals of uncertainty are equal. For an even number of experiments, this yields the following construction (considering the resolution).

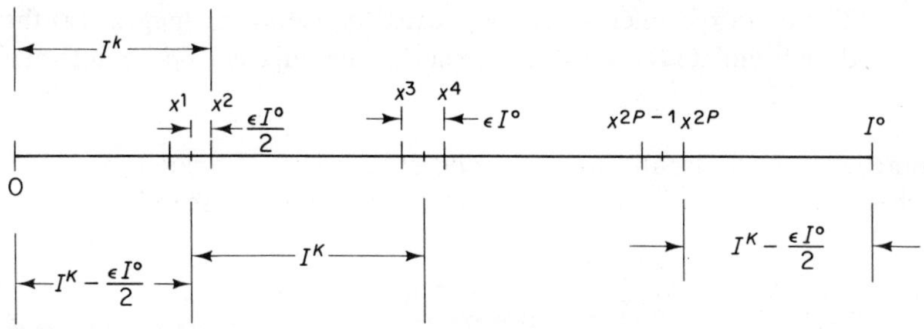

Each pair is a distance ϵI^0 (the resolution) apart from one another, and the pair *centers* are all I^K *apart*. Thus

$$x^2 = I^K; \quad x^4 - x^2 = I^K; \quad \ldots; \quad x^{2P} - x^{2P-2} = I^K$$

If we sum the sequence above:

$$x^{2P} = PI^K$$

since $I^0 = x^{2P} = I^K - \epsilon I^0$ from the figure by substitution. $I^0 - PI^K = I^K - \epsilon I^0$. Thus

$$\boxed{\frac{I^0}{I^K} = \frac{1 + P}{1 + \epsilon}} \qquad \text{(reduction ratio)}$$

or, in terms of δ:

$$\frac{I^0}{I^K} = P + I - \delta$$

Odd plan

Again, for the minimax plan we should place the experiments such that all final intervals, I^K, are equal. Thus

$$I^2 = I^K; \quad I^4 - I^2 = I^K; \quad \ldots; \quad x^{2P} - x^{2P-2} = I^K \text{ and } I^0 - x^{2P} = I^K$$

Summing these yields

$$\boxed{\frac{I^0}{I^K} = P + 1} \qquad \text{(reduction ratio)}$$

The plan is as follows:

1. Locate the *even*-numbered experiments at

$$x^{2h} = hI^K; \qquad h = 1, 2, 3, \ldots, p$$

2. The odd experiments may be located *anywhere* as long as (a) they do not violate the resolution, and (b) the adjacent *odd* points must be $\leq I^K$.

In practice, we will usually space the odd experiments I^K apart to lessen the chances of making a mistake. One possibility is a uniform search:

$$x^j = \frac{j}{2} I^K; \qquad j = 1, 2, \ldots, K$$

4-15 COMPARISON OF EVEN VERSUS ODD PLANS

The odd plan approaches the even plan as we move x_i closer to x_{i+1} (i is odd); that is, the $(2P + 1)$st experiment becomes less useful.

If we run a search of $2P$ experiments (even plan) and one of $2P + 1$ experiments (odd plan), we can begin with an interval δI^K larger for the odd plan and get the same final interval. Or, more simply, the reduction ratio is δ larger for the odd plan. Since δ is generally very small in practice, *odd plans are rarely desirable.*

4-16 THE MAXIMUM NUMBER OF EXPERIMENTS REQUIRED FOR A SIMULTANEOUS SEARCH[1]

Note that it must be true that

$$x^1 \geq \epsilon I^0, \quad x^2 - x_1 \geq \epsilon I^0, \quad \ldots, \quad x^K - x^{K-1} \geq \epsilon I^0$$

and

$$I^0 - x^K \geq \epsilon I_0$$

Summing these conditions yields the following result:

$$I^0 \geq (K + 1)\epsilon I^0$$

or

$$K \leq \frac{1}{\epsilon} - 1$$

Since K is an *integer*:

$$\frac{1}{\epsilon} - 2 < K \leq \frac{1}{\epsilon} - 1$$

This defines the number of experiments that may be run.

[1]This applies to an even or to an odd plan.

4-17 MULTIVARIABLE ELIMINATION

After seeing how effectively the elimination techniques cut down an interval of uncertainty when there is a single independent variable, one might ask how they behave in multivariable situations. The answer is disappointing, and scant space will be devoted to them. The principal difficulty is that it is much harder to eliminate portions of a square, cube, or multidimensional generalization than it is to cut away parts of a line. In the multivariable case, the class of functions on which elimination techniques can be used is more restricted than when there is but one variable. Moreover, the region of uncertainty remaining after several eliminations may be awkward to describe mathematically. In practice, multivariable problems are usually attacked by the "hill-climbing" methods o˜ gradient optimization rather than by elimination techniques. Since, however, the data gathered to implement a hill-climbing procedure can often be used to eliminate part of the region, elimination techniques are discussed briefly here.

At present there are two principal multivariable elimination methods. The first, called the *contour tangents* procedure, closes in more rapidly on the optimum but produces irregularly shaped regions of uncertainty hard to visualize in multidimensional space. The second, called the *multivariable dichotomous* procedure, gives regions of uncertainty which are hypercubes (the multidimensional generalization of the square) of predictable size, so that it is almost the only multivariable technique with measurable effectiveness. But this efficiency, although measurable, is not very high.

The descriptions to follow are shortened by treating only cases where the objective is differentiable and the first partial derivatives can be computed directly and accurately. In circumstances where such derivatives are not available, one can adapt these methods in ways described at the beginning of Chapter 5.

4-18 CONTOUR TANGENTS

Both multivariable elimination schemes are based on properties associated with a first-order approximation of the objective function in the neighborhood of a point x^0, which is a vector of feasible values of the independent variables. Let the first partial derivatives $\partial y/\partial x_i$ be evaluated and assembled into a *gradient vector* ∇y.

$$\nabla y \equiv \left(\frac{\partial y}{\partial x_1}, \ldots, \frac{\partial y}{\partial x_N}\right) \qquad (4\text{-}44)$$

Let $\Delta \mathbf{x}$ be a vector of perturbations about $\mathbf{x}^0$.

$$\Delta \mathbf{x} \equiv (\Delta x_1, \ldots, \Delta x_N) \equiv (x_1 - x_1^0, \ldots, x_N - x_N^0) \qquad (4\text{-}45)$$

Then a first-order approximation of the change Δy in the objective resulting from this perturbation is

$$\Delta y = \nabla y \, \Delta \mathbf{x}' = \sum_{n=1}^{N} \frac{\partial y}{\partial x_n} \Delta x_n \tag{4-46}$$

Suppose that y is to be maximized; then in the neighborhood of $\mathbf{x}^0$, perturbations that improve the objective satisfy the inequality

$$\Delta y = \nabla y \, \Delta \mathbf{x}' > 0$$

Whereas undesirable perturbations are those satisfying

$$\Delta y = \nabla y \, \Delta \mathbf{x}' < 0$$

The boundary between these two groups of perturbations is the set of $\Delta \mathbf{x}$ satisfying

$$\Delta y = \nabla y \, \Delta \mathbf{x}' = 0 \tag{4-47}$$

Since at $\mathbf{x}^0$ the components of ∇y are known constants, this equation is linear and describes a hyperplane of $N - 1$ dimensions passing through $\mathbf{x}^0$. This hyperplane is called the *contour tangent* because it is tangent to the contour—the curved set of points $\mathbf{x}^0$ satisfying

$$y(\mathbf{x}) = y(\mathbf{x}^0)$$

that is, having the same value of y as at $\mathbf{x}^0$. Figures 4-9 through 4-12 show contours and contour tangents for cases with two independent variables.

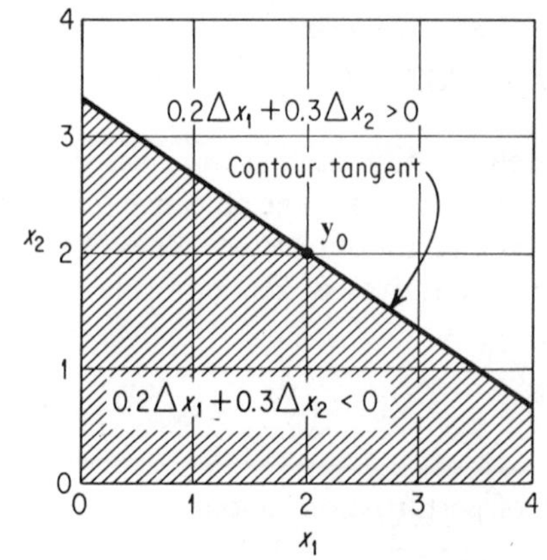

Figure 4-9. Projection of contour tangent.

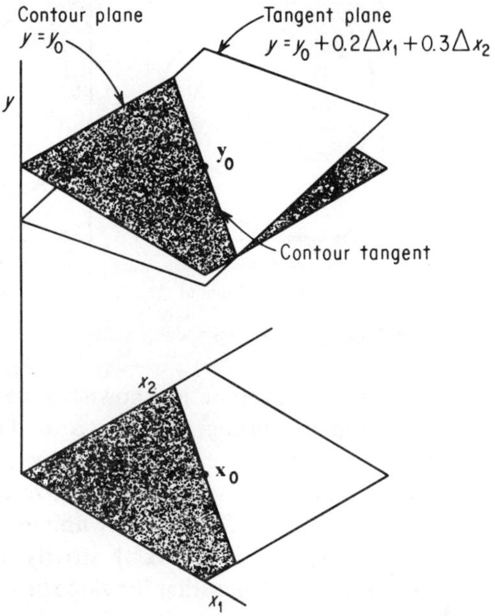

Figure 4-10. Contour tangent in space.

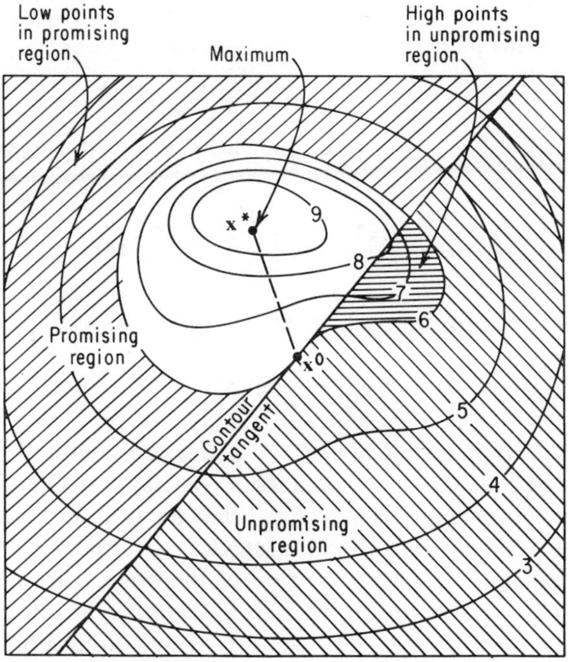

Figure 4-11. Strongly unimodal function.

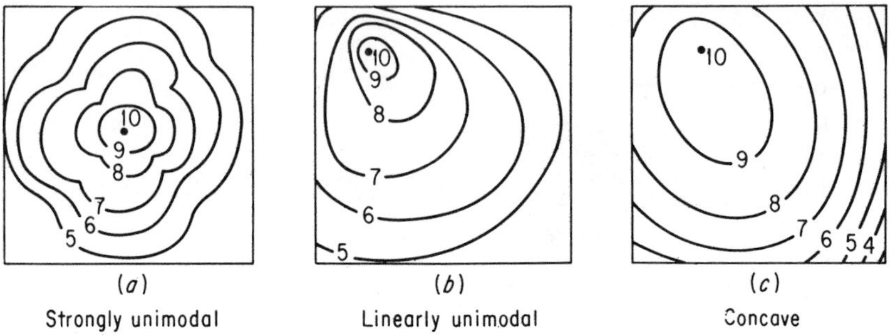

Figure 4-12. Strongly unimodal functions.

Suppose that the objective function is known to be differentiably unimodal on the straight line passing through the point $\mathbf{x}^0$ and the peak at $\mathbf{x}^*$. Since all points on this line are given by $\mathbf{x}^0 + \lambda(\mathbf{x}^* - \mathbf{x}^0)$, where λ is a parameter taking on any real values (positive, negative, or zero), the assumption is that the function $y[\mathbf{x}^0 + \lambda(\mathbf{x}^* - \mathbf{x}^0)]$ is strictly unimodal in the single variable λ. This means that $y[\mathbf{x}^0 + \lambda(\mathbf{x}^* - \mathbf{x}^0)]$ strictly increases with increasing λ (up to $\lambda = 1$), and in particular that the slope at $\mathbf{x}^0$ is positive.

$$\frac{\partial y[\mathbf{x}^0 + \lambda(\mathbf{x}^* - \mathbf{x}^0)]}{\partial \lambda} > 0 \tag{4-48}$$

But since along this path

$$\Delta \mathbf{x} = \lambda(\mathbf{x}^* - \mathbf{x}^0) \tag{4-49}$$

differentiation of Eq. (4-46) with respect to λ gives, after substitution of Eq. (4-49),

$$\frac{\partial y[\mathbf{x}^0 + \lambda(\mathbf{x}^* - \mathbf{x}^0)]}{\partial \lambda} = \nabla y(\mathbf{x}^* - \mathbf{x}^0)' \tag{4-50}$$

Thus Eqs. (4-48) and (4-50) together establish that

$$\nabla y(\mathbf{x}^* - \mathbf{x}^0)' > 0 \tag{4-51}$$

This means that the maximum must be above the contour tangent, and Eq. (4-51) can be added to the original set of constraints defining the feasible region $\mathfrak{F}$. Doing this eliminates all points below the contour tangent because they violate Eq. (4-51). Figure 4-11 illustrates this situation.

This elimination of a large fraction of the original feasible region is prudent only if one can be sure that the objective is differentiably unimodal on the straight line from $\mathbf{x}^0$ to $\mathbf{x}^*$. Fortunately, many objective functions have this property at every point, in which case they are said to be *strongly unimodal.* Such functions (Figs. 4-11, 4-12, and 4-13) lend themselves to an iteration of the elimination procedure which is called the method of *contour tangents* (Wilde, 1963; 1964, chap. 4). One determines the contour tangent at a point **x** in the interior of the feasible region, as at point 1 in Fig. 4-13, and cuts down the region of uncertainty. Then another contour tangent is measured at a point inside the remaining region (point 2 in Fig. 4-13). Repetition of this procedure gives successively smaller regions irregular in shape. Five applications of the method are illustrated in Fig. 4-13.

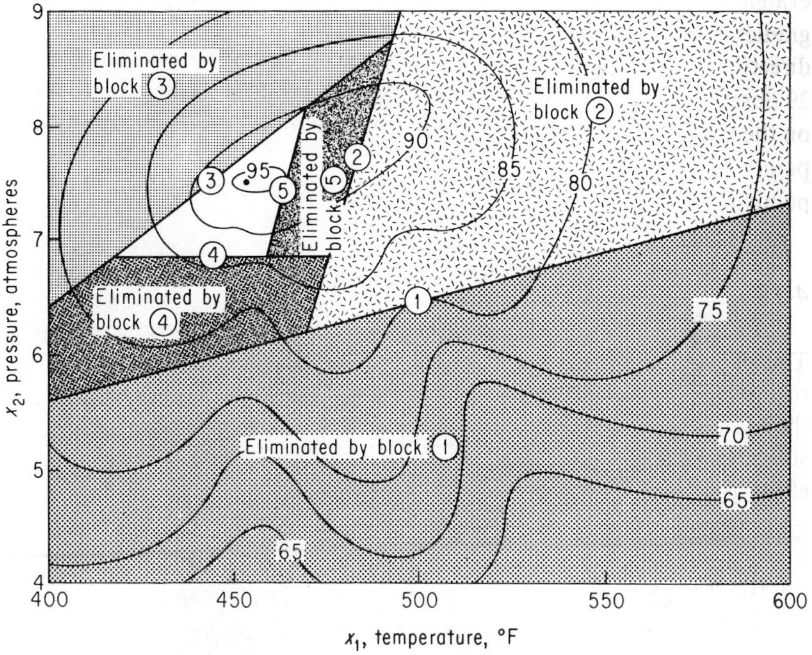

Figure 4-13. Contour tangent elimination.

The new point inside a cut-down region might be chosen in several ways (four are given in Wilde, 1964), but discussion of such details is not warranted here because the method has not really been developed to the point of usefulness. Although the contour tangents procedure should reduce the region rapidly, it gives regions of such irregular shape that the area remaining cannot be predicted in advance. Thus there is no way to specify the "optimal" location of a new point.

The only quantitative result to date concerning the method of contour tangents is that of Newman, who locates each new point at the centroid of the region remaining. He requires the objective function to be strictly unimodal on every straight line, not just on those rays passing through the maximum. Such functions are called *quasi-concave* because each contour encloses a convex set of points (Arrow and Enthoven). Newman also considered the variant of the problem in which the function can be evaluated only on a finite set of points on a square lattice with k points in each dimension. He showed that, in the two-dimensional case, application of the contour tangents method using centroids will take less than

$$10 \ln (k + 1) + 6 \tag{4-52}$$

evaluations of the function, where measurement of the function y and its gradient ∇y is counted as three evaluations in all. For 99 grid points in each direction (9801 points in all), the method could find the best after fewer than 29 observations (see Exercise 4-4). Newman has indicated that upper bounds on the number of experiments grow as the logarithm of the number of lattice points k in each dimension. This is comforting because the total number of points is k^N.

4-19 SUMMARY: PART I

This part has dealt with direct optimization methods, which narrow down the region containing the optimum until it is acceptably small. Although the theory is almost complete for unimodal functions of a single variable, it is still primitive in the multivariable case. As far as economy of experimental effort is concerned, sequential search plans demonstrate great superiority over simultaneous ones. Indistinguishability and its effect on resolution problems were also discussed. Lacking prior knowledge of resolution, one can use the Fibonacci scheme, historically the first of all plans to be developed. The golden section method can be used when the number of observations is not specified in advance.

When the interval containing the optimum can be established in advance, the Fibonacci and golden section search schemes can be utilized. A viable alternative when an unbracketed search is required was discussed in Section 4-10. Once the optimum is bounded, extrapolation and interpolation methods can also be employed (DSC/Powell searches).

Multidimensional elimination schemes depend on the properties of contour tangents when special forms of unimodality can be assumed. The contour tangent method chooses a point and finds the orientation of the contour tangent through it. Elimination methods can often be used in conjunction with direct climbing techniques, which will now be developed.

PART II: Direct Climbing Procedures

How shall I reach the top?
No time for thus reflecting!
Start to climb!

FRIEDRICH NIETZSCHE, *Excelsior*, 1882

Direct elimination procedures having been developed, it is now time to follow Nietzsche's advice and "start to climb." Any direct optimization scheme using past information to generate better points is called a *climbing* procedure. From another point of view, the problem is to reach a specified minimum acceptable level of performance in as few trials as possible. Geometrically speaking, we would like to climb as quickly as possible, even though the only information we have about the mountain comes from the past experiments we have run.

Each climbing experiment has two purposes: (1) to attain an improved value of the objective, and (2) to give information useful for locating future experiments where desirable values are likely to be found. Throughout the search we must continually be deciding whether to climb or to explore. If we expend all our experiments on exploration, we may guess where the peak is but have no measurements near it. On the other hand, it would seem shortsighted to try to reach the top without exploring at all, for we might well end far from the peak through ignorance of the behavior of the function. We need a master search plan, or strategy, that properly combines achievement with exploration.

The character of the strategies considered here will change as the search progresses. At the beginning when nothing at all is known about the function, we must explore in some small region, chosen at random, so that we may place our next experiments where the objective is higher. In the middle of a search, after having left the very low regions behind, we try to climb as fast as possible, exploring only when strictly necessary to guide our successive jumps. Toward the end of the search, when we are finally near the top, extensive exploration may be needed to attain any increase in elevation, the slope of the response surface often being slight near the maximum. Another reason for surveying rather closely the region near a supposed maximum is to check whether it really is a maximum or not.

Thus, multidimensional search strategy, like chess strategy, seems to have three phases: opening game, middle game, and end game. The opening moves set the stage; the middle ones push for advantages; the final ones strive to reach the goal. We shall see that the various search schemes differ from each other only in the middle phase, for each strategy always begins and ends with an exploration. This strategic approach, in which the experimenter changes tactics as the search moves on, seems to work well in practice, according to empirical studies by Flood and Leon, and of Lapidus et al.

Since climbing methods are designed to cope with problems involving many independent variables, a thorough understanding of multivariable functions is in order. There are two principal ways of studying such functions. The more general method involves expressing the dependent variable in terms of the independent variables, using algebraic equations. According to this approach, all analysis of the function is carried out by studying the equations. A second technique is to represent the function graphically, using Cartesian coordinates. This latter method often gives deep insight into the nature of the function because it brings to bear the geometric intuition and visual experience more or less well developed in all of us. Unfortunately, graphical techniques are not practical when there are more than two independent variables. Thus we face the unhappy choice between geometric methods which are intuitively appealing but powerless to handle large numbers of variables, and algebraic methods which are generally applicable but difficult to visualize.

A compromise is definitely in order. We shall develop an algebraic apparatus of general applicability, each algebraic concept being illustrated geometrically for the special case in which there are exactly two independent variables. In this three-dimensional context, such geometric ideas as tangent, contour, gradient, curvature, and perpendicularity are easier to visualize than their algebraic counterparts. Once the algebra for two independent variables is understood, it can be extended quite simply to the general case, using the geometric concepts as a guide. Or, from another point of view, the geometric relations holding in three dimensions can be generalized readily in terms of their algebraic analogs.

After the algebra and geometry of multidimensional spaces have been developed, we turn to describing and analyzing specific climbing procedures. The *gradient* method, going back to Cauchy, is treated first because most of the other techniques build upon it. The undesirable tendency of gradient techniques to oscillate can be overcome by *acceleration* procedures (Forsythe and Motzkin), which themselves have developed into the *ridge-following* methods of Hooke and Jeeves, Rosenbrock, Gelfand and Tsetlin, Humphrey, and Bear. An important technique, the *deflected gradient* procedure of Fletcher and Powell, not only is good at following ridges, but also has ideal behavior on quadratic objectives, a property known as *quadratic convergence*. All the methods described in this part involve breaking a multidimensional search problem down into sequences of unidimensional optimizations solvable by the powerful elimination techniques presented in Part I.

An optimization problem arising often in empirical and numerical work is the *least-squares* problem, in which one wishes to minimize a sum of the squares of a set of functions. Gauss devised a highly effective *quadratic approximation* scheme for such problems, which Levenberg has crossed with the gradient method to give the rapidly converging *damped ascent* procedure.

The "end game" of all climbing strategies involves checking the character of any stationary point found. This involves fitting a quadratic equation to the function and applying tests which determine the nature of the solution vector. This part also gives an example showing the pitfalls of inadequate analysis of stationary points. Considerations of the second-order behavior of the objective in the neighborhood of maxima led G. E. P. Box to propose a method for continually monitoring an optimum in large systems plagued by measurement errors. This *evolutionary operations* idea led Spendley, Hext, and Himsworth to develop the *simplicial* procedure for finding and following an optimum which shifts as time passes.

Several procedures for adapting climbing procedures of the pattern search variety to constrained optimization problems have been developed by Wood, by Glass and Cooper, and by Klingman and Himmelblau. The classical methods and the Jacobian approach of Chapter 2 should also prove useful when there are equality constraints, for once the constrained gradient has been estimated, any of the unconstrained methods can be used to select new, improved points.

4-20 MULTIVARIABLE ALGEBRA, GEOMETRY, AND GRAPHICAL INTERPRETATIONS

Knowledge of the behavior of functions of many variables is useful, not only for visualizing direct experimental optimization methods, but also for understanding many other topics in applied science. Indeed, one would be hard put to find in the complicated modern world a problem of any importance involving a function of a single variable. Most of the simple problems solvable by the graphical techniques so characteristic of basic engineering and economic analysis have already been solved. We therefore need to know something about the behavior of systems of many variables, especially since high-speed computers are now available to handle the calculations. Since multidimensional geometry is rarely discussed in college analytic geometry and calculus courses, at least in the United States, we must devote some time here to this elementary but exceedingly practical topic. Fortunately, the few simple concepts needed can be developed quickly and without much effort.

It is difficult, using only line drawings on the flat pages of a book, to describe solid objects, but this is exactly what we must do before we can discuss the geometry of curves and surfaces in space. The isometric projection technique, with which most engineers are already familiar, is employed because of its simplicity and pictorial qualities. *Isometric projection* consists of representing three axes, mutually perpendicular in space, by a vertical line

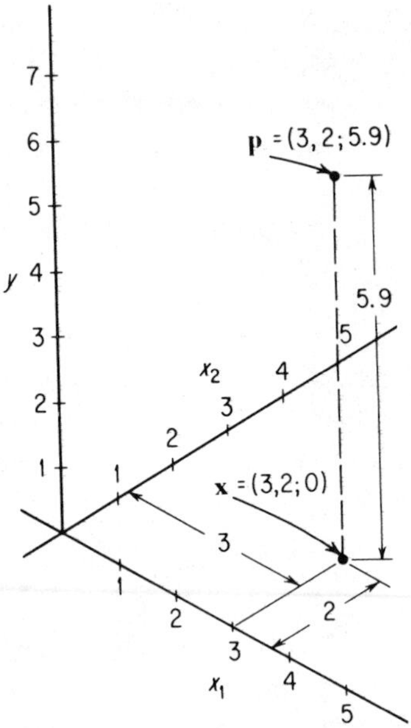

Figure 4-14. Isometric representation of a point in space.

and two slanted lines, as shown in Fig. 4-14. The slanted lines stand for the axes that are horizontal in space; the vertical line corresponds to the axis that is vertical in space. The independent variables, denoted x_1 and x_2, are plotted parallel to these horizontal axes, and values of the dependent variable y (the objective function) are measured vertically above the x_1x_2-plane.

In Fig. 4-14 the point **p** (notice the use of **boldface** type to denote points or vectors) has the coordinates $x_1 = 3$, $x_2 = 2$, and $y = 5.9$. Thus any ordered triplet of values $(x_1, x_2; y)$ can be represented graphically by a point in an isometric projection. The set of points corresponding to all possible ordered triplets forms a space of three dimensions. There are exactly *three* dimensions because at least three independent measurements, one for each coordinate, must be known to fix the location of a point.

Suppose that y is a function of two independent variables x_1 and x_2.

$$y = y(x_1, x_2)$$

In this case not all of the points in the three-dimensional space satisfy the relation. Those that do will lie on a surface floating in the space as in Fig. 4-15, where the particular function depicted is

$$y = 5 - 0.2(x_1 - 3)^2 + 0.1(x_2 + 1)^2 \tag{4-53}$$

The lines $x_1 = 0, 1, 2, 3, 4$ and $x_2 = 0, 1, 2, 3, 4$ have been drawn in the x_1x_2-plane ($y = 0$) and projected onto the surface to make it easier to visualize. Notice that the point **p**, already shown in Fig. 4-14, satisfies Eq. (4-53)

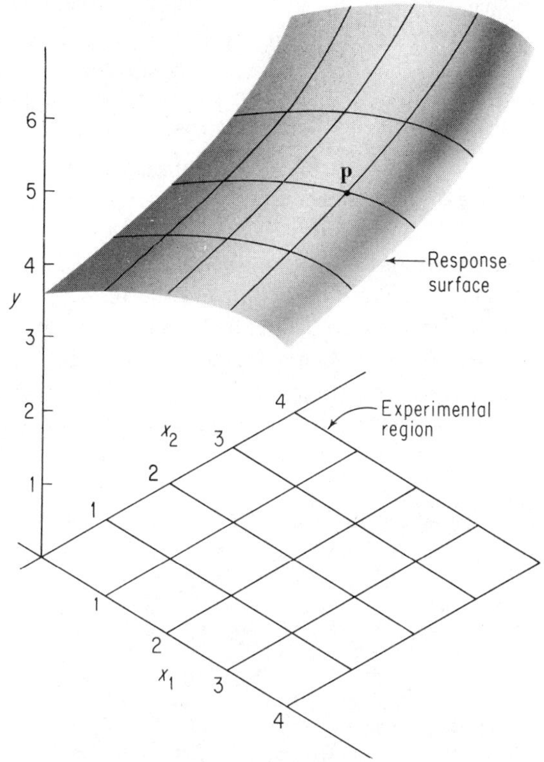

Figure 4-15. Isometric representation of a response surface.

and therefore lies on the surface. It is marked with a heavy dot at the intersection of $x_1 = 3$ with $x_2 = 2$ on the surface.

Biologists and statisticians, for whom y is often the response of a living organism to environmental factors x_1 and x_2, have come to call surfaces of this sort *response surfaces*, a name that found its way into engineering terminology because of the industrial applications of the work of G. E. P. Box (1954). The response surface of Fig. 4-15 is a two-dimensional object, because only two coordinates are needed to specify a point on it. For example, we may locate point **p** simply by stating that it lies on the response surface and then giving the two coordinates $x_1 = 3$ and $x_2 = 2$. This corresponds to the algebraic operation of finding y at the point **p** by substituting the values of x_1 and x_2 into Eq. (4-53).

In algebraic terminology the difference between the number of variables and the number of independent equations relating them is called the *number of degrees of freedom*. For the response surface there are three variables (x_1, x_2, and y) linked by Eq. (4-53), leaving two degrees of freedom. Since the number of degrees of freedom equals the number of coordinates that must be fixed to determine a point, it is the same as the number of dimensions.

Therefore, the space of all possible points $(x_1, x_2; y)$ not necessarily satisfying Eq. (4-53) has three degrees of freedom, or dimensions, whereas the response surface has only two. Geometrically speaking, a surface is a two-dimensional object embedded in a three-dimensional space. For that matter, so is the x_1x_2-plane on which $y = 0$, and the vertical plane on which $x_1 = 0$.

The intersection of two surfaces is a curve, which is one-dimensional because each surface is the representation of one equation relating the three variables. Points on the intersection belong to both surfaces and therefore must satisfy both equations, leaving only one degree of freedom. The space curves of Fig. 4-15 are at the intersections of the response surface with the equally spaced vertical planes $x_1 = 0, 1, 2, 3, 4$ and $x_2 = 0, 1, 2, 3, 4$.

By similar reasoning we can conclude that a point which is at the simultaneous intersection of three surfaces must be considered a zero-dimensional object, for it can have no degrees of freedom. The point $\mathbf{p}$ in Fig. 4-15 is at the intersection of the response surface with the two planar surfaces $x_1 = 3$ and $x_2 = 2$.

Figure 4-16 shows another way of depicting a response surface graph-

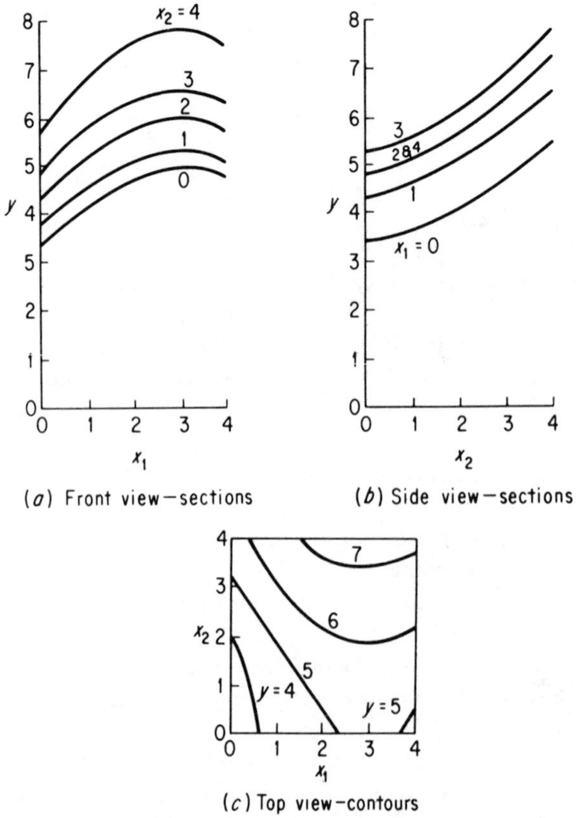

Figure 4-16. Orthogonal projection (sections) of response surface.

ically. A front view of the surface as seen from the negative side of the x_2-axis is given in Fig. 4-16(a). It shows the five space curves on the surface for $x_2 = 0, 1, 2, 3$, and 4. Similarly, the right-side view in Fig. 4-16(b) consists of the projections of the space curves for $x_1 = 0, 1, 2, 3$, and 4 onto the yx_2-plane, as they appear from the positive side of the x_1-axis. The curves showing in these views are the same as the ones in Fig. 4-15.

The curves in the top view, Fig. 4-16(c), are the projections onto the x_1x_2-plane of intersections of the response surface with the horizontal planes on which $y = 4, 5, 6$, and 7. These curves, known as *contours*, are also shown in the isometric projection of Fig. 4-17. In Fig. 4-16(c) these contours are all viewed from above. Readers who have studied mechanical drawing will recognize this technique of describing a solid object by three mutually perpendicular views as the method of *orthogonal projection.*

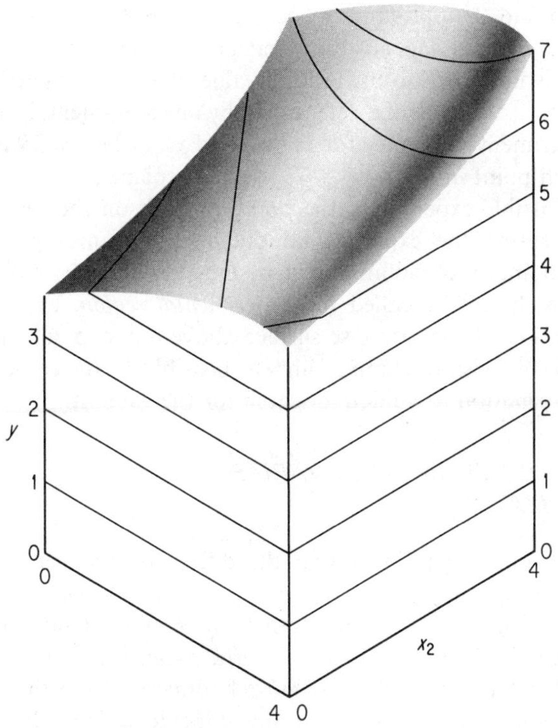

Figure 4-17. Contours in isometric representation.

Let us now generalize some of these geometric notions to situations where there are more than two independent variables. Suppose that the objective y is a function of N variables $x_1, x_2, \ldots, x_N$.

$$y = y(x_1, \ldots, x_N) \tag{4-54}$$

To represent this relation graphically one would need $N + 1$ axes all mutually perpendicular, which is physically impossible in our three-dimensional world. But even though we cannot construct such a coordinate system physically, we can certainly imagine a space of more than three dimensions if we are willing to extend the geometric concept of dimensionality by using the algebraic idea of degrees of freedom. It is customary to use the prefix *hyper-* to indicate any extension of ordinary solid geometry to spaces with more than three dimensions. Thus the term $(N + 1)$-*dimensional hyperspace* refers to the set of all possible points $(x_1, \ldots, x_N; y)$. In this terminology the set of points satisfying the single Eq. (4-54) would be an N-dimensional *hypersurface* embedded in an $(N + 1)$-dimensional hyperspace. Similarly, we may conceive of the intersection of the response hypersurface with any hyperplane on which y is constant as a *hypercontour*, an object of $N - 1$ dimensions.

The multivariable search problem can now be stated in geometric terms. It is desired to find the optimum (to be specific, say the maximum) value of some objective y which depends on N independent variables $x_1, x_2, \ldots, x_N$. This function is unknown to us, but the value of y for any particular set of values of the $x_1, \ldots, x_N$ can be determined by an experiment, as in Fig. 4-14, where the experiment conducted for $x_1 = 3$ and $x_2 = 2$ gives 5.9 for the value of y. Thus each point in the $x_1, x_2, \ldots, x_N$ hyperplane (where $y = 0$) corresponds to a possible experiment, the point above it on the response hypersurface representing the experimental outcome. Commonly, the possible experimental points are confined to a bounded portion of the $x_1, \ldots, x_N$ hyperplane which will be called the *experimental region*. Each experiment gives the elevation of the response surface above a new point in the experimental region. We wish to climb as high as possible on the response surface, using past information to guide the search for the summit.

4-21 DIFFICULTIES IN MULTIVARIABLE OPTIMIZATION

At first glance one might think that the difference between multivariable search problems and the single-variable ones already analyzed is only one of degree, that with a little extra calculation one could extend single-variable methods to multivariable problems. Unfortunately, this is not true; multivariable problems having a structure entirely different from that of a single-variable problem. Bellman refers to the difficulties engendered by these differences as "the curse of dimensionality." This "curse" takes two forms in the problems of interest here.

One deleterious effect of multidimensionality is that we will be unable to find a measure of search effectiveness that does not depend on the experimenter's luck in some way. Recall that the minimax methods developed for unimodal functions of a single variable gave the same length of final interval of uncertainty no matter where the peak happened to be. To see why the measures of effectiveness used previously are not valid in multidimensional problems, consider the two experiments **a** = (3, 2) and **b** = (1, 1) used to find, in the experimental region $0 \leq x_1 \leq 4$, $0 \leq x_2 \leq 4$, the maximum of the function given in Eq. (4-53). The values of y, respectively 5.9 and 4.6, are shown in Fig. 4-18. The function is unimodal above the straight line passing

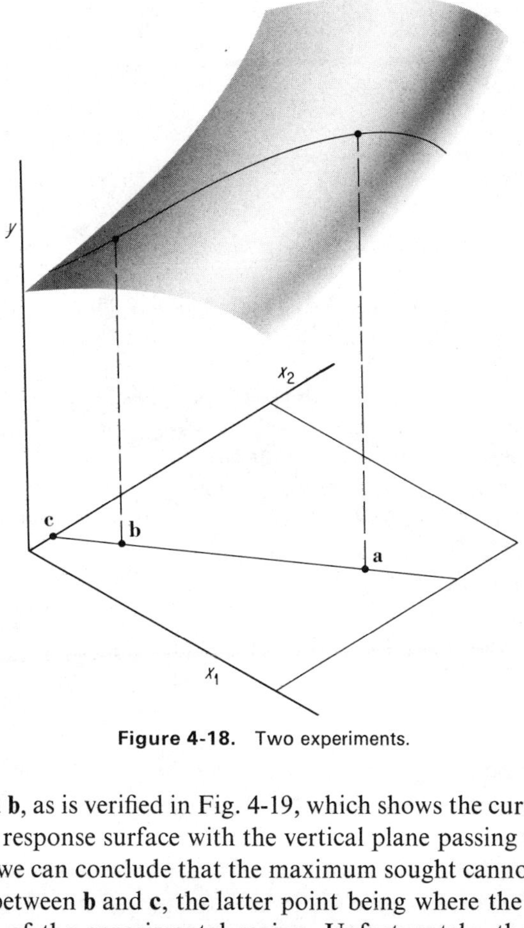

Figure 4-18. Two experiments.

through **a** and **b**, as is verified in Fig. 4-19, which shows the curve at the intersection of the response surface with the vertical plane passing through **a** and **b**. Therefore, we can conclude that the maximum sought cannot be above the line segment between **b** and **c**, the latter point being where the line intersects the boundary of the experimental region. Unfortunately, the line segment eliminated is negligible compared to the two-dimensional experimental region remaining. Thus multidimensionality overpowers the search techniques that were so effective when we had only lines to deal with.

The other difficulty is the very vastness of multidimensional spaces. This is a subtle point, difficult to visualize. To overcome our lack of multidimensional experience, let us attempt to imagine what happens as we pass from a one-dimensional line to a two-dimensional square and then to a three-dimensional cube.

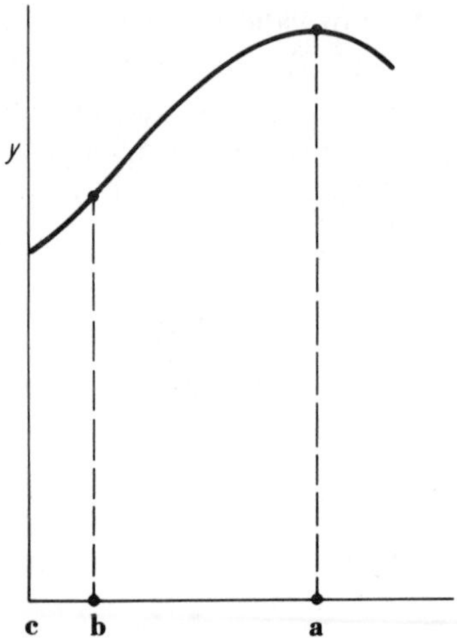

Figure 4-19. Vertical section along line through **a** and **b**.

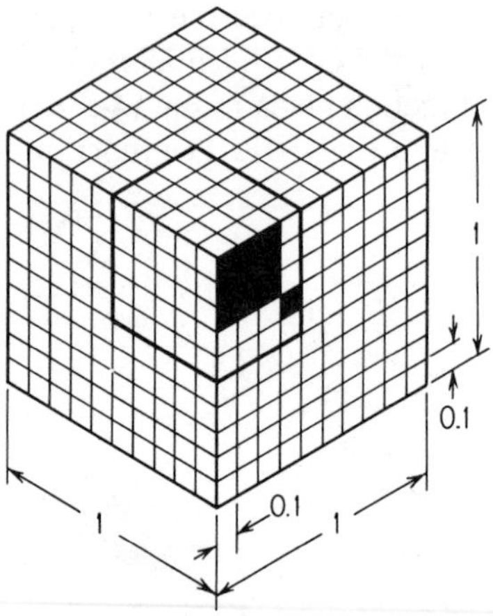

Figure 4-20. The second curse of dimensionality.

Consider a line of unit length, the top edge of the cube in Fig. 4-20. Suppose that we had a search method which would leave us with a final interval of uncertainty of 0.1, only one-tenth of the initial length. Such a technique would in effect select, from among 10 equally sized segments of the line, the interval containing the optimum, and this interval looks relatively small compared to the original one of unit length.

Next imagine a unit square divided into squares, each measuring 0.10 unit on the side, as one of the faces of the cube of Fig. 4-20. The darkened region shown comprises 10 of these 100 squares—again, one-tenth of the original region. Somehow, this tenth of a square looks larger than the tenth of the line.

This effect is more marked in the unit cube of Fig. 4-20, which has been divided into 1000 cubes each measuring 0.10 unit on a side. To encompass 10% of this region we must take 100 cubes, in Fig. 4-20 a volume measuring $0.4 \times 0.5 \times 0.5$. We see that a 10% interval of uncertainty, which looked small enough on a line, appears rather large on a cube. If we were dealing with a function of 50 variables, the hypercube containing only 10% of the volume would measure $(0.1)^{1/50} = 0.91$ unit on a side!

This phenomenon, which may appear paradoxical at first, arises from our application of a first-degree measure, percentage, to a high-degree quantity, volume, and its multidimensional generalization. Thus, even if we had a search method guaranteed to cut down the multidimensional volume of uncertainty to a fixed fraction, this percentage would have to be extremely small before the ranges of the individual independent variables would be significantly reduced. This difficulty, which might be called the *vastness* of hyperspace, has been pointed out by Hooke and Jeeves (1958). Thus multidimensionality hurts us in two ways: it takes away the a priori measure of effectiveness that led to the powerful unidimensional minimax techniques, and it forces us to seek regions of uncertainty that are but tiny fractions of the original experimental region.

4-22 OPENING GAMBIT: ESTIMATING FIRST DERIVATIVES

Let us begin the discussion of search strategy by studying the problem of locating the first few experiments. Since we have no advance information about the objective function, the opening trials must be purely exploratory. We shall see that all they tell us is our elevation on the response surface and the way we need to move in order to go up from our initial group of experiments.

To avoid being too abstract at the very beginning, consider the particular problem of finding the maximum, in the square experimental region $0 \leq x_1 \leq 4$, $0 \leq x_2 \leq 4$, of an unknown function $y(x_1, x_2)$ of two independent variables, x_1 and x_2. Since nothing is known about the function, one point in the experimental region is as good a place to begin as any other, but to be specific, let us start at the exact center, that is, at $(2.0, 2.0) \equiv \mathbf{x}_0$. Its first coordinate will be written x_{01}; its second, x_{02}.

$$\mathbf{x}_0 = (x_{01}, x_{02})$$

The outcome of the experiment $\mathbf{x}_0$ will be denoted by y_0. Suppose that this result happens to be 5.70. The point $(x_{01}, x_{02}; y_0)$, which will be abbreviated $\mathbf{y}_0$, will be on the response surface directly above $\mathbf{x}_0$, as shown in Fig. 4-21.

Not much has been gained by this initial trial, for we still have no idea where to put the next experiment. Suppose, however, that we knew the general slope of the response surface in the neighborhood of the point $\mathbf{y}_0$. Then we would have at least a rough notion of which combination of changes in x_1 and x_2 might bring an increase in y. To find the slope in the direction parallel to the x_1-axis we need only run an experiment $\mathbf{x}_1 = (x_{11}, x_{12})$, whose second coordinate x_{12} is the same as for the initial trial. That is, we set $x_{12} = x_{02}$. The first coordinate x_{11} should, however, be slightly different from x_{01}. In order to get an accurate estimate of the slope we must place $\mathbf{x}_1$ as close as possible to $\mathbf{x}_0$, allowing just enough distance between them to make the outcome y_1 distinguishable from y_0. Let us take, in this example,

$$\mathbf{x}_1 = (2.1, 2.0)$$

and suppose that the result is

$$y_1 = 5.72$$

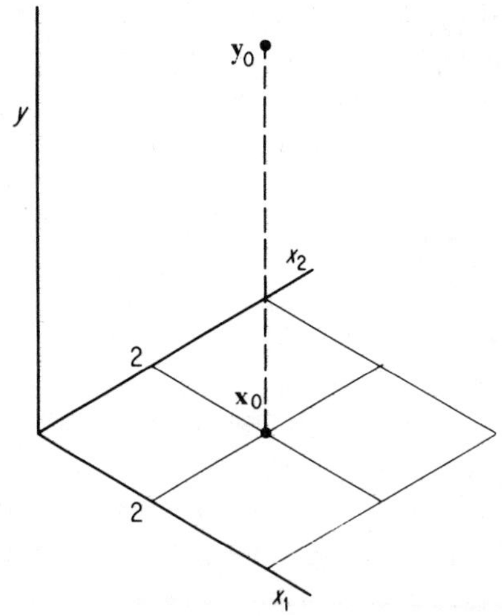

Figure 4-21. Result of first experiment.

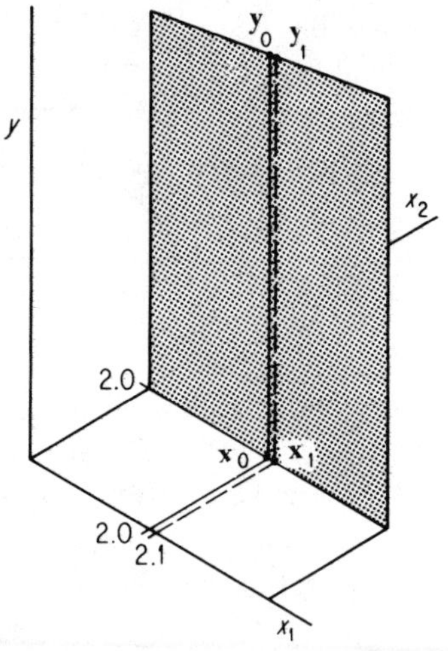

Figure 4-22. Approximate tangent to response surface at $\mathbf{y}_0$ in plane $x_2 = 2.0$.

Then the situation after the first two experiments is that shown in Fig. 4-22. The straight line through the points $\mathbf{y}_0$ and $\mathbf{y}_1$ lies entirely in the vertical plane $x_2 = 2.0$ and is approximately tangent to the response surface at $\mathbf{y}_0$. Its slope is given by

$$\left(\frac{\partial y}{\partial x_1}\right)_0 \approx \frac{y_1 - y_0}{x_{11} - x_{01}} = \frac{5.72 - 5.70}{2.1 - 2.0} = 0.2$$

where the subscript 0 indicates that the partial derivative is approximated at the point $\mathbf{y}_0$.

It would seem reasonable to locate the next experiment where x_1 is larger than at $\mathbf{x}_0$ and $\mathbf{x}_1$, since the objective y seems to be increasing in this direction. One could indeed conduct a one-dimensional search to find the high point of the response surface in the plane $x_2 = 2$. From this high point we could then carry out another one-dimensional search in the x_2-direction, holding the coordinate x_1 at the value obtained on the previous search. This procedure could be continued, holding one variable constant and adjusting the other until the maximum is found. Unfortunately, as Box and Wilson have pointed out, this method, described more fully in Section 4-24, will fail to find the maximum of a function shaped like the rising ridge of Fig. 4-23. The initial search along the line $x_2 = 2$ will find a summit at the point (3, 2) where the line crosses the ridge. Attempts to search for a higher point along the line $x_1 = 3$ would be doomed to failure, since (3, 2) itself is higher than any other point on that line. Therefore, the search would terminate at (3, 2), far from the true maximum at (0, 4). The contours of this ridge are shown in Fig. 4-24.

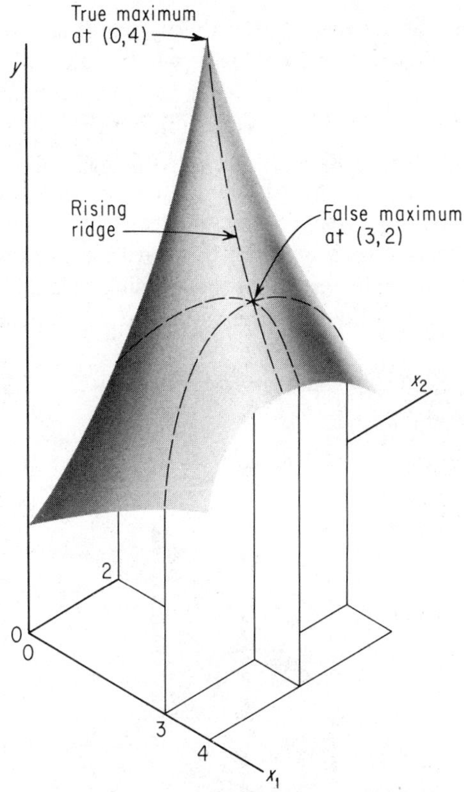

Figure 4-23. Response surface with rising ridge—isometric view.

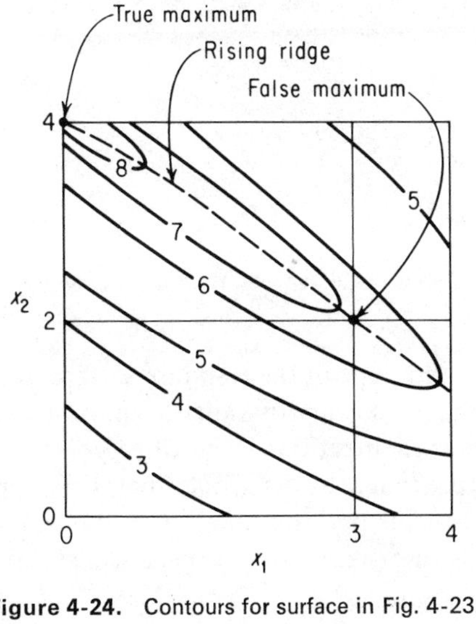

Figure 4-24. Contours for surface in Fig. 4-23.

It seems then that we should alter the x_2 coordinate at the same time as we change the coordinate for x_1. To decide whether x_2 should be increased or decreased, we must perform another exploratory experiment very close to $\mathbf{x}_0$, this time varying the x_2-coordinate while holding the x_1-coordinate constant at x_{01}. Let us place an experiment at (2.0, 1.9), to be designated $\mathbf{x}_2$. The decision to decrease the x_2 coordinate rather than to increase it for this exploratory run was, of course, completely arbitrary. If we suppose the outcome to be 5.67, then the situation after this third exploratory experiment will be as shown in Fig. 4-25. The straight line passing through points $\mathbf{y}_0$ and $\mathbf{y}_2$, approximately tangent to the response surface at $\mathbf{y}_0$, has a slope given by

$$\left(\frac{\partial y}{\partial x_2}\right)_0 \approx \frac{y_2 - y_0}{x_{22} - x_{02}} = \frac{5.67 - 5.70}{1.9 - 2.0} = 0.3$$

This tangent lies in the vertical plane $x_1 = 2.0$.

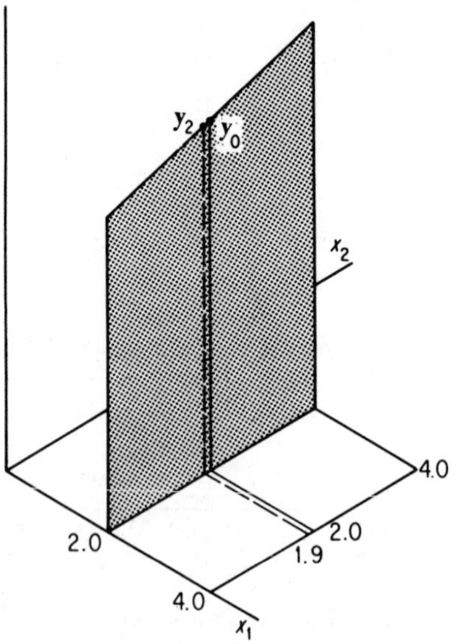

Figure 4-25. Approximate tangent to response surface at $\mathbf{y}_0$ in plane $x_1 = 2.0$.

We now know the slope of the response surface in two particular directions—parallel to the x_1-axis and parallel to the x_2-axis. What can we infer about the slope in other directions? The three points $\mathbf{y}_0$, $\mathbf{y}_1$, and $\mathbf{y}_2$ on the response surface are enough to determine the plane approximately tangent to the surface at $\mathbf{y}_0$. Let us find the equation of this tangent plane.

A plane in this three-dimensional space will satisfy an equation of the form

$$y(x_1, x_2) = m_0 + m_1 x_1 + m_2 x_2 \tag{4-55}$$

where m_0, m_1, and m_2 are constants. Notice that since a plane has two degrees of freedom, it is a special kind of surface, as our geometric experience readily verifies. The particular character of a plane follows from the special form of Eq. (4-55), which expresses y as a *linear* function of the independent variables x_1 and x_2. The word "linear" comes from the geometric fact that the intersection of two planes is a straight line. If any of the terms in Eq. (4-55) involved powers of x_1 and x_2 not equal to unity, or product of x_1 with x_2, then the equation would be *nonlinear* and the surface *curved.*

Knowing that the equation of the tangent plane must take the form of Eq. (4-55), we need evaluate only the three constants m_0, m_1, and m_2. One way to do this is to substitute into Eq. (4-55) the values of x_1, x_2, and y for the three exploratory experiments performed. This would give the following three independent equations in the three unknown coefficients m_0, m_1, and m_2:

$$y_0 = m_0 + m_1 x_{01} + m_2 x_{02} \tag{4-56a}$$

$$y_1 = m_0 + m_1 x_{11} + m_2 x_{12} \tag{4-56b}$$

$$y_2 = m_0 + m_1 x_{21} + m_2 x_{22} \tag{4-56c}$$

In the specific case under consideration, these equations would be

$$5.70 = m_0 + 2.0m_1 + 2.0m_2 \tag{4-57a}$$

$$5.72 = m_0 + 2.1m_1 + 2.0m_2 \tag{4-57b}$$

$$5.67 = m_0 + 2.0m_1 + 1.9m_2 \tag{4-57c}$$

It remains only to solve these equations simultaneously for m_0, m_1, and m_2.

In determining the equation of the tangent plane, it is convenient to deal only with the deviations of x_1, x_2, and y from the original point $\mathbf{y}_0$. For the experiment $\mathbf{x}_i = (x_{i1}, x_{i2})$, let us define the deviations

$$\Delta x_{i1} = x_{i1} - x_{01} \tag{4-58a}$$

$$\Delta x_{i2} = x_{i2} - x_{02} \tag{4-58b}$$

$$\Delta y_i = y_i - y_0 \tag{4-58c}$$

In our example, $\Delta x_{i1} = x_{i1} - 2.0$, $\Delta x_{i2} = x_{i2} - 2.0$, and $\Delta y_i = y_i - 5.70$. This translation of coordinates eliminates the need to determine the intercept constant m_0, for by subtracting Eq. (4-56a) successively from Eq. (4-56b) and (4-56c) and applying Eqs. (4-58) we obtain

$$\Delta y_i = m_1\,\Delta x_{i1} + m_2\,\Delta x_{i2} \qquad \text{for } i = 1, 2 \tag{4-59}$$

A further simplification results because $\mathbf{x}_1$ was chosen to make $\Delta x_{12} = 0$ and

$\mathbf{x}_2$ was chosen to make $\Delta x_{21} = 0$. For these special choices it is clear that

$$m_1 = \frac{\Delta y_1}{\Delta x_{11}} = 0.2 \approx \left(\frac{\partial y}{\partial x_1}\right)_0$$

$$m_2 = \frac{\Delta y^2}{\Delta x_{22}} = 0.3 \approx \left(\frac{\partial y}{\partial x_2}\right)_0$$

and therefore that the equation of the plane tangent to the response surface at $\mathbf{y}_0$ is

$$\Delta y = 0.2\,\Delta x_1 + 0.3\,\Delta x_2 \tag{4-60}$$

This tangent plane is shown in Fig. 4-26. The general equation of the tangent plane, obtained by combining Eqs. (4-55) and (4-58), is

$$\Delta y = m_1\,\Delta x_1 + m_2\,\Delta x_2 = \nabla y\,\Delta \mathbf{x}' \tag{4-61}$$

where $\nabla y \equiv (\partial y/\partial x_1, \partial y/\partial x_2)$ is the *gradient* of y.

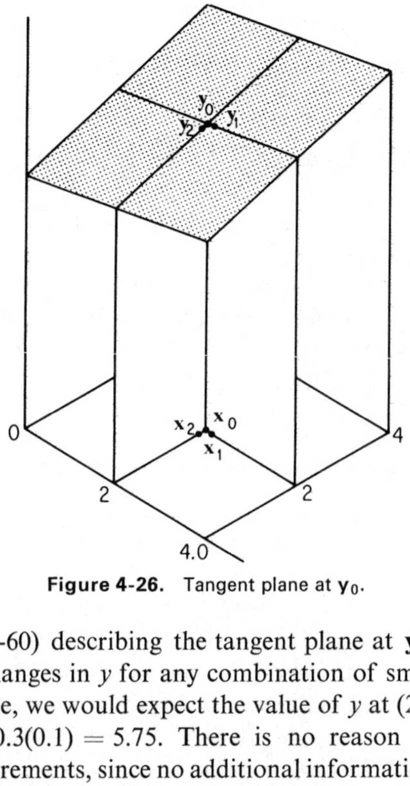

Figure 4-26. Tangent plane at $\mathbf{y}_0$.

Given Eq. (4-60) describing the tangent plane at $\mathbf{y}_0$, we could, if we wished, estimate changes in y for any combination of small deviations in x_1 and x_2. For example, we would expect the value of y at (2.1, 2.1) to be about $5.70 + 0.2(0.1) + 0.3(0.1) = 5.75$. There is no reason then to make any further slope measurements, since no additional information would be gained. Hence, in accordance with our geometric intuition, only three experiments are needed to determine the tangent plane at $\mathbf{y}_0$. Any three, not just the particular ones we chose, would have been acceptable as long as they did not all lie on the same straight line in the x_1x_2-plane. The geometric reason for this restriction is that to determine a plane through a point one must have at least two different lines through the point. From an algebraic standpoint, three points on the same line would lead to Eqs. (4-59) that would not be linearly independent and which therefore could not give a unique solution for m_1 and m_2.

To guide our location of future experiments, we shall use the tangent plane as an approximate representation of the response surface in the neighborhood of $\mathbf{y}_0$. It would be well at this point to examine the limitations on this approximation. Let us assume that the objective function y is a continuous function of x_1 and x_2 with continuous first partial derivatives everywhere in the experimental region. (This hypothesis will be weakened later.) Then we may expand y in a Taylor's series about the point $\mathbf{y}_0$ as follows:

$$\Delta y = \left(\frac{\partial y}{\partial x_1}\right)_0 \Delta x_1 + \left(\frac{\partial y}{\partial x_2}\right)_0 \Delta x_2 + 0(\Delta x)^2 \qquad (4\text{-}62)$$

where $(\partial y/\partial x_i)_0$ is the partial derivative of y with respect to x_i, evaluated at the point $\mathbf{y}_0$, and $0(\Delta x)^2$ represents terms of second order and higher. For sufficiently small deviations, these terms of higher order are negligible compared to the first-order, or linear terms, which have precisely the same form as the tangent plane Eq. (4-61). Hence the coefficients m_1 and m_2 can be interpreted as approximations to the first derivatives $(\partial y/\partial x_1)_0$ and $(\partial y/\partial x_2)_0$, or conversely, the linear terms of Taylor's expansion can be considered to be the equation of the tangent plane at $\mathbf{y}_0$. It follows that in the immediate vicinity of $\mathbf{y}_0$ the tangent plane approximates very closely the behavior of the objective y.

Generalization of the results of this section to functions of many independent variables is easy using the algebraic concepts already developed. Let $y(x_1, x_2, \ldots, x_N)$ be a function of N independent variables, let $\mathbf{x}_0 = (x_{01}, x_{02}, \ldots, x_{0N})$ be the original experiment, and let y_0 be its outcome. Define the deviations

$$\begin{aligned} \Delta x_j &= x_j - x_{0j} \qquad \text{for } j = 1, 2, \ldots, N \\ \Delta y &= y - y_0 \end{aligned} \qquad (4\text{-}63)$$

and let m_j be the derivative $(\partial y/\partial x_j)$ evaluated at y_0. The linear approximation for y at y_0 is

$$\Delta y = \sum_{j=1}^{N} m_j \,\Delta x_j = \nabla y \,\Delta x' \qquad (4\text{-}64)$$

To evaluate the coefficients m_j, one must perform N experiments (not including $\mathbf{x}_0$) and solve the N simultaneous equations

$$\Delta y_j = \sum_{j=1}^{N} \Delta x_{ij} m_j \qquad \text{for } i = 1, 2, \ldots, N \tag{4-65}$$

for the constants m_j, since the Δx_{ij} and Δy_i are given for each experiment. These computations can be simplified by choosing $\Delta x_{ij} = 0$ for all $i \neq j$, in which case Eq. (4-65) gives simply

$$m_i = \frac{\Delta y_i}{\Delta x_{ii}} \qquad \text{for } i = 1, 2, \ldots, N \tag{4-66}$$

Once the m_j are known, one can state that the combinations of Δx_j giving increased y near $\mathbf{y}_0$ must satisfy the inequality

$$\nabla y \, \Delta \mathbf{x}' = \sum_{j=1}^{N} m_j \, \Delta x_j > 0 \tag{4-67}$$

Geometrically speaking, if one considers the $(N + 1)$-dimensional space of y and the x_j, Eq. (4-64) describes the N-dimensional hyperplane tangent to the response surface at y_0. In terms of the experimental region, which is part of a space of N dimensions, the boundary between the upward region (that is, of increasing y) and the downward region is the $(N - 1)$-dimensional hyperplane satisfying the equation

$$\nabla y \, \Delta \mathbf{x}' = \sum_{j=1}^{N} m_j \, \Delta x_j = 0 \tag{4-68}$$

4-23 GRADIENT METHODS

In the preceding section, as well as throughout the early chapters, the vector of first partial derivatives $(\partial y/\partial x_1, \ldots, \partial y/\partial x_N) \equiv \nabla y$ has been given the name *gradient*. This vector gets its name because it points in the direction in which the response surface has the steepest slope. To see why this is so, consider the N-dimensional hypersphere of radius r, centered about the point $\mathbf{x}$. Points $\mathbf{x} + \partial\mathbf{x}$ on this sphere satisfy

$$\sum_{j=1}^{N} (\partial x_j)^2 = |\partial \mathbf{x}|^2 = r^2 \tag{4-69}$$

The first-order approximation of the objective function in the neighborhood of $\mathbf{x}$ gives the value of the objective function at various points on the sphere as

$$\Delta y = \nabla y \, \partial \mathbf{x}' \tag{4-70}$$

Let us seek the point on the hypersphere where Δy is maximum. At this point the following Lagrangian must be stationary:

$$L = \nabla y \, \partial \mathbf{x}' - \lambda[|\partial \mathbf{x}|^2 - r^2] \tag{4-71}$$

Hence the maximizing perturbation $\partial \mathbf{x}^*$ satisfies

$$\nabla L = \nabla y - 2\lambda \, \partial \mathbf{x}^* = 0$$

whence

$$\partial \mathbf{x}^* = (2\lambda)^{-1} \nabla y \tag{4-72}$$

Since λ, the Lagrange multiplier, is a constant, the geometric interpretation of Eq. (4-72) is that the optimal perturbation vector $\partial \mathbf{x}^*$ points in the same direction as the gradient vector. The constraint Eq. (4-69) gives λ:

$$|\partial \mathbf{x}^*|^2 = (2\lambda)^{-2} |\nabla y|^2 = r^2$$

whence

$$\lambda = \frac{|\nabla y|}{2r} \tag{4-73}$$

$$\partial \mathbf{x}^* = \frac{r \, \nabla y}{|\nabla y|} \tag{4-74}$$

and

$$\Delta y^* = r \, |\nabla y| \tag{4-75}$$

The vector $\nabla y / |\nabla y|$ is called the *normalized gradient*. All these quantities are shown graphically in Fig. 4-27 for $\nabla y = (0.2, 0.3)$, the two-dimensional example of Section 4-21.

The *gradient method* for seeking a maximum is to determine the gradient at a point $\mathbf{x}_0$. The set of points in the gradient direction is given by

$$\Delta \mathbf{x}_0 = \rho \, \nabla y(\mathbf{x}_0) \equiv \rho \, \nabla y_0 \tag{4-76}$$

where ρ is a normalized hypersphere radius given by

$$\rho \equiv \frac{r}{|\nabla y|} \tag{4-77}$$

Positive values of the normalized radius ρ give locally increasing values of y, so the value of ρ maximizing Δy is found either by a one-dimensional (Fibonacci) search or, when possible, by direct differentiation. The latter alternative involves substituting Eq. (4-76) into the objective function, differentiating with respect to ρ, setting the derivative to zero, and solving for the

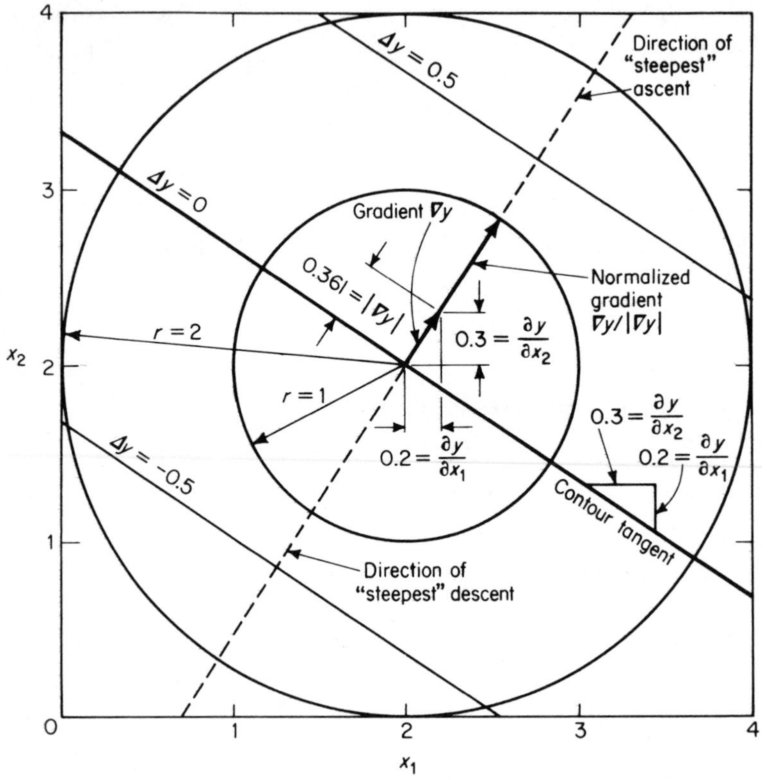

Figure 4-27. The gradient for $\Delta y = 0.2\,\Delta x_1 + 0.3\,\Delta x_2$.

minimizing value ρ^*. Thus one finds ρ^* satisfying

$$\left.\frac{\partial y(\mathbf{x} + \rho\,\nabla y)}{\partial \rho}\right|_{\rho=\rho^*} = 0 \tag{4-78}$$

At the new point $\mathbf{x}_1$, one evaluates a new gradient and iterates the gradient climbing procedure. That is,

$$\mathbf{x}_1 = \mathbf{x}_0 + \rho_0^*\,\nabla y_0 \tag{4-79}$$

$$\Delta\mathbf{x}_1 = \rho_1^*\,\nabla y_1 \tag{4-80}$$

and so on. Figure 4-28 shows the progress of this method toward the minimum in the chemical plant design problem of Chapter 2. The technique, first proposed by Cauchy in connection with solving simultaneous linear equations, was exhumed a century later by Courant for application to problems in mathematical physics. Its application to problems in industrial statistics has been due largely to the efforts of Box and Wilson.

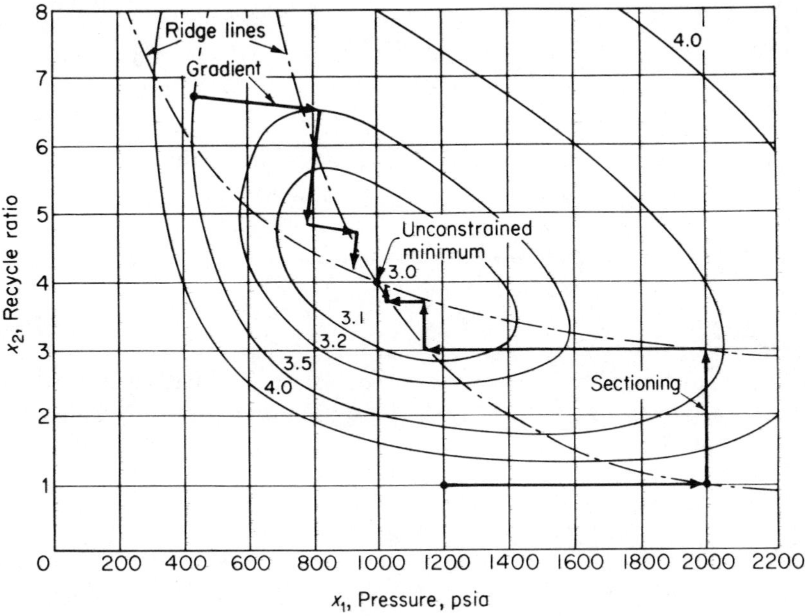

Figure 4-28. Section and gradient methods.

4-24 SCALE AND REPRESENTATION

It is instructive to study the effects of linear transformations of the independent variables, that is, changes of scale and rotation of coordinates, on the behavior of the gradient method. The effect is so marked that often a preliminary transformation of the original problem is justified by the improvement in performance it brings. Investigation of this phenomenon in the case of objectives with ellipsoidal contours suggests ways of speeding up convergence simply by thoughtful choice of the functional representative of the objective. Briefly stated, in terms as yet ill defined, the principles for preliminary preparation of a problem are three: remove interactions between independent variables, make the contours as spherical as possible by symmetric choice of scales of measurement, and represent the objective by a function well approximated by a low-order Taylor expansion in the neighborhood of the optimum. These suggestions were first made by Buehler, Shah, and Kempthorne in 1961.

First, consider the consequences of a nonsingular linear transformation of the independent variables $\mathbf{x}$. Let $\mathbf{z}$ be the N-component column vector obtained by transforming the N-component column vector $\mathbf{x}$, using the transformation whose matrix is the N by N nonsingular matrix $\mathbf{L}$. Then,

$$\mathbf{z} = \mathbf{L}\mathbf{x} \tag{4-81}$$

Since the transformation is nonsingular, each vector $\mathbf{z}$ is the image of a unique $\mathbf{x}$ which can be found by using the inverse matrix $\mathbf{L}^{-1}$.

$$\mathbf{x} = \mathbf{L}^{-1}\mathbf{z} \tag{4-82}$$

Let the gradient method be applied in the image space of the $\mathbf{z}$. The points in the gradient direction are given by the chain rule as

$$\begin{aligned}\Delta\mathbf{z} &= \rho_z \nabla y(\mathbf{z}) \\ &= \rho_z \frac{\partial \mathbf{x}}{\partial \mathbf{z}} \frac{\partial \mathbf{y}}{\partial \mathbf{x}} = \rho_z (\mathbf{L}^{-1})' \nabla y(\mathbf{x})\end{aligned} \tag{4-83}$$

where ρ_z is the appropriate parameter in $\mathbf{z}$-space. This path may be transformed back into the $\mathbf{x}$ space to see how its points compare with those along the gradient there.

$$\Delta\mathbf{x} = \mathbf{L}^{-1}\,\Delta\mathbf{z} = \rho_z \mathbf{L}^{-1}(\mathbf{L}^{-1})' \nabla y(\mathbf{x}) \tag{4-84}$$

Since the gradient path in the $\mathbf{x}$ space is

$$\Delta\mathbf{x} = \rho_x \nabla y(\mathbf{x}) \tag{4-85}$$

the two paths are the same if and only if

$$\mathbf{L}^{-1}(\mathbf{L}^{-1})' = (\rho_z^{-1}\rho_x)\mathbf{I} \tag{4-86}$$

where $\mathbf{I}$ is the N by N unit matrix. In geometric terms this means that unless the linear transformation is orthogonal, the two directions of "steepest" ascent will be different. All other types of transformation, such as unequal changes of scale, are forbidden.

The effect of compressing the x_2-axis by a factor of 2 in the example illustrated in Fig. 4-27 is shown in Fig. 4-29. Here

$$\mathbf{L} = \begin{pmatrix} 1 & 0 \\ 0 & \frac{1}{2} \end{pmatrix}$$

and the inverse is

$$\mathbf{L}^{-1} = \begin{pmatrix} 1 & 0 \\ 0 & 2 \end{pmatrix}$$

In the image space,

$$\begin{aligned}\Delta y &= 0.2\,\Delta x_1 + 0.3\,\Delta x_2 \\ &= 0.2\,\Delta z_1 + 0.3(2\,\Delta z_2) \\ &= 0.2\,\Delta z_1 + 0.6\,\Delta z_2\end{aligned}$$

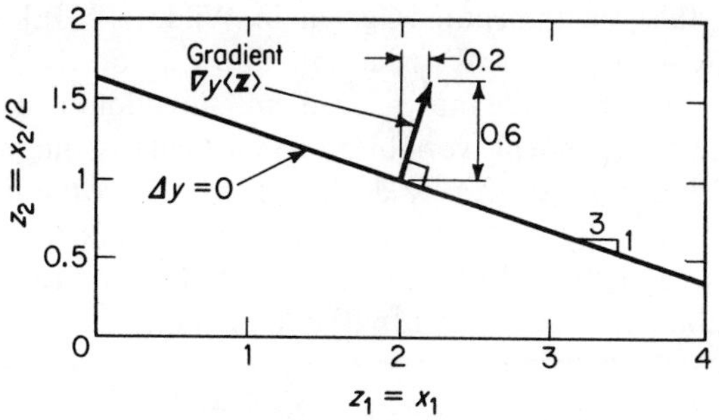

Figure 4-29. The gradient for $z_1 = x_1$; $z_2 = x_2/2$; $\Delta y = 0.2\,\Delta z_1 + 0.6\,\Delta z_2$.

whence

$$\nabla y(z) = (0.2, 0.6)'$$

as predicted by Eq. (4-83). Transformation of the path

$$\Delta z = p_z(0.2, 0.6)'$$

back into the **x** space gives

$$\Delta x_1 = \Delta z_1 = 0.2p_z$$
$$\Delta x_2 = 2\,\Delta z_2 = 1.2p_z$$

which is clearly not the gradient direction, as shown in Fig. 4-30. [A fuller

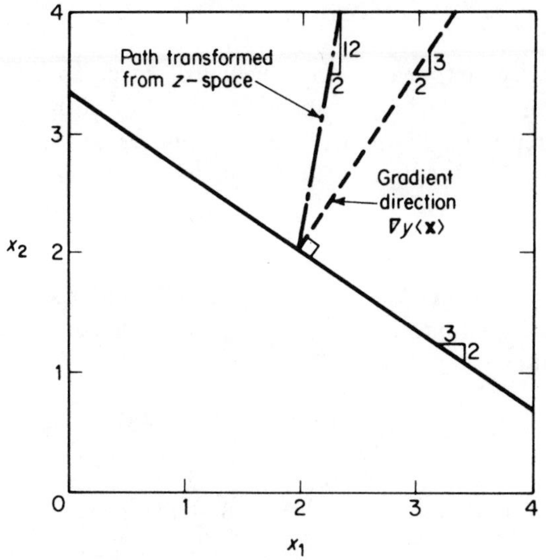

Figure 4-30. Inverse transformation.

discussion of this phenomenon is given in Wilde (1964).] The important thing to realize at the moment is that *any* direction on the high side of the contour tangent can be made into a "gradient" direction by a linear transformation of the independent variables. A particularly important class of problems which can be effectively dealt with are *quadratic functions*. These problems arise naturally in economic portfolio selection procedures, approximation techniques, and as theoretical examples of convergence properties. A simple quadratic function is given by the following:

$$y = \mathbf{z}'\mathbf{z} = \sum_{i=1}^{N} z_i^2$$

Note that there are no cross-product terms $z_1 z_2 \ldots$, which means that the change in y resulting from adjusting a given variable z_i does not depend on what values the other variables z_j $(j \neq i)$ happen to have. When this happens, the variables are said to be *noninteracting*. Moreover, the equality of all coefficients of the quadratic terms implies that the contours are spherical, that is, completely symmetrical. At the optimum, a unit change in one variable produces the same effect on the objective as the same change in any other variable. Although in practice one rarely has an objective that is precisely quadratic, experience and preliminary study often suggest ways to reduce interaction between variables and to achieve approximate symmetry in the neighborhood of the optimum. Such preparation of the problem is, in view of the potentially increased convergence rate, often well worth the effort.

As an illustration, consider the function

$$y = -2x_1^2 - x_2^2 - 3x_3^2 - 2x_4^2 \tag{4-87}$$

Suppose that we wish to search along the line of steepest ascent from the point $\mathbf{x}_0 = (-1, 0, 3, -2)$. The linear approximation to y at this point is obtained by direct differentiation in this artificial example, since the function y has been given in advance. This approximation, which corresponds to Eq. (4-70), is

$$\Delta y = 4\,\Delta x_1 + 0\,\Delta x_2 - 18\,\Delta x_3 + 8\,\Delta x_4$$

with $\Delta x_1 \equiv x_1 + 1$, $\Delta x_2 \equiv x_2$, $\Delta x_3 \equiv x_3 - 3$, and $\Delta x_4 \equiv x_4 + 2$ in the usual manner. Equation (4-76) gives the parametric equations of the line of steepest ascent as

$$\Delta x_1 = 4\rho; \qquad \Delta x_2 = 0; \qquad \Delta x_3 = -18\rho; \qquad \Delta x_4 = 8\rho$$

or

$$x_1 = -1 + 4\rho; \quad x_2 = 0; \quad x_3 = 3 - 18\rho; \quad x_4 = -2 + 8\rho \tag{4-88}$$

The four equations in the five variables have one degree of freedom, as we would expect for a line. A typical point on the line of steepest ascent is specified by selecting any positive ρ, say $\rho = 2$. The corresponding point is $(7, 0, -33, 14)$. The reader may verify that the point $(-5, 0, 21, -10)$ is on the line of steepest *descent*. (What would be the corresponding ρ?)

Since all the points on the line have been expressed in terms of a single parameter, the values that the function y takes along the line can also be made a function of ρ alone. Equations (4-87) and (4-88) together give

$$y = -2(-1 + 4\rho)^2 - 3(3 - 18\rho)^2 - 2(-2 + 8\rho)^2$$

The value of ρ giving maximum y on the line can now be obtained by differentiating the preceding equation with respect to ρ:

$$\frac{dy}{d\rho} = 404 - 2264\rho = 0$$

where $\rho = 0.179$ gives the highest value of y on the line of steepest ascent. The corresponding point is $(-0.28, 0, -0.22, -0.57)$.

The line of steepest ascent will always be characterized parametrically by one equation for each of the N coordinates. There being $N + 1$ variables in all (counting ρ), the set of points represented always has one degree of freedom. Since the equations are linear, the points must lie on a straight line in space. Parameterization enables us to work with one dimension instead of N coordinates, simplifying things both algebraically and conceptually. In practice, when the function is not known in advance, the high point on a gradient line can be found by the powerful unidimensional techniques developed in the previous section. Thus ρ may be used as the independent variable for a Fibonacci search. Most of the multivariable search methods we shall study will involve similarly breaking the problem down into a sequence of unidimensional searches.

A great advantage of gradient methods, and one not widely recognized, is that they will inherently stay away from saddle points. Zellnick, Sondak, and Davis found that their gradient search computer program avoids saddles so dependably that the only way they could test their subroutine for exploring the neighborhood of a pass was to start the search there.

Figure 4-31, in which gradient lines (dashed) are superimposed on the contours of a bimodal response surface, suggests why. Only one gradient line out of the infinite number possible actually passes through the saddle. The other lines all lead directly to one peak or the other. Hence the possibility of a gradient method's stumbling upon a saddle is remote, although the prudent experimenter should still check at the end of a search to see whether he has, by chance, found a saddle instead of a peak.

The simpler but related "sectioning" method, described by Friedman and Savage, involves altering only one variable at a time, holding all the

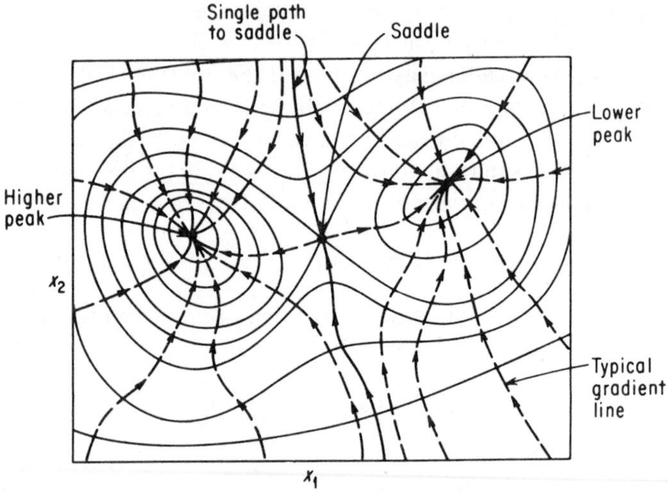

Figure 4-31. Gradient lines in the presence of a saddle.

others constant. One searches for the high point on the straight line described by $\Delta x_i = 0$ for all $i \neq j$, where j is the identifying index of the variable adjusted. Once this high point is found, x_j is fixed and some other variable altered. This procedure is continued until no further improvement is possible. Since each leg of the search will be parallel to one of the coordinate axes, for two independent variables the sectioning search path will resemble the staircase shown in Fig. 4-28. Figures 4-32(a) and 4-32(b) demonstrate how dependent the performance of the one-at-a-time method is on the shapes of the contours. It is highly effective for circles or ellipses having their major and minor axes parallel to the coordinate axes. This would mean that the independent variables are noninteracting. But when the major and minor axes are tilted, as in Fig. 4-32(b), the method is forced to change direction many times before reaching the optimum. The method fails completely when the response surface has a sharp ridge as in Fig. 4-32(c), for, being unable to move diagonally, it cannot find any higher places once it reaches the ridge where the contours come to a point. Buehler, Shah, and Kempthorne concluded that the method, uncombined with other techniques, is not suitable unless the experimenter knows in advance that such ridges are absent. Figure 4-33 shows what would happen if (fortuitously) all contours were spherical.

4-25 LEAST SQUARES

Before studying climbing methods for general unimodal functions, let us examine the important special case where the objective function is the sum of the squares of a set of functions. That is, there are M nonlinear functions $\phi_m(x)$ $(m = 1, \ldots, M)$ of the N independent variables x such that

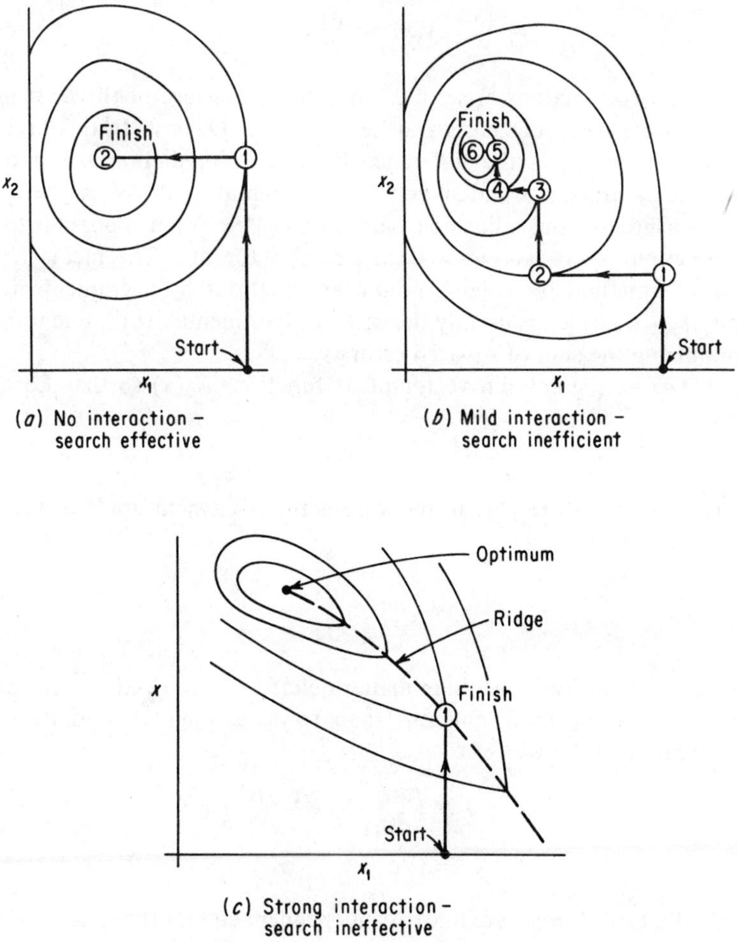

Figure 4-32. Sectional search.

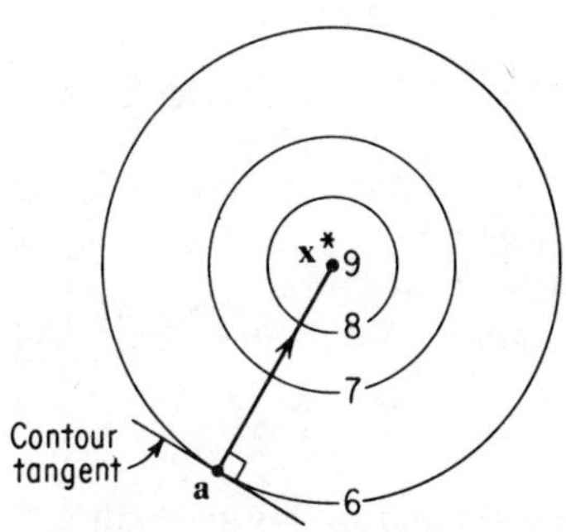

Figure 4-33. Circular contours.

$$y(x) = \sum_{m=1}^{M} \phi_m^2 \tag{4-89}$$

Such objective functions arise often in practice because of the *least-squares* curve-fitting method of Gauss and Legendre (see Davies). In this procedure each $\phi_m(x)$ represents the difference between a prediction based on the adjustable parameters $\mathbf{x}$ and an actual experimental result. When there are as many measurements as adjustable parameters ($M = N$), it is possible to drive each function to zero, and $y^* = 0$ since $\phi_m^2 \geq 0$ for all $\mathbf{x}$. This fact is the basis of Booth's method for solving nonlinear equations by a gradient method. When $M > N$, which is usually the case in experimental work, one wants the $\mathbf{x}$ minimizing the sum of squared errors y.

Let $\boldsymbol{\phi}$ be the column vector of M functions $\phi_m(\mathbf{x})$ so that Eq. (4-89) becomes

$$y = \boldsymbol{\phi}'\boldsymbol{\phi} \tag{4-90}$$

Differentiation with respect to the $\mathbf{x}$ gives the N-element gradient vector

$$\begin{aligned} \nabla y &= \left(\frac{\partial y}{\partial \mathbf{x}}\right) \\ &= 2\mathbf{J}\boldsymbol{\phi} \end{aligned} \tag{4-91}$$

where $\mathbf{J}'$ is the M by N Jacobian matrix defined by Eq. (2-81). In the present context, $\mathbf{J}$, having more columns than rows, is singular, and its various representations follow:

$$\begin{aligned} \mathbf{J} &= \left(\frac{\partial(\phi_1, \ldots, \phi_M)}{\partial(x_1, \ldots, x_N)}\right)' \\ &= (\nabla\phi_1, \ldots, \nabla\phi_M) \end{aligned} \tag{4-92}$$

Assume that the N rows are linearly independent so that the rank of $\mathbf{J}$ equals N. Gauss noticed that if the $\phi_m(\mathbf{x})$ are all linear functions of the $\mathbf{x}$, so that $y(\mathbf{x})$ is quadratic, the Jacobian matrix does not change from one point to another. Thus he suggested approximating the gradient at a point $\mathbf{x} + \Delta\mathbf{x}$ as follows [compare this with Eq. (4-91)]:

$$\nabla y(\mathbf{x} + \Delta\mathbf{x}) \approx 2\mathbf{J}(\mathbf{x})\boldsymbol{\phi}(\mathbf{x} + \Delta\mathbf{x}) \tag{4-93}$$

An approximation for $\boldsymbol{\phi}(\mathbf{x} + \Delta\mathbf{x})$ is obtained from the linear terms of the Taylor expansion about $\mathbf{x}$.

$$\begin{aligned} \boldsymbol{\phi}(\mathbf{x} + \Delta\mathbf{x}) &\approx \boldsymbol{\phi}(\mathbf{x}) + \frac{\partial\boldsymbol{\phi}'}{\partial\mathbf{x}}\Delta\mathbf{x} \\ &\approx \boldsymbol{\phi} + \mathbf{J}'\,\Delta\mathbf{x} \end{aligned} \tag{4-94}$$

Combination of Eqs. (4-93) and (4-94) gives an estimate of the gradient at $\mathbf{x} + \Delta\mathbf{x}$.

$$\nabla y(\mathbf{x} + \Delta\mathbf{x}) \approx 2\mathbf{J}(\boldsymbol{\phi} + \mathbf{J}'\,\Delta\mathbf{x}) = 2(\mathbf{J}\boldsymbol{\phi} + \mathbf{J}\mathbf{J}'\,\Delta\mathbf{x}) \tag{4-95}$$

Since the N rows of $\mathbf{J}$ have been assumed linearly independent, the N by N matrix $\mathbf{JJ}'$ is nonsingular and has an inverse $(\mathbf{JJ}')^{-1}$. Therefore, one can solve Eq. (4-95) for the correction $\Delta\mathbf{x}$, which drives all components of the gradient to zero, and obtain

$$\Delta\mathbf{x} = -(\mathbf{JJ}')^{-1}\mathbf{J}\boldsymbol{\phi} \tag{4-96}$$

The point $\mathbf{x} + \Delta\mathbf{x}$ is where Gauss's approximation predicts the minimum to be, since the gradient ∇y would vanish there. Gauss's procedure is to measure the true values of $\boldsymbol{\phi}(\mathbf{x} + \Delta\mathbf{x})$ and $\mathbf{J}(\mathbf{x} + \Delta\mathbf{x})$ to see whether the gradient $2\mathbf{J}\boldsymbol{\phi}$ really vanishes. If it does, a stationary point has been found; if not, the information is used to generate a new approximation and further correction. If the functions of ϕ are linear in $\mathbf{x}$, so that y is quadratic in $\mathbf{x}$—which is the case in the "linear regression" problem of statistics (Davies)—then Gauss's method finds the minimum in one move.

Harkins studied the behavior of Gauss's method on the following test function devised by Rosenbrock:

$$y = 100(x_2 - x_1^2)^2 + (1 - x_1)^2 \tag{4-97}$$

Figure 4-34 shows the response surface to have a shallow curved valley. Starting at the point $(-1.2, 1)$, Gauss's method found the minimum at $(1, 1)$ after two function evaluations. Since other methods to be described in this chapter required from 10–50 times as much effort, Gauss's procedure seems to be about the best available for least-squares problems. It is also interesting that both the gradient and the sectioning methods stopped far short of finding the minimum, as shown in Fig. 4-34.

Powell (1965) has devised a variation of Gauss's method which does not require knowledge of the derivatives, equivalent information being generated during unidimensional searches. Powell's method required 70 function evaluations to minimize Rosenbrock's function.

To illustrate Gauss's method numerically, the computations needed to make the first step from the starting point $(-1.2, 1)$ are given. Here $\phi_1 = 10(x_2 - x_1^2) = -4.4$, $\phi_2 = 1 - x_1 = 2.2$, and

$$\mathbf{J} = \begin{pmatrix} -20x_1 & -1 \\ 10 & 0 \end{pmatrix} = \begin{pmatrix} 24 & -1 \\ 10 & 0 \end{pmatrix}$$

Therefore,

$$\mathbf{JJ}' = \begin{pmatrix} 577 & 240 \\ 240 & 100 \end{pmatrix}$$

and

$$(\mathbf{JJ}')^{-1} = \begin{pmatrix} 1 & -2.4 \\ -2.4 & 5.77 \end{pmatrix}$$

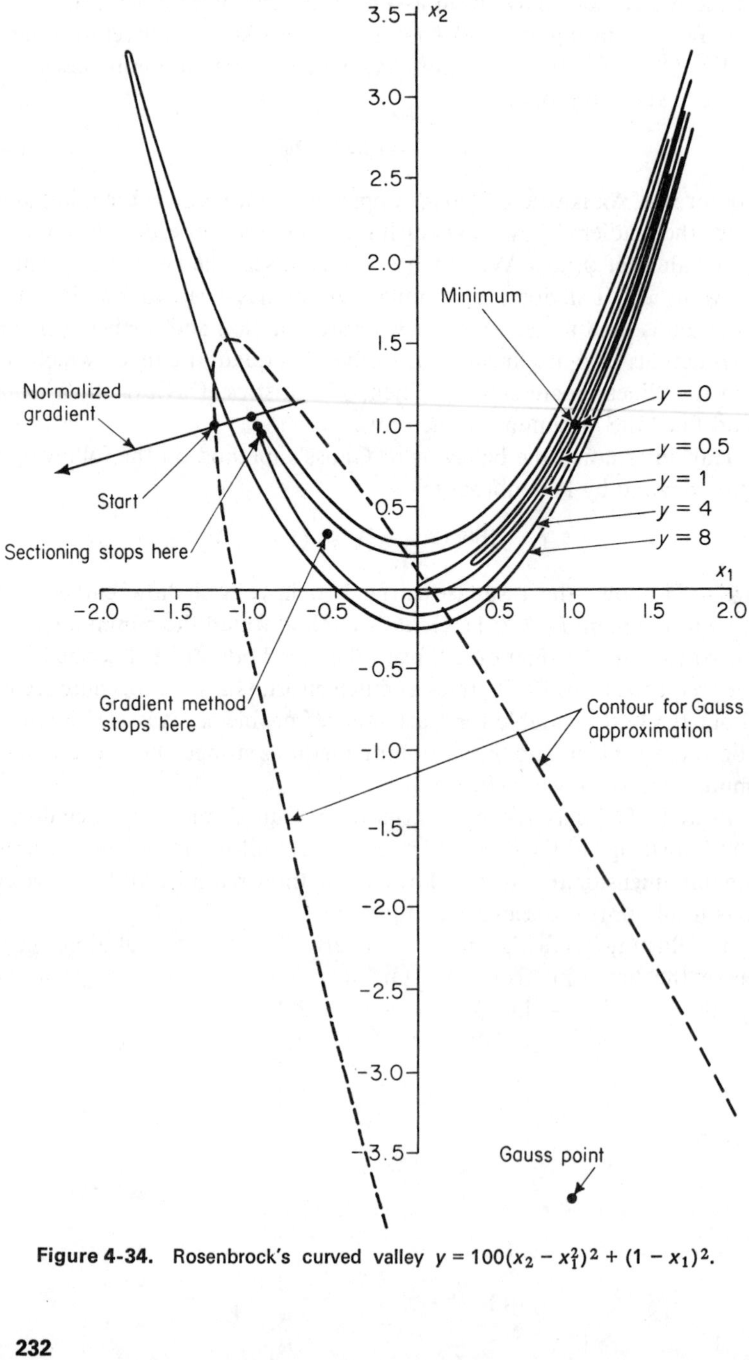

Figure 4-34. Rosenbrock's curved valley $y = 100(x_2 - x_1^2)^2 + (1 - x_1)^2$.

The step is given by

$$\Delta \mathbf{x} = -(\mathbf{J}\mathbf{J}')^{-1}\mathbf{J}\boldsymbol{\phi} = (2.2, -4.84)'$$

The new point is therefore

$$\mathbf{x} + \Delta \mathbf{x} = (1.0, -3.84)'$$

as shown in Fig. 4-34. Gauss's method has found the correct value of x_1, but because of the high-order behavior of the objective, it has overshot badly on x_2. The dashed contour, that for the straight ridged quadratic approximation, shows that it is the curvature of the ridge that causes the trouble. In fact, the objective function actually *increases* from the point $(-1.2, 1)$ to the point $(1, -3.84)$. Still the method recovers rapidly and converges faster than any of the others tried. Incidentally, the nonlinear transformation $z_1 = x_2 - x_1^2$, $z_2 = 1 - x_1$, would have straightened out the ridge so that either Gauss's method, or even the gradient method, would have found the minimum after one step.

4-26 ACCELERATION ALONG A RIDGE

When the objective function is too general for least-squares representation, extensions of the gradient method can be used to search for the peak. The various methods have in common a propensity for finding a "ridge," to use geographic imagery, and following it upward until it reaches the summit. This seems to be effective because the ridges of many objective functions encountered in practice tend to point toward the peak.

Consider the objective function with concentric ellipsoidal contours shown in Fig. 4-35. If, as at point **a**, the first point for a gradient search happens to be precisely on one of the axes of the system of ellipses, the gradient line will pass right through the peak and the search will be over in one ascent. Otherwise, the search will follow a zigzag course such as the one from $\mathbf{p}_0$ to $\mathbf{p}_2$ to $\mathbf{p}_3$ to **b**, and so on. It is interesting to notice that, in principle,

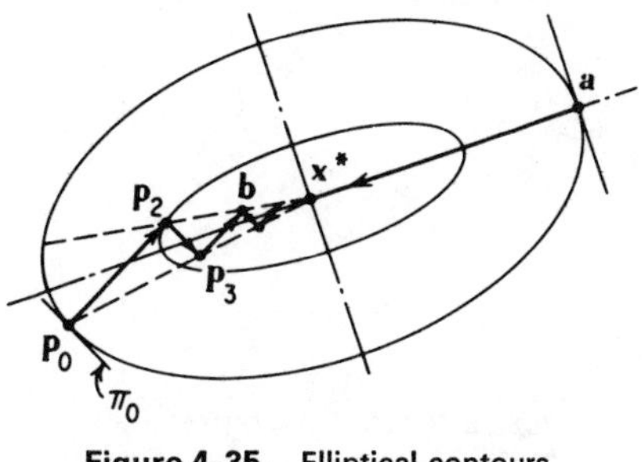

Figure 4-35. Elliptical contours.

gradient search will not reach the peak in a finite number of steps because the steps shorten as the maximum is approached. The peak can, however, be approached as closely as desired, and if the starting point is not too near the major axis, the neighborhood of the peak is attained rapidly.

Notice that the crooked path is bounded by two straight lines which intersect at the peak. This suggests that the search from point $\mathbf{p}_3$ be conducted, not in the gradient direction toward **b**, but along the straight line from $\mathbf{p}_0$ through $\mathbf{p}_3$. In this way, the peak would be located exactly after three unidimensional searches: first from $\mathbf{p}_0$ to $\mathbf{p}_2$ along the gradient at $\mathbf{p}_0$, then from $\mathbf{p}_2$ to $\mathbf{p}_3$ along the gradient at $\mathbf{p}_2$, and finally from $\mathbf{p}_3$ along the line through $\mathbf{p}_0$ and $\mathbf{p}_3$. This sort of acceleration of gradient search was first proposed by Forsythe and Motzkin. It is the two-dimensional version of what will later be called *gradient partan*. In this exposition we have identified the starting point as $\mathbf{p}_0$ rather than $\mathbf{p}_1$ for reasons that will become clear when we generalize the technique to many dimensions.

In a manner of speaking, the acceleration path can be said to "follow a ridge." But the geographic concept of "ridge" is not sufficiently precise for our purposes, so we must define the special idea of a *resolution ridge* in order to discuss ridge-following methods unambiguously. Before introducing this new concept, let us see what is unsuitable about the geographic one.

Consider two men viewing a mountain from two different positions, as in Fig. 4-36. Mr. A would say that the points on curve *A* would be on the

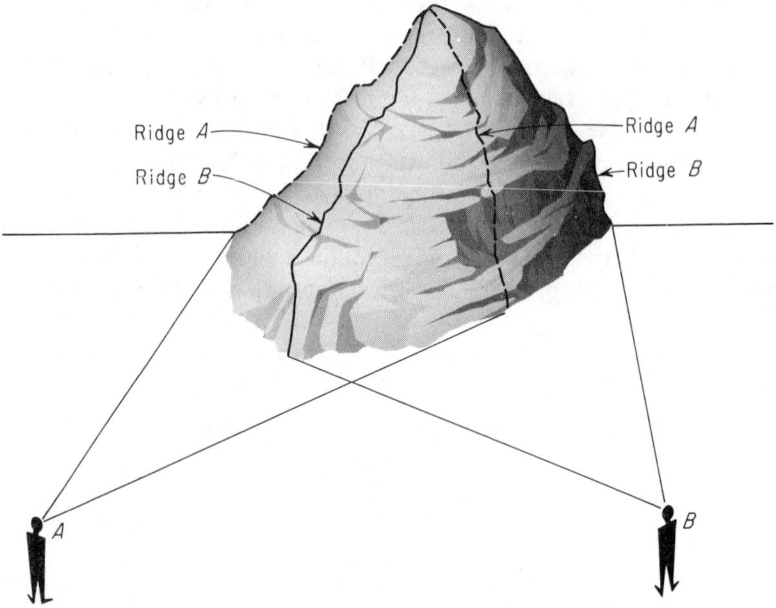

Figure 7-22. Geographic "ridges."

"ridge," since from where he stands these points form the profile of the mountain. On the other hand, Mr. B would see a different "ridge." In fact, any point on the mountain that can be seen at all will, from some viewpoint, be on a "ridge." If there are some points that appear to be a ridge when seen from many different angles, they will often be called the "crests of the ridge," and if the points are all about the same elevation the whole mountain may be called a *ridge*. Thus the geographic idea of "ridge," although helping us picture the formation or response surface with which we are dealing, is not precise enough to be useful in analyzing ridge-climbing techniques.

We prefer to define a *principal ridge* as the locus of points where the sectioning method will stop before reaching the optimum. Hence any point on a *principal ridge* will be the best attainable in any direction parallel to the coordinate (hence "principal") axes. Such a ridge is shown in Fig. 4-32(c); the other two response surfaces in Fig. 4-32 do not have principal ridges, according to this definition.

A ridge of this sort can occur only where the contour lies entirely in one quadrant (for two variables; the multidimensional generalization of the quadrant is called an *orthant*). This can happen only if the contour comes to a sharp point, meaning that the first derivatives are discontinuous there. Since, practically speaking, observations must be a finite distance from each other to be distinguishable, this can cause sectioning to stick on the ridge even when the contours are smooth. Let ϵ_i (> 0) be the closest distance between two points on a line parallel to the x_i axis for which a difference between the results can be detected. That is, ϵ_i, the *resolution* in the x_i direction, is such that for all $x_1, \ldots, x_i, \ldots, x_n$ it can be asserted that

$$y(x_1, \ldots, x_i, \ldots, x_n) \neq y(x_1, \ldots, x_i + \epsilon_i, \ldots, x_N)$$

A point $(x_1, \ldots, x_i, \ldots, x_N)$ will be said to be on a *principal resolution ridge* if it is above all points $(x_1, \ldots, x_i + \epsilon_i, \ldots, x_N)$ and $(x_1, \ldots, x_i - \epsilon_i, \ldots, x_N)$. If a point is *below* all such neighboring points, it will be said to be in a *principal resolution valley*. Principal resolution ridges are important in maximization problems, and they may be regions rather than lines. As shown in Fig. 4-37, a principal resolution ridge becomes narrower as the resolution is made finer. With sufficient resolution, a ridge may even vanish.

The sectioning method will not reach the peak if it runs into a principal resolution ridge. On the other hand, the contour tangent elimination method discussed in Part I of this chapter is not particularly confounded by a ridge; it will merely lead to a long narrow region of uncertainty. Gradient methods can safely navigate a ridge, but unless one fortunately steers right up the ridge, there is likely to be much inefficient zigzagging, as shown in Fig. 4-35. The techniques about to be described will find the trend of a principal resolution ridge and move rapidly along it.

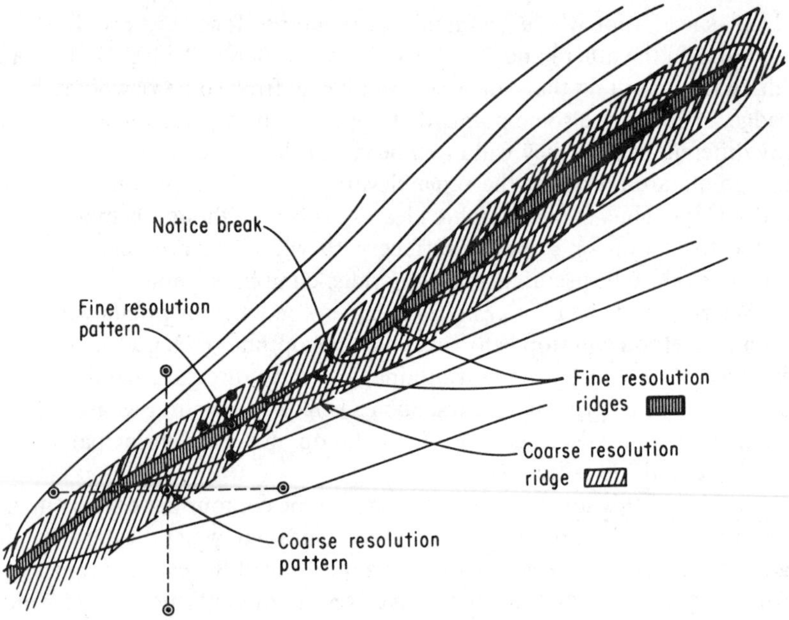

Figure 4-37. Coarse and fine resolution ridges.

Suppose that one conducts two sectioning searches, each starting from a different point, and suppose that each search encounters the principal resolution ridge at a different point. Consider the consequences of finding the best point on the line of search determined by the two points. If the resolution ridge is straight, this acceleration step will find a point close to the true optimum. This is the basic idea behind the "principle of nonlocal search" of Gelfand and Tsetlin. They do not restrict themselves to sectioning searches, however, and, in general, other procedures might be used to find the resolution ridge.

4-27 PATTERN SEARCH

An easily programmed accelerated climbing technique with ridge-following properties is the *pattern search* method of Hooke and Jeeves (1961). Their technique is based on the hopeful conjecture that any set of moves, that is, adjustments of the independent variables, which have been successful during early experiments, will be worth trying again. This strategy is successful on straight ridges because an early pattern of moves can succeed only if it lies along the crest. Hence, further moves in the same direction will be worthwhile if the ridge is straight.

Although the method starts cautiously with short excursions from the starting point, the steps grow with repeated success. Subsequent failure indicates that shorter steps are in order, and if a change in direction is required, the technique will start over again with a new pattern. In the vicinity of the peak the steps become very small, to avoid overlooking any promising direction.

In visualizing what is meant by a "pattern," it is helpful to think of an arrow, its base at one end and its head at the other. The search begins at a base point $\mathbf{b}_1$, which may be chosen arbitrarily; as yet the pattern has not been established. The experimenter chooses a step size δ_i for each independent variable x_i $(i = 1, 2, \ldots, N)$. Let $\boldsymbol{\delta}_i$ be the vector whose ith component is δ_i, all the rest being zero. After measuring the criterion at the initial base $\mathbf{b}_1$, one takes an observation at $\mathbf{b}_1 + \boldsymbol{\delta}_1$. If this new point is better than the base, we call $\mathbf{b}_1 + \boldsymbol{\delta}_1$ the *temporary head* $\mathbf{t}_{11}$, where the double subscript shows that we are developing the first pattern and that we have already perturbed the first variable x_1. Now $\mathbf{b}_1 + \boldsymbol{\delta}_1$ may not be as good as $\mathbf{b}_1$, in which case we forget $\mathbf{b}_1 + \boldsymbol{\delta}_1$ and try $\mathbf{b}_1 - \boldsymbol{\delta}_1$. If this new point is better than $\mathbf{b}_1$, we make it the temporary head; otherwise, $\mathbf{b}_1$ is designated temporary head. In summary, when we are maximizing,

$$\mathbf{t}_{11} = \begin{cases} \mathbf{b}_1 + \boldsymbol{\delta}_1 & \text{if } y(\mathbf{b}_1 + \boldsymbol{\delta}_1) > y(\mathbf{b}_1) & \text{(4-98a)} \\ \mathbf{b}_1 - \boldsymbol{\delta}_1 & \text{if } y(\mathbf{b}_1 - \boldsymbol{\delta}_1) > y(\mathbf{b}_1) > y(\mathbf{b}_1 + \boldsymbol{\delta}_1) & \text{(4-98b)} \\ \mathbf{b}_1 & \text{if } y(\mathbf{b}_1) > \max\,[y(\mathbf{b}_1 + \boldsymbol{\delta}_1), y(\mathbf{b}_1 - \boldsymbol{\delta}_1)] & \text{(4-98c)} \end{cases}$$

In Fig. 4-38, Eq. (4-98b) governs.

Perturbation of x_2, the next independent variable, is now carried out in a similar manner, this time about the temporary head $\mathbf{t}_{11}$ instead of the original base $\mathbf{b}_1$. In general, the jth temporary head $\mathbf{t}_{1j}$ is obtained from the preceding one, $\mathbf{t}_{1,j-1}$, as follows:

$$\mathbf{t}_{1j} = \begin{cases} \mathbf{t}_{1,j-1} + \boldsymbol{\delta}_j & \text{if } y(\mathbf{t}_{1,j-1} + \boldsymbol{\delta}_j) > y(\mathbf{t}_{1,j-1}) & \text{(4-99a)} \\ \mathbf{t}_{1,j-1} - \boldsymbol{\delta}_j & \text{if } y(\mathbf{t}_{1,j-1} - \boldsymbol{\delta}_j) > y(\mathbf{t}_{1,j-1}) > y(\mathbf{t}_{1,j-1} + \boldsymbol{\delta}_j) & \text{(4-99b)} \\ \mathbf{t}_{1,j-1} & \text{if } y(\mathbf{t}_{1,j-1}) > \max\,[y(\mathbf{t}_{1,j-1} + \boldsymbol{\delta}_j), y(\mathbf{t}_{1,j-1} - \boldsymbol{\delta}_j)] & \text{(4-99c)} \end{cases}$$

This expression covers all $1 \leq j \leq N$ if we adopt the convention that

$$\mathbf{t}_{10} \equiv \mathbf{b}_1$$

In Fig. 4-38, Eq. (4-99a) applies for $j = 2$. When all the variables have been perturbed, the last temporary head point $\mathbf{t}_{1N}$ is designated the *second base point* $\mathbf{b}_2$.

$$\mathbf{t}_{1N} \equiv \mathbf{b}_2$$

The original base point $\mathbf{b}_1$ and the newly determined base point $\mathbf{b}_2$ together establish the first pattern. Reasoning that if a similar exploration

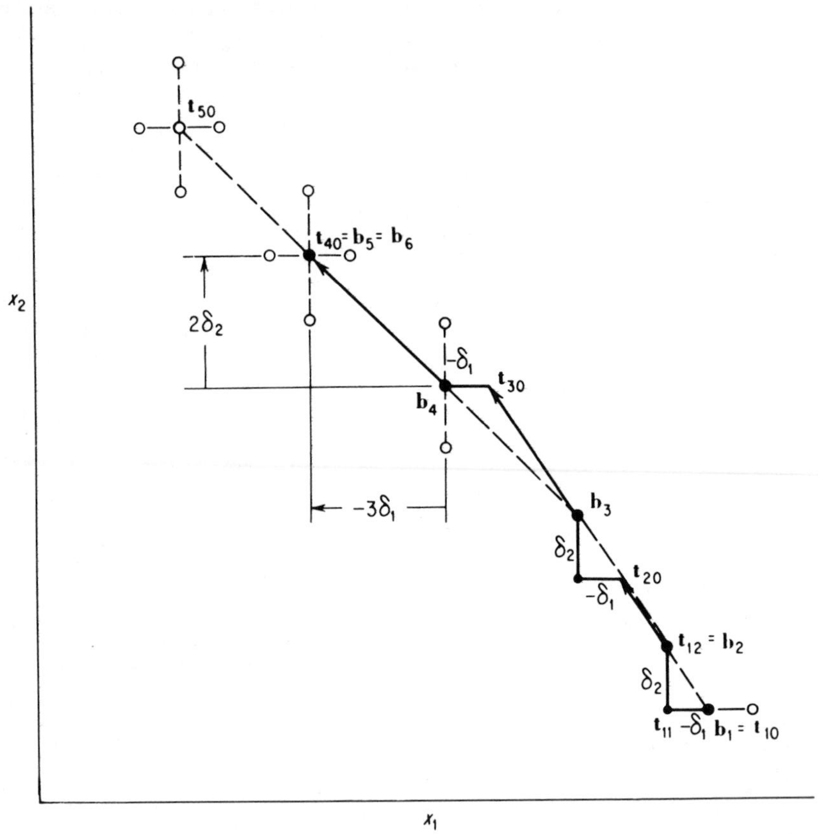

Figure 4-38. Finding a ridge.

were conducted from $\mathbf{b}_2$, the results are likely to be the same, we skip the local excursions and extend the arrow from $\mathbf{b}_1$ to $\mathbf{b}_2$, immediately doubling its length. This establishes a new temporary head $\mathbf{t}_{20}$ for the second pattern based at $\mathbf{b}_2$. This initial temporary head is given by

$$\mathbf{t}_{20} \equiv \mathbf{b}_1 + 2(\mathbf{b}_2 - \mathbf{b}_1) = \mathbf{b}_2 + (\mathbf{b}_2 - \mathbf{b}_1) = 2\mathbf{b}_2 - \mathbf{b}_1$$

The double subscript 20 indicates that we are building a second pattern and that we have not yet begun to perturb the variables. A local exploration about $\mathbf{t}_{20}$ is now carried out to correct the tentative second pattern if necessary, as shown in Fig. 4-38. The logical equations governing establishment of the new temporary heads $\mathbf{t}_{21}, \mathbf{t}_{22}, \ldots, \mathbf{t}_{2N}$ will be similar to Eqs. (4-99), the only difference being that the first subscript will be 2 instead of 1. The reconnaissance is completed when all the variables have been perturbed, and the last temporary head, $\mathbf{t}_{2N}$, is designated the third base point $\mathbf{b}_3$, if, as in Fig. 4-38, the outcome there is better than at $\mathbf{b}_2$.

As before, a new temporary head $\mathbf{t}_{30}$ is established by extrapolating from $\mathbf{b}_2$ through $\mathbf{b}_3$.

$$\mathbf{t}_{30} = 2\mathbf{b}_3 - \mathbf{b}_2$$

In Fig. 4-38 the new base $\mathbf{b}_3$ is collinear with $\mathbf{b}_2$ and $\mathbf{b}_1$, indicating that the direction of the pattern is not to be changed. Notice that the repeated success in this direction causes the pattern to grow, because

$$\mathbf{b}_3 - \mathbf{b}_2 = 2(\mathbf{t}_{20} - \mathbf{b}_2) = 2(\mathbf{b}_2 - \mathbf{b}_1)$$

The procedure is iterated for the third pattern. Suppose that perturbation of x_2 fails to produce any improvement over temporary head $\mathbf{t}_{31}$, as in Fig. 4-38, but that $\mathbf{t}_{31}$ is still a better point than $\mathbf{b}_3$. Then,

$$\mathbf{b}_4 = \mathbf{t}_{32} = \mathbf{t}_{31}$$

and the pattern will veer to the left, still growing in length.

For the fourth pattern, imagine that none of the perturbations about the initial temporary head $\mathbf{t}_{40}$ improve the outcome, but that $y(\mathbf{t}_{40}) > y(\mathbf{b}_4)$. Then,

$$\mathbf{b}_5 = \mathbf{t}_{42} = \mathbf{t}_{41} = \mathbf{t}_{40}$$

and the pattern will maintain its direction and length without any growth. The fourth pattern, $\mathbf{b}_5 - \mathbf{b}_4$, has components $(-3\delta_1, 2\delta_2)$, representing the cumulative effect of three successful steps in the negative x_1 direction (that is, to the left) and two in the positive x_2 direction (that is, upward).

Suppose that none of the temporary heads, $\mathbf{t}_{50}$, $\mathbf{t}_{51}$, or $\mathbf{t}_{52}$, is any better than the fifth base, $\mathbf{b}_5$, as in Fig. 4-38. Then $\mathbf{b}_6 = \mathbf{b}_5$ and the pattern is destroyed. Since this could mean we are either at the peak or crossing a resolution ridge, new maneuvers are in order.

Unable to continue the old pattern from $\mathbf{b}_5$ even by modifying it, we must abandon it entirely and try to build a new one using $\mathbf{b}_5$ as the base point, but designating it $\mathbf{b}_6$, since we now are working with the sixth pattern. We start all over again, making $\mathbf{b}_6$ the initial temporary head $\mathbf{t}_{60}$ for a local exploration. If this scouting expedition locates a better point, then we can begin a new pattern. But if, as in Fig. 4-39, no better point is found, then the steps must be shortened in an attempt to break the resolution ridge, if there is one. In Fig. 4-39 we have cut the steps in half and are able to obtain improvement, which starts us off on a fresh, albeit tiny pattern.

After a few minor modifications of direction and rapid growth in size, the pattern coincides with the trend of the ridge from the ninth through eleventh pattern moves and hence holds its length constant. At $\mathbf{b}_{12}$ the ridge starts to curve, and the pattern swerves sharply to follow the crest, shortening up as necessary until $\mathbf{b}_{16}$, where the ridge straightens out again. The sixteenth pattern lengthens rapidly on the straight ridge, but the seventeenth pattern fails to find a better point. Again the pattern must be destroyed.

As before, we retreat to the last successful base point, in this case $\mathbf{b}_{17}$ (now designated $\mathbf{b}_{18}$) and try to establish a nineteenth pattern by shortening the exploratory steps. When this fails to resolve the ridge, we retrench further, again with no improvement. The search terminates when the step sizes fall below a preselected minimum, as after the second reduction in Fig. 4-39. In our case, $\mathbf{b}_{17}$ is actually at the maximum, at least as far as we can tell with the finest resolution available.

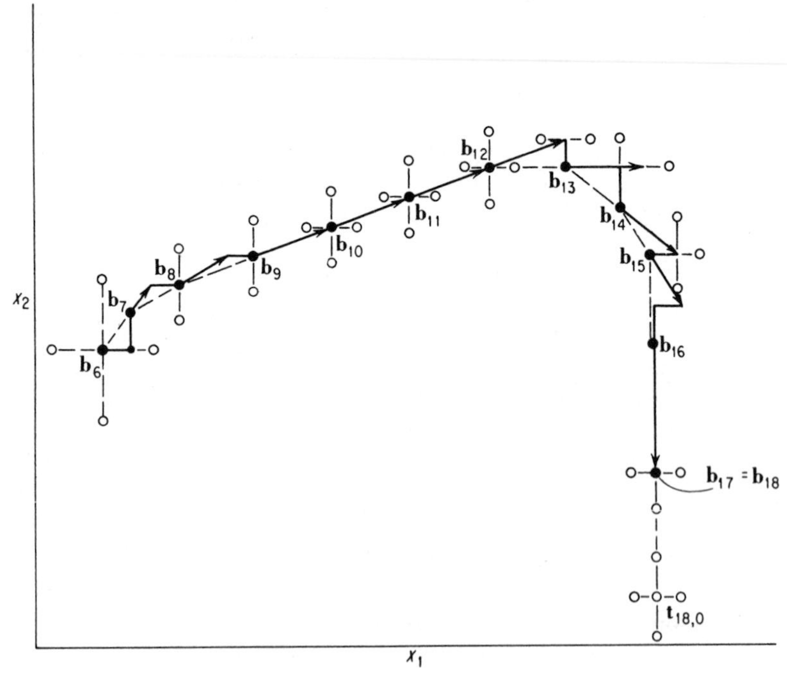

Figure 4-39. Following a ridge.

This example shows how smoothly pattern search finds the trend of a ridge and follows it to the top. Although performance of the method does not depend on the choice of scale, it certainly would be sensitive to the step size selected and the speed at which the grid is reduced to resolve a ridge. In the original form described here, pattern search occasionally has difficulty establishing a new pattern after arriving at the resolution ridge. To correct this shortcoming, Wood has incorporated a quadratic strategy into the routine that uses second-order information to detect the trend of the resolution ridge and restart the pattern. These techniques are developed in Section 4-27.

Hooke and Jeeves found empirically, in a curve-fitting problem involving neutron reactor, that the computation time for pattern search increased only as the *first* power of the number variables. This is striking because, with classical minimization techniques, the computations grow with the *cube* of the dimensionality. This phenomenon may be rationalized by observing that a ridge is really a one-dimensional object, since it may be characterized by a single parameter. Thus the empirically observed efficiency of pattern search may be due precisely to its ability to follow a ridge and reduce the effective dimensionality of the problem.

Rosenbrock has devised a ridge-following procedure which has proved effective at finding the minimum of his test function

$$y = 100(x_2 - x_1^2)^2 + (1 - x_1)^2$$

having its low point at (1, 1) in the shallow curving valley shown in Fig. 4-32. His *method of rotating coordinates* differs from pattern search mainly in the way it carries out local explorations. Instead of perturbing each of the original variables independently as in pattern search, Rosenbrock rotates the coordinate system so that one axis points along the direction of the ridge as estimated by the previous trial. The other axes are arranged in directions normal to the first. Naturally, excursions in these normal directions are effective in correcting the estimate of the trend of the ridge, as indicated in Fig. 4-34.

Instead of taking a fixed step in each direction, Rosenbrock in effect tries to find the optimum point on each line. This procedure continuously adjusts what in pattern search would be the step size. The combined rotation of the ridge-tracking vector and scale adjustment proves extremely effective on the test function. Table 4-2 shows where several schemes, all started at the point (−1.2, 1), ended after 200 moves. The pattern search trials were recorded by Wood in his 1962 report. Gauss's method, which happens to be

TABLE 4-2

Performance of Various Search Schemes after 200 Trials on $y = 100(x_2 - x_1^2)^2 + (1 - x_1)^2$

Method	x_1	x_2	y
Sectioning	−0.970	0.945	3.882
Gradient	−0.605	0.371	2.578
Ordinary pattern	------	-----	0.803
Pattern with adjusted steps	-----	-----	0.0103
Rotating coordinates	0.995	0.991	0.000022
Optimum	1	1	0

applicable to Rosenbrock's function, gets close to the minimum with one-tenth of the effort.

This completes discussion of the "middle-game" strategies for organizing first-order information into climbing procedures. By acceleration and by following ridges, they scale the response surface rapidly in steeply sloped regions and bring the investigator to the vicinity of a stationary point. There, where the tangent plane is almost horizontal and the first derivatives become small, a change in tactics is called for, which is the topic of Section 4-28.

4-28 EXPLORATION NEAR A STATIONARY POINT

Both the beginning and the end of a climb involve local exploration, the former being a simple linear study near an arbitrary point; the latter, a nonlinear exploration of the vicinity of the optimum. Thus a fairly simple extension of the concepts already developed enables us to dispose of the end-game tactics in this section.

Box and Wilson have remarked that often an experimenter is not satisfied merely with locating the optimum; he also needs to know how the objective function behaves at points nearby. Since the tangent plane will be horizontal at a peak, curvature, asymmetry, and other nonlinear effects become important there, and the investigator is led to fit quadratic or higher-degree expressions to the unknown function. Even an investigator totally unconcerned about the objective's behavior near the summit would be reckless not to examine the supposed optimum closely, for there may actually be better points nearby. We shall study an example where the fitting of a quadratic expression signals the presence of higher ground that otherwise would have been overlooked.

First, we measure the curvature by considering only the first- and second-degree terms of the Taylor expansion in the region of interest. If the function is asymmetric, cubic terms might be necessary, but we will find this out later when we check our quadratic predictions against actual observations. Box (1954) has discussed the dangers of trying to approximate a high-degree expression with a lower-degree polynomial. Since the work required to construct a nonlinear approximation grows rapidly with its degree, choice of a good algebraic representation for the function may well justify the initial research and reflection needed to obtain it, as pointed out in Section 4-19.

We shall develop the principles of nonlinear exploration by studying a specific example. An investigator wishes to find the combination of temperature and chemical reactor volume for which the maximum profit, considering value of product less operating and construction costs, is obtainable. He has constructed a complicated mathematical model relating profit (y, \$/day) to the *logarithm* of temperature in degrees centigrade (x_1, dimensionless) and the reactor volume (x_2, cubic feet). The logarithm was chosen because theoretical considerations, namely, the Arhennius relation of chemical kinetics (McCutcheon, Seltz, and Warner), indicate that reaction rate is an exponential function of temperature. Use of the logarithm is more likely to give a representation which can be fitted by a low-degree polynomial.

Suppose that, according to a pattern search, the optimum appears to be near the point (2.45, 8.5), where the profit is \$756 per day, the highest measured. The daily profits at the four nearest points are as follows: $y(2.49, 8.5) = 574$; $y(2.45, 8.9) = 742$; $y(2.41, 8.5) = 646$; $y(2.45, 8.1) = 702$. In the experimental region obtained by plotting x_2 against x_1 the five points form a cross. As we saw in Section 4-27 many search strategies give this sort of pattern in the neighborhood of a candidate for the optimum.

Consider the Taylor series for a function of two variables with the terms of higher than second-degree neglected.

$$\Delta y = m_1\,\Delta x_1 + m_2\,\Delta x_2 + \tfrac{1}{2}[m_{11}(\Delta x_1)^2 + 2m_{12}(\Delta x_1)(\Delta x_2) + m_{22}(\Delta x_2)^2] \tag{4-100}$$

This expression has five constants, but there are available only four points different from the base point, which we shall take to be (2.45, 8.5) in the center of the cross. We must either take another observation or throw out one term. As a first trial, let us neglect the interaction term involving both Δx_1 and Δx_2. Thus we would have

$$\Delta y = m_1\,\Delta x_1 + m_2\,\Delta x_2 + \tfrac{1}{2}[m_{11}(\Delta x_1)^2 + m_{22}(\Delta x_2)^2] \tag{4-101}$$

This approximation will be used to estimate the location of the true optimum. A check measurement there will tell us whether the approximation is good enough; if it is not, we shall use the check point to evaluate the constants when the interaction term is included.

The crosslike arrangement permits great simplification in the computations. Consider first the constants m_1 and m_{11} associated with the variable x_1. Let the base point (2.45, 8.5) be designated $\mathbf{x}_0$; let the point (2.49, 8.5) to its right be $\mathbf{x}_{11}$; and let (2.41, 8.5) on the left be $\mathbf{x}_{12}$. With Δy_{ij} denoting $y(\mathbf{x}_{ij}) - y(\mathbf{x}_0)$ and $\Delta \mathbf{x}_{ij}$ being the distance from the base to the point $\mathbf{x}_{ij}$, we may write Eq. (4-101) for $\mathbf{x}_{11}$ and $\mathbf{x}_{12}$ as follows:

$$\Delta y_{11} = m_1\,\Delta \mathbf{x}_{11} + \tfrac{1}{2}m_{11}(\Delta \mathbf{x}_{11})^2 \tag{4-102a}$$

$$\Delta y_{12} = m_1\,\Delta \mathbf{x}_{12} + \tfrac{1}{2}m_{11}(\Delta \mathbf{x}_{12})^2 \tag{4-102b}$$

But $\Delta \mathbf{x}_{12} = -\Delta \mathbf{x}_{11}$; hence the latter equation becomes

$$\Delta y_{12} = -m_1\,\Delta \mathbf{x}_{11} + \tfrac{1}{2}m_{11}(\Delta \mathbf{x}_{11})^2 \tag{4-102c}$$

Adding (4-102a) to (4-102c) gives

$$m_{11} = \frac{\Delta y_{11} + \Delta y_{12}}{(\Delta \mathbf{x}_{11})^2} \tag{4-103a}$$

Subtracting (4-102c) from (4-102a) gives

$$m_1 = \frac{\Delta y_{11} - \Delta y_{12}}{2\,\Delta \mathbf{x}_{11}} \tag{4-103b}$$

In general, if there are N independent variables arranged so that there are points $\mathbf{x}_{i1}$ and $\mathbf{x}_{i2}$ such that

$$\mathbf{x}_{i1} = \mathbf{x}_0 + \Delta \mathbf{x}_{i1}\mathbf{e}_i \quad \text{and} \quad \mathbf{x}_{i2} = \mathbf{x}_0 - \Delta \mathbf{x}_{i1}\mathbf{e}_i$$

where $\mathbf{e}_i$ is the ith unit vector, then

$$m_{ii} = \frac{\Delta y_{i1} + \Delta y_{i2}}{(\Delta \mathbf{x}_{i1})^2} \tag{4-104}$$

and

$$m_i = \frac{\Delta y_{i1} - \Delta y_{i2}}{2\,\Delta \mathbf{x}_{i1}} \tag{4-105}$$

In the example at hand, the approximation would be

$$\Delta y = -900\,\Delta x_1 + 50\,\Delta x_2 - 90{,}000(\Delta x_1)^2 - 200(\Delta x_2)^2 \tag{4-106}$$

This quadratic expression may be differentiated partially with respect to Δx_1 and Δx_2, the derivatives set equal to zero, and the two equations solved simultaneously to give the coordinate changes Δx_1^0 and Δx_2^0 to reach the apparent optimum. Thus

$$\frac{\partial\,\Delta y}{\partial\,\Delta x_1} = -900 - 180{,}000\,\Delta x_1 = 0$$

whence

$$\Delta x_1^0 = -0.005$$

and, similarly,

$$\Delta x_2^0 = 0.125$$

In general, for this cruciform pattern,

$$\Delta x_i^0 = -\frac{m_i}{m_{ii}} = \frac{(\Delta y_{i2} - \Delta y_{i1})\,\Delta \mathbf{x}_{i1}}{2(\Delta y_{i2} + \Delta y_{i1})} \tag{4-107}$$

Since the expression fit indicates that the optimum is indeed quite close to $\mathbf{x}_0$, we may be tempted to conclude that the optimum has been found. It is prudent, however, to test another point first. Although the predicted optimum may seem like a good place to test, it is so close to the point $\mathbf{x}_0$ that it would be of little value in subsequent surface fitting should the prediction be invalid. Thus we shall try the point (2.41, 8.9), obtained by *decreasing* x_1 an amount $\Delta\mathbf{x}_{11}$ and *increasing* x_2 by $\Delta\mathbf{x}_{21}$. This puts the new point not only in the same quadrant as the predicted optimum, but also where it will make future computations convenient. Equation (7-81) forecasts that at the new point the profit will be \$636 per day—\$120 less than at the best point so far. Evaluation of this point shows the profit to be only \$492 per day, indicating that the variables are interacting so strongly that the simple model should not be used.

Equation (4-101) having failed, we are forced to use Eq. (4-100), with its additional interaction term $m_{12}\,(\Delta x_1)(\Delta x_2)$. Fortunately, the calculations are quite simple, for, since the interaction term will still be zero in equations such as (4-102), the expressions (4-104) and (4-105) for m_i and m_{ii} remain valid. In fact, the numerical values of m_1, m_2, m_{11}, and m_{22} already computed will be unchanged in the new expression. To evaluate m_{12} we need merely write Eq. (4-100) for the new point $\mathbf{x}^{12}$, substituting the values of the known constants.

$$\begin{aligned}\Delta y^{12} = -264 &= -900(-0.04) + 50(0.4)\\ &\quad -90{,}000(0.04)^2 - 200(0.4)^2 + m_{12}(-0.04)(0.4)\end{aligned}$$

whence $m_{12} = 9000$, clearly not negligible.

Before finding the optimum according to the new expression, let us indicate how to handle interaction when there are more than two independent variables. For N variables there will be $N(N-1)/2$ interaction terms, each requiring a point. To evaluate the coefficient m_{ij}, place a new point at

$$\mathbf{x}_{ij} \equiv \mathbf{x}_0 \pm \Delta x_i \mathbf{e}_i \pm \Delta x_j \mathbf{e}_j$$

where the $\pm$ indicates that the sign is arbitrary. Equation (4-100), when written for this point, will have m_{ij} as its only unknown, since all the constants m_i and m_{ii} have already been determined.

Next, we reestimate the location of the optimum by differentiating Eq. (4-100), which in this case is now

$$\Delta y = -900\,\Delta x_1 + 50\,\Delta x_2 - 90{,}000(\Delta x_1)^2 + 9000(\Delta x_1)(\Delta x_2) - 200(\Delta x_2)^2 \tag{4-108}$$

Solution of the two simultaneous equations resulting gives

$$\Delta x_1^0 = -0.01 \quad \text{and} \quad \Delta x_2^0 = -0.10$$

A test at the corresponding point $\mathbf{x}^0 = (2.44, 8.4)$ verifies the slight profit improvement of \$2 per day predicted by Eq. (4-108). Hence we shift our coordinate system to put $\mathbf{x}^0$ at the origin, defining

$$\Delta\bar{x}_1 \equiv x_1 - x_1^0 = \Delta x_1 + 0.01 \tag{4-109a}$$

$$\Delta\bar{x}_2 \equiv x_2 - x_2^0 = \Delta x_2 + 0.10 \tag{4-109b}$$

To obtain a quadratic expression valid in the vicinity of this apparent optimum, we could invoke Taylor's theorem, using Eq. (4-108) to give us the first and second derivatives we need. But we know that at the optimum the first derivatives vanish, and the values of the second derivatives of a quadratic expression are necessarily the same no matter where they are evaluated. Therefore,

$$\Delta\bar{y} \equiv y - y^0 = y - 756 = -90{,}000(\Delta\bar{x}_1)^2 + 9000(\Delta\bar{x}_1)(\Delta\bar{x}_2) - 200(\Delta\bar{x}_2)^2 \tag{4-110}$$

The right member is a *homogeneous quadratic form* in two variables because each term is of second degree. In Section 2-09 it was shown how to identify the character of a stationary point by completing the square on the quadratic terms, which in this case gives

$$\begin{aligned}\Delta\bar{y} &= -90{,}000(\Delta\bar{x}_1)^2 + 9000(\Delta\bar{x}_1)(\Delta\bar{x}_2) - 200(\Delta\bar{x}_2)^2 \\ &= -[90{,}000(\Delta\bar{x}_1)^2 - 9000(\Delta\bar{x}_1)(\Delta\bar{x}_2) + 225(\Delta\bar{x}_2)^2] + 25(\Delta\bar{x}_2)^2 \\ &= -(300\,\Delta\bar{x}_1 - 15\,\Delta\bar{x}_2)^2 + 25(\Delta\bar{x}_2)^2\end{aligned} \tag{4-111}$$

This expression is indefinite because the two squared terms have different signs. Hence the point $\mathbf{x}^0$ is not a peak; it is a saddle. Notice that the expression would have been different if we had eliminated x_2 first. This does not matter because we are interested merely in the arrangement of signs, which is known as the *signature* of the function (Birkhoff and Mac Lane).

Since $\mathbf{x}^0$ is in reality a saddle, there is the possibility of further improvement of the profit. To do this we must choose a combination of $\Delta\bar{x}_1$ and $\Delta\bar{x}_2$ which will make Eq. (4-111) positive. Let us, for the sake of convenience, simply nullify the first term by taking

$$\Delta\bar{x}_2 = 20\,\Delta\bar{x}_1 \tag{4-112}$$

Since the second term must always be positive, any pair of values satisfying Eq. (4-112) will be satisfactory: we shall choose $\Delta\bar{x}_1 = 0.02$ and $\Delta\bar{x}_2 = 0.4$. Equation (4-111) predicts $\Delta\bar{y} = 4$ at this point; hence, if this value is verified, we should search in this direction. Notice that if opposite signs were chosen, the improvement would be the same, which indicates that there are two possible peaks to be examined.

If the experiments do not confirm the predictions of the equation, then the observer is forced to consider fitting a cubic equation, but we shall not go into this matter here. It is important to notice that the direction of improvement was found from data which at first glance would suggest that no better profit was possible. Directions of improvement can be found only if the interaction term is taken into account; the simpler no-interaction model is unsuitable because it involves only squared terms. Figure 4-40 shows the contours of the objective function.

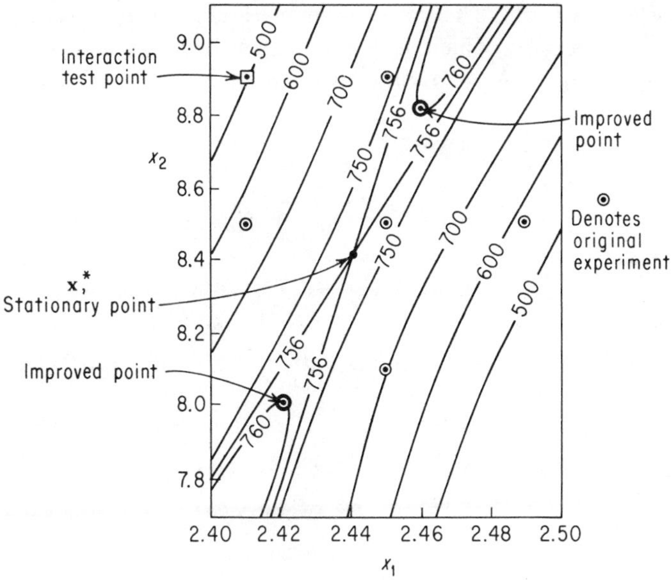

Figure 4-40. The objective function.

4-29 *EVOLUTION AND THE SIMPLICIAL METHOD*

In most large-scale industrial operations any observations are obscured by appreciable experimental error. One must make many repeated measurements to distinguish between true changes in optimum conditions and the spurious fluctuations caused by random noise. Since experimentation with a full-sized manufacturing plant is a costly business, monitoring a moving optimum would be out of the question but for the large amount of data available for small changes in the independent variables. Hence, if the experimenter is patient and uses good statistical techniques in gathering and analyzing information, he or she may be able to make the plant slowly follow shifts in optimum conditions.

This idea of letting a system adapt to changing conditions without unduly upsetting it dynamically or sacrificing profits is due to Box (1957), who named the procedure *evolutionary operations*, perhaps in the spirit of the Darwin Centennial being celebrated at that time. The bookkeeping details of evolutionary operations are not within the scope of this text, which assumes no statistical background on the part of the reader. The method deserves attention, however, for it has proved itself valuable in industrial optimization. Its principles are very close to those we have described for fitting a nonlinear function to the response surface near the optimum.

The concept of evolutionary operations inspired Spendley, Hext, and Himsworth to devise a simple system for approaching a nearby optimum and following it if it drifts. Their scheme involves placing observations on the vertices of a *simplex*, which is the N-dimensional generalization of the equilateral triangle ($N = 2$) and the regular tetrahedron ($N = 3$). To describe algebraically where to put the measurements, define the quantities (Spendley, Hext, and Himsworth)

$$p_N \equiv \frac{\sqrt{N+1} - 1 + N}{N\sqrt{2}} \tag{4-113}$$

$$q_N \equiv \frac{\sqrt{N+1} - 1}{N\sqrt{2}} \tag{4-114}$$

Then the $N + 1$ vertices of a simplex with unit edge are specified by

$$\mathbf{x}_0 = 0 \tag{4-115}$$

$$\mathbf{x}_1 = (p_N, q_N, q_N, \ldots, q_N) \tag{4-116}$$

$$\mathbf{x}_2 = (q_N, p_N, q_N, \ldots, q_N) \tag{4-117}$$

$$\vdots$$

$$\mathbf{x}_N = (q_N, \ldots, q_N, p_N) \tag{4-118}$$

That is, the nth component of $\mathbf{x}_n$ is p_N, the others being q_N ($n \neq 0$). For $N = 2$, the three points are

$$\mathbf{x}'_0 = (0, 0)$$

$$\mathbf{x}'_1 = \frac{\sqrt{3} + 1, \sqrt{3} - 1}{2\sqrt{2}} = (0.965, 0.259)$$

$$\mathbf{x}'_2 = \frac{\sqrt{3} - 1, \sqrt{3} + 1}{2\sqrt{2}} = (0.259, 0.965)$$

They are shown graphically in Fig. 4-41. Notice that they are a unit distance apart (Exercise 4-12).

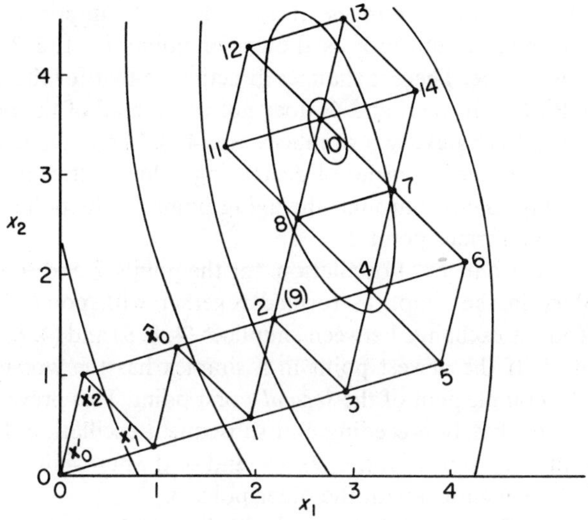

Figure 4-41. Simplicial search.

Next consider $N + 1$ *complementary points* $\hat{\mathbf{x}}_n$, defined by

$$\hat{\mathbf{x}}_n = \sum_{j=0}^{N} \mathbf{x}_j - \left(\frac{N+2}{N}\right)\mathbf{x}_n \tag{4-119}$$

For $N = 2$, the three complementary points are

$$\hat{\mathbf{x}}_0' = \sum_{j=0}^{2} \mathbf{x}_j - 2(0, 0) = (1.224, 1.224)$$
$$\hat{\mathbf{x}}_1' = (1.224, 1.224) - 2(0.965, 0.259)$$
$$= (-0.706, 0.706)$$
$$\hat{\mathbf{x}}_2' = (1.224, 1.224) - 2(0.259, 0.965)$$
$$= (0.706, -0.706)$$

Let the typical point $\mathbf{x}_n$ be deleted from the original set. The remaining N points, together with the complementary point $\hat{\mathbf{x}}_n$, form a new regular unit simplex as in Fig. 4-41. Suppose that the objective function is measured at each point $\mathbf{x}_n$ in the original simplex, and let the index of the worst value be w. That is, if y is to be maximized, then

$$y_w \equiv y(\mathbf{x}_w) \equiv \min_{0 \le n \le N} [y(\mathbf{x}_n)] \tag{4-120}$$

The strategy is to measure next at $\hat{\mathbf{x}}_w$, the point complementary to the worst one in the original set. Then one finds the worst point in the new set, measures at its complement, and continues. If started in a nonoptimal region, this procedure will migrate upward, as shown for points $\hat{\mathbf{x}}_0$, 1, 2, 3, 4, 5, and 6, until it crosses a ridge. There it changes direction so that for the first time the current simplex (points 4, 7, and 6) does not contain all of the most recently measured points. The next two simplices are (4, 8, 7) and (4, 8, 9), the point numbers being arranged in order of decreasing value. Notice that the procedure is beginning to circle around the aging point 4. The point 9 is, in fact, identical with the former point 2.

Here there is danger of oscillation, for the points 7 and 9 are mutually complementary in the simplices formed together with points 8 and 4. To prevent an endless exchange between simplices (4, 8, 9) and (4, 8, 7), a second rule is invoked. If the newest point in a simplex has the worst value, then proceed to the complement of the *second* worst point. This prevents return to the simplex immediately preceding and subsequent oscillation. In the example, the second worst point in (4, 8, 9) is 8, and so the new simplex is (4, 9, 10), where point 10 is identical with the older point 3.

Here again the antioscillation rule applies, and since the next simplex (4, 5, 10) would be identical with the older one (4, 5, 3), the procedure would circle the point 4 forever if something were not done. Spendley, Hext, and Himsworth give a stopping rule which terminates the search whenever a given point has remained in the number of successive simplices given by Table 4-3. These numbers seem low to us, for in Fig. 4-41 the rule would

TABLE 4-3

Stopping Ages for Simplicial Method

Number of independent variables	2	3	4	5	6	7	8	9	10	15	20	30
Maximum age	3	5	7	9	11	13	16	18	21	36	52	92

stop the search at the simplex (4, 6. 5) before it has had a chance to go up the ridge. They probably do not intend to stop the search and declare point 4 the optimum, which it certainly is but rather to make a quadratic approximation to determine the character of the apparent stationary point. The end-game procedure would in this case predict greener pastures in directions of increased x_2 that would justify points 7 and 8.

This time, when point 9 gives such poor results that the antioscillation rule would be called for, the quadratic prediction is used to suggest a new point instead. This directs the search to point 10, a radical move because the point 4, best until now, is left behind. In Fig. 4-41 this gamble pays off, since point 10 is really better than point 4 and the search can resume. Simplices (10, 8, 11), (10, 11, 12), and (10, 12, 13) follow from the usual logic. At this juncture, point 10 has aged sufficiently to justify another shift to end-game strategy, which would indeed verify that (10, 12, 13) contains the optimum. Spendley, Hext, and Himsworth give several second-order experimental designs for the end game which build upon simplices.

This technique is called the *sequential simplex method* by its inventors. To avoid confusing it with Dantzig's simplex method for linear programming problems, we prefer to call it the *simplicial method.* Since it does not accelerate its steps, it is probably too slow for ordinary optimization problems free from experimental error. It does fit in well with Box's evolutionary operations philosophy when there is experimental error present. In this case the stopping rule is replaced by the dictum that any over-age point be rechecked, since it may be spuriously high, owing to sampling error. In evolutionary operations there is no need for a stopping rule, since measurement continues forever. The simplicial method will circle around an optimum, gathering more information until a quadratic fit suggests trying a point off the beaten path. In Fig. 4-35, dropping the stopping rule would generate simplices (10, 13, 14), (10, 7, 14), and finally (10, 8, 7) again. Here the cycle would be completed, and the procedure would continue circling about point 10 until the optimum moved outside the hexagon of points 7, 8, 11, 12, 13, and 14.

4-30 QUADRATIC CONVERGENCE

A quadratic function $q(\mathbf{x})$ of N variables $\mathbf{x}$ has in it one constant term, N linear terms, and $N(N+1)/2$ quadratic terms—$(N+1)(N+2)/2$ in all. If values of a quadratic objective function are known at $(N+1)(N+2)/2$ points, then the coefficients can be found by solving the resulting linear simultaneous equations, provided that the points have been located so as to make the equations linearly independent. This done, one could predict the precise location of the optimum by setting the first derivatives to zero.

Next consider an objective that is a monotonic function $m(q)$ of a quadratic function $q(\mathbf{x})$ of N variables $\mathbf{x}$.

$$y(\mathbf{x}) = m[q(\mathbf{x})] \tag{4-121}$$

Such an objective is called *quasi-quadratic,* its contours being N-dimensional ellipsoids. If the form of the function $m(q)$ is known, then it is, in principle, possible to invert it because of its monotonicity. Application of this inverse $m^{-1}(q)$ gives

$$\begin{aligned} m^{-1}[y(\mathbf{x})] &= m^{-1}\{m[q(\mathbf{x})]\} \\ &= q(\mathbf{x}) \end{aligned} \tag{4-122}$$

In this case one can determine the $(N+1)(N+2)/2$ coefficients of $q(\mathbf{x})$ by evaluating $y(\mathbf{x})$, and consequently $m^{-1}[y(\mathbf{x})]$, at $(N+1)(N+2)/2$ points.

Then one could optimize $q(\mathbf{x})$ directly, the monotonicity of $m(q)$, ensuring that the same point x^* would optimize $y(\mathbf{x})$.

In practice, these ideal situations rarely arise. Usually, $m(q)$ either is not known in advance or is too complicated to be inverted. Yet quadratic or quasi-quadratic functions are useful as standards against which to compare the performances of competing climbing techniques. For example, Section 4-26 showed that on a quasi-quadratic function the gradient method may not achieve the optimum in a finite number of steps, whereas, for $N = 2$, the accelerated gradient technique will always reach the optimum after three straight-line unidimensional searches (see Fig. 4-35). Any climbing procedure capable of finding the optimum of a (quasi-) quadratic function after measuring N gradients is said to *converge (quasi-) quadratically* or to have *(quasi-) quadratic convergence.* These properties are of considerable theoretical importance, and are often referred to when convergence is being discussed or algorithmic convergences compared.

4-31 DEFLECTED GRADIENTS

The Hooke–Jeeves, simplicial, and related methods are useful when only response information is available. The *deflected gradient method* of Fletcher and Powell appears more efficient when accurate gradients are relatively easy to obtain. The latter situation arises when the objective is implicit, making function evaluations so costly that the extra effort needed to find the gradient is relatively small. The deflected gradient method requires N gradient measurements and subsequent unidimensional searches. The searches, although in directions of locally improving values of the objective, are rarely exactly along the gradient—hence the name "deflected" gradient. Fletcher and Powell developed the procedure from some earlier ideas of Davidon and gave proofs of its quadratic convergence. It required only 18 iterations (gradient measurements) to find the optimum for Rosenbrock's function. To compare this with other procedures, one should multiply this number by 3, the number of function evaluations needed to measure a gradient by perturbation in two dimensions. The 54 or so equivalent evaluations are not as low as the two needed by Gauss's method, but certainly better than the 200 needed by partan and the simple ridge-following techniques.

To understand the strategy involved, consider the quadratic objective

$$y = y_0 + c'\mathbf{x} + \tfrac{1}{2}\mathbf{x}'\mathbf{Q}\mathbf{x} \tag{4-123}$$

One can easily show that the step $\Delta\mathbf{x}^*$ needed to go to the optimum $\mathbf{x}^*$ from a point $\mathbf{x}_0$ where the gradient $\nabla y(\mathbf{x}_0)$ ($\equiv \nabla y_0$) is known is

$$\Delta\mathbf{x}^* \equiv (\mathbf{x}^* - \mathbf{x}_0) = -\mathbf{Q}^{-1}\,\nabla y_0 \tag{4-124}$$

In practice, however, $\mathbf{Q}$ is not known in advance, which makes the foregoing relation unusable. This analysis does suggest, however, that one proceed in a direction which is not along the gradient. Fletcher and Powell suggest using an N by N matrix $\mathbf{H}_0$ such that

$$\mathbf{x}_1 - \mathbf{x}_0 \equiv \Delta\mathbf{x}_1 = -\mu_1\mathbf{H}_0 \nabla y_0 \tag{4-125}$$

The matrix $\mathbf{H}_0$ is chosen to be positive-definite, so that the new direction will give local improvement of y, whereas μ_1 is the search parameter chosen to optimize $y(\Delta\mathbf{x}_1)$ along the line of search defined by Eq. (4-125). If the investigator has some idea in which direction the optimum might be, he should select $\mathbf{H}_0$ accordingly. In the absence of such information he might as well choose the N by N unit matrix $\mathbf{I}$, it being the simplest to deal with. This means that the first step is an ordinary gradient optimization after all.

The deflected gradient method generates a sequence of points $\mathbf{x}_1, \ldots,$ $\mathbf{x}_n$ such that the gradient ∇y_n at each point $\mathbf{x}_n$ is orthogonal (perpendicular) to all preceding steps $\Delta\mathbf{x}_1, \ldots, \Delta\mathbf{x}_n$.

$$(\nabla y_n)' \, \Delta\mathbf{x}_i = 0; \qquad i = 1, \ldots, n; \quad n = 1, \ldots, N \tag{4-126}$$

The Nth gradient, ∇y_N, must therefore be orthogonal to N vectors $\Delta\mathbf{x}$, which are constructed to be linearly independent. Hence ∇y_N must vanish, indicating that $\mathbf{x}_N$ is the minimum sought in this ideal quadratic case. Moreover, a sequence of N by N positive definite matrices $\mathbf{H}_1, \ldots, \mathbf{H}_N$ is generated such that the last one is precisely the inverse of the originally unknown Hessian matrix $\mathbf{Q}$.

$$\mathbf{H}_N = \mathbf{Q}^{-1} \tag{4-127}$$

In ideal circumstances one need not compute this matrix, but in practice ∇y_N may not vanish, say because of accumulated roundoff errors, and an $(N + 1)$th step using $\mathbf{H}_N$ would be justified. In this case

$$\begin{aligned}\Delta\mathbf{x}_{N+1} \equiv \mathbf{x}_{N+1} - \mathbf{x}_N &\equiv -\mu_{N+1}\mathbf{H}_N \nabla y_N \\ &= -\mu_{N+1}\mathbf{Q}^{-1}\nabla y_N \\ &= \mu_{N+1}(\mathbf{x}^* - \mathbf{x}_N)\end{aligned}$$

Thus, if $\mu_{N+1} = 1$,

$$\mathbf{x}_{N+1} = \mathbf{x}^* \tag{4-128}$$

and the search terminates after $N + 1$ steps. Further correction for rounding errors can be made by minimizing with respect to μ_{N+1}, which in practice may differ from unity.

When y is not quadratic, the procedure will not end so soon, but after N steps the deflection matrices $\mathbf{H}$ usually become increasingly better estimates of the curvature at the minimum, as measured by the inverse of the Hessian matrix of second derivatives there. Thus the method converges rapidly as soon as it gets close enough to make a quadratic approximation valid. Following is a detailed description of the method and its behavior in ideal quadratic circumstances.

After the nth step has been taken ($n = 1, \ldots, N$), the location of the nth point $\mathbf{x}_n$ is known. The gradient ∇y_n is determined there, as are two new N by N matrices $\mathbf{A}_n$ and $\mathbf{B}_n$ used to form the new deflection matrix $\mathbf{H}_n$ from the preceding one, $\mathbf{H}_{n-1}$.

$$\mathbf{H}_n = \mathbf{H}_{n-1} + \mathbf{A}_n + \mathbf{B}_n \tag{4-129}$$

The role of the $\mathbf{A}_n$ is to generate the inverse $\mathbf{Q}^{-1}$ in N steps. In fact, the $\mathbf{A}_n$ will be chosen so that their sum will be $\mathbf{Q}^{-1}$.

$$\sum_{n=1}^{N} \mathbf{A}_n = \mathbf{Q}^{-1} \tag{4-130}$$

The sequence of matrices $\mathbf{B}_n$ is intended to cancel out the initial assumption for $\mathbf{H}_0$. Their sum is therefore the negative of $\mathbf{H}_0$.

$$\sum_{n=1}^{N} \mathbf{B}_n = -\mathbf{H}_0 \tag{4-131}$$

To see that Eqs. (4-130) and (4-131) do give a sequence converging to $\mathbf{Q}^{-1}$, write

$$\begin{aligned}
\mathbf{H}_N &= \mathbf{H}_{N-1} + \mathbf{A}_N + \mathbf{B}_N \\
&= \mathbf{H}_{N-2} + (\mathbf{A}_N + \mathbf{A}_{N-1}) + (\mathbf{B}_N + \mathbf{B}_{N-1}) \\
&\ \ \vdots \\
&= \mathbf{H}_0 + \sum_{n=1}^{N} \mathbf{A}_n + \sum_{n=1}^{N} \mathbf{B}_n = \mathbf{Q}^{-1}
\end{aligned} \tag{4-132}$$

It remains to find how to compute the $\mathbf{A}_n$ and $\mathbf{B}_n$ from information generated along the way. The principal agent for this is the *gradient difference vector* $\mathbf{g}_n$, defined as the difference between the gradients at the beginning and at the end of the nth step.

$$\mathbf{g}_n \equiv \nabla y_n - \nabla y_{n-1} \tag{4-133}$$

The gradient difference can be related to the unknown matrix $\mathbf{Q}$ by differentiating the objective function to obtain the gradients.

$$\begin{aligned}
\mathbf{g}_n &= (\mathbf{c} + \mathbf{Q}\mathbf{x}_n) - (\mathbf{c} + \mathbf{Q}\mathbf{x}_{n-1}) \\
&= \mathbf{Q}\,\Delta\mathbf{x}_n
\end{aligned} \tag{4-134}$$

This furnishes N linear equations, not enough to determine all the N^2 elements of $\mathbf{Q}$. One can see that N such gradient differences would be needed, and any of the previous methods would yield an estimate of $\mathbf{Q}$ after N steps if gradient differences were tabulated and the resulting N^2 equations were solved. The advantage of the deflected gradient method is that, in effect, it solves these equations in a stepwise fashion that does not require storage of data from all previous steps. Updating the matrices $\mathbf{A}_n$ and $\mathbf{B}_n$ is all that is needed.

To find $\mathbf{A}_n$, write, using Eqs. (4-130) and (4-134),

$$\begin{aligned}\Delta\mathbf{x}_n &= \mathbf{I}\,\Delta\mathbf{x}_n = \mathbf{Q}^{-1}\mathbf{Q}\,\Delta\mathbf{x}_n \\ &= \sum_{i=1}^{N} \mathbf{A}_i\mathbf{Q}\,\Delta\mathbf{x}_n \\ &= \sum_{i=1}^{N} \mathbf{A}_i\mathbf{g}_n\end{aligned} \tag{4-135}$$

But if $\mathbf{A}_n$ is to depend only upon information generated at the nth step, then only the nth term of the sum should be nonzero. That is,

$$\Delta\mathbf{x}_n = \mathbf{A}_n\mathbf{g}_n \tag{4-136a}$$

and

$$\mathbf{A}_i\mathbf{g}_n = 0 \qquad \text{for } i \neq n \tag{4-136b}$$

Multiplication of $\Delta\mathbf{x}_n$ by unity (written in a complicated way) gives

$$\Delta\mathbf{x}_n = \Delta\mathbf{x}_n\left(\frac{\Delta\mathbf{x}_n'\mathbf{g}_n}{\Delta\mathbf{x}_n'\mathbf{g}_n}\right) = \left(\frac{\Delta\mathbf{x}_n\,\Delta\mathbf{x}_n'}{\Delta\mathbf{x}_n'\mathbf{g}_n}\right)\mathbf{g}_n$$

which by comparison with Eq. (4-136) gives

$$\mathbf{A}_n = \frac{\Delta\mathbf{x}_n\,\Delta\mathbf{x}_n'}{\Delta\mathbf{x}_n'\mathbf{g}_n} \tag{4-137}$$

The inductive proof of orthogonality Eq. (4-136b), given by Fletcher and Powell, will not be detailed here. It depends on the fact that the gradient at $\mathbf{x}_n$ must be orthogonal to the preceding step, $\Delta\mathbf{x}_n$, because y is optimized along the line of search. That is,

$$(\nabla y_n)'\,\Delta\mathbf{x}_n = 0 \tag{4-138}$$

Success of the deflected gradient method depends on the fact that the steps $\Delta\mathbf{x}_1, \Delta\mathbf{x}_2, \ldots, \Delta\mathbf{x}_n$ are related to $\mathbf{H}_n$ by

$$\mathbf{H}_n\mathbf{Q}\,\Delta\mathbf{x}_i = \Delta\mathbf{x}_i; \qquad i = 1, 2, \ldots, n \tag{4-139}$$

In particular, at the last step ($n = N$)

$$\mathbf{H}_N\mathbf{Q}\,\Delta\mathbf{x}_i = \Delta\mathbf{x}_i \tag{4-140}$$

which can be true for N linearly independent steps $\Delta\mathbf{x}_i$ only if $\mathbf{H}_N\mathbf{Q}$ is the unit matrix. This proves that

$$\mathbf{H}_N = \mathbf{Q}^{-1} \tag{4-141}$$

In matrix jargon, the steps $\Delta\mathbf{x}_i$ are called *eigenvectors* of $\mathbf{H}_n\mathbf{Q}$, each with unit eigenvalue. Arranging things so that Eq. (4-139) holds gives a way to compute the sequence of $\mathbf{B}_n$. Equations (4-129), (4-134), and (4-136a) give

$$\begin{aligned}\mathbf{H}_n\mathbf{Q}\,\Delta\mathbf{x}_n &= \mathbf{H}_n\mathbf{g}_n \\ &= \mathbf{H}_{n-1}\mathbf{g}_n + \mathbf{A}_n\mathbf{g}_n + \mathbf{B}_n\mathbf{g}_n \\ &= \mathbf{H}_{n-1}\mathbf{g}_n + \Delta\mathbf{x}_n + \mathbf{B}_n\mathbf{g}_n\end{aligned} \tag{4-142}$$

Equations (4-139) and (4-142) imply that

$$\mathbf{B}_n\mathbf{g}_n = -\mathbf{H}_{n-1}\mathbf{g}_n \tag{4-143}$$

The obvious solution ($\mathbf{B}_n = -\mathbf{H}_{n-1}$) must be ruled out because of its uninteresting consequences, but there is a nontrivial possibility obtained by multiplying the right member of Eq. (4-143) by unity, written in a way even more complicated than before. Observe that for any N vector $\mathbf{z}$, it is true that

$$1 = \frac{\mathbf{z}'\mathbf{g}_n}{\mathbf{g}_n'\mathbf{z}} \qquad (\mathbf{g}_n'\mathbf{z} \neq 0) \tag{4-144}$$

Equations (4-143) and (4-144) together give

$$\mathbf{B}_n\mathbf{g}_n = -\frac{\mathbf{H}_{n-1}\mathbf{g}_n\mathbf{z}'\mathbf{g}_n}{\mathbf{g}_n'\mathbf{z}} \tag{4-145}$$

whence, by comparing matrices,

$$\mathbf{B}_n = -\frac{\mathbf{H}_{n-1}\mathbf{g}_n\mathbf{z}'}{\mathbf{g}_n'\mathbf{z}} \tag{4-146}$$

Notice that the numerator is a (singular) N by N matrix, whereas the denominator is a scalar. Since $\mathbf{H}_n$, $\mathbf{H}_{n-1}$, and $\mathbf{A}_n$ are all required to be symmetric, the vector $\mathbf{z}$ must be selected by make $\mathbf{B}_n$ symmetric. The correct choice is

$$\mathbf{z} = \mathbf{H}_{n-1}\mathbf{g}_n \tag{4-147}$$

which gives

$$\mathbf{B}_n = -\frac{\mathbf{H}_{n-1}\mathbf{g}_n\mathbf{g}_n'\mathbf{H}_{n-1}'}{\mathbf{g}_n'\mathbf{H}_{n-1}\mathbf{g}_n} \tag{4-148}$$

The reader may wish to test his or her understanding of the deflected gradient method by applying it to the following problem.

$$\text{Minimize } y = 2x_1^2 + x_2^2 + 3x_3^2 \tag{4-149}$$

starting at the point (—1, 1, —1), where $\Delta y = (-4, 2, -6)$. Table 4-4, prepared by B. A. Williams while a graduate student at Stanford, shows the computations involved.

TABLE 4-4

Deflected Gradient Method Example

$\mathbf{x}_0 = (-1, 1, -1)$ $y\langle \mathbf{x}_0 \rangle = 6.0000$ $\nabla y_0 = (-4, 2, -6)$ $\mathbf{H}_0 = \begin{pmatrix} 1 & & \\ 0 & 1 & \\ 0 & 0 & 1 \end{pmatrix}$

	$n=1$	$n=2$	$n=3$
$\mathbf{H}_{n-1}\nabla \mathbf{y}_{n-1}$	-4.0000 2.0000 -6.0000	-1.0612 1.2653 0.6122	0.2641 0.5281 -0.0587
μ_n	0.1944	0.3116	0.4106
$\mathbf{x}_n$	-0.2222 0.6111 0.1667	0.1084 0.2169 -0.0241	0.0000 0.0000 0.0000
$y\langle \mathbf{x}_n \rangle$	0.5556	0.0723	0.0000
$\nabla \mathbf{y}_n$	-0.8889 1.2222 1.0000	0.4337 0.4337 -0.1446	0.0000 0.0000 0.0000
$\mathbf{g}_n$	0.3111 -0.7778 7.0000	1.3226 -0.7885 -1.1446	-0.4337 -0.4337 0.1446
$\mathbf{A}_n$	0.0556 -0.0278 0.0139 0.0833 -0.0417 0.1250	0.1131 -0.1349 0.1608 -0.0652 0.0778 0.0377	0.0813 0.1627 0.3253 -0.0181 -0.0361 0.0040
$\mathbf{B}_n$	-0.1633 0.0408 -0.0102 -0.3673 0.0918 -0.8265	0.6387 0.3553 -0.1976 0.3233 -0.1798 -0.1637	-0.1980 -0.3961 -0.7922 0.0440 0.0880 -0.0098
$\mathbf{H}_n$	0.8923 0.0130 1.0037 -0.2840 0.0502 0.2985	0.3667 0.2334 0.9669 -0.0259 -0.0519 0.1724	0.2500 0.0000 0.5000 0.0000 0.0000 0.1667

On nonquadratic functions convergence cannot be guaranteed any more than it can for any other climbing methods. Yet Fletcher and Powell have had success with it even on helical ridges and with functions of up to 50 variables. In all cases the number of steps increased only linearly with the number of variables. Progress toward the optimum, although slow during early iterations when data are being gathered for estimates of $\mathbf{Q}^{-1}$, often speeds up remarkably as the quadratic approximation gets better. It appears to be the best method available when accurate gradients are relatively easy to obtain. Powell (1964) has devised a variation not requiring explicit evaluation of the gradients, equivalent information being generated during the unidimensional searches. This modification required 140 function evaluations to minimize Rosenbrock's function, a performance comparable to that of partan.

This completes the description of climbing techniques for finding interior optima. In Chaptor 5 we will discuss the consequences of boundary solutions generated by constraint relationships.

4-32 SUMMARY: PART II

The climbing methods begin by measuring a gradient and end with estimating curvature by second-order approximations. The most effective techniques have ridge-following properties, and one of them, gradient partan, also has quasi-quadratic convergence. When gradients are relatively easy to measure, or when the objective function is particularly difficult to measure, the deflected gradient procedure with its quadratic convergence seems best. For least-squares problems, Levenberg's method is highly effective. In the presence of experimental error, and for continual monitoring of a shifting optimum, the simplicial method is suitably cautious and easy to implement. Constraints can be handled by straightforward adaptations of the Jacobian procedures of nonlinear programming, by local linearization, or by methods that repel the search path away from the constraints. Many climbing methods could be combined with elimination techniques if one wanted to know the region of uncertainty. The climbing procedures, originally developed for unconstrained objectives, may also be useful in nonlinear programming computer codes because of their ability to adjust several variables at once.

No matter how sophisticated the climbing procedure used, one should always be careful in setting up the problem. By removing interaction, by scaling variables to make contours approximately circular, and by choosing representations easy to approximate quadratically, one can cut down the searching effort significantly. There is no need to make the problem unnecessarily difficult.

With absolutely perfect preparation of the problem, the number of function evaluations needed is proportional to N, the number of independent variables. Imperfect scaling drives the number up to N^2, even for quasi-quadratic functions, which is about as good as one could hope for. Linear programming computations increase as N^3, so all the procedures described so far must be confined to problems with a number of independent variables much smaller than those encountered in most practical problems. The next two chapters show how to break very large optimization problems down into pieces of a size solvable by the methods developed so far. When this is done, the computation effort grows only linearly, not quadratically or cubically, with the number of pieces.

BIBLIOGRAPHY

ARROW, K. J., and A. C. ENTHOVEN, "Quasi-concave programming," *Econometrica*, **29**, 4 (October 1961) 779–800.

AVRIEL, M., and D. J. WILDE, "Optimal search for a maximum with sequences of simultaneous function evaluations," *Man. Sci.*, **12**, 9 (May 1966), 722–731.

———, and D. J. WILDE, "Optimality proof for the symmetric Fibonacci search technique," *Fibonacci Quart. J.*, **4**, 3 (October 1966), 265–269.

BAER, R. M., "Note on an extremum locating algorithm," *Comp. J.*, **5**, 3 (1962), cited in Flood and Leon.

BALL, W. W. R., *A Short Account of the History of Mathematics* (Macmillan, London, 1888).

BEIGHTLER, C. S., G. D. BOUCHEY, E. L. DRAPER, and B. V. KOEN, "Multiple foil activation spectrum determination using a numerical direct search technique," *Amer. Nucl. Soc. Trans.*, **14**, 2 (1971).

BEIGHTLER, C. S., L. G. MITTEN, and G. L. NEMHAUSER, "A short table of *z*-transforms and generating functions," *Operations Res.*, **9**, 4 (August 1961), 574–578.

BELL, E. T., *The Development of Mathematics* (McGraw-Hill, New York, 1940).

BELLMAN, R., *Dynamic Programming* (Princeton Univ., Princeton, N.J., 1957), p. 34.

BERGE, CLAUDE, *Topological Spaces*, E. M. Patterson, trans. (Oliver & Boyd, Edinburgh and London, 1963; French ed., 1959), pp. 199–200.

BIRKHOFF, G., and S. MAC LAND, *A Survey of Modern Algebra*, 3rd ed. (Macmillan, New York, 1965).

BOOTH, A. D., "An application of the method of steepest descents to the solutions of systems of nonlinear simultaneous equations," *Quart. J. Mech. Appl. Math.*, **2**, 4 (1949), 460.

BOOTH, A. W., and T. I. PETERSON, "Nonlinear estimation," IBM 704 program WL NLI (1960), cited in Harkins.

BOX, G. E. P., "The exploration and exploitation of response surfaces," *Biometrics*, **10** (1954), 16.

———, "Evolutionary operation: a method for increasing industrial productivity," *Appl. Stat.*, **6** (1957), 81–101.

———, and K. B. WILSON, "On the experimental attainment of optimum conditions," *J. Roy. Stat. Soc.*, **B13** (1951), 1.

BOX, M. J., "A comparison of several current optimization methods, and the use of transformations in constrained problems," *Comp. J.*, **9** (1966), 67.

BROOKS, S. H., "A comparison of maximum-seeking methods," *Operations Res.*, **7**, 4 (July 1959), 430–457.

BUEHLER, R. J., B. N. SHAH, and O. KEMPTHORNE, "Some properties of steepest ascent and related procedures for finding optimum conditions," Iowa State Univ. Statistical Laboratory, Ames, Iowa (April 1961).

CAMPEY, I. G., and D. G. NICKOLS: "Simplex Minimization," Imperial Chemical Industries Ltd., August 1961.

CAUCHY, A. L., "Méthode générale pour la résolution des systèmes d'équations simultanées," *Compt. Rend. Acad. Sci. Paris*, **25** (1847), 536–538. Also in *Oeuvres complètes d'Augustin Cauchy*, **10** (Gauthier-Villars, Paris, 1901), pp. 399–406.

COXETER, H. S. M., "The golden section, phyllotaxis, and Wythoff's game," *Scripta Math.* (1954), 135–143.

DAVIDON, W. C., "Variable metric method for minimization," AEC Research and Development Report Anl-5990, December 1959; cited in Fletcher and Powell.

DAVIES, D., The Use of Davidon's Method in Nonlinear Programming, *ICI Ltd. Report MSDH/68/110*, August 1968; *Document N69–33235* available from CFSTI, Springfield, Va.

———, Review of Constrained Optimization, Clearinghouse for Federal Scientific and Technical Information, *Document N69–36898*, September 30, 1968.

———, "Some practical methods of optimization: notes for the NATO Summer School, on integer and nonlinear programming," June 8–20, 1969, Academic Press, New York (in press). (Discussion of Rosenbrock's and Davidon's methods extended to linear and nonlinear inequality constraints.)

DAVIES, O. L., *The Design and Analysis of Industrial Experiments* (Oliver & Boyd. London, 1956).

FIBONACCI, LEONARDO, *Algebra et almuchabala* (*liber abbaci*) (1202), cited in Bell, p. 160.

FLETCHER, R., and M. J. D. POWELL, "A rapidly convergent descent method for minimization," *Comp. J.*, **6**, 2 (1963), 163–168.

FLOOD, M. M., and A. LEON, "A universal adaptive code for optimization (GROPE)," in Lavi and Vogl, pp. 101–130.

FORSYTHE, G. E., and T. S. MOTZKIN, "Acceleration of the optimum gradient method," preliminary report (abstract) *Bull. Amer. Math. Soc.*, **57** (1951), 304–305, cited in Shah, Buehler, and Kempthorne.

FRIEDMAN, M., and L. S. SAVAGE, "Planning experiments seeking maxima," in *Selected Techniques of Statistical Analysis*, C. Wisenhart, M. W. Hastay, and W. A. Wallis, eds. (McGraw-Hill, New York, 1947), cited in Shah, Buehler, and Kempthorne.

GAUSS, C. F., *Werke*, **4** (Göttingen, 1821), cited in O. L. Davies, p. 578.

GELFAND, I. M., and M. L. TSETLIN, "The principle of nonlocal search in automatic optimization systems," *Doklady Akad. Nauk, SSSR*, **137**, 2 (March 1961), 295–298.

GENOCCHI, ANGELO, and G. PEANO, *Calcolo Differenziale e Principii di Calcolo Integrale* (1884), app., prob. 133–136; cited in Hancock, pp. 33 *et seq.*

GIRARD, A., *L'arithmétique de Simon Stévin de Bruges* (Leiden, Holland, 1634), p. 667, cited in Coxeter, p. 139.

GLASS, H., and L. COOPER, "Sequential search: a method for solving constrained optimization problems," *J. Assoc. Comp. Mach.*, **12** (1965), 71.

HARKINS, A., "The use of parallel tangents in optimization," *Optimization Techniques* (see Blakemore and Davis 1964, pp. 35–40).

HESTENES, M. R., and E. STIEFEL, "Method of conjugate gradients for solving linear systems," *J. Res. Natl. Bur. Standards*, **49** (1952), 409–436.

HIMSWORTH, F. R., "Empirical methods of optimization," *Trans. Inst. Chem. Engrs.*, **40** (1962), 345–349.

HOOKE, R., and T. A. JEEVES, "Comments on Brooks' discussion of random methods," *Operations Res.*, **6**, 6 (November 1958), 881–882.

———, "Progress report on Opcon," *Control Eng.*, **6** (November 1959), 124.

———, and T. A. JEEVES, "Direct search solution of numerical and statistical problems," *J. Assoc. Comp. Mach.*, **8**, 2 (April 1961), 212–229.

HORN, FRIEDRICH (private communication, 1965).

HUMPHREY, W. E., "A general minimizing routine-Minfun," Univ. of California, Lawrence Radiation Laboratory, Berkeley; internal memorandum (September 1962), 9 pp., cited in Flood and Leon.

JOHNSON, S. M., "Optimal search for a maximum is Fibonaccian," RAND Corporation Report P-856 (1956); see also Bellman.

KELLEY, H. J., "Method of gradients," in G. Leitman, ed., *Optimization Techniques with Applications to Aerospace Systems* (New York, Academic Press, 1962), pp. 205–254.

KIEFER, J., "Sequential minimax search for a maximum," *Proc. Amer. Math. Soc.*, **4** (1953), 502–506.

KLINGMAN, W. R., and D. N. HIMMELBLAU, "Nonlinear programming with the aid of a multiple-gradient summation technique," *J. Assoc. Comp. Mach.*, **11**, 4 (October 1964), 400–415.

LAPIDUS, L., E. SHAPIRO, S. SHAPIRO, and R. E. STILLMAN, "Optimization of process performance," *Amer. Inst. Chem. Engrs. J.*, **7**, 2 (June 1961), 288–294.

LAVI, A., and T. P. VOGL, eds., *Recent Advances in Optimization Techniques* (Wiley, New York, 1966).

LEGGE, J., *The Chinese Classics*, Vol. 1 (Trübner and Co., London, 1861).

LEITMAN, G., ed., *Optimization Techniques with Applications to Aerospace Systems* (New York, Academic Press, 1962).

LEVENBERG, K., "A method for the solution of certain non-linear problems in least squares," *Quart. Appl. Math.*, **2** (1944), 164–168.

LUCAS, E., "Note sur l'application des séries récurrents à la recherche de la loi de distribution des nombres premiers," *Compt. Rend. Acad. Sci. Paris*, **82** (1876), 165–167, cited in Coxeter.

MCCUTCHEON, T. P., H. SELTZ, and J. C. WARNER, *General Chemistry* (Van Nostrand, Princeton, N.J., 1939), p. 236.

MOOD, A. M., *Introduction to the Theory of Statistics* (McGraw-Hill, New York, 1950).

NEWMAN, D. J., "Location of the maximum on unimodal surfaces," *J. Assoc. Comp. Mach.*, **12**, 3 (July 1965), 395–398.

NIETZSCHE, FRIEDRICH, "Excelsior" (1882), from *Joyful Wisdom*, Thomas Commons, trans. Poetry versions by Paul V. Cohn and Maude D. Petrie (F. Ungar, New York, 1964).

OLIVER, L. T., and D. J. WILDE, "Symmetric sequential minimax search for a maximum," *Fibonacci Quart. J.*, **2**, 3 (October 1964), 169–175.

PHILLIPS, H. B., *Vector Analysis* (Wiley, New York, 1933), pp. 35 *et seq.*

PISANO, LEONARDO (FIBONACCI), *Scritti*, Vol. 1 (1857), pp. 282–284, cited in Coxeter.

POWELL, M. J. D., "An efficient method for finding the minimum of a function of several variables without calculating derivatives," *Comp. J.*, **7** (1964), 155–162.

———, "A method for minimizing a sum of squares of nonlinear functions without calculating derivatives," *Comp. J.*, **7** (1965), 303–307.

PYTHAGORAS, cited in Turnbull.

RAPHSON, JOSEPH, *Analysis aequationem universalis*, cited in Cajori, p. 203.

ROSENBROCK, H. H., "An automatic method for finding the greatest or least value of a function," *Comp. J.*, **3**, 3 (October 1960), 175–184.

SCHEEFER, LUDWIG, "Über die Bedeutung der Begriffe Maximum and Minimum in der Variationsrechnung," *Math. Ann.*, **26** (1886), 197–208, cited in Hancock, Chap. 4.

SHAH, B. V., R. J. BUEHLER, and O. KEMPTHORNE, "Some algorithms for minimizing a function of several variables," *J. Soc. Ind. Appl. Math.*, **12**, 1 (March 1964), 74–92.

SOUTHWORTH, R., and S. DELEEUW, *Digital Computation and Numerical Methods* (McGraw-Hill, New York, 1965).

SPENDLEY, N., G. R. HEXT, and F. R. HIMSWORTH, "Sequential application of simplex design in optimization and evolutionary operations," *Technometrics*, **4**, 4 (November 1962), 441–459.

SWANN, W. H., "Report on the development of a new direct search method of optimization," *Imperial Chemical Industries Ltd., Central Instr. Lab. Res. Note 64/3*, 1964.

TURNBULL, H. W., *The Great Mathematicians* (New York Univ. Press, New York, 1961), pp. 8–15. (Thanks to M. Manoff for this reference.)

WILDE, D. J., "Optimization by the method of contour tangents," *Amer. Inst. Chem. Engrs. J.*, **9**, 2 (March 1963), 186–190.

———, *Optimum Seeking Methods* (Prentice-Hall, Englewood Cliffs, N.J., 1964).

———, "A multivariable dichotomous optimum seeking method," *Inst. Elec. Electron. Engrs. G-AC Trans. AC-10*, **1** (January 1965), 85–87.

———, "Jacobians in constrained nonlinear optimization," *Operations Res.*, **13**, 5 (September 1965), 848–856.

———, "Objective function indistinguishability in unimodal optimization," in Recent Advances in Optimization Techniques, *Proc. IEEE—OSA Conf. Optimization*, A. Lavi and T. P. Vogl, eds. (Wiley, New York, 1966), pp. 341–350.

WOOD, E. F., "Application of direct search to the solution of engineering problems," *Westinghouse Scientific Paper 64–1210–1–P*, 1960.

ZELLNICK, H. E., N. E. SONDAK, and R. S. DAVIS, "Gradient search optimization," *Chem. Eng. Progr.*, **58**, 8 (1962), pp. 35–41.

EXERCISES

4-1. Prove that, in a Fibonacci search, $x_1^1 = I_1^2$ implies that $x_1^2 = I_1^3$.

4-2. Prove Eq. (4-36).

4-3. Prove Eq. (4-37). (Proof given in Wilde, 1966.)

4-4. Verify Newman's upper bound for $k = 99$, setting up the lattice in the first quadrant. (For definiteness, assume that the contours are circles centered on the origin.) Suggest a smaller bound.

4-5. Prove that the line $x_2 = \frac{1}{2}$ is tangent to a contour only at $\mathbf{u}^1$, the high point, in Fig. 4-10.

4-6. Prove that all convex functions are unimodal.

4-7. Find a cubic function of x that is unimodal but not convex for $0 \leq x \leq 1$.

4-8. Search for the maximum of the function $y = 3 + 6x - 4x^2$ in the interval $0 \leq x \leq 1$ with four sequential experiments. Space all experiments at least 0.05 unit apart. What is the length of the final interval? What is the highest value of y attained?

4-9. Conduct the same search as in Exercise 4-8 with four lattice experiments. What is the highest value of y attained?

4-10. Could the preceding search have been conducted efficiently with five experiments? Why?

4-11. A medical research team is seeking the American with the highest concentration of exogen in the blood stream. Exogen concentration is known to be a unimodal function depending only on the date of birth of the individual, which is known to the team through U.S. Census Bureau records. What would be the least number of Americans that the team could examine and still find the one with the highest exogen content?

4-12. Draw an isometric projection and the three orthogonal views for the following functions in the region $-2 \leq x_1 \leq 2$, $-2 \leq x_2 \leq 2$. Indicate all the maxima, minima, and saddles in the region.

(a) $y = x_1 x_2$

(b) $y = x_1^2 + x_2^2$

(c) $y = \exp(-x_1^2 + x_2^2)$
(d) $y = \ln(x_1^2 + x_1x_2 - 2x_2^2)$
(e) $y = x_1^3 + x_2^3$
(f) $y = x_1^2x_2^2$

4-13. Find the equation of the contour tangent through the point (1, −2, 3) for the functions
(a) $y = x_1^2 + x_2^2 + x_3^2$
(b) $y = \exp(x_1^2 + x_3^2 - x_1 + 3x_2 + 2)$
(c) $y = \ln(x_1^2 + x_1x_2 + x_2^2)$
(d) $y = x_1^3 + x_2^3$

4-14. Find the equation of the contour tangent through the point (−2, 1, −1, 4), using the following data:

x_1	x_2	x_3	x_4	y
−2	1	−1	4	10.0
−1.9	1	−1	4	10.3
−2	1.1	−1	4	9.8
−2	1	−0.9	4	9.7
−2	1	−1	4.1	10.4

Estimate the value of y in Problem 4-13 at the point

$$(-1.5, 0.5, -0.5, 4.5)$$

4-15. Use the combination algorithm of Box, Davies, and Swann on Example Problem 4-3 of Section 4-13.

4-16. Construct a quadratic approximation to the function $y = \exp(2x_1^2 + 2x_2^2 + x_1 - 5x_2 + 10)$ in the vicinity of (0, 1). Where does the approximation predict the minimum to be? Compare the predicted value at this point with the actual value.

4-17. From the accompanying data, construct both a noninteracting and an interacting quadratic approximation. Compare the stationary points predicted by these approximations. Is the stationary point a maximum, minimum, or saddle?

x_1	x_2	y
1.0	4.0	5.85
1.0	3.0	5.85
1.0	2.0	6.00
2.0	4.0	6.10
2.0	3.0	6.10
2.0	2.0	6.10
3.0	4.0	5.85
3.0	3.0	6.05

4-18. A function y depends on four independent variables x_1, x_2, x_3, and x_4, and the following table gives the measured value of y at eight different points.

(a) Give the coordinates of any point on the line of steepest ascent passing through (0, 1, −1, 3). [Do not give (0, 1, −1, 3) as an answer.]

(b) Give the coordinates of any point [except (0, 1, −1, 3)] in the contour tangent hyperplane passing through (0, 1, −1, 3).

x_1	x_2	x_3	x_4	y
0	1	−1	3	5
1	1	−1	3	7
2	1	−1	3	9
−1	2	−1	3	2
0	−1	−1	3	7
0	1	1	3	7
0	1	−1	2	5
0	2	0	3	5

4-19. It is desired to find the point where an unknown function y is maximum on the line between the two points (1, −1, 0, 2) and (−5, −1, 3, 1).

(a) Assuming perfect resolution and unimodality of the function on the line, give the coordinates of the points where you would measure the function next, assuming that you are going to conduct a total of five new experiments in sequence.

(b) What is the final interval of uncertainty of the coordinate x_1?

4-20. Use the sectioning method and the gradient method to find the minimum of the following functions, starting from the points given. (Make no more than four linear searches.)

(a) $y = x_1^2 + 3x_2^2 + 2x_3^2$; (2, −2, 1)

(b) $y = 2x_1^2 + 2x_1x_2 + 5x_2^2$; (2, −2)

4-21. Prove that the points given in Fig. 4-39 are a unit distance apart. Using the results in Table 4-4 for the deflected gradient example, verify Eqs. (4-30), (4-31), (4-136b), (4-140), and (4-141).

4-22. On the functions of Exercise 4-12, use the following procedures, making no more than five moves:

(a) Pattern search (set $\delta_1 = \delta_2 = 0.1$ and use second-order check at apparent optimum)

(b) Deflected gradients

Constrained Nonlinear Optimization 5

The dragon exceeds the proper limits;
there will be occasion for repentance.

The Book of Changes (China, ca. 1200 B.C.,
Translated by J. LEGGE)

The ancient *Book of Changes* (*Yî King* in Cantonese; *I Ching* in Mandarin) is a manual for divining the future from tortoise shells and the stalks of a sacred flower. Its cryptic and ambiguous fortune-telling recipes were refined by Confucius and his disciples into proverbs, an example being the quotation beginning this chapter. With oriental subtlety this maxim reminds us that most human endeavors have bounds which are dangerous or impossible to violate. Policies and designs, like dragons, must not exceed the proper limits, lest later there be "occasion for repentance." To be useful, optimization theory must recognize such bounds and show how to handle them.

In Chapter 2 the question of finding an optimum in a feasible region defined by equations and inequalities was raised but was only partly settled. In Chapter 4 we dealt exclusively with unconstrained optimization procedures. When there are no equations, but only strict inequalities, satisfied at the optimum, it can be found by setting first derivatives to zero. When a known set of equations, but no inequalities, constrains the optimum, it can be located by a similar indirect approach involving decision derivatives. More generally, however, one cannot predict which inequality constraints will be "tight" (satisfied as strict equalities) and which "loose" (satisfied as strict inequalities at the optimum. Optimization problems of this sort, which have come to be called *mathematical programming* problems, will appear throughout the rest of the book. The totally linear case has been dealt with independently in Chapter 3. This section shows what can be done when the functions are nonlinear. Chapter 6 will exploit nonlinearities of a special kind.

Solving a mathematical programming problem is like looking for the high point inside a fenced-in field on the *side* of a mountain. The peak could in principle be found by classical methods, but it lies outside the fence. If the mountain is unimodal, as in Fig. 5-1, the high point in the field is up against a fence, perhaps even in a corner where two fences meet. Thus an explorer climbs upward until he strikes the fence and then continues along the fence in a rising direction until he reaches a local optimum, as along the path *abc*. A more fortunate track starts at *d*, hits a fence at *e*, leaves it again at *f*, encounters a new one at *g*, and follows it to the true maximum at *h*. Notice that strictly rising paths do not always lead to the true maximum in a constrained region, even when the objective function is unimodal.

Not knowing against which segments of the boundary the optimum lies

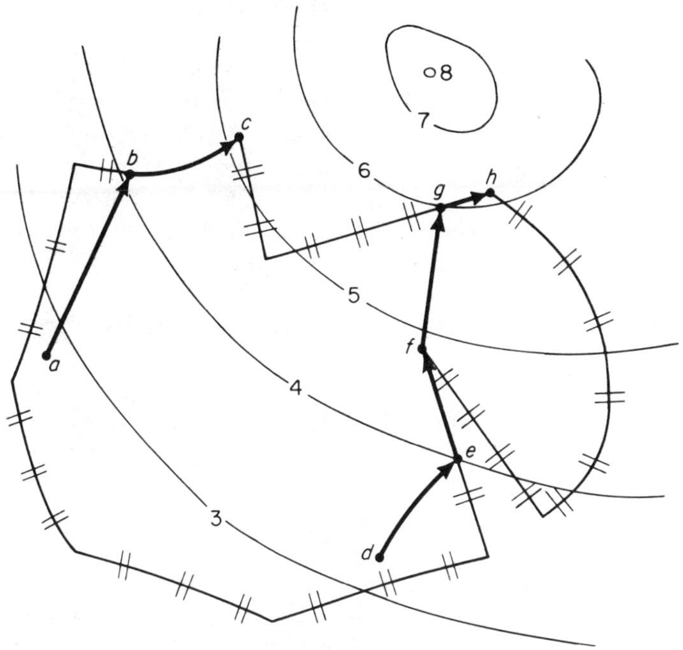

Figure 5-1. Hill climbing inside a fence.

forces one to proceed from one trial point to another in several jumps instead of one. But for each step the critical constraints can usually be identified, which permit indirect methods to be used, at least after some modification. Thus many mathematical programming procedures consist of a sequence of indirect optimizations in subregions of the feasible domain, guided by a master strategy for estimating new directions in which to proceed and the constraints most likely to be encountered. This section extends the indirect methods for use in these semidirect techniques. The key idea is to develop new constrained derivatives by manipulating differentials as in Chapter 2.

5-01 EQUALITY CONSTRAINTS

In most operations research problems of any size and importance, the variables are required to satisfy side conditions expressed as equations. These arise, for example, whenever conservation of mass, energy, or momentum is involved, or when forces or moments must be kept in equilibrium. Such equations, or *equality constraints*, are henceforth written in the *null form* $\mathbf{e}(\mathbf{x}) = \mathbf{0}$; when there are several, say E, equality constraints, they may be written as an E-vector of the constraint functions: $(e_1(\mathbf{x}), \ldots, e_E(\mathbf{x}))^T \equiv \mathbf{e}(\mathbf{x}) = \mathbf{0}$. This form is not unique, because if $e(x)$ is a scalar constraint function, then so is $k(e(\mathbf{x}))^{\alpha}$ for any constants k and α. This ambiguity, especially in the sign, is one of the difficulties with which the theory to follow attempts to cope.

In principle, for every equality-constrained optimization problem, there exists an equivalent unconstrained problem having fewer variables that can be solved by the methods already developed. There are fortunate but rare situations where this equivalent unconstrained problem can be constructed directly. More often, this is inconvenient or impossible. But even then, the first and second derivatives of the equivalent unconstrained problem can, as will be shown, be computed from those of the original objective and constraint functions.

Consider, then, the problem of minimizing an objective function $y(\mathbf{x})$ subject to a vector of strict null equalities $\mathbf{e}(\mathbf{x}) = \mathbf{0}$, where $y(\mathbf{x})$ and $\mathbf{e}(\mathbf{x})$ are continuously differentiable functions, possibly negative, of the nonnegative variables $\mathbf{x} \geq \mathbf{0}$. Once this *equality-constrained* optimization problem is understood, inequality constraints will be studied, beginning in Section 5-10.

5-02 LINEAR EQUALITIES

Many important concepts needed to handle equality constraints are illustrated easily when all the constraints are linear. Thus let $\mathbf{E}$ be an E by N matrix of constants, and let $\boldsymbol{\phi}$ be a column vector of E constant components, so that the constraint equation is $\mathbf{e}(\mathbf{x}) \equiv \mathbf{E}\mathbf{x} - \boldsymbol{\phi} = \mathbf{0}$.

Let S be the *rank* of **E**, which is the maximum number of linearly independent rows or columns of **E**. Of course, this rank cannot exceed the number of rows or the number of columns; that is, $S \leq \min(E, N)$. The rank is easily found by Gauss elimination. As an example, let

$$\mathbf{E} \equiv \begin{pmatrix} 1 & 0 & 1 & 1 & 0 \\ 0 & 1 & 1 & 0 & -2 \\ 1 & -1 & 0 & 1 & 2 \end{pmatrix}; \qquad \boldsymbol{\phi} = \begin{pmatrix} 3 \\ 2 \\ 1 \end{pmatrix}$$

Gauss elimination from the left gives the next two matrices:

$$\begin{pmatrix} 1 & 0 & 1 & 1 & 0 \\ 0 & 1 & 1 & 0 & -2 \\ 0 & -1 & -1 & 0 & 2 \end{pmatrix}\begin{pmatrix} 3 \\ 2 \\ -2 \end{pmatrix}; \qquad \begin{pmatrix} 1 & 0 & 1 & 1 & 0 \\ 0 & 1 & 1 & 0 & -2 \\ 0 & 0 & 0 & 0 & 0 \end{pmatrix}\begin{pmatrix} 3 \\ 2 \\ 0 \end{pmatrix}$$

The number of nonvanishing rows is 2, the rank of **E**. Thus in this example the first two variables, x_1 and x_2, can be expressed as linear functions of the remaining variables: $x_1 = 3 - x_3 - x_4$; $x_2 = 2 - x_3 + 2x_5$. When, as in this example, the rank (2) is less than the number of rows (3), the constraint system is said to be *singular*. The number of variables (or columns) N minus the rank S is called the number of *degrees of freedom*, D; 3 in this case.

$$D \equiv N - S \tag{5-1}$$

The number of degrees of freedom is the number of variables which, when given numerical values, determine the values of all the others needed to satisfy the constraints. Thus, if x_3, x_4, and x_5 are assigned numerical values, the remaining variables, x_1 and x_2, are fixed.

The S variables solved for (x_1 and x_2 in the example) are called the *solution variables*; the other D are called the *decision variables*. In the literature the solution variables are also known by the less descriptive names "state" or "basic" variables. This regrouping of variables may be regarded as a partitioning of the original vector **x** into an S-vector **s** of solution variables and a D-vector **d** of decision variables. In the example, $\mathbf{x} \equiv (\mathbf{s}^T, \mathbf{d}^T)^T$; $\mathbf{s} \equiv (x_1, x_2)^T$; $\mathbf{d} \equiv (x_3, x_4, x_5)^T$. Usually, a renumbering and rearrangement of these variables may be needed. Thus x_2 and x_4 could have been the two solution variables, in which case $\mathbf{s} = (x_2, x_4)^T$ and $\mathbf{d} = (x_1, x_3, x_5)^T$. Moreover, not all sets of S variables can be the solution variables. In the example, x_1 and x_4 cannot together form a solution set because there is no way of solving for x_1 independently of x_4 as a function of the other variables (try it). Nor can x_2 and x_5 form a solution set.

A systematic way to identify a solution set is to find an S^2 nonsingular submatrix $\mathbf{S}_s$, called the *solution* or *basis matrix*. Then **E** can be rearranged and partitioned as follows:

$$\mathbf{E} \equiv \begin{pmatrix} \mathbf{S}_s & \mathbf{S}_d \\ \mathbf{R}_s & \mathbf{R}_d \end{pmatrix} \tag{5-2}$$

so that the original constraints become

$$\mathbf{Ex} - \boldsymbol{\phi} = \begin{pmatrix} \mathbf{S}_s\mathbf{s} + \mathbf{S}_d\mathbf{d} \\ \mathbf{R}_s\mathbf{s} + \mathbf{R}_d\mathbf{d} \end{pmatrix} - \begin{pmatrix} \boldsymbol{\phi}_s \\ \boldsymbol{\phi}_r \end{pmatrix} \tag{5-3}$$

Here the vector $\boldsymbol{\phi}$ of constants has been partitioned into $\boldsymbol{\phi}_s$, corresponding to the *solution* or *basic constraints* having full row rank S, and $\boldsymbol{\phi}_r$, corresponding to the redundant constraints remaining, whose matrix is labeled $\mathbf{R}$. In the example,

$$\mathbf{S}_s = \begin{pmatrix} 1 & 0 \\ 0 & 1 \end{pmatrix}; \quad \mathbf{S}_d = \begin{pmatrix} 1 & 1 & 0 \\ 1 & 0 & -2 \end{pmatrix}; \quad \mathbf{R}_s = (1, -1);$$

$$\mathbf{R}_d = (0 \quad 1 \quad 2); \quad \boldsymbol{\phi}_s = \begin{pmatrix} 3 \\ 2 \end{pmatrix}; \quad \text{and} \quad \boldsymbol{\phi}_r = (1)$$

The point of all this is that the solution matrix $\mathbf{S}_s$ has been constructed to be nonsingular so that the solution variables can always be expressed as a linear function of the decision variables by $\mathbf{s} = \mathbf{S}_s^{-1}\mathbf{e}_s - \mathbf{S}_s^{-1}\mathbf{S}_d\mathbf{d} \equiv \mathbf{k} + \mathbf{Dd}$, where $\mathbf{k}$ is a particular solution and the columns of $\mathbf{D}$ are a set of D homogeneous solutions to $\mathbf{S}_s\mathbf{s} = \boldsymbol{\phi}_s - \mathbf{S}_d\mathbf{d}$. As in the example, $\mathbf{k} = (3, 2)^T$ and $\mathbf{D} = \begin{pmatrix} -1 & -1 & 0 \\ -1 & 0 & +2 \end{pmatrix}$ can be found directly by Gauss elimination.

Although these definitions forbid some choices of solution and decision variables, more than one choice usually remains. The same goes for deciding which constraints to make basic and which to make redundant. Thus any one of the three equations in the example could be regarded as redundant, leaving the other two for the basic set of solution constraints. There are, in fact, 24 ways to choose a 2^2 submatrix of $\mathbf{E}$ having rank 2, and each way corresponds to a different combination of solution and decision variables, with different redundant constraints. This leaves much latitude for choosing a solution basis with a matrix easy to invert, a matter which Section 5-08 discusses in detail.

5-03 NONLINEAR EQUALITIES

The concepts of the preceding section must be modified when any of the constraint equalities are nonlinear, as in the example following.

$$\begin{aligned} e_1(\mathbf{x}) &= x_1 + x_3 + x_4 - 3 = 0 \\ e_2(\mathbf{x}) &= x_2^3/3 + x_3^2/2 - x_5^4/2 - \tfrac{1}{3} = 0 \\ e_3(\mathbf{x}) &= x_1 - x_2 + x_4^2/2 + 2x_5 - \tfrac{5}{2} = 0 \end{aligned}$$

At a given point $\mathbf{x}$, the Taylor expansion for these equations is

$$\partial \mathbf{e} \sim \frac{\partial \mathbf{e}}{\partial \mathbf{x}} \partial \mathbf{x} = \begin{pmatrix} 1 & 0 & 1 & 1 & 0 \\ 0 & x_2^2 & x_3 & 0 & -2x_5^3 \\ 1 & -1 & 0 & x_4 & 2 \end{pmatrix} \partial \mathbf{x} \equiv \mathbf{e}_x \, \partial \mathbf{x}$$

The matrix $\partial \mathbf{e}/\partial \mathbf{x} \equiv \mathbf{e}_x$ changes the values of its elements from point to point. Consider, for example, a point $\mathbf{x}_1$ where $\mathbf{e}(\mathbf{x}_1) = \mathbf{0}$, say at $x_1 = (1, 1, 1, 1, 1)^T$. Then

$$(\mathbf{e}_x)_1 = \begin{pmatrix} 1 & 0 & 1 & 1 & 0 \\ 0 & 1 & 1 & 0 & -2 \\ 1 & -1 & 0 & 1 & 2 \end{pmatrix}$$

which is the same as the matrix analyzed in the preceding section. Thus at the particular point $\mathbf{x}_1$, where the rank is 2, one can plausibly speak of $5 - 2 = 3$ degrees of freedom with respect to *perturbations* $\partial \mathbf{x}$. That is, in the neighborhood of $\mathbf{x}_1$, specifying three of the perturbations will automatically determine the values of the two others needed to keep the constraint functions at zero to first order.

Yet the situation changes drastically very nearby. At the point $\mathbf{x}_2 = (0.98, 1.00, 1.02, 1.00, 1.01)^T$, which also satisfies the three nonlinear equations, the matrix of first derivatives is

$$(\mathbf{e}_x)_2 = \begin{pmatrix} 1 & 0 & 1 & 1 & 0 \\ 0 & 1 & 1.02 & 0 & -2.10 \\ 0.98 & -1 & 0 & 1 & 2.02 \end{pmatrix}$$

which has rank 3. Here there are only two degrees of freedom and decision variables, whereas there are three solution variables—any trio not including both x_2 and x_5.

The example demonstrates that the number of degrees of freedom can change from point to point in nonlinear systems, so it is not a global property as it is for linear equations. One can, however, speak of the *minimum* number of degrees of freedom as a global property, being simply the difference between the number of variables and the number of equations. Any point where the number of degrees of freedom exceeds this minimum will be called a *degenerate point* and handled specially by methods described in Section 5.15.

When there are no degrees of freedom, the number of solutions is finite, although there may be more than one, as for $e(x) = x^2 - 4x + 3 = (x - 1)(x - 3) = 0$. At any such solution, there are no nonzero perturbations satisfying the constraints, because $\partial \mathbf{e} \sim \mathbf{e}_x \, \partial \mathbf{x} = \mathbf{0}$ has the unique solution $\partial \mathbf{x} = \mathbf{0}$. Hence any solution to a zero-degree-of-freedom problem is called an *isolated point*.

It is strange, but true, that many engineering design problems are intentionally formulated to have isolated points. That is, equations arising from simplifying assumptions are appended to the original set, and certain variables given fixed numerical values, until the number of variables just equals the number of independent equations. This reduces the design problem to one of solving the constraint equations, which has the advantage that no objective function need be formulated. The drawback of this popular procedure is that it artificially rules out designs that might be better than that at the isolated point. It therefore behooves the designer skilled in optimization to regain degrees of freedom, and consequently opportunities for improvement, by either deleting unneeded "simplifying" equations or replacing them with inequalities.

5-04 SOLVABLE CONSTRAINTS

Optimization is straightforward when the constraint equations can be solved directly for the solution variables. In this case the solution variables can be eliminated from the objective, which then depends only on the decision variables and is not constrained. This new unconstrained function can be differentiated for optimization by any of the methods already described. This method of *direct substitution* is so straightforward that the reader may be tempted to skip the example of it to be developed here. Please do not; certain concepts needed later will be defined in the context of this simplest of cases. Pay special attention to the *solution function*, which is the expression for the solution variables as functions of the decision variables and its derivatives with respect to the decisions. An important feature of the *differential method*, to be employed later, is that the derivatives of a solution function can be computed even when the constraints cannot be solved.

As a simple example with two solution variables and two decisions, consider finding the point on the elliptic paraboloid $x_1 = 4x_2^2 - 2x_2x_3 + x_3^2$ closest to the point $(1, 1)^T$ and equidistant from the x_1 and x_4 axes. The equation of the elliptic paraboloid, in standard null form for a constraint, is

$$e_1(\mathbf{x}) = x_1 - 4x_2^2 + 2x_2x_3 - x_3^2 = 0$$

Points equidistant from the x_1 and x_4 axes satisfy

$$e_2(\mathbf{x}) = x_1 - x_4 = 0$$

These are solvable for the linear variables x_1 and x_4, which are designated the solution variables s_1 and s_2, respectively, leaving x_2 and x_3 as the decision variables d_1 and d_2. Solving the constraints gives

$$s_1 = 4d_1^2 - 2d_1d_2 + d_2^2$$
$$s_2 = 4d_1^2 - 2d_1d_2 + d_2^2$$

Henceforth the E-vector $\mathbf{s}$, whose components are $s_1, \ldots, s_E$, is to be considered a function of the decision vector $\mathbf{d}$, which is emphasized by calling it the *solution function*, symbolized $\mathbf{s}(\mathbf{d})$. This solution function can of course be differentiated with respect to $\mathbf{d}$, so the first and second *solution derivatives* $\mathbf{s}_d$ and $\mathbf{s}_{dd}$ are defined. In the example,

$$\mathbf{s}_d = \begin{pmatrix} 8d_1 - 2d_2, & -2d_1 + 2d_2 \\ 8d_1 - 2d_2, & -2d_1 + 2d_2 \end{pmatrix}; \qquad \mathbf{s}_{dd} = \begin{pmatrix} 8 & -2 \\ -2 & 2 \end{pmatrix}$$

To be minimized is the distance $[(d_1 - 1)^2 + (d_2^2 - 1)^2 + (s_1 - 1)^2 + (s_2 - 1)^2]^{1/2}$ from $(1, 1, 1, 1)^T$ to any point $(\mathbf{d}^T, \mathbf{s}^T)^T$ on the intersection of the elliptic paraboloid with the plane equidistant from the s_1 and s_2 axes. For simplicity, the objective function $y(\mathbf{d}, \mathbf{s})$ is taken as the square of this distance, for, being an increasing function of the distance, it is minimum at the same point as where the distance is least.

$$y(\mathbf{d}, \mathbf{s}) \equiv (d_1 - 1)^2 + (d_2 - 1)^2 + (s_1 - 1)^2 + (s_2 - 1)^2$$

Now use the solution function to eliminate the solution variables.

$$\begin{aligned} y(\mathbf{d}, \mathbf{s}(\mathbf{d})) &= (d_1 - 1)^2 + (d_2 - 1)^2 + [s_1(d) - 1]^2 + [s_2(d) - 1]^2 \\ &= (d_1 - 1)^2 + (d_2 - 1)^2 + 2(4d_1^2 - 2d_1d_2 + d_2^2 - 1)^2 \\ &= z(\mathbf{d}) \end{aligned}$$

Notice the new symbol $z(\mathbf{d})$ for the reduced objective function (see Chapter 2), from which the solution variables have been eliminated. This function is *not* constrained and depends only on the decision variables. Since it therefore can be optimized by the methods of Chapter 2, it is called the *unconstrained* (*or reduced*) *objective function*.

Stationary points of $z(\mathbf{d})$ must satisfy $(z_d) = \mathbf{0}^T$, so

$$\left(\frac{\partial z}{\partial d_1}\right) = [4(4d_1^2 - 2d_1d_2 + d_2^2 - 1)(8d_1 - 2d_2) + 2(d_1 - 1)] = 0$$

$$\left(\frac{\partial z}{\partial d_2}\right) = [4(4d_1^2 - 2d_1d_2 + d_2^2 - 1)(-2d_1 + 2d_2) + 2(d_2 - 1)] = 0$$

These cubic equations cannot be solved directly, so the Newton–Raphson method must be employed. The details will be exhibited for later comparison with the case where the constraint is not solved.

Begin by fixing the decision variables at unity, the base case. Thus $(\mathbf{d})_0 = (1, 1)^T$ and $(z_d)_0 = (48, 0)$. The vector z_d is called the *first decision derivative*. Other names for this vector are "reduced gradient" and "gradient projection." The base case is not a stationary point in $\mathbf{d}$-space because the first decision derivative vector does not vanish there. Notice that the values of the solution variables in the base case, although easily computed from

$(\mathbf{s})_0 = \mathbf{s}((\mathbf{d})_0) = (3, 3)^T$, are not needed to find a stationary point. However, the Newton–Raphson method requires the matrix z_{dd}, called the *second decision derivative.*

$$\left(\frac{\partial^2 z}{\partial d_1^2}\right)_0 = 4[(4d_1^2 - 2d_1d_2 + d_2^2 - 1)(8) + (8d_1 - 2d_2)^2]_0 + 2 = 210$$

$$\left(\frac{\partial^2 z}{\partial d_1\, \partial d_2}\right)_0 = 4[(4d_1^2 - 2d_1d_2 + d_2^2 - 1)(-2) + (-2d_1 + 2d_2)(8d_1 - 2d_2)]_0$$
$$= -16$$

$$\left(\frac{\partial^2 z}{\partial d_2^2}\right)_0 = 4[(4d_1^2 - 2d_1d_2 + d_2^2 - 1)(2) + (-2d_1 + 2d_2)^2]_0 + 2 = 18$$

$$(z_{dd})_0 = \begin{pmatrix} 210 & -16 \\ -16 & 18 \end{pmatrix}$$

The second decision derivative matrix is positive definite, so the Newton–Raphson procedure has a good chance of finding a better point. The Newton–Raphson equation is

$$(z_{dd})_0(\Delta\mathbf{d})_0 = -(z_d)_0$$

whence $(\Delta\mathbf{d})_0 = (-0.245, -0.218)^T$ and $(\mathbf{d})_1 = (0.755, 0.782)^T$. The predicted reduction in z is

$$(\Delta a)_0 = \tfrac{1}{2}(z_d)_0(\Delta d)_0 = \tfrac{1}{2}[48(-0.245) + 0(-0.218)] = -5.88$$

The actual reduction is better:

$$(\Delta z)_0 = (z)_1 - (z)_0 = 1.12 - 8.00 = -6.88$$

The reader is strongly advised to make another iteration, for the numbers will be used in the next section, which shows how to generate the decision derivatives without solving the constraints.

5-05 UNSOLVABLE CONSTRAINTS

In the example just solved, the first and second decision derivatives had to be evaluated at every iteration in order to use the Newton–Raphson method. This section demonstrates that the constraints need not be solved explicitly for the solution function $\mathbf{s}(\mathbf{d})$ to compute the decision derivatives. This is important in practice, for often either the solution function has no closed form, or its use to eliminate solution variables makes the unconstrained objective too complicated. Therefore, let us solve the preceding example without ever using the solution function $\mathbf{s}(\mathbf{d})$. Computations will be made in the neighborhood of the point $(\mathbf{d})_1 = (0.755, 0.782)^T$ generated in the preceding section. There the solution vector (not the solution function) is $(\mathbf{s})_1 = (1.710, 1.710)^T$, found easily in this case by solving the constraints, evaluated at $(\mathbf{d})_1$. In general, as will be demonstrated in Section 5-07, it can be obtained numerically by using the N–R method to drive the constraint functions to zero.

Consider now the first decision derivative z_d. Differentiation of z with respect to $\mathbf{d}$ gives

$$z_d \equiv \left(\frac{\partial z(\mathbf{d})}{\partial \mathbf{d}}\right) = \left(\frac{\partial y(\mathbf{d}, \mathbf{s}(\mathbf{d}))}{\partial \mathbf{d}}\right) = \left(\frac{\partial y}{\partial \mathbf{d}}\right) + \left(\frac{\partial y}{\partial \mathbf{s}}\right)\left(\frac{\partial \mathbf{s}}{\partial \mathbf{d}}\right) = y_d + y_s \mathbf{s}_d \tag{5-4}$$

Thus only first derivatives are needed to obtain z_d. The first derivatives of the original objective function y are

$$(y_d)_1 = (2(d_1 - 1), 2(d_2 - 1))_1 = (-0.490, -0.436)$$
$$(y_s)_1 = (2(s_1 - 1), 2(s_2 - 1))_1 = (1.42, 1.42)$$

Here evaluation of y_s would seem to require the solution function $\mathbf{s}(\mathbf{d})$. But at a given feasible point $(\mathbf{d})$, the corresponding value of the solution vector $(\mathbf{s}) = \mathbf{s}(\mathbf{d})$ is all that is needed, not the entire function $\mathbf{s}(\mathbf{d})$.

The derivative $\partial \mathbf{s}/\partial \mathbf{d}$ must also be evaluated at the point in question. To obtain it, differentiate the constraint equations to obtain

$$\frac{\partial \mathbf{e}(\mathbf{d}, \mathbf{s}(\mathbf{d}))}{\partial \mathbf{d}} = \frac{\partial \mathbf{e}}{\partial \mathbf{d}} + \frac{\partial \mathbf{e}}{\partial \mathbf{s}}\frac{\partial \mathbf{s}}{\partial \mathbf{d}} = \mathbf{0} = \mathbf{e}_d + \mathbf{e}_s \mathbf{s}_d \tag{5-5}$$

This matrix is null because the constraints must remain tight after the perturbation. In this expression, $\mathbf{e}_d$ and $\mathbf{e}_s$ are known. In the example

$$(\mathbf{e}_d)_1 = \begin{pmatrix} -8d_1 + 2d_2, & 2d_1 - 2d_2 \\ 0 & 0 \end{pmatrix}_1 = \begin{pmatrix} -4.475 & -0.054 \\ 0 & 0 \end{pmatrix}$$

Only the matrix $\mathbf{s}_d$ is unknown in this equation.

$$(\mathbf{e}_d)_1 + (\mathbf{e}_s)_1(\mathbf{s}_d)_1 = \mathbf{0} = \begin{pmatrix} -4.475 & -0.054 \\ 0 & 0 \end{pmatrix} + \begin{pmatrix} 1 & 0 \\ 1 & -1 \end{pmatrix} \begin{pmatrix} \frac{\partial s_1}{\partial d_1} & \frac{\partial s_1}{\partial d_2} \\ \frac{\partial s_2}{\partial d_1} & \frac{\partial s_2}{\partial d_2} \end{pmatrix}_1$$

In scalar form, this gives in the example four linear equations in the four unknown elements of $\mathbf{s}_d$.

$$\left(\frac{\partial s_1}{\partial d_1}\right)_1 = 4.475; \quad \left(\frac{\partial s_1}{\partial d_2}\right)_1 = 0.054; \quad \left(\frac{\partial s_1}{\partial d_1}\right)_1 - \left(\frac{\partial s_2}{\partial d_1}\right)_1 = 0;$$
$$\left(\frac{\partial s_1}{\partial d_2}\right)_1 - \left(\frac{\partial s_2}{\partial d_2}\right)_1 = 0$$

Therefore,

$$(\mathbf{s}_d)_1 = \begin{pmatrix} 4.475 & 0.054 \\ 4.475 & 0.054 \end{pmatrix}$$

This can be expressed in terms of the inverse matrix $\mathbf{e}_s^{-1}$:

$$\mathbf{s}_d = -\mathbf{e}_s^{-1}\mathbf{e}_d$$

In the example $\mathbf{e}_s = \mathbf{e}_s^{-1}$ by a convenient coincidence, so

$$(\mathbf{s}_d)_1 = -\begin{pmatrix} 1 & 0 \\ 1 & -1 \end{pmatrix}\begin{pmatrix} -4.475 & -0.054 \\ 0 & 0 \end{pmatrix} = \begin{pmatrix} 4.475 & 0.054 \\ 4.475 & 0.054 \end{pmatrix}$$

Substitution of this into the expression for z_d gives

$$\begin{aligned}(z_d)_1 = (y_d + y_s\mathbf{s}_d)_1 &= (-0.490, -0.436) + (1.42, 1.42)\begin{pmatrix} 4.475 & 0.054 \\ 4.475 & 0.054 \end{pmatrix} \\ &= (-0.490, -0.436) + (12.71, 0.153) \\ &= (12.22, -0.283)\end{aligned}$$

which, of course, matches the result derived by differentiating z_d directly.

In principle, the feasible stationary points of y could be found by solving simultaneously the constraint equations together with those obtained by setting z_d to zero. However, this would require the constraints to be solvable, which may not be true in more complicated circumstances. This approach is a disguised form of the widely known "Lagrange" procedure to be discussed in Section 5-09.

5-06 SECOND DERIVATIVES

To obtain second derivatives, differentiate the column vector z_d^T again with respect to **d**, keeping in mind that all first derivatives except $\mathbf{s}_d$ depend on **s** as well as **d**.

$$z_{dd} \equiv \frac{\partial z_d^T}{\partial \mathbf{d}} = \frac{\partial}{\partial \mathbf{d}}(y_d^T + \mathbf{s}_d^T y_s^T) \tag{5-6}$$

The derivative of the first term is

$$\frac{\partial}{\partial \mathbf{d}} y_d^T(\mathbf{d}, \mathbf{s}) = y_{dd} + \frac{\partial}{\partial \mathbf{s}} y_d^T \frac{\partial \mathbf{s}}{\partial \mathbf{d}} = y_{dd} + y_{ds}\mathbf{s}_d \tag{5-7}$$

The derivative of the second term requires expressing it as a column vector of sums.

$$\mathbf{s}_d^T y_s^T = \left(\sum_{i=1}^{S} \frac{\partial s_i}{\partial d_1}\frac{\partial y}{\partial s_i}, \ldots, \sum_{i=1}^{S} \frac{\partial s_i}{\partial d_D}\frac{\partial y}{\partial s_i}\right)^T$$

The jth-row element is, consequently,

$$\sum_{i=1}^{S} \frac{\partial s_i}{\partial d_j}\frac{\partial y}{\partial s_i}$$

and its derivative with respect to a typical decision d_l is

$$\frac{\partial}{\partial d_l}\left(\sum_{i=1}^{S} \frac{\partial s_i}{\partial d_j}\frac{\partial y}{\partial s_i}\right) = \sum_{i=1}^{S}\left[\frac{\partial s_i}{\partial d_j}\left(\frac{\partial^2 y}{\partial s_i \partial d_l} + \sum_{k=1}^{S} \frac{\partial^2 y}{\partial s_i \partial s_k}\frac{\partial s_k}{\partial d_l}\right) + \frac{\partial^2 s_i}{\partial d_j \partial d_l}\frac{\partial y}{\partial s_i}\right] \tag{5-8}$$

This is the element in row j and column l of the matrix z_{dd}. Arrangement of these elements and expression of the sums as matrix products gives

$$z_{dd} = y_{dd} + y_{ds}\mathbf{s}_d + \mathbf{s}_d^T y_{sd} + \mathbf{s}_d^T y_{ss}\mathbf{s}_d + \sum_{i=1}^{S} \frac{\partial y}{\partial s_i} s_{idd} \tag{5-9}$$

where $s_{idd} \equiv \partial^2 s_i/\partial \mathbf{d}^2$; $i = 1, \ldots, S$. All quantities except the s_{idd} are known at a given point. In the example

$$y_{dd} = \begin{pmatrix} 2 & 0 \\ 0 & 2 \end{pmatrix}; \quad y_{ds} = y_{sd}^T = \begin{pmatrix} 0 & 0 \\ 0 & 0 \end{pmatrix}; \quad y_{ss} = \begin{pmatrix} 2 & 0 \\ 0 & 2 \end{pmatrix}$$

These matrices can conveniently be regarded as submatrices of the matrix y_{xx}, rearranged so that the first rows and columns correspond to the decisions; the rest, to the solution variables. Thus, in the example,

$$y_{xx} \equiv \left(\begin{array}{cc|cc} \dfrac{\partial^2 y}{\partial d_1^2} & \dfrac{\partial^2 y}{\partial d_1\,\partial d_2} & \dfrac{\partial^2 y}{\partial d_1\,\partial s_1} & \dfrac{\partial^2 y}{\partial d_1\,\partial s_2} \\ \dfrac{\partial^2 y}{\partial d_2\,\partial d_1} & \dfrac{\partial^2 y}{\partial d_2^2} & \dfrac{\partial^2 y}{\partial d_2\,\partial s_1} & \dfrac{\partial^2 y}{\partial d_2\,\partial s_2} \\ \hline \dfrac{\partial^2 y}{\partial s_1\,\partial d_1} & \dfrac{\partial^2 y}{\partial s_1\,\partial d_2} & \dfrac{\partial^2 y}{\partial s_1^2} & \dfrac{\partial^2 y}{\partial s_1\,\partial s_2} \\ \dfrac{\partial^2 y}{\partial s_2\,\partial d_1} & \dfrac{\partial^2 y}{\partial s_2\,\partial d_2} & \dfrac{\partial^2 y}{\partial s_2\,\partial s_1} & \dfrac{\partial^2 y}{\partial s_2^2} \end{array}\right) = \begin{pmatrix} y_{dd} & y_{ds} \\ y_{sd} & y_{ss} \end{pmatrix} \tag{5-10}$$

With y_{xx} arranged this way, the matrices needed can be written in a single matrix equation in which $\mathbf{I}$ represents the D^2 unit matrix.

$$(\mathbf{I}, \mathbf{s}_d^T) y_{xx} \begin{pmatrix} \mathbf{I} \\ \mathbf{s}_d \end{pmatrix} = y_{dd} + y_{ds}\mathbf{s}_d + \mathbf{s}_d^T y_{sd} + \mathbf{s}_d^T y_{ss}\mathbf{s}_d \equiv Y_{dd} \tag{5-11}$$

In the example,

$$(Y_{dd})_1 = \begin{pmatrix} 2 & 0 \\ 0 & 2 \end{pmatrix} + \mathbf{0}\mathbf{s}_d + \mathbf{s}_d^T\mathbf{0} + \begin{pmatrix} 4.475 & 4.475 \\ 0.054 & 0.054 \end{pmatrix}\begin{pmatrix} 2 & 0 \\ 0 & 2 \end{pmatrix}\begin{pmatrix} 4.475 & 0.054 \\ 4.475 & 0.054 \end{pmatrix}$$
$$= \begin{pmatrix} 82.1 & 0.97 \\ 0.97 & 2.01 \end{pmatrix}$$

Hence Y_{dd} is an abbreviation, and now

$$z_{dd} = Y_{dd} + \sum_i \frac{\partial y}{\partial s_i} s_{idd} \tag{5-12}$$

Notice, incidentally, that since $\mathbf{s}_d^T y_{sd} = (y_{ds}^T \mathbf{s}_d)^T$, only the first matrix product need be computed. It remains to find the unknown quantity

$$\sum_i \frac{\partial y}{\partial s_i} s_{idd}.$$

To do this, differentiate a typical vector $(\partial e_h/\partial \mathbf{d})^T$ with respect to $\mathbf{d}$ and set the result equal to the null matrix, since the constraint function e_h must remain zero to second as well as first order.

$$\frac{\partial}{\partial \mathbf{d}}\left(\frac{\partial e_h}{\partial \mathbf{d}}\right)^T = \frac{\partial}{\partial \mathbf{d}}[e_{hd}^T + \mathbf{s}_d^T e_{hs}^T] = \mathbf{0}; \qquad h = 1, \ldots, E \tag{5-13}$$

Since this has exactly the same form as the expression for y, manipulations similar to those earlier give

$$E_{hdd} + \sum_{i=1}^{E} \frac{\partial e_h}{\partial s_i} s_{idd} = 0; \qquad h = 1, \ldots, E \tag{5-14}$$

where

$$E_{hdd} \equiv (\mathbf{I}, \mathbf{s}_d^T) e_{hxx} \begin{pmatrix} \mathbf{I} \\ s_d \end{pmatrix} \quad \text{and} \quad e_{hxx} \equiv \begin{pmatrix} e_{hdd} & e_{hds} \\ e_{hds}^T & e_{hss} \end{pmatrix}; \qquad h = 1, \ldots, E \tag{5-15}$$

In the example

$$E_{1dd} = \begin{pmatrix} -8 & 2 \\ 2 & -2 \end{pmatrix} \quad \text{and} \quad E_{2dd} = \begin{pmatrix} 0 & 0 \\ 0 & 0 \end{pmatrix}$$

The latter matrix is null in this example because the second constraint is completely linear. The two matrix equations are therefore, at $(\mathbf{d})_1$,

$$\begin{pmatrix} -8 & 2 \\ 2 & -2 \end{pmatrix} + (s_{1dd})_1 = \begin{pmatrix} 0 & 0 \\ 0 & 0 \end{pmatrix}$$

and

$$\begin{pmatrix} 0 & 0 \\ 0 & 0 \end{pmatrix} + (s_{1dd})_1 - (s_{2dd})_1 = \begin{pmatrix} 0 & 0 \\ 0 & 0 \end{pmatrix}$$

It follows that

$$(s_{1dd})_1 = (s_{2dd})_1 = \begin{pmatrix} 8 & -2 \\ -2 & 2 \end{pmatrix}$$

With the second derivatives of the solution functions known, the second decision derivatives can now be evaluated.

$$\begin{aligned} (z_{dd})_1 &= \left(Y_{dd} + \sum_i \frac{\partial y}{\partial s_i} s_{idd}\right)_1 \\ &= \begin{pmatrix} 82.1 & 0.97 \\ 0.97 & 2.01 \end{pmatrix} + 1.42\begin{pmatrix} 8 & -2 \\ -2 & 2 \end{pmatrix} + 1.42\begin{pmatrix} 8 & -2 \\ -2 & 2 \end{pmatrix} \\ &= \begin{pmatrix} 104.8 & -4.71 \\ -4.71 & 7.69 \end{pmatrix} \end{aligned}$$

Energetic readers can verify these numbers independently by differentiating the unconstrained objective twice with respect to **d**.

The Newton–Raphson move resulting generates the new point $(\mathbf{d})_2 = (0.637, 0.746)^T$, where $(\mathbf{s})_2 = (1.229, 1.229)^T$, and the predicted reduction is $(\Delta a)_1 = \frac{1}{2}[12.2(-0.118) - 0.28(-0.036)] = -0.72$. The actual change is $(\Delta y)_1 = (y)_2 - (y)_1 = 0.30 - 1.12 = -0.82$.

The objective is now close to zero, an obvious lower bound, but the reader is advised to try another iteration from $(\mathbf{d})_2$ to gain experience with the procedure. As a check, here are the coordinates of the next three points, together with their objective values and decision derivatives.

$$\mathbf{x}_3 = (1.114, 0.593, 0.837, 1.114)^T; \quad y_3 = 0.218; \quad (z_d)_3 = (0.591, -0.103)$$

$$\mathbf{x}_4 = (1.085, 0.573, 0.890, 1.085)^T; \quad y_4 = 0.209; \quad (z_d)_4 = (0.105, -0.0032)$$

$$\mathbf{x}_5 = (1.079, 0.569, 0.897, 1.079)^T; \quad y_5 = 0.209; \quad (z_d)_5 = (0.008, 0.001)$$

Although the end appears to be near, since y_4 and y_5 are the same, it is wise to construct a local lower bound with the known decision derivatives. This local bounding is made possible by the positive definiteness of the curvature matrix z_{dd} in the neighborhood of the local minimum, for this implies that the first-order part of any Taylor expansion must underestimate the substituted objective function.

$$\partial z = z_d \partial \mathbf{d} + \tfrac{1}{2} \partial \mathbf{d}^T z_{dd} \partial \mathbf{d} > z_d \partial \mathbf{d} \qquad (5\text{-}16)$$

At least three points are needed, since there are two decisions in addition to the objective function. The last three points would give inequalities which unfortunately would be unbounded with respect to d_2, all coefficients of d_2 being positive. Thus another point is needed having a negative value of $\partial z/\partial d_2$. This is obtained, not with a NR step, but rather by applying the last step again to obtain $\mathbf{x}_6 = (1.073, 0.565, 0.904, 1.073)^T$, where $y_6 = 0.209$ and $(z_d)_6 = (-0.082, 0.005)$. The inequalities constructed from the best three points $\mathbf{x}_4$, $\mathbf{x}_5$, and $\mathbf{x}_6$ are

$$y - 0.105d_2 + 0.0032d_3 \geq 0.152$$
$$y - 0.008d_2 - 0.001d_3 \geq 0.204$$
$$y + 0.082d_2 - 0.005d_3 \geq 0.251$$

Minimizing y subject to these inequalities, a simple linear programming problem, gives a lower bound on y^* where all three inequalities are active. Solution of the simultaneous equations gives $y = 0.208$, only 1/2% below the objective value at the best points. This bounds the local minimum closely enough:

$$0.208 \leq y^* \leq 0.209$$

This section has demonstrated that both first and second decision derivatives of the unconstrained objective function can be computed at any feasible point, without solving the constraints. The Newton–Raphson method therefore becomes easily available for finding constrained minima, provided that feasible points can be found. But this task is also readily accomplished by using the derivatives of the solution functions, as the next section shows.

5-07 *REGAINING FEASIBILITY*

A major difficulty of equality-constrained optimization problems is that the solution variables must be readjusted for feasibility with every move. This section shows, for the example, how the NR method can be used to regain feasibility. Even faster convergence to a feasible solution can be achieved by employing the second derivatives of the solution functions as generated in the preceding section.

Consider the problem of finding feasible values of the solution variables at the point $\mathbf{d}_2 = (0.576, 0.635)^T$. At this point the constraint functions read

$$(e_1)_2 = (s_1 - 4d_1^2 + d_1d_2 - d_2^2)_2 = (s_1)_2 - 1.365 = 0$$
$$(e_2)_2 = (s_1 - s_2)_2 = (s_1)_2 - (s_2)_2 = 0$$

These expressions are easily solved for $(s_1)_2 = (s_2)_2 = 1.365$, but this would not be simple if the constraints were nonlinear in the solution variables. Consequently, these results will be obtained without solving the constraint equations.

Since the numerical value of $(\mathbf{s}_d)_1$ is known at the preceding point $\mathbf{x}_1$, a first approximation to the move $(\Delta\mathbf{s})_1$ is obtained from the first-order Taylor expansion of $\mathbf{s}$ as a function of $\mathbf{d}$.

$$\Delta\mathbf{s}' \sim (\mathbf{s}_d)_1(\Delta\mathbf{d})_1 = \begin{pmatrix} 5.076 & 0.768 \\ 5.076 & 0.768 \end{pmatrix}\begin{pmatrix} -0.152 \\ -0.113 \end{pmatrix} = \begin{pmatrix} -0.858 \\ -0.858 \end{pmatrix}$$

This would give estimates $(\mathbf{s}')_2^T \equiv [(\mathbf{s})_1 + \Delta\mathbf{s}']^T = (2.135, 2.135) + (-0.858, -0.858) = (1.277, 1.277)$. Because of the nonlinearity, this linear approximation is slightly off, which can be detected by evaluating the constraint functions at this new point $(d_1, d_2, s_1, s_2) = (0.576, 0.635, 1.277, 1.277)$. There $e_1 = (s_1')_2 - 4(d_1)_2^2 + (d_1)_2(d_2)_2 - (d_2)_2^2 = -0.088 \neq 0$; $e_2 = (s_1')_2 - (s_2')_2 = 0$. The Newton–Raphson method would call for writing a first-order Taylor expansion of the constraints as a function of $\mathbf{s}$, and then adjusting $\mathbf{s}$ to remove the error predicted.

$$\partial e_1 \sim \left(\frac{\partial e_1}{\partial s_1}\right)_2' \partial s_1 + \left(\frac{\partial e_1}{\partial s_2}\right)_2' \partial s_2 = \partial s_1 = 0.088$$

$$\partial e_2 \sim \left(\frac{\partial e_2}{\partial s_1}\right)_2' \partial s_1 + \left(\frac{\partial e_2}{\partial s_2}\right)_2' \partial s_2 = \partial s_1 - \partial s_2 = 0$$

This gives the appropriate step $\Delta s_1 = \Delta s_2 = 0.088$, whence the error drops to zero and the feasible point $\mathbf{x}_2 = (d_1, d_2, s_1, s_2)_2^T = (0.576, 0.635, 1.365, 1.365)^T$ is generated. Newton–Raphson converged in one step because the constraints are linear in the solution variables, which is a good reason for choosing the linear variables to be the solution variables.

Actually, the second derivatives of the solution functions $\mathbf{s}$ were also available at $\mathbf{x}_1$. Consequently, a *quadratic* prediction of the solution variables would be

$$\begin{aligned}(\Delta s_1)_1'' &\approx (s_{1d})_1(\Delta\mathbf{d})_1 + \tfrac{1}{2}(\Delta\mathbf{d})_1^T(\mathbf{s}_{1dd})_1(\Delta\mathbf{d})_1 \\ &= (5.076, 0.768)(-0.152, -0.113)^T \\ &\quad + \tfrac{1}{2}(-0.152, -0.113)\begin{pmatrix} 8 & -1 \\ -1 & 2 \end{pmatrix}\begin{pmatrix} -0.152 \\ -0.113 \end{pmatrix} = -0.858 + 0.088 \\ &= -0.770 \quad [= (\Delta s_2)_1'']\end{aligned}$$

Notice that the quadratic term 0.088 equals the correction needed for the linear approximation, which of course is the first term. Thus further Newton–Raphson correction is not even needed. This demonstrates that the quadratic correction is worth making, especially since the second derivatives needed are available anyway. In this example, in which the solution functions are exactly quadratic, there was in fact no discrepancy for the Newton–Raphson method to correct.

Now that the usefulness of finding the second derivatives of the solution functions with respect to the decisions has been demonstrated, it may be well to show that they can always be computed, not just in the special example given. Recall that for a typical constraint e_h, $\sum_i (\partial e_h/\partial s_i)s_{idd} = -e_{hdd}$ for $h = 1, \ldots, E$. Since this amounts to E linear matrix equations in the E unknown matrices s_{idd}, the system can, in general, be solved by Gauss elimination, just as in the example.

An alternative form using the inverse of the matrix $\mathbf{e}_s$ will be used in Section 5-09. Let the inverse matrix $\mathbf{e}_s^{-1}$ be symbolized by $\mathbf{s}_e$.

$$\mathbf{e}_s^{-1} \equiv \mathbf{s}_e \tag{5-17}$$

The elements of $\mathbf{s}_e$ can be written $\partial s_g/\partial e_h$ for row g and column h. Then since $\mathbf{s}_e$ and $\mathbf{e}_s$ are inverse ($\mathbf{s}_e\mathbf{e}_s = \mathbf{e}_s\mathbf{s}_e = \mathbf{I}$), the system of linear matrix equations can be multiplied on the right by $\mathbf{s}_e$ to obtain the solution

$$\sum_h \frac{\partial s_g}{\partial e_h} \sum_i \frac{\partial e_h}{\partial s_i} s_{idd} = s_{gdd} = -\sum_h \frac{\partial s_g}{\partial e_h} e_{hdd}; \qquad g = 1, \ldots, E \tag{5-18}$$

In practice, one does not have to compute the inverse $\mathbf{s}_e$ to find the $\mathbf{s}_{gdd}$. They can be calculated directly by Gauss elimination of the matrix equations.

5-08 CHOOSING AND ORDERING SOLUTION VARIABLES

The analyst has great latitude in deciding which variables should be decision and which solution variables. Wise choices based on the structure and solvability of the equations can greatly clarify an optimization problem and reduce computation. This section gives simple guides for selecting solution variables and the order in which the constraint equations should be solved.

First, consider the question of when a single nonlinear equation can be solved for one of its variables in closed form. An equation $e(\mathbf{x}) = 0$ can be solved for a specific variable, say x_1, whenever x_1, or some monotonic transformation of it such as $\ln x_1$, x_1^α, or $\exp(x_1)$, appears only to first or second power in the equation. If, for example,

$$e(\mathbf{x}) = x_1^{2.4} - x_1^{1.2}x_2^{0.8}e^{3x_3}(1 + x_2)^{-1} + 5$$

then $e(\mathbf{x})$ is solvable for x_1 and x_3, but not x_2, in closed form. A circumflex ($\wedge$) will indicate a variable for which an equation is not solvable in closed form in the functional expression for the equality constraint function; $e(\mathbf{x}) = e(x_1, \hat{x}_2, x_3)$ in the example.

Consider now an example having eight equalities in 13 variables, which comes from an air-conditioning system design problem of W. F. Stoecker in *Design of Thermal Systems* (New York: McGraw-Hill, 1971), pp. 230–233.

$$e_1 \equiv v_0[3.71(10^{-3})(\omega/L)]^{-0.58} + 0.0183 + V^{-0.8}[0.0119F + 6.32(10^{-3})] - 1$$

$$e_2 \equiv C - \omega^{2.725}L^{-1.725}(243 + 60.8F)$$

$$e_3 \equiv C' + [L(1 - R)][5.84(10^{-3}) + 2.26(10^{-4})\omega F + 3.62(10^{-3})\omega]$$

$$e_4 \equiv D - v_0 L\{V^{-1}[5.58(10^{-5})F + 4.71(10^{-5})] - [1.38(10^{-5})F + 1.166(10^{-3})]\}$$

$$e_5 \equiv G(0.0987V - e^D) + e^D - 1$$

$$e_6 \equiv T(G - 2) + 115 - 75G$$

$$e_7 \equiv q - 2.165(10^{-4})VG + 288.7VTG$$

$$e_8 \equiv P - V^{18.5}(0.05032\omega L + 1.129\omega + 2.74)(10^{-11}) + 0.0965V$$

The solvability of this bewilderingly practical array of equations is expressed by the functional forms: $e_1(V, \omega, L, F)$, $e_2(C, \omega, L, F)$, $e_3(C', L, R, \omega, F)$, $e_4(D, v_0, L, V, F)$, $e_5(G, V, D)$, $e_6(T, G)$, $e_7(q, V, G, T)$, and $e_8(P, \hat{V}, \omega, L)$. Notice that in the entire array, only e_8 cannot be solved in terms of any one of its variables. Table 5-1 exhibits the equation structure more clearly.

TABLE 5-1

Equation Structure

Variable

Equation e_i: i	v_0	ω	L	C'	C	F	R	D	G	T	Q	P	V
1	1	1	1			1							1
2		1	1		1	1							
3		1	1	1		1	1						
4	1		1			1		1					1
5								1	1				1
6									1	1			
7									1	1	1		1
8		1	1									1	∧
			d_1			d_2	d_3	d_4					d_5

This structure can be used to decide which variables to call decision variables, and further, in what order the equations should be solved to minimize, or even avoid, times when several must be solved simultaneously. First scan the columns to see if any are identical; in the example, that for C' is the same as that for R. To prevent singularity, no more than one of these variables can be solution variables. With this in mind, select five columns to represent the five decision variables. At any point, the numerical values of the decisions will be given, the corresponding values of the solution variables being computed from these values. Hence it is plausible to choose as decisions those variables appearing in many of the constraints, for once these are fixed, the remaining expressions should be relatively easy to solve. Four of the variables ω, L, F, and V, each appear in four or more equations, whereas the other variables appear in three or fewer equations. Moreover, e_8 is not solvable in V, which is another good reason to let V be a decision variable. If possible, any variable required to be discrete, as are D, F, R, and L, should not be a solution variable, since, being dependent on the fluctuating decisions, they cannot be held at discrete values. Thus a good trial choice of decision variables would be D, F, R, and L, because they are discrete, together with V, because it cannot be solved for in one of the equations. The bottoms of the columns for these variables in Table 5-1 are marked with their new identifications as decision variables, numbered arbitrarily. With these columns deleted, the square structure matrix for the solution variables is shown in Table 5-2.

Since, when the decision variables are given, constraint e_4 has only one solution variable v_0 remaining, the value of v_0 can be found immediately. Similarly, G can be found by solving e_5. These variables v_0 and G are thus identified as the first two solution variables s_1 and s_2, respectively, and the

TABLE 5-2

Decisions Deleted

Equation e_i	i	v_0	ω	C'	C	G	T	q	P	
	1	1	1							
	2		1		1					
	3		1							
	4	1								f_1
	5					1				f_2
	6					1	1			
	7					1	1	1		
	8		1						1	
		s_1				s_2				

corresponding equations, relabeled f_1 and f_2, are understood to be solved for these solution variables as functions of the decision variables. Thus $e_4(v_0) = 0$ is rearranged as $v_0 - f_1(\mathbf{d}) \equiv s_1 - f_1(\mathbf{d}) = 0$. Since v_0 and G are now fixed by **d**, these values are assigned to all other equations containing them. This permits deleting not only rows 4 and 5, but also columns v_0 and G, leaving Table 5-3.

TABLE 5-3

Solution Variable Deletion

Equation e_i

i	ω	C'	C	T	q	P	
1	1						f_3
2	1		1				
3	1	1					
6				1			f_4
7				1	1		
8	1					1	
	s_3			s_4			

Rows 1 and 6 each contain a single entry, so the corresponding variables ω and q are designated s_3 and s_4, respectively. All rows of the matrix remaining after deletion of these columns and rows contain single elements, so the remaining solution variables can be numbered arbitrarily. As shown in Table 5-4, all these operations could have been carried out on the original structure matrix by labeling the lower and right margins as the single elements in a row are identified and circled.

TABLE 5-4

Summary

Original equation / Renumbered equation

i	v_0	ω	L	C'	C	F	R	D	G	T	q	P	V	
1	1	(1)	1			1							1	f_3
2		1	1		(1)	1								f_6
3		1	1	(1)		1	1							f_5
4	(1)		1			1		1					1	f_1
5								1	(1)				1	f_2
6									1	(1)				f_4
7									1	1	(1)		1	f_7
8		1	1									(1)	$\wedge$	f_8
	s_1	s_3	d_1	s_5	s_6	d_2	d_3	d_4	s_2	s_4	s_7	s_8	d_5	

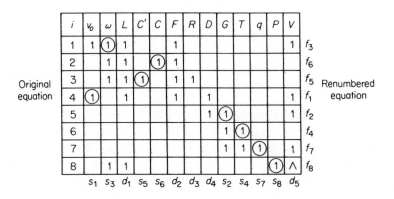

Now renumber the solved equations according to the new ordering. Then the equation set looks like this:

$$\begin{aligned}
f_1 &= s_1 && + g_1(d_1, d_2, d_4, d_5)\\
f_2 &= s_2 && + g_2(d_4, d_5)\\
f_3 &= s_3 && + g_3(d_1, d_2; s_1)\\
f_4 &= s_4 && + g_4(s_2)\\
f_5 &= s_5 && + g_5(d_1, d_2, d_3; s_3)\\
f_6 &= s_6 && + g_6(d_1, d_2; s_3)\\
f_7 &= s_7 && + g_7(d_5; s_2, s_4)\\
f_8 &= s_8 && + g_8(d_1, d_5; s_3)
\end{aligned}$$

This system, having a unit diagonal with nothing below it and little above, can be solved for $\mathbf{s}$ very rapidly for any given vector of decision values. Moreover, the perturbations $\partial\mathbf{s}$ at any point, as functions of decision perturbations $\partial\mathbf{d}$ there, are found from the easily solvable matrix equation $\partial\mathbf{f} = \mathbf{0} = \mathbf{f}_s\,\partial\mathbf{s} + \mathbf{f}_d\,\partial\mathbf{d} = (\mathbf{I} + \mathbf{g}_s)\,\partial\mathbf{s} + \mathbf{g}_d\,\partial\mathbf{d}$, where $\mathbf{I}$ is the unit matrix.

This section has shown how to reduce computation by intelligent selection of decision variables and of the order in which the equations are solved. Handled one at a time, simultaneous equations are very easy to solve.

5-09 CONSTRAINED STATIONARY POINTS

This section presents a more usual formulation of the equality-constrained optimization problem in which, unlike that of earlier sections, derivatives of the solution function are neither generated nor used. This alternative has both practical and theoretical utility.

Let any feasible decision point $\mathbf{d}_\dagger$ be called a *constrained stationary point* if the first decision derivatives vanish there, i.e., $(z_d)_\dagger = \mathbf{0}^T$. At $\mathbf{d}_\dagger$, the S-vector $(y_s\mathbf{e}_s^{-1})_\dagger$ is unique and computable; let it be abbreviated by

$$\boldsymbol{\lambda}^T \equiv (\lambda_1, \ldots, \lambda_S) \equiv (y_s\mathbf{e}_s{}^{-1})_\dagger \equiv (y_s\mathbf{s}_e)_\dagger \tag{5-19}$$

Then since $\mathbf{s}_d = -\mathbf{s}_e\mathbf{e}_d$,

$$(z_d)_\dagger = (y_d)_\dagger - (y_s\mathbf{s}_e\mathbf{e}_d)_\dagger = (y_d - \boldsymbol{\lambda}\mathbf{e}_d)_\dagger = \mathbf{0}^T \tag{5-20}$$

Now the partitioning of $\mathbf{x}$ into decision and solution variables is not unique in general, so the numerical values of y_s and $\mathbf{e}_s$ certainly change for different partitions. Yet the vector $\boldsymbol{\lambda}$ does not! To see why, consider that for any partition, the vector equation represents D equations in the D decisions, S solution variables, and S components of $\boldsymbol{\lambda}$. In addition, the decision and solution variables must satisfy the S nonsingular constraints because $\mathbf{d}^0$ is feasible. Thus the $D + S$ equations in the $D + 2S$ variables have exactly S degrees of freedom, provided that the solution basis is nonsingular as assumed. Hence the S components of λ are uniquely determined, independent of which set of solution variables is chosen.

It follows that at a stationary point, but only there, the differentiation can be made with respect to any and all of the independent variables so that

$$z_x - \boldsymbol{\lambda}^T \mathbf{e}_x = \mathbf{0}^T \tag{5-21}$$

This represents an equation for each of the $D + S$ independent variables. Since $y_x - \lambda^T \mathbf{e}_x = \frac{\partial}{\partial \mathbf{x}}(y - \boldsymbol{\lambda}^T \mathbf{e})$, constrained stationary points are often described in terms of the *Lagrangian function*

$$y - \boldsymbol{\lambda}^T \mathbf{e} \equiv L(\mathbf{x}, \boldsymbol{\lambda}) \tag{5-22}$$

Then since for stationarity $(L_x)_\dagger = (y_x - \boldsymbol{\lambda}^T \mathbf{e}_x)_\dagger = \mathbf{0}^T$, and for feasibility $(L_\lambda)^0 = -(\mathbf{e}_x^T)^0 = \mathbf{0}^T$, it follows that constrained stationary points for $y(\mathbf{x})$ occur at the *un*constrained stationary points of $L(\mathbf{x}, \boldsymbol{\lambda})$. Even though the Lagrangian L has more variables than $y(\mathbf{x})$, it is sometimes easier to find the unconstrained stationary points of L than the constrained ones of y, especially when the objective and constraints are linear or quadratic.

The Lagrangian function for the paraboloid example of Section 5-04 is

$$\begin{aligned} L = y - \boldsymbol{\lambda}\mathbf{e} = (x_1 - 1)^2 + (x_2 - 1)^2 + (x_3 - 1)^2 + (x_4 - 1)^2 \\ - \lambda_1(x_1 - 4x_2^2 + 2x_2x_3 - x_3^2) - \lambda_2(x_1 - x_4) \end{aligned}$$

The stationary points of L must satisfy

$$\left(\frac{\partial L}{\partial x_1}\right)_\dagger = 2(x_1 - 1)_\dagger - \lambda_1 - \lambda_2 = 0$$

$$\left(\frac{\partial L}{\partial x_2}\right)_\dagger = 2(x_2 - 1)_\dagger - \lambda_1(-8x_2 + 2x_3)_\dagger = 0$$

$$\left(\frac{\partial L}{\partial x_3}\right)_\dagger = 2(x_3 - 1)_\dagger - \lambda_1(2x_2 - 2x_3) = 0$$

$$\left(\frac{\partial L}{\partial x_4}\right)_\dagger = 2(x_4 - 1)_\dagger - \lambda_2(-1) = 0$$

Elimination of λ_1 and λ_2, which appear linearly, gives the nonlinear equations

$$2(x_1 + x_4 - 2)(x_3 - 4x_2) = x_2 - 1$$
$$2(x_1 + x_4 - 2)(x_2 - x_3) = x_3 - 1$$

These must be solved simultaneously with the two constraints, which would require a numerical procedure like the Newton–Raphson method. This procedure, the "method of undetermined (Lagrange) multipliers," is useful when the constraints are linear, but for nonlinear constraints as in the example, it is primarily of theoretical interest. Other chapters in this book employ the Lagrangian mainly to establish important theoretical results on which practical procedures are based.

So much for constraints which are strict equalities. Many of these results remain valuable in the more general case where some or all constraints are inequalities, a situation to be considered throughout the remainder of the chapter.

5-10 RESTRICTED DECISIONS

In many problems, some decision variables are bounded above and/or below by constants. In a minimization (maximization) problem, any decision variable d bounded in this way is said to be *restricted* at a given point $(\mathbf{x})$ if (1) d is at its lower (upper) bound, and (2) the value $(\partial z/\partial d)$ of its first decision derivative $\partial z/\partial d$ evaluated at $(\mathbf{x})$ is positive (negative). As in preceding sections, parentheses, with or without subscript, distinguish a particular value of a variable from the variable itself. Thus at any particular feasible point $(\mathbf{x})$, there is a vector $(\mathbf{r})$ of restricted decision variables satisfying these two conditions.

$$\mathbf{r} \geq (\mathbf{r}); \qquad (z_r) > \mathbf{0}^T \tag{5-23}$$

Every remaining decision variable u, said to be *unrestricted*, satisfies either $u > (u)$ or $(z_u) \leq 0$.

This formulation leads to an important necessary condition for a constrained minimum. The vector $\partial \mathbf{r}$ of feasible perturbations for the restricted variables must be nonnegative ($\partial \mathbf{r} \geq \mathbf{0}$), since by definition $\mathbf{r} \geq (\mathbf{r})$, the vector $(\mathbf{r})$ by definition equaling the vector of constant lower bounds. Hence to first order $(z_r)^* \, \partial r \geq 0$, and the objective can only be made worse (that is, increased) if the restricted variables are feasibly perturbed at $(\mathbf{x})^*$. To first order, then,

$$\partial z \sim (z_r)^* \, \partial \mathbf{r} + (z_u)^* \, \partial \mathbf{u} \geq (z_u)^* \, \partial \mathbf{u} \tag{5-24}$$

But the perturbations $\partial \mathbf{u}$ are not restricted in sign for the unrestricted variables $\mathbf{u}$, so $(\mathbf{x})^*$ can be locally minimum only if $(z_u)^* = \mathbf{0}^T$, so that $\partial z \geq 0$. This proves:

> ***Necessary Condition for Constrained Minima:*** *If a feasible nonsingular point* $(\mathbf{x})^*$ *is a local minimum, then* $(z_u)^* = \mathbf{0}^T$.

Any feasible point where this necessary (although not sufficient) condition holds is called a *quasi-minimum*. This concept plays almost the same role in constrained problems as does that of a stationary point in unconstrained ones. The only difference is that although an unconstrained maximum is certainly a stationary point, a constrained maximum is not a quasi-minimum, because none of its decision derivatives can be positive. The dagger subscript will be used to indicate quasi-minima, it being understood that if there are no restricted decisions, such a point is stationary. The necessary condition for a local constrained minimum can now be expressed entirely in words as follows: *If a feasible nonsingular point is a local minimum, then it is a quasi-minimum.*

This formulation reduces the constrained minimization problem to one of nullifying only the *unrestricted* decision derivatives. Thus the gradient or Newton–Raphson methods can be used directly on the unrestricted variables while the restricted variables are held fixed at their bounds. It is, however, necessary at every new point to recompute all the first decision derivatives to see which decisions are restricted there. This approach capitalizes on the fact that in many optimization problems, most decisions are restricted at the minimum. Significantly, the second derivatives of the Newton–Raphson method are needed only for the few unrestricted decisions remaining.

Since all restricted decisions are held fixed, only that part of the Taylor expansion for $z(\mathbf{r}, \mathbf{u})$ dependent on the unrestricted variables is pertinent for perturbation analysis. Hence a Newton–Raphson move $\Delta\mathbf{u}$ is the solution to $z_{uu}\,\Delta\mathbf{u} = -z_u$, and only the Hessian with respect to $\mathbf{u}$ is needed in:

> ***Sufficient Condition for Constrained Minima:*** *If* $(z_{uu})_\dagger$ *is positive definite at a quasiminimum* $(\mathbf{d})_\dagger = (\mathbf{r}^T, \mathbf{u}^T)^T_\dagger$, *then* $(\mathbf{x})_\dagger$ *is a local minimum.*

At first, the concept of a restricted variable may seem too specialized to be of much practical value, even though many problems do have variables bounded by constants. Yet the next section shows that understanding restricted variables simplifies the solution of any problem having inequality constraints.

5-11 INEQUALITY CONSTRAINTS

A problem with inequalities is readily converted into an equality-constrained problem having restricted variables. Let the inequality constraints $\mathbf{i}(\mathbf{x})$ be rearranged so that they may be written in *nonnegative vector form* as follows:

$$\mathbf{i}(\mathbf{x}) \geq \mathbf{0} \tag{5-25}$$

Then at any particular feasible point $(\mathbf{x})$, each inequality constraint function can be evaluated to see whether it is positive or zero. Those which are positive are said to be *loose* at (x) because they are not violated for suitably small perturbations of the **x** variable. Consequently, they can be ignored in the perturbation analysis, which will, however, need to weigh effects on those *tight* constraints which achieve their lower bounds of zero at (x). Let the vector of tight constraints, originally written

$$\mathbf{t}(\mathbf{x}) \geq \mathbf{0}; \qquad (\mathbf{t}) = \mathbf{0}$$

where $(\mathbf{t}) = \mathbf{0}$ is the set of tight constraints which achieve their lower bounds at zero, be replaced by:

$$\begin{aligned} \mathbf{t}(\mathbf{x}) - \mathbf{v} &= \mathbf{0} \\ \mathbf{v} &\geq \mathbf{0}; \qquad (\mathbf{t}) = \mathbf{0} \end{aligned} \tag{5-26}$$

The new variables **v** are called *tight slack variables* because if they were positive they would measure the differences between the actual values of the constraint functions and their lower bounds at zero. By construction, this difference is zero at (x).

The problem now has T new variables and T new equality constraints, so the number of degrees of freedom is unchanged. Since these tight slack variables are at their bounds, they cannot be used as solution variables, which must be free to change in either direction in response to decision variable perturbations. Hence they must be treated as decisions. This forces T of the original variables to become dependent solution variables, and, of course, these must be selected to avoid singularity and make the set of T constraint equations as easy to solve as possible.

To simplify the presentation, assume that at first there are no strict equality constraints in the original problem. Then if a nonsingular solution set can be found, S, the number of solution variables, equals T, the number of tight slack variables. Setting the derivative with respect to **t** of the equality-constraint equations $\mathbf{t}(\mathbf{s}, \mathbf{d}) - \mathbf{v} = \mathbf{0}$ to zero gives the T^2 system $\mathbf{t}_s\mathbf{s}_t - \mathbf{I} = \mathbf{0}$, whence $\mathbf{t}_s\mathbf{s}_t = \mathbf{I}$, and so

$$\mathbf{s}_t = \mathbf{t}_s^{-1} \tag{5-26}$$

Similar differentiation with respect to the decisions **d** (excluding the new slack variables) gives the T by D system $\mathbf{t}_s\mathbf{s}_d + \mathbf{t}_d = \mathbf{0}$, whence $\mathbf{s}_d$ is the T by D matrix solution of the T^2 system.

$$\mathbf{t}_s\mathbf{s}_d = -\mathbf{t}_d \tag{5-27}$$

There are two pertinent expressions each for z_t and z_d, derivatives central to this differential method. The first set uses the solution derivatives

$\mathbf{s}_t$ and $\mathbf{s}_d$ as follows. Since y does not depend on $\mathbf{t}$, $y_t = \mathbf{0}^T$, so the slack derivative vector is

$$z_t = y_s \mathbf{s}_t \tag{5-28}$$

This is simpler than the expression for the first decision derivative vector,

$$z_d = y_d + y_s \mathbf{s}_d \tag{5-29}$$

A second set of expressions does not involve $\mathbf{s}_t$ and $\mathbf{s}_d$, thus avoiding matrix inversion. Since $\mathbf{t}_s$ is inverse to $\mathbf{s}_t$, the two matrices can be multiplied in either order to obtain the unit matrix: $\mathbf{t}_s \mathbf{s}_t = \mathbf{s}_t \mathbf{t}_s = \mathbf{I}$. Hence multiplication of the slack derivative vector z_t on the right by $\mathbf{t}_s$ gives $z_t \mathbf{t}_s = y_s \mathbf{s}_t \mathbf{t}_s = y_s$. Subsequent transposition expresses the unknown z_t as a column vector solution of a T^2 set of linear equations

$$\mathbf{t}_s^T z_t^T = y_s^T \tag{5-30}$$

But be careful; a common mistake in using this equation is to forget to transpose the square matrix $\mathbf{t}_s$.

An alternative formula for the decision derivatives similarly free of solution derivatives is obtained by eliminating $y_s = z_t \mathbf{t}_s$ from $z_d = y_d + y_s \mathbf{s}_d = y_d + z_t \mathbf{t}_s \mathbf{s}_d$. Since it has been shown that $\mathbf{t}_s \mathbf{s}_d = -\mathbf{t}_d$, it follows that

$$z_d = y_d - z_t \mathbf{t}_d \tag{5-31}$$

which is computed by vector-matrix multiplication after z_t has been calculated. The latter forms can be used early in the search when second derivatives are not calculated, provided that the solution derivatives are not needed to generate new feasible points.

The next section works through an example showing how these derivatives are employed to generate improvement. Roughly speaking, one first computes the slack derivatives z_t. Since the corresponding tight slack variables all have values zero, any of them having a nonnegative slack derivative will be restricted and therefore must be held tight during the next move. Conversely, any with a negative slack derivative will be unrestricted and can therefore be made positive to improve the objective. This is equivalent to loosening the corresponding constraint; the procedure will be demonstrated in the next section's example.

Only if $z_t \geq \mathbf{0}^T$, in which case all tight slack variables are restricted, does the vector z_d of decision derivatives involving the original problem variables need be computed, which is why z_t was calculated first. If any of the derivatives for the unrestricted decisions $\mathbf{u}$ are not zero, a Newton–Raphson step will be in order requiring second constrained derivatives z_{uu}. These are computed by the formulas of Section 5-06, modified by deleting the restricted decisions, which must be held at zero. Thus, if y'_{xx} and t'_{ixx} for $i = 1, \ldots, T$ represent second derivative matrices of the objective and constraints with the $\mathbf{x}$ vector rearranged to have unrestricted decisions before the solution variables, restricted decisions being deleted, then the formulas of Section 5-06 would involve the solution derivatives $\mathbf{s}_u$ and s_{iuu} with respect to the unrestricted decisions. Thus the key formulas of Section 5-07 in the present context would read

$$Y_{uu} = (\mathbf{I}, \mathbf{s}_u^T) y'_{xx} (\mathbf{I}, \mathbf{s}_u^T)^T \tag{5-32}$$

$$T_{iuu} = (\mathbf{I}, \mathbf{s}_u^T) t_{iuu} (\mathbf{I}, \mathbf{s}_u^T)^T; \qquad i = 1, \ldots, T \tag{5-33}$$

$$z_{uu} = Y_{uu} + \sum \left(\frac{\partial y}{\partial s_i}\right) s_{iuu} \tag{5-34}$$

where the s_{iuu} are the solutions to $\sum_i (\partial t_h/\partial s_i) s_{iuu} = -T_{huu}$. In making the Newton–Raphson move, only the subvector z_u and submatrix z_{uu} for the unrestricted decisions should be used in the Newton–Raphson formula,

$$z_{uu}\, \Delta\mathbf{u} = -z_u^T \tag{5-35}$$

Such a move must be checked for feasibility and then again for optimality.

All of this is becoming too complicated to imagine without an example. After the next section shows how these derivatives are computed and used, Section 5-13 gives a more formal set of rules governing the successive improvement procedure based on these constrained derivatives. The singular case, which occurs in the example, is discussed in Section 5-14.

Although nonlinear, the example is simple enough that second solution derivatives are not used to regain feasibility. Hence formulas are derived here for computing z_{dd} (or z_{uu}) directly from the T_{iuu}. Let $\partial s_i/\partial t_h$ be the element in row i and column h of $\mathbf{s}_t \equiv \mathbf{t}_s^{-1}$ so that the second solution derivatives s_{iuu} can be written explicitly as $s_{iuu} = -\sum_h (\partial s_i/\partial t_h) T_{huu}$. Then substitution for s_{iuu} in the expression for z_{uu} gives

$$\begin{aligned} z_{uu} &= Y_{uu} + \sum_i \left(\frac{\partial y}{\partial s_i}\right) \sum_h \left(\frac{-\partial s_i}{\partial t_h}\right) T_{huu} \\ &= Y_{uu} - \sum_h \left(\frac{\partial z}{\partial t_h}\right) T_{huu} \end{aligned} \tag{5-36}$$

since $y_s \mathbf{s}_t = z_t$, as shown earlier in this section. This important formula is used frequently in the next example.

5-12 INEQUALITY EXAMPLE

The ideas of the sections preceding and immediately following are here illustrated with an example involving four positive variables and four inequality constraints, three of them nonlinear. You will notice that the hardest part is not performing the calculations, all easily done on a portable calculator, but rather keeping track of which constraints are tight and which variables are decisions, for this information changes from point to point. The problem is to minimize $y(\mathbf{x})$, where

$$y(\mathbf{x}) \equiv (x_1 - 1)^2 + (x_2 - 1)^2 + (x_3 - 1)^2 + (x_4 - 1)^2$$

subject to $\mathbf{x} \geq \mathbf{0}$ and

$$\begin{aligned} i_1(\mathbf{x}) &\equiv x_1 - 4x_2^2 + 2x_2x_3 - x_3^2 \geq 0 \\ i_2(\mathbf{x}) &\equiv -x_1 + x_4 \geq 0 \\ i_3(\mathbf{x}) &\equiv x_1 - x_2^2 + x_4 \geq 0 \\ i_4(\mathbf{x}) &\equiv x_1 - 2x_2 - 2x_3^2 + x_4 \geq 0 \end{aligned}$$

Start at the base case $\mathbf{x}_0 \equiv (1, 1, 1, 1)^T$. Evaluation of the inequality constraint functions gives $\mathbf{i}_0 = (-2, 0, 1, -2)^T$. Hence $\mathbf{x}_0$ is infeasible, both i_1 and i_4 being negative. At $\mathbf{x}_0$, i_2 is tight because it is satisfied as a strict equality, and i_3 is loose because it is positive there.

The first task is to find a feasible point, in this case where the three constraints i_1, i_2, and i_4 are all zero, if such a point exists. This gives three equations in the four unknown coordinates, leaving one degree of freedom, so let one of the coordinates, say x_2, be set to the base case value of unity; $(x_2)_1 = 1$. This gives the equation system

$$\begin{aligned} (x_1)_1 + 2(x_3)_1 - (x_3^2)_1 &= 4 \\ -(x_1)_1 + (x_4)_1 &= 0 \\ (x_1)_1 - 2(x_3^2)_1 + (x_4)_1 &= 2 \end{aligned}$$

Solution by the Newton–Raphson method, or in this case by direct algebraic manipulation, yields $\mathbf{x}_1 = (3.250, 1.000, 1.500, 3.250)^T$. The remaining constraint, i_3, when evaluated at $\mathbf{x}_1$, is found to equal 5.500, so it is loose and $\mathbf{x}_1$ is proven feasible.

Incidentally, if these equations had been found to have had no solution for $(x_2)_1 = 1$, x_2 would have to have been changed and a new system solved. Then if no obvious value of x_2 produced a feasible solution, x_2 would have to remain a variable while the function $i_1^2 + i_2^2 + i_3^2 + i_4^2$ is minimized with respect to **x** using Gauss's least-squares method of Section 4-24. If the minimum is zero, a feasible solution has been generated. But if the minimum is positive, then no feasible solution exists.

At $\mathbf{x}_1$ the tight constraints are $\mathbf{t}_1 = (i_1, i_2, i_4)_1^T \equiv (t_1, t_2, t_4)_1^T$, the tight constraint subscripts matching those of the original inequalities to reduce confusion. Since three constraints are tight, three solution variables must be chosen; let them be $s_1 \equiv (x_1, x_3, x_4)_1^T \equiv (s_1, s_3, s_4)_1^T$, where again the subscripts refer to the original variables. The remaining variable x_2 is designated a decision: $\mathbf{d}_1 \equiv x_2 \equiv (d_2)_1$.

Next, the slack derivatives are computed from $(\mathbf{t}_s)_1^T (z_t^T)_1 = (y_s^T)_1$. Here

$$(\mathbf{t}_s)_1 = \begin{pmatrix} \partial t_1/\partial x_1 & \partial t_1/\partial x_3 & \partial t_1/\partial x_4 \\ \partial t_2/\partial x_1 & \partial t_2/\partial x_3 & \partial t_2/\partial x_4 \\ \partial t_4/\partial x_1 & \partial t_4/\partial x_3 & \partial t_4/\partial x_4 \end{pmatrix} = \begin{pmatrix} 1 & -1 & 0 \\ -1 & 0 & 1 \\ 1 & -6 & 1 \end{pmatrix}$$

and

$$(y_s)_1 = (2(x_1 - 1), 2(x_3 - 1), 2(x_4 - 1)) = (4.5, 1.0, 4.5)$$

Hence $(z_t)_1$ is the solution to (note transpositions)

$$\begin{pmatrix} 1 & -1 & 1 \\ -1 & 0 & -6 \\ 0 & 1 & 1 \end{pmatrix} \begin{pmatrix} \partial z/\partial t_1 \\ \partial z/\partial t_2 \\ \partial z/\partial t_4 \end{pmatrix}_1 = \begin{pmatrix} 4.5 \\ 1.0 \\ 4.5 \end{pmatrix}$$

Therefore, $(z_t)_1 = (14, 7, -2.5)$. Only the signs are important. The positive signs on $(\partial z/\partial t_1)_1$ and $(\partial z/\partial t_2)_1$ indicate that i_1 and i_2 should stay tight during the next move. On the other hand, t_4 should be loosened because $(\partial z/\partial t_4)_1$ is negative.

In the region into which the move will be made, there will be two degrees of freedom, since only two constraints will be tight. The point of departure is still $\mathbf{x}_1$, but computation of the derivatives there must use a solution basis having only i_1 and i_2 active. Let a prime designate this point, but with the new basis (i_1, i_2), the new solution set $(\mathbf{s})_1' \equiv (x_1', x_4')_1^T \equiv (s_1', s_4')^T$, and the new decision set $(\mathbf{d})_1' \equiv (x_2', x_3')_1^T \equiv (d_2', d_3')^T$. The slack derivatives are the solutions to

$$\begin{pmatrix} 1 & -1 \\ 0 & 1 \end{pmatrix} \begin{pmatrix} \partial z/\partial t_1 \\ \partial z/\partial t_2 \end{pmatrix}_1' = \begin{pmatrix} 4.5 \\ 4.5 \end{pmatrix}$$

whence $(z_t)_1' = (9.0, 4.5)$. The two constraints should therefore remain tight during the next move, their slack derivatives being positive.

The new values of the decision variables must now be determined. The first decision derivatives are

$$(z_d)'_1 = (y_d)'_1 - (z_t)'_1(\mathbf{t}_s)'_1 = (0, 1) - (9.0, 4.5)\begin{pmatrix}-5 & -1\\ 0 & 0\end{pmatrix} = (45, 10)$$

The large values of these decision derivatives suggest that $\mathbf{x}'_1$ is too far from the optimum to justify second-order methods. Hence the new point $\mathbf{x}_2$ will be generated by following the negative constrained gradient until a constraint is encountered. Thus $(\Delta\mathbf{d})'_1 \equiv -\alpha_1(45, 10)^T$, where α_1 is the stepping parameter. The least upper bound on α_1 is $1/45 = 0.0222$, where x_2 vanishes, for no other constraint is violated for smaller α_1. Although strictly speaking, $x_2 = 0$ is not feasible, since x_2 is required to be positive, the value can be made arbitrarily small, say $\epsilon > 0$. So take $(x_2)_2 = \epsilon$ and $(x_3)_2 = 1.5 - 10(0.0222) = 1.278$. Since $(i_1)_2 = 0$, $(x_1)_2$ is the solution to $(i_1)_2 = 0 = (x_1)_2 - 4\epsilon^2 - 2\epsilon(1.278) - (1.278)^2$, so $(x_1)_2 = 1.633$. Since $(i_2)_2 = (-x_1 + x_4)_2 = 0$, $(x_4)_2 = 1.633$, and the new point is $\mathbf{x}_2 = (1.633, \epsilon, 1.278, 1.633)^T$, where $y_2 = 1.88 < 10 = y_1$.

At $\mathbf{x}_2$ there are now three tight constraints, counting the nonnegativity condition on x_2. Yet only i_1 and i_2 need be considered, as long as x_2 is made a decision variable. Then if $(\partial z/\partial d_2)_2 \geq 0$, x_2 need only be treated as a restricted decision and held at zero for the next move. If, on the other hand, $(\partial z/\partial d_2)_2 < 0$, x_2 can be made positive without restriction.

Since at $\mathbf{x}_2$ the same constraints are active as at $\mathbf{x}'_1$ (the nonnegativity condition on x_2 does not count), the same pattern of solution and decision variables will be used. Then

$$\begin{pmatrix}1 & -1\\ 0 & 1\end{pmatrix}(z_t)_2 = \begin{pmatrix}1.266\\ 1.266\end{pmatrix}$$

so $(z_t)_2 = (2.532, 1.266) > \mathbf{0}^T$, implying that these constraints should remain tight during the next move and that decision derivatives are needed. They are

$$(z_d)_2 = (-2, 0.556) - (2.532, 1.266)\begin{pmatrix}2.556 & -2.556\\ 0 & 0\end{pmatrix} = (-8.472, 7.028)$$

Thus x_2 is not restricted, its decision derivative being negative, and another gradient move can be made: $(\Delta\mathbf{d})_3 = -\alpha_2(-8.472, 7.028)^T$. The maximum value of α_2 is found by direct search to be where $i_4 = 0$. The correct value is $\alpha_2 = 0.0742$, giving the new point $\mathbf{x}_3 = (1.201, 0.628, 0.757, 1.201)^T$, where $y_3 = 0.278 < 1.88 = y_2$ and $(i_3)_3 > 0$.

At $\mathbf{x}_3$ now, $\mathbf{t}_3 = (i_1, i_2, i_4)_3^T$; let $\mathbf{s}_3 = (x_1, x_3, x_4)_3^T$ and $\mathbf{d}_3 = (x_2)_3$. The constrained gradient computations follow for checking on your calculator.

$$\begin{pmatrix} 1 & -1 & 1 \\ -0.258 & 0 & -3.028 \\ 0 & 1 & 1 \end{pmatrix} (z_t)_3 = \begin{pmatrix} 0.402 \\ -0.486 \\ 0.402 \end{pmatrix}$$

Therefore, $(z_t)_3 = (0.582, 0.291, 0.111) > \mathbf{0}^T$, so these constraints will remain active for the next move. The decision derivative is $(z_d)_3 = -0.744 - (0.582, 0.291, 0.111)(-3.510, 0, -2)^T = 1.521$.

This time the simple gradient move, which would involve decreasing x_2, is not bounded by the remaining constraint i_3, so second derivatives will be computed for a Newton–Raphson step. First, $(\mathbf{s}_d)_3$.

$$\begin{pmatrix} 1 & -0.258 & 0 \\ -1 & 0 & 1 \\ 1 & -3.028 & 1 \end{pmatrix} (\mathbf{s}_d)_3 = -\begin{pmatrix} -3.510 \\ 0 \\ -2 \end{pmatrix}$$

whence $(\mathbf{s}_d)_3 = (4.44, 3.59, 4.44)^T$. Next, the matrices.

$$(Y_{dd})_3 = (1, 4.44, 3.59, 4.44)(2)(1, 4.44, 3.59, 4.44)^T = 106.5$$

$$(T_{1dd})_3 = (1, 4.44, 3.59, 4.44)\left(\begin{array}{c|ccc} -8 & 0 & 2 & 0 \\ \hline 0 & 0 & 0 & 0 \\ 2 & 0 & -2 & 0 \\ 0 & 0 & 0 & 0 \end{array}\right)\begin{pmatrix} 1 \\ 4.44 \\ 3.59 \\ 4.44 \end{pmatrix} = -19.4$$

$$(T_{2dd})_3 = \mathbf{0}$$

$$(T_{4dd})_3 = (1, 4.44, 3.59, 4.44)\left(\begin{array}{c|ccc} 0 & 0 & 0 & 0 \\ \hline 0 & 0 & 0 & 0 \\ 0 & 0 & -4 & 0 \\ 0 & 0 & 0 & 0 \end{array}\right)\begin{pmatrix} 1 \\ 4.44 \\ 3.59 \\ 4.44 \end{pmatrix} = -51.6$$

Then the constrained curvature matrix $(z_{dd})_3$ is

$$\begin{aligned}(z_{dd})_3 &= (Y_{dd} - \textstyle\sum (\partial z/\partial t_i) T_{idd})_3 \\ &= 106.5 - (0.582)(-19.4) - (0.111)(-51.6) = 123.5\end{aligned}$$

The Newton–Raphson step is simply $(\Delta\mathbf{d})_3 = -(z_d/z_{dd})_3 = -1.521/123.5 = -0.012$. Therefore, $(d_2)_4 \equiv (x_2)_4 = 0.616$ and $\mathbf{x}_4 = (1.152, 0.616, 0.732, 1.152)^T$, where $y_4 = 0.265 < 0.278 = y_3$ and $(i_3)_4 > 0$.

The step and improvement are now getting small enough to consider plans for ending the search. The procedure to be used here is appropriate only when, as in the example, there is exactly one degree of freedom near the local optimum. Special though this may seem, problems often do have but few degrees of freedom, being nearly constraint-bound.

The idea is to assume that z_{dd} stays positive and the same constraints remain tight near $\mathbf{x}^*$, so that the single decision variable can be bracketed into such a small interval that one lower bound constructed from a decision derivative will end the search. In the example, the next point will be obtained, not by repeating the full Newton–Raphson method, but instead by applying the same step as before to the latest point, giving $(d_2)_5 = (d_2)_4 + (\Delta d_2)_3 = 0.604$, where $\mathbf{x}_5 = (1.105, 0.604, 0.708, 1.105)^T$ and $y_5 = 0.264 < 0.265 = y_4$. The good news is that the objective has improved; the bad news is that the optimum has still not been bracketed, for this only happens when the best value is between two worse ones, assuming that z_{dd} stays positive throughout the interval. Consequently, another step is made to $(d_2)_6 = (d_2)_5 + (\Delta d_2)_3 = 0.592$, where $y_6 = 0.273 > 0.264 = y_5$, and the locally optimal coordinate for d_2 is bracketed between that for $\mathbf{x}_6$ and $\mathbf{x}_4$: $0.592 < (x_2)^* < 0.616$.

All this has been based on an assumption which must now be tested—that the slack derivatives remain nonnegative throughout the interval. This turns out *not* to be true at $\mathbf{x}_6$, where $(z_t)_6 = (-0.257, -0.128, 0.248)$. Thus there are better points near $\mathbf{x}_6$ obtainable merely by loosening i_1 and i_2. But rather than look for improvement near $\mathbf{x}_6$, let us test the still better $\mathbf{x}_5$ to see if the assumption holds there. Fortunately, it does, for $(z_t)_5 = (0.009, 0.004, 0.206) > \mathbf{0}^T$, so it is reasonable by continuity to assume that constraints i_1, i_2, and i_4 stay tight between $\mathbf{x}_4$ and $\mathbf{x}_5$. One could now look for a point near $\mathbf{x}_5$ that, being worse, would bracket d_2, but there is a simpler alternative—compute the decision derivative $(z_d)_5$, for if it is negative (positive), then lesser (greater) values of d_2 cannot be minimal near $\mathbf{x}_5$. It happens that $(z_d)_5 = -0.351 < 0$, so $(x_2)^*$ is successfully bracketed: $0.604 < (x_2)^* < 0.616$.

A lower bounding function may now be constructed at the best point $\mathbf{x}_5$, for if z_{dd} stays positive, then

$$\Delta y \geq (z_d)_5\, \Delta x_2 = -0.351(0.616 - 0.604) = -0.004$$

Thus $0.264 - 0.004 = 0.260$ is a valid lower bound on y: $0.260 \leq y^* \leq 0.264$. The gap is less than 2‰, but to cover all eventualities, suppose this is not within the threshold desired. Then the decision derivative at $\mathbf{x}_4$ can be computed to give another lower bounding function as in Section 5-06. To avoid this extra calculation, use the value of $(z_d)_3$ as a bound on $(z_d)_4$, valid as long as $z_{dd} > 0$: $(z_d)_4 < (z_d)_3 = 1.521$, whence

$$y \geq 0.265 + 1.521(x_2 - 0.616) = 1.521x_2 - 0.672$$

This, together with the inequality

$$y \geq -0.351x_2 + 0.476$$

constructed at $\mathbf{x}_2$ gives a lower bound of 0.261 at $x_2 = 0.613$. Thus y^* is now bounded to within slightly more than 1%:

$$0.261 \leq y^* \leq 0.264$$

So much has happened that a recapitulation is in order. At the base case $\mathbf{x}_0$, one constraint was tight and two were violated, so a feasible point $\mathbf{x}_1$ was found by solving these three constraints as equations, one of the four coordinates being set to the base-case value of unity. There one tight constraint had a negative slack derivative, so the constraint was deleted from the basis; the slack derivatives were recomputed with only the other two constraints active. The resulting two-dimensional decision gradient being large, a move along the negative gradient was made until one of the decision variables vanished at $\mathbf{x}_2$. There the gradient changed sign (and magnitude) so that following it would encounter a third constraint, the one in fact that had been dropped at $\mathbf{x}_1$. Thus slack derivatives were computed at $\mathbf{x}_2$ with three constraints active. The resulting one-dimensional decision gradient indicated increasing the variable that had vanished at $\mathbf{x}_2$. A negative gradient move would have been unbounded, so second derivatives were calculated for a Newton–Raphson step to $\mathbf{x}_3$. There the slack derivatives indicated that the three tight constraints should remain active, and another Newton–Raphson step was made to $\mathbf{x}_4$. The improvement being slight and the decision space being one-dimensional, further steps aimed to bracket the local minimum and construct a local lower bound. This required only two more points $\mathbf{x}_5$ and $\mathbf{x}_6$, of which $\mathbf{x}_5$ had the lowest value of any of the seven measured. The objective function there was within 1% of the unknown true local minimum, so the search was terminated. Overall, the objective function was reduced almost two orders of magnitude from the base-case value. Rules governing this differential procedure are given in the next section.

5-13 DIFFERENTIAL OPTIMIZATION PROCEDURE

Having digested an example, the reader may now be ready for a general framework guiding the differential solution of a nonlinear inequality constrained optimization problem. The rules following cover the situations occurring most of the time, but exceptional situations involving degeneracy (singularity) are deferred to the next section.

0. **Start.** Select a base case.
   a. Evaluate all constraint functions.
   b. If point is feasible, go to 1.
   c. If point is infeasible, set all infeasible constraint functions to zero and solve the resulting simultaneous equations (numerically or analytically) with as many variables fixed at their base values as there are degrees of freedom.

d. If solution is found, go to 0(a).
e. If no solution is found, minimize the sum of the squared infeasible constraint functions, using Gauss's method of least squares described in Section 4-24.
f. If this minimum is zero, the solution is feasible; go to 0(a).
g. If this minimum is positive, no feasible solution exists for this constraint set. Guess again on a base case or give up; maybe the original constraints are inconsistent.

1. **Slack derivatives**
   a. Select a nondegenerate set of solution variables, excluding any variables at bounds. The remaining variables are decisions.
   b. If 1(a) can be done, point is nondegenerate; go to 1(d).
   c. If 1(a) is impossible, point is degenerate. This exceptional situation is discussed in Section 5-14.
   d. Compute slack derivatives z_t for tight constraints by solving $\mathbf{t}_s^T z_t^T = y_s^T$, the matrix $\mathbf{t}_s$ and vector y_s being known numerically.
   e. If $z_t \geq \mathbf{0}^T$, go to 2.
   f. If any $\partial z/\partial t_i < 0$, delete constraint i from the solution basis and return to 1(a).

2. **Decision gradient**
   a. Compute first decision derivatives z_d from $z_d = y_d - z_t \mathbf{t}_d$, all quantities on the right being known numerically.
   b. Determine which decisions are unrestricted and assemble them into a vector $\mathbf{u}$.
   c. Move along the negative gradient: $\Delta u = -\alpha z_u^T$ with $\alpha > 0$ as large as possible without violating any constraints. If α is finite, go to 2(e).
   d. If α is unbounded, go to 3.
   e. Determine $\mathbf{u} + \Delta\mathbf{u}$ and solve active constraints to determine values of solution variables $\mathbf{s}$ at new point. (As these instructions are written, solution derivatives are not available, so algebraic manipulation or the Newton–Raphson method must be used. Alternatively, the methods of Section 5-07 can be used to obtain quadratic estimates if one wishes to compute the first and second solution derivatives.)
   f. If the solution variables are unique, evaluate the objective y at the new point and go to 2(h).
   g. If solution variables $\mathbf{s}$ do not exist, go to 2(j); if they are multiple, choose those giving the best value of the objective.
   h. If the objective has improved, add any newly tight constraints to the basis and go to 1.
   i. If objective has not improved, go to 3 if point has not already been through that step. If it has, go to 2(j).
   j. Move along negative gradient $\Delta\mathbf{u} = -\beta z_u^T$ with β chosen to improve $y(\beta)$. Return to 2(e).

3. **Second decision derivatives**
   a. Compute $\mathbf{s}_u$ by solving $\mathbf{t}_s \mathbf{s}_u = -\mathbf{t}_u$, where $\mathbf{t}_s$ and $\mathbf{t}_u$ are known numerically.
   b. Compute $Y_{uu} = (\mathbf{I}, \mathbf{s}_u^T) y'_{xx} (\mathbf{I}, \mathbf{s}_u^T)^T$ and $T_{iuu} = (\mathbf{I}, \mathbf{s}_u^T) t'_{ixx} (\mathbf{I}, \mathbf{s}_u^T)^T$ for all active constraints.
   c. Compute $z_{uu} = Y_{uu} - \sum_i (\partial z / \partial t_i) T_{iuu}$.
   d. Compute Newton–Raphson step $\Delta \mathbf{u}$ as solution of $z_{uu} \Delta \mathbf{u} = -z_u^T$, meanwhile checking to see if z_{uu} is positive-definite.
   e. If z_{uu} is not positive-definite, go to 2(j).
   f. If z_{uu} is positive-definite, go to 2(e).

This is neither an algorithm nor a computer code; rather, it should be called a "procedure," or even a set of guidelines. Some of the instructions are vague, especially 2(j), which calls for any move along the negative gradient that will improve the objective. A designer or analyst with a hand calculator or an interactive computer, who can keep in mind several steps at once, may prefer a short step when feasibility is a problem, or a long one if a line minimization program (see Chapter 4) is handy. Nothing in the procedure tells what to do if z_{uu} is not positive-definite; the informed engineer will, however, be alerted to the possibilities of multiple optima when this happens. No termination routine is in the procedure, this being left to the designer's discretion.

Roughly speaking, the procedure first tries to find a feasible solution. Then at each subsequent feasible point the slack derivatives are computed to see if any constraints should be loosened. Early moves are along the negative decision gradient, this phase ending as soon as such moves overshoot or become unbounded. Then the second decision derivatives are used for Newton–Raphson steps. From point to point, work is required to find the values of the solution variables resulting from a decision move, and the procedure may well bog down here. Any anomalies in the second-order search cause the procedure to revert to first-order moves, short enough to guarantee cautious improvement if any is possible. By now, the reader knows not to attempt termination unless z_{uu} is positive-definite. And if possible, a local lower bound should be constructed to preclude overlooking evasive directions of improvement. Degeneracy, the final theoretical loose end, is disposed of in the next section.

5-14 DEGENERACY

All the theory so far has depended on having the matrix $\mathbf{t}_s$ nonsingular. It may happen, however, that no choice of solution variables at a point can produce such a nondegenerate basis. This section shows how to find directions of improvement, if any, when this happens. It also derives necessary conditions for a local minimum under these circumstances, in which the vanishing of decision derivatives and nonnegativity of slack derivatives is no longer necessary.

The method involves deleting constraints from the basis until the remaining matrix is nonsingular. Then, constrained derivatives can be computed as before. However, any moves must be restricted by the deleted, or *hidden* constraints, given the mnemonic symbol $\mathbf{h}(\mathbf{x})$. To do this, a first-order approximation to the hidden constraints is used to construct an inequality restricting the feasible region. This requires decision derivatives $\mathbf{h}_d$ of the hidden constraint functions themselves, computed from formulas entirely analogous to those for constrained derivatives of the objective function. Incidentally, to reduce this theory's consumption of letters of the alphabet, the vector of active constraints, after deletion of the hidden ones, will retain the symbol $\mathbf{t}(\mathbf{x})$, even at the risk of some confusion. After all, degeneracy does not happen very often, anyway.

As an example, consider the point $\mathbf{x}_2 = (1.633, 0, 1.278, 1.633)^T$ in the problem of Section 5.12. There the three tight inequalities are:

$$i_1 \equiv x_1 - 4x_2^2 + 2x_2x_3 - x_3^2 \geq 0; \qquad i_2 \equiv -x_1 - x_4 \geq 0;$$
$$i_4 \equiv x_1 - 2x_2 - 2x_3^2 + x_4 \geq 0$$

The next point, x_3, will therefore have $(i_1)_3 = (i_2)_3 = (i_4)_3 = 0$. There is only one degree of freedom left, and it must be assigned to x_2 because $(x_2)_2 = 0$. Thus $\mathbf{d}_2' \equiv (x_2)_2'$ and $\mathbf{s}_2' \equiv (x_1, x_3, x_4)^T$.

The slack derivative vector $(z_t)_2'$ is the solution to

$$\begin{pmatrix} 1 & -1 & 1 \\ -2.556 & 0 & -5.112 \\ 0 & 1 & 1 \end{pmatrix} (z_t^T)_2' = \begin{pmatrix} 1.266 \\ 0.556 \\ 1.266 \end{pmatrix}$$

Attempting to solve this set of equations would show that the matrix is singular, being of rank 2. The reader can verify this by adding the first and third rows, obtaining a multiple of the second row. Therefore, $\mathbf{x}_2'$ is a degenerate point requiring special measures.

The remedy is to delete i_4 from the basis, using it instead to restrict the move to feasible directions. Doing this requires using the slack derivatives at $\mathbf{x}_2$ (without the prime) to construct a linear approximation to the effects on i_4, now hidden and designated h_4, of perturbing the decisions d_2 and d_3. Let $\bar{h}_4$ represent the hidden constraint function h_4 with the solution variables eliminated. This is analogous to the relation between z and y, and so the constrained derivative formulas for $\bar{h}_4$ have the same form as for z, namely,

$$(\mathbf{t}_s)_2^T \left(\frac{\partial \bar{h}_4}{\partial \mathbf{t}}\right)_2^T = \left(\frac{\partial \bar{h}_4}{\partial \mathbf{s}}\right)_2^T \quad \text{or} \quad \begin{pmatrix} 1 & -1 \\ 0 & 1 \end{pmatrix} \begin{pmatrix} \partial \bar{h}_4/\partial t_1 \\ \partial \bar{h}_4/\partial t_2 \end{pmatrix} = \begin{pmatrix} 1 \\ 1 \end{pmatrix}$$

whence $(\partial\bar{h}_4/\partial t_1)_2 = 2$ and $(\partial\bar{h}_4/\partial t_2)_2 = 1$. Next, compute

$$\left(\frac{\partial\bar{h}_4}{\partial\mathbf{d}}\right)_2 = \left(\frac{\partial\bar{h}_4}{\partial\mathbf{d}}\right)_2 - \left(\frac{\partial h_4}{\partial\mathbf{t}}\right)_2(\mathbf{t}_d)_2 = (0, -5.226) - (2, 1)\begin{pmatrix}2.556 & -2.556\\ 0 & 0\end{pmatrix}$$
$$= (-5.226, 0)$$

Thus for feasibility to first order, the move $\Delta\mathbf{d} \equiv \Delta\mathbf{u}$ must satisfy $(\partial\bar{h}_4/\partial u_2)_2\,\Delta u_2 + (\partial\bar{h}_4/\partial u_3)_2\,\Delta u_3 = -5.23\,\Delta u_2 \geq 0$. For improvement of the objective, it must also satisfy

$$\Delta z = \left(\frac{\partial z}{\partial u_2}\right)_2 \Delta u_2 + \left(\frac{\partial z}{\partial u_3}\right)_2 \Delta u_3 = -7.98\,\Delta u_2 + 1.96\,\Delta u_3 < 0$$

Hence increasing u_2, which is profitable, will force infeasibility, so u_2 must be held at zero. Then u_3 can be decreased to improve the objective without altering h_4, at least to first order.

Inspection of i_4 shows that such a move would be unbounded, so second-order information is needed to limit the step. But the Newton–Raphson method cannot be used because the direction of the move has already been set. However, in this special situation where x_2 must be held constant, it can be treated just like a restricted decision, leaving only x_3 unrestricted. Thus the usual computation, with x_2 deleted from the analysis, gives $(z_{uu})_2$, now the single element $(\partial z/\partial u_3)_2 = 33.2$. Application of the Newton–Raphson method then gives $(\Delta u_3)_2 = -7.03/33.2 = -0.212$. Hence $(x_3)_3 = 1.066$, and of course $(x_2)_2 = 0$. Solution of the two tight constraints gives $(x_1)_3 = (x_4)_3 = 1.136$, and by a happy coincidence arising from the nonlinearities, the hidden constraint is just tight at the new point: $(i_4)_3 = 1.136 - 2(1.066)^2 + 1.136 = 0$. The objective has improved, for $y_3 = 1.04 < 1.88 = y_2$. As an exercise, the reader can make the next iteration from this new point, which happens also to be degenerate.

It is instructive to consider the more general situation where x_2 need not be held at zero. For the sake of argument, suppose that $(\partial\bar{h}_4/\partial u_3)_2$ were -0.1 instead of zero, so that for feasibility, $-5.23\,\Delta u_2 - 0.1\,\Delta u_3 \geq 0$. Assume that the same improvement inequality holds: $-7.98\,\Delta u_2 + 1.96\,\Delta u_3 < 0$. Any of several vectors satisfying both inequalities can be a feasible direction of improvement; for instance, $(\Delta\mathbf{u})_2 = \alpha_2(0.1, -6)^T$ with $\alpha_2 > 0$. In this case $(z_{uu})_2$, computed for practice in the next paragraph, gives a second-order prediction of the change in y.

$$\Delta y \approx (z_u)(0.1, -6)^T\alpha_2 + \tfrac{1}{2}\alpha_2^2(0.1, -6)(z_{uu})_2(0.1, -6)^T$$
$$= -43.0\alpha_2 + \tfrac{1}{2}\alpha_2^2(1170)$$

Therefore, $\alpha_2 = 43.0/1170 = 0.037$, which determines the next step.

The calculation of $(z_{uu})_2$ is detailed now because it is the first two-dimensional example of the second-order formulas of Section 5-11. First solve for $\mathbf{s}_u$ in $\mathbf{t}_s\mathbf{s}_u = -\mathbf{t}_u$.

$$\begin{pmatrix} 1 & 0 \\ -1 & 1 \end{pmatrix}\begin{pmatrix} \partial s_1/\partial \mathbf{u} \\ \partial s_4/\partial \mathbf{u} \end{pmatrix}_2 = -\begin{pmatrix} 2.556 & -2.556 \\ 0 & 0 \end{pmatrix}; \qquad (\mathbf{s}_u)_2 = \begin{pmatrix} -2.566 & 2.566 \\ -2.566 & 2.566 \end{pmatrix}$$

Then

$$(Y_{uu})_2 = \left(\begin{array}{cc|cc} 1 & 0 & -2.556 & 2.556 \\ 0 & 1 & -2.556 & 2.556 \end{array}\right)\left(\begin{array}{cc|cc} 2 & 0 & 0 & 0 \\ 0 & 2 & 0 & 0 \\ \hline 0 & 0 & 2 & 0 \\ 0 & 0 & 0 & 2 \end{array}\right)\left(\begin{array}{cc} 1 & 0 \\ 0 & 1 \\ \hline -2.556 & -2.556 \\ 2.556 & 2.556 \end{array}\right)$$

$$= \begin{pmatrix} 28.1 & 26.1 \\ 26.1 & 28.1 \end{pmatrix}$$

$$(t_{1uu})_2 = \left(\begin{array}{cc|cc} -8 & 2 & 0 & 0 \\ 2 & -2 & 0 & 0 \\ \hline 0 & 0 & 0 & 0 \\ 0 & 0 & 0 & 0 \end{array}\right)$$

so

$$(T_{1uu})_2 = \begin{pmatrix} -8 & 2 \\ 2 & -2 \end{pmatrix}; \qquad (T_{2uu})_2 = \mathbf{0};$$

$$(z_{uu})_2 = \begin{pmatrix} 28.1 & 26.1 \\ 26.1 & 28.1 \end{pmatrix} - 2.532\begin{pmatrix} -8 & 2 \\ 2 & -2 \end{pmatrix} = \begin{pmatrix} 48.4 & 21.1 \\ 21.1 & 33.2 \end{pmatrix}$$

At a degenerate point, the conditions of Section 5.10 are not necessary for minimality, since they were derived for unhampered perturbations of the unrestricted decisions. Instead, it is necessary that there be no feasible directions of improvement.

> ***Necessary Condition for a Degenerate Constrained Minimum:*** *If* $\mathbf{h}(\mathbf{x})$ *is the vector of hidden constraints deleted to make the basis nondegenerate, and if* $(\mathbf{h})^* = \mathbf{0}$ *at a local minimum* $\mathbf{x}^*$*, then there exist no solutions* $\Delta\mathbf{u}$ *to the inequalities*
>
> $$\left(\frac{\partial z}{\partial \mathbf{u}}\right)^* \Delta\mathbf{u} < 0 \quad \text{and} \quad \left(\frac{\partial \bar{h}}{\partial \mathbf{u}}\right)^* \Delta\mathbf{u} \geq 0$$
>
> *where* $\mathbf{u}$ *is the vector of unrestricted decisions.*

The first inequality gives directions of improvement, and the others, one for each hidden constraint, give directions feasible to first order. If, for example, at $\mathbf{x}_2$ the derivative $(\partial \bar{h}_4/\partial u_3)_2$ were positive, say 0.1, so that for feasibility $-5.23\ \Delta u_2 + 0.1\ \Delta u_3 \geq 0$, then $\Delta u_3 \geq 52.3\ \Delta u_2$, which would be inconsistent with the improvement inequality $-7.98\ \Delta u_2 + 1.96\ \Delta u_3 < 0$ that implies $\Delta u_3 < 4.07\ \Delta u_2$. To first order at least, such a point would appear to be a local minimum, although the conclusion of the theorem is not sufficient for minimality. At a *non*degenerate point, where no constraints are deleted, the theorem generates the requirement that $(\partial z/\partial \mathbf{u})^* = \mathbf{0}^T$, as derived in Section 5-10.

Many nonlinear optimization algorithms now in use depend on what mathematicians call *constraint qualifications*, which are a priori assumptions about the functions. If true, these constraint qualifications would guarantee that no degeneracy can occur. The previous example shows, however, that degeneracy can and does happen all too naturally in optimization problems, which only rarely can be *proven* to satisfy the constraint qualifications, even when by chance they do. Rather than wishing degeneracy away, the analyst armed with the theory of this section can deal with it whenever it intrudes. In fact, analysts having trouble with unfamiliar canned optimization programs are advised to look for degeneracy. Even when $\mathbf{t}_s$ is almost, although not quite, singular, convergence can be speeded by taking a smaller but well-conditioned basis, handling the deleted constraints by the methods of this section.

5-15 SENSITIVITY ANALYSIS

Not infrequently the analyst must modify a design or operating plan in the face of new information. When this happens, it is helpful to know what impact such changes might have on the objective, as well as on the problem variables. Fortunately, the derivatives generated by the differential procedure make available quick and accurate predictions of these effects without forcing the analyst to solve the problem all over again.

Suppose, in the example, that an adjustment in the decision variable d_2 of up to 5% ($\Delta \mathbf{d} = 0.030$) may be necessary once construction has begun. Since the decision derivative $\partial z/\partial d_2$ must be close to zero at the satisfactory design accepted, the objective should be insensitive to small changes in d_2, and it is. However, the decision change is large enough that second-order effects might be important; in fact, $\Delta y \approx \frac{1}{2}\Delta \mathbf{d}^T(z_{dd})\ \Delta \mathbf{d} = \frac{1}{2}(123.5)(0.030)^2 = 0.056$, which is 21% of the minimum value. This rather large number warns the designer that the design is sensitive to the decision variable, which therefore needs close supervision during fabrication. If one wishes no more than 5% change in the *objective*, then the decision must be controlled to within $[2(0.05)(0.264)/123.5]^{1/2} = 0.015$, which is 2.5% of the optimal value.

Suppose, then, that, to be on the safe side, measures are taken to hold the decision within 2% of its optimal value. If a change must be made, the solution variables must also adjust to maintain feasibility. Estimation of these effects requires the solution derivatives $\partial \mathbf{s}/\partial \mathbf{d}$. In the example, a 2% change in d_2 would require estimated (to first order) changes of $\Delta s \sim (\mathbf{s}_d)\,\Delta \mathbf{d} = (4.44, 3.59, 4.44)^T(0.604)(0.02) = (0.054, 0.043, 0.054)^T$, respectively, 5%, 6%, and 5% of their optimal values. A better estimate would be available if the second derivatives s_{idd} had been computed, but the first-order one is good enough for assessing the sensitivity of the design. In the example we see that the objective is rather sensitive to d_2 because the three solution variables require large adjustments to maintain feasibility.

The constraints themselves may not be precise when the time comes to build the device. For instance, in the example it may happen that the second constraint may indeed require a slight excess, say 0.001, of x_1 over x_4. This is equivalent to increasing the tight slack variable t_2 by 0.001, as the constraint equation $-x_1 + x_4 + v_2 = 0$ shows. The effect of this on the optimized objective is found in the first-order Taylor expansion of z as a function of v and $\mathbf{d}$.

$$\partial z^* \sim z_t^* \,\partial \mathbf{v} = 0.009 \partial v_1 + 0.004 \partial v_2 + 0.206 \partial v_4$$

Thus the coefficient $\partial z/\partial v_2 = 0.004$ measures the sensitivity of the objective to changes in constraint t_2. In this case, the sensitivity is quite small, the objective increasing only $(0.004)(0.001) = 4 \times 10^{-6}$ in response to the small tightening of the second constraint. In this example, constraint i_4 has 50 times more impact, a fact the designer may find important if there is any uncertainty in that constraint. In practice, constraints tend to become more restrictive as time passes and the world gets more crowded, and the components of z_d, sometimes called *sensitivity coefficients*, can warn of potential diseconomies or poor performance. On the other hand, they can point to possible improvement through making constraints *less* restrictive, say by purchasing sturdier materials or employing better surveys. Thus in the example, the designer would want to reexamine constraint i_4, the most sensitive, to see if and how it could be relaxed.

A short flight of imagination converts the current geometric example, in which the objective is a squared distance, into an economic one in which the objective is money. Suppose that the problem is interpreted as locating an electrical power source, subject to the geometric constraints. Then the squared distance can be regarded as proportional to the cost of transmission losses. With this interpretation, the sensitivity coefficient $\partial z/\partial t_4$ tells how much the designer can afford to spend on a unit relaxation of t_4. In the example, this amounts to 0.206 economic unit per unit of slack variable, at least for small changes.

Effects on the solution variables resulting from constraint changes are measured by the inverse matrix $\mathbf{s}_t = \mathbf{t}_s^{-1}$, which did not need computation in this example. Decision variables of course need no adjustment to maintain feasibility after a constraint shift; the solution variables perform this duty.

Giving interpretations to the various derivatives generated at a local optimum is a healthy intellectual exercise. It shows which variables and constraints, being sensitive, will bear close supervision or even reexamination. Also it measures quantitatively the interdependence of the device and other system components, as communicated through the variables and constraints.

5-16 PENALTY FUNCTION METHODS

An algorithm developed by Fiacco and McCormick for solving constrained optimization problems is based on an approach quite different from those previously described. In order to minimize an objective function, $y(\mathbf{x})$, subject to linear/nonlinear inequality/equality constraints of the form

$$f_k(\mathbf{x}) - b_k \leq 0; \qquad k = 1, 2, \ldots, M \tag{5-37}$$

this method seeks the *unconstrained* minimum of a new objective function defined by the following problem.

$$\text{Minimize } p(\mathbf{x}, s, r) = y(\mathbf{x}) + sr\left[\sum_{k=1}^{M} \phi(f_k(\mathbf{x}) - b_k)\right] \tag{5-38}$$

where

$$\begin{aligned} y(\mathbf{x}) &= \text{original objective function} \\ s &= \pm 1 \\ r &= \text{monotonically increasing or decreasing scale factor} \\ \phi(f_k(\mathbf{x}) - b_k) &= \text{penalty function involving each original constraint} \end{aligned}$$

The nature of s, r, and $\phi(f_k(\mathbf{x}) - b_k)$ in this general formulation depends upon whether or not one wishes to begin with a feasible solution vector and proceed toward optimality or start with an infeasible solution vector and proceed toward both optimality and feasibility. These two objectives, which utilize different numerical procedures, have been investigated extensively by Fiacco and McCormick and others, and are called respectively *interior* and *exterior point* penalty methods. These methods are both most easily explained through the use of numerical examples. The treatment that follows will be based upon the work of Gottfried and Weisman.

Consider the following simple but illustrative example.

$$\text{Minimize } y(x) = x^2 - 10x \tag{5-39}$$

subject to

$$f_1(x) - b_1 = x - 3 \leq 0 \tag{5-40}$$

Suppose that we reformulate the problem such that the constraint is made part of the objective function. Further, we should ensure that two essential conditions are satisfied. First, if an optimal solution exists that requires the constraint to be tight, then any perturbation away from constraint equality should be heavily penalized. Second, if an optimal solution dictates that the constraint be loose at optimality, then the problem should behave as if it were unconstrained by that constraint. Both conditions are satisfied by the following formulation.

$$\text{Minimize } p(x, r, s) = x_2 - 10x + sr(x - 3)^2 \tag{5-41}$$

This formulation corresponds to Eq. (5-38) with $\phi(f_k(\mathbf{x}) - b_k) \equiv (x - 3)^2$.

Let us now discuss the nature of s and r. Note that if we set $s \equiv 0$, then the (modified) objective function would be equivalent to the original unconstrained optimization problem. If the constraint is "loose," $(x - 3) < 0$, then we are not in constraint violation and s should be zero. If the constraint is "violated," $(x - 3) > 0$, then s should be positive to incur penalties. Obviously, if $x - 3 \equiv 0$, then the value of s does not affect our solution. If s is positive unity, given any positive value for r, the (modified) objective function is increased or "penalized" by any value of $x - 3 \neq 0$. Since we choose to square all deviations away from the constraint boundary, we "penalize" any such deviations regardless of whether they are positive or negative. The factor r can be interpreted as a scale factor that controls the actual effect of constraint violations. All cases can be dealt with by setting $s = 1$ if the constraint is active in problem solution and $s = 0$ otherwise. The choice of magnitude for r will be addressed shortly. To start, consider the constraint inactive by choosing $s = 0$ and $r = 1$. The solution for this simple example is by inspection $x_0 = 5$. Since this violates the constraint, it is clear that $f(\mathbf{x})$ is binding at optimality and we should set s to a positive quantity. With $r = 1$, we obtain $(s \equiv 1)$:

$$\text{Minimize } p(x, 1, 1) = x^2 - 10x + (x - 3)^2$$

The optimal solution to this new problem is given by $x_0 = 4$. It seems logical at this point that the optimal solution should be $x_0 \equiv 4$. However, by examining the original problem, we can easily verify (for this simple example) that the true optimal solution is given by $x^* \equiv 3$. A moment's reflection will allow us to rationalize this result. Note that for the solution $x_0 = 4$, the constraint is violated; yet we sought to penalize such violations by setting $r = 1$. The result simply implies that our penalty is not *severe* enough to produce any further adjustment than $\Delta x_0 = 4 - 5 = -1$. For this simple example, we can easily minimize Eq. (5-41) and obtain the optimal solution in terms of r, the scale factor on our penalty (with $s \equiv 1$). This yields

$$x^* = \frac{6r + 10}{2r + 2}$$

Note that this result clearly shows that as $r \longrightarrow \infty$, $x^* \longrightarrow 3$, the desired result. An algorithmic procedure is now evident. The optimal solution can be approached through a sequence of unconstrained optimization problems using a monotonically increasing sequence of r's. Note that almost any "penalty function" ϕ could be used, including $[f_k(\mathbf{x}) - b_k]^4$ or $[f_k(\mathbf{x}) - b_k]$. However, the choice of $[f_k(\mathbf{x}) - b_k]^2$ is convenient since it penalizes plus or minus deviations in a quadratic fashion. This gives rise to application of any convenient unconstrained optimization procedure, and derivatives of quadratic functions, being linear, are readily available. Conversely, if an absolute-value penalty function is used, solution procedures are limited to those using only functional evaluations. In theory, many constrained optimization problems can be solved numerically in this manner by using the following formulation. Minimize

$$G(\mathbf{x}, s, r) = y(x) + r \sum_{j=1}^{M} s_j \phi_j [f_j(x) - b_j]^2 + \sum_{k=M+1}^{\bar{M}} \phi_k [f_k(\mathbf{x}) - b_k]^2 \qquad (5\text{-}42)$$

where $f_k(\mathbf{x}) - b_k$; $k = M + 1, \ldots, \bar{M}$, are given *equality* constraints. The scale factor r should be negative for maximization. Fiacco and McCormick have investigated this formulation for a wide range of penalty functions, and variations of the scale factor r. Minimization of this function through a sequence of unconstrained optimizations using a monotonic sequence of r values will approach a quasioptimum, often the optimal constrained solution. Note that a major disadvantage is that this formulation generates points *exterior* to the feasible solution space, and only at termination is a feasible solution obtained. This is characteristic of exterior penalty function techniques.

To combat the disadvantage above, consider an alternative formulation which would ensure that our solution vector remain feasible. Suppose that we structure our problem in the following form.

$$\text{Minimize } y(\mathbf{x})$$

subject to

$$f_k(\mathbf{x}) - b_k \geq 0$$

Fiacco and McCormick have shown that instead of employing Eq. (5-42), either of the following formulations will generate a sequence of *feasible* vectors to the original problem for a strictly *decreasing* sequence of r values:

$$p(\mathbf{x}, r) = y(\mathbf{x}) + r \sum_{k=1}^{M} \frac{1}{f_k(\mathbf{x}) - b_k} \qquad (5\text{-}43)$$

or

$$p(\mathbf{x}, r) = y(\mathbf{x}) - r \sum_{k=1}^{M} \ln [f_k(\mathbf{x}) - b_k] \qquad (5\text{-}44)$$

Under certain minor restrictions, as r approaches zero, the solution vector approaches $\mathbf{x}^*$, the optimal constrained solution. The important restrictions are that $y(\mathbf{x})$ and $f_k(\mathbf{x})$ all be continuously twice differentiable, and that for each positive value of r, the function $p(\mathbf{x}, r)$ be strictly convex. Consider the same (simplified) example previously presented [Eqs. (5-39) and (5-40)], now rewritten as

$$\text{Minimize } y(x) = x^2 - 10x$$

subject to

$$f_1(x) - b_1 = -x + 3 \geq 0$$

In order to generate a set of *feasible* solutions, consider the following penalty function:

$$p(x, r) = x^2 - 10x - r[\ln (3 - x)]$$

For this example, we will apply the calculus to obtain

$$\frac{\partial[p(x, r)]}{\partial x} = 2x - 10 + \frac{r}{3 - x} = 0$$

and

$$\mathbf{x}^*(r) = \frac{8 - \sqrt{4 + 2r}}{2}$$

where the negative sign is chosen since the constraint requires that $x \leq 3$. The minimum is given by $x^* = 3$ and can be generated through a monotonically decreasing sequence of r:

Iteration	*r*	*x*(r)*
1	10	1.550
2	5	2.130
3	2.5	2.500
4	1.0	2.780
5	0.50	2.880
6	0.25	2.970
7	0.10	2.980
8	0.05	2.990
9	0.01	2.998

In summary, this method is based on transforming a constrained minimization problem into a sequence of unconstrained minimization problems. The algorithm develops both primal-feasible and dual-feasible points, so that the dual solution vector is generated explicitly. The primal objective function, $y(\mathbf{x})$, is monotonically decreased as r decreases (or increases), and a subproblem of the original problem is solved with each unconstrained minimization.

This subproblem provides information for estimating the limiting (solution) values of the problem variables as well as the $\mathbf{x}$ that minimizes the augmented objective function for some smaller value of r.

In dealing with more general, complicated, nonlinear programming problems, Fiacco and McCormick give a rationale for computing the initial value of r consistent with attempting to reduce the effort of minimizing $p(\mathbf{x}, r)$; the initial value, r_1, is selected so as to minimize the magnitude of the gradient of $p(\mathbf{x}, r)$, evaluated at the starting point, $\mathbf{x}_0$. This leads to an analytic determination of r_1 which the authors state has worked well in practice.

The major computational effort required in this algorithm is concerned with obtaining the sequence of points that minimize the penalty function for various values of r. First-order (steepest ascent) gradient methods have generally converged too slowly to be adequate for minimizing p, and a second-order optimum gradient method has proved to be much more reliable and efficient. In this second-order method, a point, $\mathbf{x}_{j+1}$, is obtained from the previous point, $\mathbf{x}_j$, by descending the mapped gradient of the objective function, p, evaluated at $\mathbf{x}_j$. The mapping is obtained by premultiplying the gradient of p by the inverse of the matrix of second partial derivatives of p, evaluated at the point x_j. The point x_{j+1} is found by approximating the point where p is minimized on the mapped gradient vector using the direct search methods of Chapter 4. This becomes the next point in the iteration, and the process is then repeated until the minimum is approached. The authors give criteria for terminating convergence toward the minimum as well as for selecting the rate of reduction of the parameter r. Also given are a theoretical basis for developing the trajectory of minima of P as a function of r. A detailed computer algorithm is available from Fiacco and McCormick.

5-17 GENERALIZED LAGRANGIAN MULTIPLIERS

In a 1963 article, Hugh Everett introduced the *generalized Lagrange multiplier* (GLM) method as a computational procedure for the solution of mathematical programming problems. The method can be described briefly as follows. Consider the mathematical program:

Program A

$$\text{Maximize } y(\mathbf{x})$$

subject to

$$f_j(\mathbf{x}) \leq b_j; \qquad j = 1, \ldots, M$$
$$\mathbf{x} \in S$$

where S is a subset of R^n on which $y(\mathbf{x})$ and f_j, $j = 1, \ldots, M$, are real-valued.

The generalized Lagrangian function associated with program A is

$$F(\mathbf{x}, \boldsymbol{\lambda}) = y(\mathbf{x}) - \sum_{j=1}^{M} \lambda_j f_j(\mathbf{x}) \tag{5-45}$$

Everett proved that if $\mathbf{x}^*(\hat{\boldsymbol{\lambda}}) \equiv \hat{\mathbf{x}}$ is the solution to the problem of maximizing the *unconstrained* Lagrangian $F(\mathbf{x}, \hat{\boldsymbol{\lambda}})$, over all $\mathbf{x} \in S$, for a specific choice $\hat{\boldsymbol{\lambda}}$ of the $\boldsymbol{\lambda}$, then $\mathbf{x}$ also maximized $y(\mathbf{x})$ over all $\mathbf{x} \in S$ subject to the constraints

$$f_j(\mathbf{x}) \leqq f_j(\hat{\mathbf{x}}); \qquad j = 1, \ldots, M \tag{5-46}$$

Hence program A can be solved by finding the set of multipliers $\boldsymbol{\lambda}^0$ such that $f_j(\mathbf{x}^0) \leqq b_j$ and $\lambda_j^0 b_j = \lambda_j^0 f_j(\mathbf{x}^0), j = 1, \ldots, M$, where $\mathbf{x}^0$ maximizes $F(\mathbf{x}, \boldsymbol{\lambda}^0)$ over S.

These statements may be formalized as follows.

Everett's condition In order to solve program A, it is *sufficient* to find nonnegative numbers (Everett's multipliers) λ_i^0 such that a corresponding solution $\mathbf{x}^*(\boldsymbol{\lambda}^0) \equiv \mathbf{x}^0$, to the problem of maximizing the unconstrained function $F(\mathbf{x}, \boldsymbol{\lambda})$:

$$y(\mathbf{x}) - \sum_{j=1}^{M} \lambda_j f_j(\mathbf{x}) \tag{5-47}$$

for $\mathbf{x} \in S$, can be found that satisfies

$$f_j(\mathbf{x}^0) \leq b_j \qquad \text{for } j = 1, \ldots, M$$

Under certain stability conditions it can be demonstrated that if the vector $\mathbf{f}(\mathbf{x}^0)$ is "close to" $\mathbf{b}$, then $y(\mathbf{x}^0)$ is "close to" the optimal objective value for program A. The unspecified part of the GLM method is the multiplier search. What we desire is an algorithm for finding a sequence of multipliers, $\{\lambda_k^i\}$, such that the values $y(x_k^i)$ converge to the optimal value in program A, where x_k^i maximizes $F(\mathbf{x}, \lambda_k^i)$ over the solution space S.

Brooks and Geoffrion have modified Everett's condition as follows:

1. If $\mathbf{x}^0$ is optimal for Eq. (5-47).
2. For $\boldsymbol{\lambda}^0 \geq 0$,

$$\left.\begin{array}{ll} \lambda_j^0 > 0 \ \text{ implies } & f_j(\mathbf{x}^0) = b_j \\ \lambda_j^0 = 0 \ \text{ implies } & f_j(\mathbf{x}^0) \leq b_j \end{array}\right\} j = 1, \ldots, M$$

If these conditions are satisfied, then $\mathbf{x}^0$ is the optimal solution to program A.

Conditions 1 and 2 are equivalent to the requirement that $(\mathbf{x}^0, \boldsymbol{\lambda}^0)$ be a *saddle point* of the Lagrangian:

$$L(\mathbf{x}, \boldsymbol{\lambda}) = y(\mathbf{x}) - \sum_{j=1}^{M} \lambda_j[f_j(\mathbf{x}) - b_j]$$

That is,

$$L(\mathbf{x}, \boldsymbol{\lambda}^0) \leq \hat{L}(\mathbf{x}^0, \boldsymbol{\lambda}^0) \leq L(\mathbf{x}^0, \boldsymbol{\lambda})$$

Thus Everett's method is essentially an iterative technique for constructing a saddle point for $L(\mathbf{x}, \boldsymbol{\lambda})$.

When program A is a linear programming problem (S the nonnegative orthant of R^N and y and all the f_j linear functions), it is easy to show that $(\mathbf{x}^0, \boldsymbol{\lambda}^0)$ satisfies conditions 1 and 2 if and only if $\mathbf{x}^0$ solves program A and $\boldsymbol{\lambda}^0$ solves the dual of A. The λ_j are then the dual variables of program A.

If program A is a *nonlinear* problem, Brooks and Geoffrion showed that a linearized version of program A can be used to get the successive values of λ_j which converge to satisfy Everett's condition. The linearized version is: For given values of $\mathbf{x}^1, \mathbf{x}^2, \ldots, \mathbf{x}^{k-1}$, find $u_1, u_2, \ldots, u_{k-1}$ which

$$\text{Maximize } \sum_{t=1}^{k-1} u_t y(\mathbf{x}^t)$$

and which satisfy

$$\sum_{t=1}^{k-1} u_t = 1$$

$$\sum_{t=1}^{k-1} u_t f_i(\mathbf{x}^t) \leq b_i; \qquad i = 1, \ldots, M$$

The dual variables of this linear program will be $\lambda_0^k, \lambda_1^k, \ldots, \lambda_m^k$, corresponding to the $M + 1$ constraints; from the simplex method, we know that all (except λ_0^k) will be nonnegative. These values λ_i^k are then substituted into the function of Eq. (5-45), and a maximizing vector $\mathbf{x}^k$ obtained. This process is repeated until a value of $\mathbf{x}^*$ is produced which is "sufficiently near" to satisfying the constraints of program A. Stopping rules are given in the original paper by Brooks and Geoffrion.

Consider now a highly simplified example to illustrate the use of Everett's method. This problem is so simple that it could be solved readily by other means; here we desire only to show the application of the iterative method that would be used on much larger problems.

$$\text{Maximize } y = 8x - x^2$$

subject to

$$x^2 + x \leq 4$$

The Lagrangian is

$$L(x, \lambda) = 8x - x^2 - \lambda[x^2 + x]$$

Although Everett's method is designed for use mainly on problems having nondifferentiable functions, here we can simply solve the equation

$$\frac{\partial L}{\partial x} = 8 - 2x - 2\lambda x - \lambda = 0$$

for the optimizing value of x:

$$x^*(\lambda) = \frac{8 - \lambda}{2 + 2\lambda} \tag{5-48}$$

In general, the value of $\mathbf{x}^*(\boldsymbol{\lambda}^k)$ is found by maximizing Eq. (5-45) using any convenient procedure.

To begin the iterative technique, two values for λ need be chosen; here we select arbitrarily, $\lambda_1^1 = 1$, $\lambda_1^2 = 2$, producing the corresponding optimal values of x: $x^1 = 1.75$ and $x^2 = 1.0$, yielding the coefficients $y(x^1) = 10.9375$, $f(x^1) = 4.8125$, $y(x^2) = 7$, and $f(x^2) = 2$, needed to form the first linear programming approximation to program A:

$$\text{Maximize } \hat{y} = 10.9375u_1 + 7u_2$$

subject to

$$u_1 + u_2 = 1$$
$$4.8125u_1 + 2u_2 \leq 4$$

The optimal values of u_1 and u_2 are of no importance; we are only interested in the value of the dual variable associated with the inequality constraint; we find: $\lambda_1^3 = 1.40$. Then, we have [from Eq. (5-48)], $x^3 = 1.375$, which produces the next linear program:

$$\text{Maximize } y = 10.9375u_1 + 7u_2 + 9.109u_3$$

subject to

$$u_1 + u_2 + u_3 = 1$$
$$4.8125u_1 + 2u_2 + 3.27u_3 \leq 4$$
$$u_1, u_2, u_3 \geq 0$$

Actually, since we are only interested in the dual variables anyway, it is much more straightforward and efficient to solve the dual programs directly. For example, the dual of this second linear program is

$$\text{Minimize } z = \lambda_0^4 + 4\lambda_1^4$$

subject to

$$\lambda_0^4 + 4.8125\lambda_1^4 \geq 10.9375$$
$$\lambda_0^4 + 2\lambda_1^4 \geq 7$$
$$\lambda_0^4 + 3.27\lambda_1^4 \geq 9.109$$

with λ_1^4 restricted to nonnegative values, while λ_0^4 is a free variable (that is, not sign-restricted). This method of solving only the dual to the linearized version of program A has the additional advantage that, in progressing from one linear program to the next, only one constraint need be added to the last linear program to produce the next linear program. When working with the primal linear programs, a term must be added to each constraint and to the objective function. This makes a computer solution more difficult than does working with the dual programs.

Here, the optimal solution is $\lambda_1^4 = 1.19$, and the next dual linear program is formed by adding the single constraint

$$\lambda_0^5 + 3.9525\lambda_1^5 \geq 10$$

to the linear program above. The iterative process terminates rapidly to yield the optimal solution: $x^* = 1.56$. It is interesting to note that Everett's technique is a special form of penalty functions in that it transforms a constrained optimization problem into an unconstrained problem.

The following example, taken from Phillips, Ravindran, and Solberg, illustrates a real-world application of this method. An astronaut's water container is to be stored in a space capsule wall. The container is made in the form of a sphere surmounted by a cone, the base of which is equal to the radius of the sphere. If the radius of the sphere is restricted to exactly 6 feet and a surface area of 450 ft² is all that is allowed, find the dimensions x_1 and x_2 such that the volume of the container is a maximum.

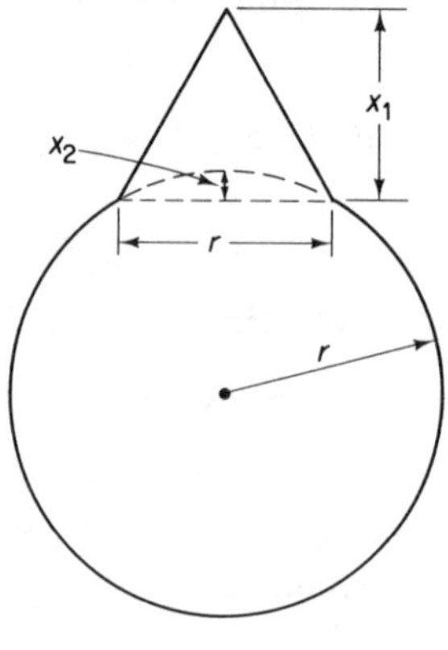

$$\text{volume of the conical top} = \frac{1}{3}\pi\frac{r^2}{4}x_1 = \frac{\pi r^2}{12}x_1$$

$$\text{volume of the cut sphere} = \frac{4}{3}\pi r^3 - \frac{1}{6}\pi x_2\left(\frac{3r^2}{4} + x_2^2\right)$$

$$\therefore \text{ volume of the capsule} = \frac{\pi}{12}r^2 x_1 + \frac{4}{3}\pi r^3 - \frac{\pi}{6}x_2\left(\frac{3r^2}{4} + x_2^2\right)$$

$$\text{surface area of cone} = \frac{\pi r}{2}\sqrt{\frac{r^2}{4} + x_1^2}$$

$$\text{surface area of spherical portion} = 4\pi r^2 - \pi\left(\frac{r^2}{4} + x_2^2\right)$$

$$\therefore \quad \text{total surface area} = \frac{\pi r}{2}\sqrt{\frac{r^2}{4} + x_1^2} + 4\pi r^2 - \pi\left(\frac{r^2}{4} + x_2^2\right)$$

From the information above, the following problem can be constructed:

$$\text{Maximize } y(x) = \frac{\pi}{12}r^2 x_1 + \frac{4}{3}\pi r^3 - \frac{\pi}{6}x_2\left(\frac{3r^2}{4} + x_2^2\right)$$

subject to

$$\frac{\pi r}{2}\sqrt{\frac{r^2}{4} + x_1^2} + 4\pi r^2 - \pi\left(\frac{r^2}{4} + x_2^2\right) = 450$$

or

$$\text{Maximize } y(x) = 9.43x_1 - 14.14x_2 - 0.52x_2^3 + 905.143$$

subject to

$$9.43\sqrt{x_1^2 + 9} - 3.14x_2^2 = 25.714$$

The Lagrangian function is given by

$$\begin{aligned}\text{maximize } L(\lambda_1, x_1, x_2) = {} & 9.43x_1 - 14.14x_2 - 0.52x_2^3 \\ & - \lambda(9.43\sqrt{x_1^2 + 9} - 3.14x_2^2)\end{aligned}$$

ITERATION 1. A common starting point is to choose $\lambda = 0$ and solve the original unconstrained problem for the solution variables. If this solution (fortuitously) satisfies the problem constraints as equalities, the problem is solved. If not, then a nonzero value of λ should be chosen and the problem solved again. For $\lambda = 0$, the solution is clearly $x_1^* = \infty$. This is logical because it will create an infinite container.

ITERATION 2. Assume that $\lambda = 3$.

$$\begin{aligned}\text{Maximize } L(x_1, x_2) = {} & 9.43x_1 - 14.14x_2 - 0.52x_2^3 \\ & - 3(9.43\sqrt{x_1^2 + 9} - 3.14x_2^2)\end{aligned}$$

The solution is

$$x_1^* = 1.06, \qquad x_2^* = 0.80$$

The constraint yields

$$9.43\sqrt{10.125} - 3.14(0.8)^2 = 27.99$$

Note that we have found the solution to a *particular* optimization problem; specifically, that problem where the right-hand side of the constraint is exactly 27.99. However, this is not the problem we wish to solve. In order to solve the problem as stated, it is necessary to modify λ iteratively as described above. The following table gives the results of successive iterations.

Iteration	λ	x_1^*	x_2^*	*Right-hand side of constraint*
1	0	∞	—	∞
2	3.0	1.06	0.80	27.99
3	2.0	1.73	1.35	26.94
4	1.5	2.68	2.79	13.52
5	1.9	1.86	1.47	26.48
6	1.8	2.00	1.61	25.89

We are sufficiently close to the answer to terminate. The optimal solution is

$$x_1^* = 2.0 \text{ ft}; \qquad x_2^* = 1.61 \text{ ft}$$

For completeness, it should be mentioned that there may not be a set of λ's that will generate $f_i(\mathbf{x}^*) \equiv b_i$. These conditions are known as *gaps*, and they are a subject of some theoretical interest. It has been found, however, that these are relatively rare in real-world problems, and there are usually multipliers for which the $f_j(\mathbf{x}^0)$ approximate the b_j closely enough for $\mathbf{x}^0$ to be a useful solution to the original problem. The procedure above will always converge to a locally optimal solution in a finite number of steps provided that there are no gaps at the optimal solution and a solution exists. Hence the real value of this technique is that it can be used on nondifferentiable functions and, in particular, on those problems in which the variables are constrained to take on only integer values.

5-18 QUADRATIC PROGRAMMING

A special case of the general nonlinear programming problem occurs when the objective function is quadratic, and all the constraints are linear. This *quadratic program* may be written as

$$\text{Minimize } y = \sum_{n=1}^{N} c_n x_n + \tfrac{1}{2} \sum_{n=1}^{N} \sum_{p=1}^{N} x_n q_{np} x_p \tag{5-49}$$

subject to

$$\sum_{n=1}^{N} a_{kn} x_n \geq b_k; \qquad k = 1, 2, \ldots, M \tag{5-50}$$

$$x_j \geq 0; \qquad j = 1, 2, \ldots, N$$

where the c_n, q_{np}, and b_k are given constants, and the matrix $\mathbf{Q} = [q_{np}]$ is positive-definite (that is, the quadratic term in y is positive for all possible values of the x_n except for all of the x_n equal to zero), and symmetric ($\mathbf{Q} = \mathbf{Q}^T$).

Although few real-world problems could be modeled accurately as quadratic programs, such programs do provide a much better approximation to the system under study than does the linear program of Chapter 3 (which frequently is used in practice for just such an approximation). In addition, it is of some interest to see how the simplex method can be adapted to solve this nonlinear problem. Wolfe was the first (1959) to show how this could be accomplished. The necessary and sufficient (Kuhn–Tucker) conditions for $\mathbf{x}^*$ to be the optimal solution to the problem of finding nonnegative x_n that minimize (5-49) subject to the constraints (5-50) are

$$\sum_{p=1}^{N} q_{np}x_p + c_n - \sum_{k=1}^{K} \lambda_k a_{kn} - v_n = 0; \qquad n = 1, \ldots, N \qquad (5\text{-}51)$$

$$\sum_{n=1}^{N} a_{kn}x_n - x_{n+k} = b_k; \qquad k = 1, \ldots, K \qquad (5\text{-}52)$$

$$v_n \geq 0; \qquad n = 1, \ldots, N \qquad (5\text{-}53)$$

$$x_n v_n = 0; \qquad n = 1, \ldots, N \qquad (5\text{-}54)$$

$$x_{n+k}\lambda_k = 0; \qquad k = 1, \ldots, K \qquad (5\text{-}55)$$

$$\lambda_k \geq 0; \qquad k = 1, \ldots, K \qquad (5\text{-}56)$$

$$x_n \geq 0; \qquad n = 1, \ldots, N + K \qquad (5\text{-}57)$$

Wolfe now sets the v_n and λ_k to zero, so that conditions (5-52), (5-57), and (5-53)–(5-56) are satisfied. His algorithm then uses the simplex method of linear programming to minimize the linear form

$$f = \sum_{n=1}^{N} \sum_{p=1}^{N} (q_{np}x_p + c_n) \qquad (5\text{-}58)$$

subject to the constraints (5-52), (5-57), and (5-53)–(5-56) and the additional constraint

$$\sum_{n=1}^{N} q_{np}x_p + c_n \geq 0; \qquad n = 1, \ldots, N \qquad (5\text{-}59)$$

where λ_k is the Lagrange multiplier associated with the kth constraint. These are necessary and sufficient conditions for solving the quadratic programming problem, provided that Q is positive definite. Wolfe first finds a basic feasible solution to the constraints (5-52). (Recall that this consists in solving the equations for K basic variables in terms of the remaining N decision variables, the latter then being set to zero.) This solution must be such that the resulting values of the basic variables are all nonnegative, and a systematic way to find such a solution is to solve a linear programming problem which formulates the sufficient conditions for the existence of the solution. One adds an artificial variable $\bar{S}_k$ to each constraint equation (5-52) to obtain a basic feasible solution, $\bar{S}_k = b_k$, with all x_n, $n = 1, \ldots, M + K$, set equal to zero. Then one minimizes the linear expression, $\sum_{k=1}^{K} \bar{S}_k$, subject to nonnegativity constraints on the $\bar{S}_k$ and x_n. If a solution to conditions (5-52) and (5-57) exists, the minimum value of $\sum_{k=1}^{K} \bar{S}_k$ will be zero, producing the desired basic feasible solution. In order to solve this problem, the simplex method must be modified so that the nonlinear constraints (5-54) and (5-55) are satisfied. This is accomplished by requiring a v_n to remain a decision variable if the corresponding x_n is a basic variable, and vice versa. This is easily accommodated through use of an appropriate *restricted basis* rule. Notice that this linear programming problem is much larger than the original quadratic problem, as it consists of $2(N + K)$ variables and $(N + K)$ constraints, plus the $N + K$ complementary slackness conditions.

5-19 DUAL PROGRAMS

Duality is a fundamental concept in mathematics and plays a particularly important role in optimization theory. The subject is so extensive that a complete coverage would require a book by itself. Here we shall be content with a brief introduction to the subject; the reader interested in a fuller treatment should consult the books and papers by Dorn, Zangwill, Duffin, Peterson, and Zener, and Beightler and Phillips, which are referenced at the end of this chapter. We shall follow the approach used by Zangwill and proceed from the notion of Lagrangian functions and the saddle-point solutions obtained for such functions.

The *primal mathematical programming problem* may be written as:

Program P

$$\text{Minimize } y(\mathbf{x})$$

subject to

$$f_i(\mathbf{x}) \geq 0; \qquad i = 1, \ldots, K$$
$$\mathbf{x} \equiv (x_1, \ldots, x_N)$$

The Lagrangian for this primal program is

$$L(\mathbf{x}, \boldsymbol{\lambda}) = y(\mathbf{x}) - \boldsymbol{\lambda}'\mathbf{f}(\mathbf{x}) = y(\mathbf{x}) - \sum_{i=1}^{K} \lambda_i f_i(\mathbf{x})$$

Any function $K(\mathbf{x}, \boldsymbol{\omega})$ is said to have a saddle point if there is a point $(\mathbf{x}^0, \boldsymbol{\omega}^0)$

such that

$$K(\mathbf{x}^0, \boldsymbol{\omega}) \leq K(\mathbf{x}^0, \boldsymbol{\omega}^0) \leq K(\mathbf{x}, \boldsymbol{\omega}^0) \tag{5-60}$$

for all $\mathbf{x}$, $\boldsymbol{\omega}$. Let

$$\hat{K}(\boldsymbol{\omega}) \equiv \min_{\mathbf{x}} [K(\mathbf{x}, \boldsymbol{\omega})] \qquad \text{(minimized over } \mathbf{x} \text{ for fixed } \boldsymbol{\omega})$$

and let

$$\bar{K}(\mathbf{x}) \equiv \max_{\boldsymbol{\omega}} [K(\mathbf{x}, \boldsymbol{\omega})] \qquad \text{(maximized over } \boldsymbol{\omega} \text{ for fixed } \mathbf{x})$$

Then

$$\bar{K}(\mathbf{x}) \geq K(\mathbf{x}, \boldsymbol{\omega}) \geq \hat{K}(\boldsymbol{\omega}) \tag{5-61}$$

for any $(\mathbf{x}, \boldsymbol{\omega})$, so that

$$\bar{K}(\mathbf{x}) \geq \hat{K}(\boldsymbol{\omega}) \tag{5-62}$$

for any $(\mathbf{x}, \boldsymbol{\omega})$.

The following two lemmas, due to Zangwill, are of fundamental importance in the development of duality theory.

Lemma 1: *The following three statements are equivalent:*

(*a*) $$\min_{\mathbf{x}} \{\max_{\boldsymbol{\omega}} [K(\mathbf{x}, \boldsymbol{\omega})]\} = \max_{\boldsymbol{\omega}} \{\min_{\mathbf{x}} [K(\mathbf{x}, \boldsymbol{\omega})]\}$$

(*b*) $$\bar{K}(\mathbf{x}^0) = \hat{K}(\boldsymbol{\omega}^0)$$

(*c*) $$K(\mathbf{x}^0, \boldsymbol{\omega}) \leq K(\mathbf{x}^0, \boldsymbol{\omega}^0) \leq K(\mathbf{x}, \boldsymbol{\omega}^0)$$

Proof: Statement (*a*) may be rewritten as

$$\min_{\mathbf{x}} [\bar{K}(\mathbf{x})] = \max_{\boldsymbol{\omega}} [\hat{K}(\boldsymbol{\omega})]$$

Let $\mathbf{x}^0$ minimize $\bar{K}(\mathbf{x})$ and $\boldsymbol{\omega}^0$ maximize $\hat{K}(\boldsymbol{\omega})$; then (*b*) holds; (*a*) implies (*b*). By Eq. (5-62),

$$\bar{K}(\mathbf{x}) \geq \hat{K}(\boldsymbol{\omega})$$

(*b*) implies

$$\min_{\mathbf{x}} [\bar{K}(\mathbf{x})] = \max_{\boldsymbol{\omega}} [\bar{K}(\boldsymbol{\omega})]$$

and hence (*a*).

From Eq. (5-61), (*b*) is equivalent to

$$\bar{K}(\mathbf{x}^0) = K(\mathbf{x}^0, \boldsymbol{\omega}^0) = \hat{K}(\boldsymbol{\omega}^0)$$

which (from the definition) can be rewritten as

$$\max_{\boldsymbol{\omega}} [K(\mathbf{x}^0, \boldsymbol{\omega})] = K(\mathbf{x}^0, \boldsymbol{\omega}^0) = \min_{\mathbf{x}} [K(\mathbf{x}, \boldsymbol{\omega}^0)]$$

which is (*c*); hence (*b*) and (*c*) are equivalent.

Lemma 1 thus provides several different ways to write a saddle point, $(\mathbf{x}^0, \boldsymbol{\omega}^0)$. Notice, however, that it is possible for a function *not* to have a saddle point if it is true that

$$\min_{\mathbf{x}} [\bar{K}(\mathbf{x})] > \max_{\boldsymbol{\omega}} [\hat{K}(\boldsymbol{\omega})]$$

The Lagrangian for the primal program P is

$$L(\mathbf{x}^0, \boldsymbol{\lambda}) = y(\mathbf{x}) - \boldsymbol{\lambda}'\mathbf{f}(\mathbf{x})$$

From Eq. (5-60), this Lagrangian possesses a saddle point if

$$L(\mathbf{x}^0, \boldsymbol{\lambda}) \leq L(\mathbf{x}^0, \boldsymbol{\lambda}^0) \leq L(\mathbf{x}, \boldsymbol{\lambda}^0)$$

By analogy to $\hat{K}$ and $\bar{K}$, we define the *primal function* to be

$$\bar{L}(\mathbf{x}) = \max_{\boldsymbol{\lambda} \geq \mathbf{0}} [L(\mathbf{x}, \boldsymbol{\lambda})]$$

and the *dual function* to be

$$\hat{L}(\boldsymbol{\lambda}) = \min_{\mathbf{x}} [L(\mathbf{x}, \boldsymbol{\lambda})]$$

From these two functions we may now define two (programming) problems: The *primal problem* is to determine the optimal $\mathbf{x}^0$, such that

$$\bar{L}(\mathbf{x}^0) = \min_{\mathbf{x}} [\bar{L}(\mathbf{x})]$$

The *dual problem* is to calculate an optimal $\boldsymbol{\lambda}^0$ such that

$$\hat{L}(\boldsymbol{\lambda}^0) = \max_{\boldsymbol{\lambda} > \mathbf{0}} [\hat{L}(\boldsymbol{\lambda})]$$

Let us now analyze the primal problem. To do this, we first evaluate the primal function:

$$\bar{L}(\mathbf{x}) = \max_{\boldsymbol{\lambda} > \mathbf{0}} \left[y(\mathbf{x}) - \sum_i \lambda_i f_i(\mathbf{x}) \right] = \begin{cases} y(\mathbf{x}) & \text{if all } f_i(\mathbf{x}) \geq 0 \\ +\infty & \text{if any } f_i(\mathbf{x}) < 0 \end{cases} \tag{5-63}$$

These results clearly hold, since if all $f_i(\mathbf{x}) \geq 0$, and since we must choose

$\lambda_i \geq 0$, then setting all $\lambda_i = 0$ would maximize $L(\mathbf{x})$. However, if any $f_i(\mathbf{x}) < 0$, then letting the corresponding $\lambda_i \longrightarrow +\infty$ would maximize the Lagrangian.

The primal problem requires that we minimize the primal function $\bar{L}(\mathbf{x})$. In this minimization process the portion where $\bar{L}(\mathbf{x}) = +\infty$ can, of course, be disregarded, and the primal problem may now be stated as

$$\min_{\mathbf{x}} [\bar{L}(\mathbf{x})] = \min_{\mathbf{x}} [y(\mathbf{x})] \text{ over all } \mathbf{x} \text{ such that } f_i(\mathbf{x}) \geq 0; \qquad i = 1, \ldots, K$$

or, in standard notation,

$$\min_{\mathbf{x}} [y(\mathbf{x})]$$

subject to

$$f_i(\mathbf{x}) \geq 0; \qquad i = 1, \ldots, K$$
$$\mathbf{x} \in E^N$$

Now consider the *dual problem*: A pair $(\mathbf{x}^A, \boldsymbol{\lambda}^A)$ is said to be *feasible* for the dual problem if

$$\hat{L}(\boldsymbol{\lambda}^A) = L(\mathbf{x}^A, \boldsymbol{\lambda}^A)$$

If $\boldsymbol{\lambda}^0$ is the optimal value of $\boldsymbol{\lambda}$ for the dual, and $(\mathbf{x}^0, \boldsymbol{\lambda}^0)$ is feasible for the dual, then the pair $(\mathbf{x}^0, \boldsymbol{\lambda}^0)$ is said to be *optimal* for the dual.

Lemma 2: *Suppose that* $\mathbf{x}^A$ *is feasible for the primal program P. Then*

(a) $$\bar{L}(\mathbf{x}^A) = y(\mathbf{x}^A)$$

Moreover, if $(\mathbf{x}^B, \boldsymbol{\lambda}^B)$ *is feasible for the dual:*

(b) $$\bar{L}(\mathbf{x}^A) = y(\mathbf{x}^A) \geq y(\mathbf{x}^B) - \sum_{i=1}^{K} \lambda_i^B f_i(\mathbf{x}^B) = L(\mathbf{x}^B, \boldsymbol{\lambda}^B) = \hat{L}(\boldsymbol{\lambda}^B)$$

Proof: (*a*) If $\mathbf{x}^A$ is feasible for the primal, then from Eq. (5-63),

$$\bar{L}(\mathbf{x}^A) = y(\mathbf{x}^A)$$

(*b*) Equation (5-62) may be rewritten in terms of $\hat{L}$ and $\bar{L}$ as

$$\bar{L}(\mathbf{x}^A) \geq \hat{L}(\boldsymbol{\lambda}^B)$$

Lemma 2 states that the dual problem always provides a *lower bound* for the primal problem. This holds even if the Lagrangian does not have a saddle point.

The dual program in general is:

Program D

$$\max_{\mathbf{x},\boldsymbol{\lambda}} z = y(\mathbf{x}) - \sum_{i=1}^{K} \lambda_i f_i(\mathbf{x}) \tag{5-64}$$

subject to

$$\boldsymbol{\lambda} \geq 0 \tag{5-65}$$

$$y(\mathbf{x}) - \sum_{i=1}^{K} \lambda_i f_i(\mathbf{x}) = \min_{\mathbf{x}} \left[y(\mathbf{x}) - \sum_{i=1}^{K} \lambda_i f_i(\mathbf{x}) \right] \tag{5-66}$$

Equation (5-66) says that $\mathbf{x}$ has been "optimized out."

Suppose now that y is strictly convex, the f_i are all concave functions, and there is at least one point $\bar{\mathbf{x}}^{(i)}$ which satisfies $f_i > 0$ (strict inequality) for each $i = 1, \ldots, K$. Then we designate the resulting *convex* primal problem as program $\mathrm{P_C}$, and observe that

$$\nabla L(\mathbf{x}, \boldsymbol{\lambda}) = 0$$

if and only if

$$L(\mathbf{x}, \boldsymbol{\lambda}) = \min_{\mathbf{x}} [L(\mathbf{x}, \boldsymbol{\lambda})]$$

So the dual problem becomes:

Program $\mathrm{D_C}$

$$\max_{\mathbf{x}} z = y(\mathbf{x}) - \sum_{i=1}^{K} \lambda_i f_i(\mathbf{x}) \tag{5-67}$$

subject to

$$\lambda_i \geq 0; \qquad i = 1, \ldots, K \tag{5-68}$$

$$\nabla y(\mathbf{x}) - \sum_{i=1}^{K} \lambda_i \nabla f_i(\mathbf{x}) = 0 \tag{5-69}$$

If either program $\mathrm{P_C}$ or $\mathrm{D_C}$ has a finite solution, then the other does also, and min y = max z. Furthermore, if $\mathbf{x}^*$ solves $\mathrm{P_C}$, then this $\mathbf{x}^*$, together with some $\boldsymbol{\lambda}^*$, solves $\mathrm{D_C}$. Conversely, if $(\mathbf{x}^*, \boldsymbol{\lambda}^*)$ is the optimal solution to program $\mathrm{D_C}$, then this same $\mathbf{x}^*$ solves program $\mathrm{P_C}$.

For example, consider the convex program

$$\text{Minimize } y = 2x_1^2 - x_1x_2 + x_2^2 - 3x_1 - 4x_2$$

subject to

$$4x_1 - x_2^2 = 12$$

Then the dual is given by:

$$\text{Maximize } z = 2x_1^2 - x_1x_2 + x_2^2 - 3x_1 - 4x_2 - \lambda[4x_1 - x_2^2 - 12]$$

subject to

$$\begin{aligned} 4x_1 - x_2 - 3 - 4\lambda &= 0 \\ -x_1 + 2x_2 - 4 - 4\lambda x_2 &= 0 \end{aligned}$$

Notice that the dual variable λ is not sign-restricted since the sole primal constraint is an equality. In general, if the ith primal constraint is an equality, then the associated dual variable, λ_i, will be a free variable; if this constraint is an inequality, then λ_i will be restricted to nonnegative values.

An important type of convex program is the *quadratic programming problem*, which was developed in Section 5-18. The program may be written as:

Program $\mathbf{P_Q}$

$$\text{Minimize } y = \frac{1}{2}\sum_{k=1}^{N}\sum_{j=1}^{N} x_k q_{kj} x_j + \sum_{j=1}^{N} c_j x_j$$

subject to

$$\sum_{j=1}^{N} a_{ij}x_j \geq b_i; \qquad i = 1, \ldots, K$$

where $q_{kj} = q_{jk}$, all j, k.

This program may be written in matrix notation by defining $\mathbf{Q} \equiv [q_{ij}]$ to be a symmetric, positive-definite matrix, and $\mathbf{A}$ the $K \times N$ matrix of constraint coefficients:

$$\text{Minimize } y = \tfrac{1}{2}\mathbf{x}'\mathbf{Q}\mathbf{x} + \mathbf{C}'\mathbf{x}$$

subject to

$$\mathbf{A}\mathbf{x} \geq \mathbf{b}$$

Then the dual program becomes

$$\text{maximize } z = \tfrac{1}{2}\mathbf{x}'\mathbf{Q}\mathbf{x} + \mathbf{C}'\mathbf{x} + \boldsymbol{\lambda}'\mathbf{b} - \boldsymbol{\lambda}'\mathbf{A}\mathbf{x}$$

subject to

$$\begin{aligned} \mathbf{x}'\mathbf{Q} + \mathbf{C}' - \boldsymbol{\lambda}'\mathbf{A} &= \mathbf{0} \\ \boldsymbol{\lambda} &\geq \mathbf{0} \end{aligned}$$

If the x_j are required to be nonnegative in the primal, then the dual constraints will be inequalities; here, the primal variables are not sign-restricted.

The dual may be further simplified by first postmultiplying the constraints by $\mathbf{x}$ to yield

$$\mathbf{x}'\mathbf{Q}\mathbf{x} + \mathbf{C}'\mathbf{x} - \boldsymbol{\lambda}'\mathbf{A}\mathbf{x} = \mathbf{0}$$

and then substituting the result into the objective function. The result is

Program $\mathbf{D}_\mathbf{Q}$

$$\text{Maximize } z = \boldsymbol{\lambda}'\mathbf{b} - \tfrac{1}{2}\mathbf{x}'\mathbf{Q}\mathbf{x}$$

subject to

$$\mathbf{A}'\boldsymbol{\lambda} - \mathbf{Q}\mathbf{x} = \mathbf{C}$$
$$\boldsymbol{\lambda} > \mathbf{0}$$

If program $\mathrm{P_Q}$ requires in addition that $\mathbf{x} \geq \mathbf{0}$, then the equality dual constraints above become

$$\mathbf{A}'\boldsymbol{\lambda} - \mathbf{Q}\mathbf{x} \leq \mathbf{C}$$

In order to illustrate the quadratic programs above, consider the following simplified example:

$$\text{Minimize } y = 2x_1^2 - 6x_1x_2 + 9x_2^2 - 18x_1 + 9x_2$$

subject to

$$x_1 + 2x_2 \leq 12$$
$$4x_1 - 3x_2 \leq 20$$
$$x_1,\ x_2 \geq 0$$

Here,

$$\mathbf{Q} = \begin{bmatrix} 4 & -6 \\ -6 & 18 \end{bmatrix}$$

and upon multiplying the constraints through by -1 to agree with the form required by program $\mathrm{P_Q}$, we have

$$\mathbf{A} = \begin{bmatrix} -1 & -2 \\ -4 & 3 \end{bmatrix}, \qquad \mathbf{b} = \begin{bmatrix} -12 \\ -20 \end{bmatrix}$$

so that the dual program, $\mathrm{D_Q}$, is:

$$\text{Maximize } z = -12\lambda_1 - 20\lambda_2 - 2x_1^2 + 6x_1x_2 - 9x_2^2$$

subject to

$$-\lambda_1 - 4\lambda_2 - 4x_1 + 6x_2 \leq -18$$
$$-2\lambda_1 + 3\lambda_2 + 6x_1 - 18x_2 \leq 9$$
$$\lambda_1, \lambda_2 \geq 0$$

The solution to the primal problem is $x_1^* = 6.300$, $x_2^* = 1.733$, which, when substituted into the dual program, produces the following linear program:

$$\text{Maximize } z = -12\lambda_1 - 20\lambda_2 - 40.90$$

subject to

$$\lambda_1 + 4\lambda_2 \geq 3.20$$
$$-2\lambda_1 + 3\lambda_2 \leq 2.40$$
$$\lambda_1, \lambda_2 \geq 0$$

which has the solution $\lambda_1^* = 0$, $\lambda_2^* = 0.80$. The reader may verify that $y(\mathbf{x}^*) = z(\boldsymbol{\lambda}^*) = -56.90$, as specified by duality theory.

In the special case for which $\mathbf{Q} = \mathbf{0}$ (y and all of the f_i are *linear functions*), programs P_Q and D_Q become the following dual linear programming problems:

Program P_L

$$\text{Minimize } y = \sum_{j=1}^{N} c_j x_j$$

subject to

$$\sum_{j=1}^{N} a_{ij} x_j \geq b_i; \qquad i = 1, \ldots, K$$
$$x_j \geq 0; \qquad j = 1, \ldots, N$$

Program D_L

$$\text{Maximize } z = \sum_{i=1}^{K} b_i \lambda_i$$

subject to

$$\sum_{i=1}^{K} a_{ij} \lambda_i \leq c_i; \qquad j = 1, \ldots, N$$
$$\lambda_i \geq 0; \qquad i = 1, \ldots, K$$

where the c_j, b_i, and a_{ij} are given numbers.

Observe that here in the fully linear case, the dual problem is no longer a function of the primal variables. In fact, as shown in Chapter 3, the solution to either of the problems above produces the optimal solution to the other problem as well. Indeed, these two programs may be written in the form of a single problem, expressed entirely in terms of inequalities, with no objective function to be optimized:

$$\sum_{i=1}^{K} b_i\lambda_i \leq \sum_{j=1}^{N} c_j x_j$$

$$\sum_{j=1}^{N} a_{ij}x_j \geq b_i; \qquad i = 1, \ldots, K$$

$$\sum_{i=1}^{K} a_{ij}\lambda_i \leq c_j; \qquad j = 1, \ldots, N$$

$$x_j \geq 0; \qquad j = 1, \ldots, N$$

$$\lambda_i \geq 0; \qquad i = 1, \ldots, K$$

If values of x_j and λ_i can be found which satisfy all the inequalities above, then these same values will be the optimal solutions to the primal program P_L and the dual program D_L, respectively.

One of the most significant developments in mathematical programming is the recent appearance of *geometric programming*, the topic covered in Chapter 6. This revolutionary new method, with its elegant solution algorithms, has made the fullest use to date of duality in solving problems with highly nonlinear constraints. Although originally applicable only to a special class of functions, recent results have shown that mathematical programming problems containing *any* algebraic functions may be converted to equivalent programs amenable to solution by geometric programming algorithms. As a result, this new technique is finding important application in many fields, most notably in engineering design.

Of fundamental importance to the numerical solution of real-world problems is the fact that, unlike other nonlinear programs, the dual geometric program is *not* a function of the primal variables. In this regard, it resembles the fully linear situation, for which the solution to either the primal or dual program gives the optimal solution to both programs. This is indeed fortunate, since in going from the primal to the dual program, a problem constrained by nonlinear inequalities is transformed into one having only linear equality constraints. Hence, the dual geometric program is much the easier of the two problems to solve, since the original (primal) problem has in general a nonconvex objective function defined over a nonconvex solution set. Owing to the special relationship between the primal and dual programs, geometric programming shares with linear programming the advantage that a full sensitivity analysis on all the input parameters can be performed with relative ease. The theory and application of geometric programming are surveyed in Chapter 6 and covered in detail in the Beightler–Phillips book referenced at the end of the chapter.

For problems that occur in practice, an interpretation of the context of the dual program is very useful in problem analysis; Dorn has given an important interpretation of certain special dual programs arising from physical problems, and we shall use his excellent description to close this section:

> The duality theorems indicate that to a certain class of minimization problems there corresponds a class of complementary maximization problems. If the original problem is cast in terms of the appropriate variables, then the variables in its dual problem will have an interpretation in terms of the physical properties of the system. To this end, the variables in any system are divided in two categories:
>
> 1. *Extensive* variables whose magnitudes increase proportionally to the size of the system (for example, mass, total cost, etc.).
> 2. *Intensive* variables whose magnitudes are independent of the size of the system (for example, pressure, price, etc.).
>
> If the original problem is derived in terms of extensive variables exclusively, then its dual will be in terms of intensive variables. Conversely, a problem in terms of intensive variables has dual variables that are extensive.
> The dual problem often sheds light on the properties of the system and provides further insight into the proper interpretation of the solutions.

BIBLIOGRAPHY

BEIGHTLER, C., "Duality theory," *Encyclopedia of Computer Science and Technology*, Vol. 7, pp. 375–393.

———, and D. T. PHILLIPS, *Applied Geometric Programming* (Wiley, New York, 1976).

———, and D. J. WILDE, "Diagonalization of quadratic forms by Gauss elimination," *Man. Sci.*, **12** (January 1966), 371–379.

BROOKS, R., and A. GEOFFRION, "Finding Everett's multipliers by linear programming," *Operations Res.*, **14** (1966), 1149–1153.

DENNIS, J. B., *Mathematical Programming and Electrical Networks* (M.I.T. Press, Cambridge, Mass., 1959).

DORN, W. S., "Non-linear programming—a survey," *Man. Sci.*, **9** (January 1963), 171–208.

DUFFIN, R. J., and L. A. KARLOVITZ, "An infinite linear program with a duality gap," *Man. Sci.*, **12** (September 1965), 122–134.

———, E. L. PETERSON, and C. ZENER, *Geometric Programming* (Wiley, New York, 1967).

EVERETT, H., "Generalized Lagrange multiplier method for solving problems of optimum allocation of resources," *Operations Res.*, **11** (1963), pp. 399–471.

FIACCO, A. V., and G. P. MCCORMICK, "Computational algorithm for the sequential unconstrained minimization technique for nonlinear programming," *Man. Sci.*, **10** (1964), 601–617.

GASS, S. I., *Linear Programming*, 2nd ed. (McGraw-Hill, New York, 1964).

GOTTFRIED, B. S., and JOEL WEISMAN, *Introduction to Optimization Theory* (Prentice-Hall, Englewood Cliffs, N.J., 1973).

HADLEY, G., *Nonlinear and Dynamic Programming* (Addison-Wesley, Reading, Mass., 1964).

KELLEY, J. E., JR., "The cutting plane method for solving convex programs," *J. Soc. Ind. Appl. Math.*, **8** (December 1960), 703–712.

KUHN, H. W., and A. W. TUCKER, "Nonlinear programming," *Proc. Second Berkeley Symp. Math. Statistics and Probability*, J. Neyman, ed. (Univ. of California Press, 1951).

KUNZ, K. S., *Numerical Analysis* (McGraw-Hill, New York, 1957).

KÜNZI, H. P., and W. KRELLE, *Nichtlineare Programmierung* (Springer-Verlag, Berlin, 1962).

PHILLIPS, D. T., A. RAVINDRAN, and J. J. SOLBERG, *Operations Research: Principles and Practice* (Wiley, New York, 1976).

WILDE, D. J., "Differential calculus in nonlinear programming," *Operations Res.*, **10** (November 1962), 764–773.

———, *Globally Optimal Design* (Wiley-Interscience, New York, 1978).

WOLFE, P., "The simplex method for quadratic programming," *Econometrica*, **27** (1959), 382–398.

———, "Accelerating the cutting plane method for nonlinear programming," *J. Soc. Ind. Appl. Math.*, **9** (September 1961), 481–488.

ZANGWILL, W. I., *Nonlinear Programming: A Unified Approach* (Prentice-Hall, Englewood Cliffs, N.J., 1969).

ZOUTENDIJK, G., *Methods of Feasible Directions* (Elsevier Publishing Co., Amsterdam, 1960).

EXERCISES

5-1. Consider an N-stage system where each stage consists of a number m_i of parallel (redundant) components having a cost c_i and a reliability r_i. This system is operable if, and only if, every stage contains at least one operable component. The allocation problem is to choose the stage redundancies (m_i's) in order to minimize the cost of achieving a given system reliability (or, *here*, to maximize the system reliability subject to a total cost constraint).

The system reliability is given by

$$R = \prod_{i=1}^{N} [1 - (1 - r_i)^{m_i}]$$

and, since the m_i that maximize the logarithm of R also maximize R, the problem can be formulated in this way:

$$\text{Maximize } \sum_{i=1}^{N} \ln [1 - (1 - r_i)^{m_i}]$$

subject to

$$\sum_{i=1}^{N} c_i m_i \leq C$$

$$m_i = 0, 1, 2, \ldots; \qquad i = 1, \ldots, N$$

where C and the r_i and c_i are given positive numbers.

Solve this using Everett's method with the Brooks–Geoffrion iterative technique, for the values $N = 4$, $C = 50.3$, and the stage component costs and reliabilities as follows:

Stage, i	*Cost, c_i*	*Reliability, r_i*
1	1.2	0.8
2	2.3	0.7
3	3.4	0.75
4	4.5	0.85

Ans.: $m_1 = 5$, $m_2 = 6$, $m_3 = 5$, $m_4 = 3$

5-2. Rework the example of Section 5-02, this time taking x_2 and x_4 as the solution variables. Why won't x_2 and x_5 work?

5-3. Make another iteration from $(\mathbf{d})_1$ for the example in Section 5-05.

5-4. As part of the derivation of Eq. (5-8), differentiate $(\partial s_i/\partial d_j)(\partial y/\partial s_i)$ partially with respect to d_l.

5-5. Derive Eq. (5-9) for $D = S = 2$.

5-6. Make another iteration from $(\mathbf{d})_2$ in Section 5-06.

5-7. Derive the functions g_1 through g_8 for the example of Section 5-08.

5-8. For the example of Section 5-12, make an iteration from the point $(1.136, 0, 1.066, 1.136)^T$.

5-9. Apply the differential procedure to the problem defined in Eqs. (5-39) and (5-40).

5-10. Apply the differential procedure to the water container problem of Section 5-17.

5-11. Minimize $y = -\exp[(x_1 - 1)^2 + (x_2 - 2)^2]$ subject to $x_1, x_2 \geq 0$ and $x_1 \geq x_2^2$; $x_2 \geq e^{-x_1}$; $2(x_1 - 1)^2 \leq x_2$; by (a) the differential procedure; (b) a penalty function method; (c) generalized Lagrange multipliers.

5-12. Show that the homogeneous quadratic form

$$q = x_1^2 - 4x_1x_2 + 6x_1x_3 + 5x_2^2 - 10x_2x_3 + 8x_3^2$$

is indefinite. For what values of the x_j can q be made arbitrarily small?

5-13. Minimize $y = 6x_1^3 - 12x_1x_2 + 3x_2^2 - 6x_1 + 7x_2$ subject to $x_1, x_2 \geq 0$ and $3x_1 + 4x_2 \leq 2$ by (a) the differential procedure; (b) a penalty function method; (c) generalized Lagrange multipliers. Start from the origin.

5-14. A company manufactures two products, each of which uses the same two raw materials. Each unit of product 1 requires 8 kg of material A and 7 kg of B; each unit of product 2 requires 3 kg of A and 6 kg of B. The maximum amount of A available each week is 1200 kg; the maximum weekly amount of B is 2100 kg. The company can sell as many units as they can make, but the selling prices depend on how many units have been sold. Thus the unit profit for product 1 is $800 - x_1 - x_2$, and the unit profit for product 2 is $2000 - x_1 - 3x_3$, where x_1 and x_2 are the weekly production rates for products 1 and 2, respectively. Find the optimal quantities of the two products to make each week for maximum weekly profit.

5-15. Use the penalty function method to solve the following problem:

$$\text{Maximize } y = 6x_1x_2 - 2x_1^2 - 9x_2^2 + 18x_1 - 9x_2$$

subject to

$$x_1 + 2x_2 \leq 12$$
$$x_1, \quad x_2 \geq 0$$

5-16. Re-solve Exercise 5-15 when the following constraint is added to the feasible set.

$$x_1^2 + x_2^2 = 9$$

5-17. (Fiacco and McCormick) Solve using penalty functions:

$$\text{Minimize } y = -x_1x_2$$

subject to

$$-x_1 - x_2^2 + 1 \geq 0$$
$$x_1 + x_2 \geq 0$$
$$x_1, \quad x_2 \geq 0$$

Geometric Programming 6

Which of you, intending to build a tower,
sitteth not down first and counteth the cost,
whether he have sufficient to finish it?

LUKE XIV, 28, C. 75

Geometric programming is a relatively new technique, developed for solving algebraic nonlinear programming problems subject to nonlinear constraints. This new technique first emerged in 1961 when Clarence Zener, then Director of Science at Westinghouse Corporation, discovered that many engineering design problems consisting of a sum of component costs could sometimes be minimized almost by inspection under suitable conditions. Coincidentally, Richard Duffin, then Professor of Mathematics at Carnegie-Mellon University, was engaged in the development of a duality theory with direct applications to nonlinear programming problems. Duffin learned of Zener's work while visiting Westinghouse Corporation and soon (March 1962) solidified Zener's discovery mathematically through application of his own recently developed theory. The original mathematical development of this new method used the arithmetic–geometric mean inequality relationship between sums and products of positive numbers. It was this theoretical development that prompted Duffin to call this technique "geometric programming."

Once the basic concepts of geometric programming had been established, Duffin and Zener began to extend and develop their new technique. Elmor Peterson, a student of Duffin's, collaborated with Zener on an electrical transformer law and with Duffin on an extension of the method to include inequality constraints. Previously, Charnes and Cooper had shown how to handle single-term inequalities. In 1967, Duffin, Peterson, and Zener produced the now-classic text, *Geometric Programming*. This landmark publication contained all the theoretical developments up to that date, along with some example problems to illustrate the technique. In addition to rigorous proofs, it contained many useful transformations and approximations for expressing optimization problems in the proper form for solution by geometric programming.

The work of Duffin, Peterson, and Zener was restricted to those problems that contained only positive coefficients for the component cost terms. Although many real-world problems in engineering design are characterized by this structure, the final developments, which released the theory of geometric programming from this restrictive environment, were already developed independently by Passy and Wilde. Since that time, many new developments have been published, extending the range of problems that can be solved by this technique. Geometric programming is now applicable to any problem involving signomials (generalized polynomials) both in the objective function and in the constraints, where inequalities of either sense are permissible. Indeed, *any* nonlinear algebraic program can be transformed into an equivalent posynomial program (with some inequalities of opposite sense). All this recent work is covered in detail in the book *Applied Geometric Programming*, by Beightler and Phillips. In this chapter we can only touch on a few of the highlights of geometric programming and summarize the important results; the interested reader should consult the Beightler–Phillips book for a full account of the entire modern theory. The book also contains numerous examples of successful applications of geometric programming to real-world problems.

We now state the generalized geometric programming problem in order to establish the terminology and notation:

$$\text{minimize } y_o(\mathbf{x}) = \sum_{t=1}^{T_o} \sigma_{ot} c_{ot} \prod_{n=1}^{N} x_n^{a_{otn}}$$

subject to

$$y_m(\mathbf{x}) = \sum_{t=1}^{T_m} \sigma_{mt} c_{mt} \prod_{n=1}^{N} x_n^{a_{mtn}} \leq \sigma_m; \qquad m = 1, 2, \ldots, M$$

$$x_n > 0; \qquad n = 1, 2, \ldots, N$$

where

$$\sigma_{mt} = \pm 1, \quad \sigma_{ot} = \pm 1, \quad \text{and} \quad \sigma_m = \pm 1$$

$$c_{mt} > 0, \quad c_{ot} > 0$$

where a_{mtn} and a_{otn} are unrestricted in sign

T_m = number of terms in the mth constraint ($m = 1, \ldots, M$)
T_o = number of terms in the objective function

In terms of engineering design formulations, the c_{mt} are economic coefficients, the x_n are design decision variables, the a_{mtn} are technological exponents of the decision variables, and the $\boldsymbol{\sigma}$ vector has as elements binary variables (± 1), whose signs represent the sign of each term and inequality in the problem statement. If all the $\boldsymbol{\sigma}$ are positive, the program is called a *posynomial* geometric programming problem; if one or more of the $\boldsymbol{\sigma}$ are negative, the program is called a *signomial* geometric programming problem.

When certain favorable properties are present in the problem formulation, problems with complex nonlinear objective functions and constraints can be solved through the solution of a set of well-defined *linear* equations! However, when these favorable properties are not present, the solution technique becomes a nontrivial problem that has been the subject of much recent research. Even in those cases, however, the technique of geometric programming provides a vehicle through which simpler formulations of the original problem can be derived.

6-01 UNCONSTRAINED POSYNOMIAL PROBLEMS

This section is concerned with *posynomial* functions, defined as

$$y \equiv y(\mathbf{x}) = \sum_{t=1}^{T} c_t p_t(\mathbf{x}) \tag{6-1}$$

where the given c_t are positive numbers, $\mathbf{x} \equiv (x_1, x_2, \ldots, x_N)'$, and the functions $p_t(\mathbf{x})$ are defined as

$$p_t(\mathbf{x}) = \prod_{n=1}^{N} x_n^{a_{tn}} \tag{6-2}$$

with the a_{tn} being any real numbers. Notice that y is *not* a polynomial since the coefficients of the terms must be positive and the exponents are not restricted to the positive integers.

Suppose that y is an objective function to be minimized; it is helpful to think of it as a total cost composed of a sum of component costs. Consider first the unconstrained problem in which it is desired to find positive values $x_1^*, x_2^*, \ldots, x_N^*$ that minimize y.

At a minimum, the first derivatives must vanish.

$$\frac{\partial y}{\partial x_k} = \sum_{t=1}^{T} c_t a_{tk} x_k^{a_{t,k}-1} \prod_{n \neq k} x_n^{a_{tn}} = 0; \qquad k = 1, \ldots, N$$

and, since all $x_n > 0$, these equations become

$$\sum_{t=1}^{T} c_t a_{tk} \prod_{n=1}^{N} x_n^{a_{tn}} = 0; \qquad k = 1, \ldots, N \tag{6-3}$$

This is a system of N simultaneous nonlinear equations in the N unknowns, x_n. Solving such a system is very difficult, and most solution methods, when they converge at all, may not find the proper root. Yet this solution of simultaneous nonlinear equations is required in the classical method for optimizing a multivariable function. This approach is a natural one, since Eq. (6-3) expresses the necessary conditions for a (local) optimum and hence must be satisfied.

Consider, instead, the following approach: rather than seek the optimal values of the independent variables first, let us find the optimal way to distribute the total cost among the various terms of the objective function. Once these optimal allocations are obtained, the minimum cost, y^*, can be found without knowledge of the optimal values of the independent variables, x_n^*.

To obtain this distribution, first define the optimal weights, w_t, by

$$w_t \equiv \frac{c_t p_t(\mathbf{x}^*)}{y^*}; \qquad t = 1, \ldots, T \tag{6-4}$$

Clearly, the weights sum to unity:

$$\sum_{t=1}^{T} w_t = 1 \tag{6-5}$$

Equations (6-3) and (6-4) give

$$\sum_{t=1}^{T} a_{tn} w_t y^* = 0; \qquad n = 1, \ldots, N$$

or, since y^* must be positive,

$$\sum_{t=1}^{T} a_{tn} w_t = 0; \qquad n = 1, \ldots, N \tag{6-6}$$

Now y^* can be found directly from knowledge of the values of the w_t, as shown from the following derivation:

$$\begin{aligned} y^* &= \prod_{t=1}^{T} (y^*)^{w_t} = \prod_{t=1}^{T} \left[\frac{c_t p_t(\mathbf{x}^*)}{w_t}\right]^{w_t} \\ &= \prod_{t=1}^{T} \left(\frac{c_t}{w_t}\right)^{w_t} \prod_{t=1}^{T} [p_t(\mathbf{x}^*)]^{w_t} \end{aligned} \tag{6-7}$$

But, from Eq. (6-6),

$$\prod_{t=1}^{T}[p_t(\mathbf{x}^*)]^{w_t} = \prod_{n=1}^{N}\prod_{t=1}^{T}(x_n^*)^{a_{tn}w_t} = \prod_{n=1}^{N}(x_n^*)^{\sum_{t=1}^{T} a_{tn}w_t} = 1$$

Therefore, Eq. (6-7) becomes

$$y^* = \prod_{t=1}^{T}\left(\frac{c_t}{w_t}\right)^{w_t} \tag{6-8}$$

so that y^* can be calculated from the coefficients c_t and the optimal weights w_t. The problem of optimizing y now becomes one of finding the values of the w_t. Notice that Eqs. (6-5) and (6-6) consist of $N + 1$ simultaneous linear equations in T unknowns. In order to make the solution method clear, let us first concentrate on an example problem for which $T = N + 1$, for then the optimal weights will be uniquely determined by the equation set [(6-5), (6-6)], as long as the coefficient matrix of these equations has full rank.

Consider a hypothetical problem in which it is desired to minimize the total inventory and production cost associated with the manufacture of a certain product. Let this total cost be

$$y = c_1x_1^{-3}x_2^{-2} + c_2x_1^3x_2 + c_3x_1^{-3}x_2^3$$

where x_1 is the tons of product manufactured during a given period, x_2 is the fraction of this production that is to be stored in inventory, and c_1, c_2, and c_3 are cost coefficients that vary with the time period and the production facilities. The first term represents the setup and downtime costs; the second, the idle time and fixed production costs; the third, the inventory and handling costs. For this problem, Eqs. (6-5) and (6-6), which are usually referred to as the "dual constraints" for reasons to be made clear later, become

$$\begin{aligned} w_1 + w_2 + w_3 &= 1 \\ -3w_1 + 3w_2 - 3w_3 &= 0 \\ -2w_1 + w_2 + 3w_3 &= 0 \end{aligned}$$

which are three independent linear equations in the three unknown weights. There are exactly enough equations to determine the optimal weights uniquely without recourse to further analysis. The solution is readily found:

$$w_1 = 0.4; \qquad w_2 = 0.5; \qquad w_3 = 0.1$$

Thus, without knowing the optimal production rate or optimal inventory level, it is nevertheless possible to state definitely that these optimal values must always be such as to make the downtime and setup costs four times as large as the inventory and handling costs, and also such as to make the idle time and fixed production costs five times as large as the inventory and handling costs. Furthermore, this optimal distribution is totally unaffected by changes in the cost coefficients c_1, c_2, and c_3. This separation of technological effects, as reflected by the exponents a_{tn}, from the economic effects, as measured by the coefficients c_t, is one of the attractive features of this approach.

As shown by Eq. (6-8), we can now find the minimum cost, still without knowing the optimal values of the decision variables, x_1 and x_2. Suppose that for a given time period and a given production facility, the unit costs are $c_1 = 60$, $c_2 = 50$, and $c_3 = 20$. Then Eq. (6-8) yields the result

$$y^* = \left(\frac{60}{0.4}\right)^{0.4}\left(\frac{50}{0.5}\right)^{0.5}\left(\frac{20}{0.1}\right)^{0.1} = 125.8$$

This minimum total cost and the optimal weights now can be used to find optimal policy, $\mathbf{x}^* = (x_1^*, x_2^*)$. For $t = 1$ (corresponding to the first term of y), Eq. (6-4) becomes

$$60x_1^{-3}x_2^{-2} = (0.4)(125.8) = 50.32$$

and for $t = 3$:

$$20x_1^{-3}x_2^{3} = (0.1)(125.8) = 12.58$$

so

$$x_1^* = 1.12 \text{ tons} \quad \text{and} \quad x_2^* = 0.944$$

Notice that the only nonlinear equations that must be solved in the geometric programming algorithm are those of Eq. (6-4), each containing only *one* term, and therefore *linear* in the logarithm of x_n. Hence, in effect, it is never necessary to solve any nonlinear equations when using this solution method. The key idea in the development of the technique was the introduction of the weights into Eq. (6-3), for without this change of variable, we would have been faced with solving a set of N simultaneous nonlinear equations. Even for the highly simplified example problem above, where $N = 2$, the solution of these equations could have presented difficulties:

$$-3c_1x_1^{-4}x_2^{-2} + 3c_2x_1^2x_2 - 3c_3x_1^{-4}x_2^3 = 0$$
$$-2c_1x_1^{-3}x_2^{-3} + c_2x_1^3 + 3c_3x_1^{-3}x_2^2 = 0$$

For the general case of N equations in N unknowns, the problem of finding a solution is, of course, more difficult by several orders of magnitude.

The original development of the geometric programming technique, and the basis for its name, relied heavily on the generalized arithmetic–geometric mean inequality:

For any positive numbers $v_1, v_2, \ldots, v_T$, and a set of positive weights $w_1, w_2, \ldots, w_T$ such that $w_1 + w_2 + \cdots + w_T = 1$,

$$w_1 v_1 + w_2 v_2 + \cdots + w_T v_T \geq v_1^{w_1} v_2^{w_2} \cdots v_T^{w_T} \tag{6-9}$$

with equality only when $v_1 = v_2 = \cdots = v_T$.

In order to use this inequality in the present analysis, we shall write the objective function in Eq. (6-1) as

$$y = \sum_{t=1}^{T} w_t \left(\frac{c_t p_t}{w_t}\right)$$

Then, from inequality (6-9),

$$\sum_{t=1}^{T} c_t p_t \geq \prod_{t=1}^{T} \left(\frac{c_t p_t}{w_t}\right)^{w_t} = \prod_{t=1}^{T} \left(\frac{c_t}{w_t}\right)^{w_t} \tag{6-10}$$

the third expression resulting from the fact that the weights are chosen to make the function p_t dimensionless with respect to the original variables x_n as in Eq. (6-7). The expression

$$\prod_{t=1}^{T} \left(\frac{c_t p_t}{w_t}\right)^{w_t} \tag{6-11}$$

will subsequently be referred to as the *predual* function, and the w_t as the dual variables. The expression

$$\prod_{t=1}^{T} \left(\frac{c_t}{w_t}\right)^{w_t} \tag{6-12}$$

will be referred to as the *dual* function. For the example problem, the predual function is

$$\left(\frac{c_1 x_1^{-3} x_2^{-2}}{w_1}\right)^{w_1} \left(\frac{c_2 x_1^3 x_2}{w_2}\right)^{w_2} \left(\frac{c_3 x_1^{-3} x_2^3}{w_3}\right)^{w_3}$$

so that the exponents of x_1 and x_2 are $-3w_1 + 3w_2 - 3w_3$, and $-2w_1 + w_2 + 3w_3$, which are the expressions

$$\sum_{t=1}^{T} a_{tn} w_t$$

that are set to zero in Eq. (6-6). Notice that for this problem,

$$\frac{c_1 p_1}{w_1} = \frac{c_2 p_2}{w_2} = \frac{c_3 p_3}{w_3} = y^*$$

as required by the arithmetic–geometric mean inequality:

$$\frac{60(1.12)^{-3}(0.944)^{-2}}{0.4} = \frac{50(1.12)^{3}(0.944)}{0.5} = \frac{20(1.12)^{-3}(0.944)^{3}}{0.1} = 125.8$$

It should be emphasized that the dimensional analysis used in developing Eqs. (6-5) and (6-6) did not include the c_t, so that the values found for w_1, w_2, and w_3 in the example above represent invariance properties of the system, and must hold for any values of the term coefficients. Suppose, for example, that c_2, the coefficient for the idle time and fixed production cost term, were to change from 50 to 200. Then, from Eq. (6-8), the new value of the minimum total cost would become

$$y^* = 125.8\left(\frac{0.5}{50}\right)^{0.5}\left(\frac{200}{0.5}\right)^{0.5} = 125.8 \cdot 2 = 251.6$$

and the new optimal policy would be determined as before from the unchanged optimal weights:

$$60x_1^{-3}x_2^{-2} = (0.4)(251.6) = 100.64$$
$$20x_1^{-3}x_2^{3} = (0.1)(251.6) = 25.16$$

so $x_1^* = 0.875$ and $x_2^* = 0.944$.

The preceding example was, as stated, special in that it had exactly one more term than variable; that is, $T = N + 1$. Thus there were as many independent linear equations [Eqs. (6-5) and (6-6)] as there were variables, w_t, so that a unique set of weights satisfied these equations. When T is greater than $N + 1$, further steps must be taken to find the optimal weights from among those satisfying Eqs. (6-5) and (6-6).

6-02 DEGREES OF DIFFICULTY

The difference between the number of variables and the number of independent linear equations is conventionally called the number of *degrees of freedom*. For the unconstrained posynomial problem, there are N orthogonality conditions [Eqs. (6-6)], one for each variable x_n, a single normality condition [Eq. (6-5)], and T weights, one for each term. Hence these equations have $T - (N + 1)$ degrees of freedom. Duffin and Zener suggest calling this quantity the number of *degrees of difficulty*, since the larger this number, the harder the problem is to solve.

In order to fix ideas, consider the following problem due to Duffin. A total of 400 yd³ of gravel must be ferried across a river. The gravel is to be shipped in an open box of length x_1, width x_2, and height x_3. The ends and bottom of the box cost \$20 per yd² to build; the sides, \$5 per yd². Runners cost \$2.50 per yd, and two are required for the box to slide on. Each round trip of the ferry costs \$0.10. The problem is to find the optimal dimensions of the box that will minimize the total costs of transportation and of the construction of the box. This total cost is evidently given by

$$y = 40x_1^{-1}x_2^{-1}x_3^{-1} + 20x_1x_2 + 10x_1x_3 + 40x_2x_3 + 5x_1 \tag{6-13}$$

where again there are no constraints save those on the nonnegativity of the x_n.

Equations (6-5) and (6-6) for this example are

$$\begin{aligned} w_1 + w_2 + w_3 + w_4 + w_5 &= 1 \\ -w_1 + w_2 + w_3 \qquad\quad + w_5 &= 0 \\ -w_1 + w_2 \qquad\quad + w_4 \qquad\quad &= 0 \\ -w_1 \qquad\quad + w_3 + w_4 \qquad\quad &= 0 \end{aligned}$$

There are not enough equations to determine the optimal weights since four linearly independent equations in five variables have no unique solution. There are, in fact, an infinity of solutions, as can be seen by solving for the first four weights in terms of the fifth.

$$\begin{aligned} w_1 &= \tfrac{2}{5} - \tfrac{1}{5}w_5; \qquad & w_2 &= \tfrac{1}{5} - \tfrac{3}{5}w_5; \\ w_3 &= \tfrac{1}{5} - \tfrac{3}{5}w_5; \qquad & w_4 &= \tfrac{1}{5} + \tfrac{2}{5}w_5 \end{aligned} \tag{6-14}$$

Any choice of w_5 gives values for w_1, w_2, w_3, and w_4, which satisfy the normality and orthogonality conditions of Eqs. (6-5) and (6-6). The problem is to select the optimal weights from among these infinite possibilities. We now derive the method for accomplishing this. Before proceeding with the derivation, however, it should be noted from Eqs. (6-14) that the optimal weights will always lie within certain fixed ranges regardless of the values of the c_t. For example, the requirement $0 \leq w_2 \leq 1$ translates into $0 \leq \frac{1}{5} - \frac{3}{5}w_5 \leq 1$, or $0 \leq w_5 \leq \frac{1}{3}$, from which it follows that $\frac{1}{3} \leq w_1 \leq \frac{2}{5}$, and $\frac{1}{5} \leq w_4 \leq \frac{1}{3}$. No matter how the unit construction or transportation costs change, this analysis shows that the ferrying cost is never less than one-third or greater than two-fifths of the total cost, for an optimal design. Thus, even in the presence of degrees of difficulty, the dual Eqs. (6-5) and (6-6) often uncover important invariance properties that might otherwise go unnoticed.

Let us substitute the solutions of the normality and orthogonality conditions, as given by Eqs. (6-14), into Eq. (6-8), where w_5 is now the only unknown variable:

$$d(w_5) = \left[\frac{40}{\frac{1}{5}(2 - w_5)}\right]^{1/5(2-w_5)} \times \left[\frac{20}{\frac{1}{5}(1 - 3w_5)}\right]^{1/5(1-3w_5)}$$
$$\times \left[\frac{10}{\frac{1}{5}(1 - 3w_5)}\right]^{1/5(1-3w_5)} \times \left[\frac{40}{\frac{1}{5}(1 + 2w_5)}\right]^{1/5(1+2w_5)} \quad (6\text{-}15)$$
$$\times \left(\frac{5}{w_5}\right)^{w_5}$$

We shall show in the derivation to follow that $d(w_5)$, called the *substituted dual function*, is maximized by the optimal weight w_5^*. Recall that the function which we wish to minimize is given by

$$y = \sum_{t=1}^{T} c_t \prod_{n=1}^{N} x_n^{a_{tn}} \quad (6\text{-}16)$$

where the c_t are positive constants, and the decision variables x_n are also constrained to be positive. If we now make the exponential transformation

$$e^{u_n} = x_n; \qquad n = 1, 2, \ldots, N \quad (6\text{-}17)$$

the problem becomes:

$$\text{Minimize } y(\mathbf{x}) = \sum_{t=1}^{T} c_t \prod_{n=1}^{N} e^{u_n a_{tn}} \quad (6\text{-}18)$$

or

$$y(\mathbf{x}) = \sum_{t=1}^{T} c_t \exp\left(\sum_{n=1}^{N} u_n a_{tn}\right) \quad (6\text{-}19)$$

Let us now study the *dual* geometric programming problem as derived in the previous discussion.

$$\text{Optimize } d(\mathbf{w}) = \prod_{t=1}^{T} \left(\frac{c_t}{w_t}\right)^{w_t} \quad (6\text{-}20)$$

subject to

$$\sum_{t=1}^{T} a_{tn} w_t = 0$$
$$\sum_{t=1}^{T} w_t = 1$$
$$w_t > 0$$

The dual constraint set obviously forms a convex region, since all constraints are linear. The nature of the dual objective function $d(\mathbf{w})$ is now determined. Rather than work with $d(\mathbf{w})$ we prefer to work instead with $z(\mathbf{w}) = \ln [d(\mathbf{w})]$. This is a legitimate transformation since all w_t variables are known to lie between zero and one, and the $\ln [d(\mathbf{w})]$ is a monotonic function in the dual variables $\mathbf{w}$. Hence

$$z(\mathbf{w}) = \ln\left[\prod_{t=1}^{T} \left(\frac{c_t}{w_t}\right)^{w_t}\right] = -\sum_{t=1}^{T} w_t \ln\left(\frac{w_t}{c_t}\right) \tag{6-21}$$

The function $z(\mathbf{w})$ is concave with respect to the weights since it is the negative of a sum of functions that are themselves convex. The dual problem, therefore, is to find a stationary point for a concave objective function subject to a set of convex constraints. Hence the function has a unique stationary point—a global maximum. Thus the dual problem is to *maximize* $z(\mathbf{w})$ subject to the normality and orthogonality conditions, with the weights w_t constrained to be positive.

Although we have chosen to define formally the nature of the dual problem, this result could have been deduced from the original arithmetic–geometric mean inequality derivation. Note that through the use of Eq. (6-9),

$$\sum_{t=1}^{T} c_t p_t \geq \prod_{t=1}^{T} \left(\frac{c_t}{w_t}\right)^{w_t} \tag{6-22}$$

It is clear intuitively that if we seek the minimum of the left-hand side of Eq. (6-22) by dealing with the functional form of the right-hand side, we will drive the inequality to *equality* by *maximizing* the right-hand side. Thus the global minimum for the primal problem equals the global maximum for the dual problem. This is important because the dual problem is usually easier to solve and, of course, offers numerous other advantages such as uncovering invariance properties.

6-03 MAXIMIZING THE DUAL FUNCTION

Now that sufficient conditions have been established, let us apply them to the example. To find the optimal weights it remains to maximize the function $d(w_5)$ with respect to the single variable w_5. It is easier to work with the logarithm

$$\begin{aligned} z(w_5) \equiv \ln [d(w_5)] = &-\frac{2 - w_5}{5} \ln \left(\frac{2 - w_5}{200}\right) \\ &-\frac{1 - 3w_5}{5} \ln \left(\frac{1 - 3w_5}{100}\right) - \frac{1 - 3w_5}{5} \ln \left(\frac{1 - 3w_5}{50}\right) \\ &-\frac{1 + 2w_5}{5} \ln \left(\frac{1 + 2w_5}{200}\right) - w_5 \ln \left(\frac{w_5}{5}\right) \end{aligned} \tag{6-23}$$

Set the first derivative to zero:

$$\frac{dz}{dw_5} = \frac{1}{5}\left[1 + \ln\left(\frac{2 - w_5}{200}\right)\right] + \frac{3}{5}\left[1 + \ln\left(\frac{1 - 3w_5}{100}\right)\right]$$
$$+ \frac{3}{5}\left[1 + \ln\left(\frac{1 - 3w_5}{50}\right)\right] - \frac{2}{5}\left[1 + \ln\left(\frac{1 + 2w_5}{200}\right)\right]$$
$$- \left[1 + \ln\left(\frac{w_5}{5}\right)\right] = 0 \tag{6-24}$$
$$= \ln\left[(2 - w_5)^{1/5}(1 - 3w_5)^{6/5}(1 + 2w_5)^{-2/5}w_5^{-1}\right]$$
$$- \ln\left[\frac{(200)^{1/5}(100)^{3/5}(50)^{3/5}}{(200)^{2/5}(5)}\right]$$

Notice that the nonlogarithmic constants cancel each other; this is always true because of the orthogonality and normality conditions. The optimal weight w_5^* must satisfy

$$(2 - w_5)^{1/5}(1 - 3w_5)^{6/5}(1 + 2w_5)^{-2/5}w_5^{-1} = 11.45 \tag{6-25}$$

The unique solution may be found with relative ease, since there is only one variable. The optimal weight is

$$w_5^* = 0.0709$$

The other weights are found from Eq. (6-14):

$$w_1^* = 0.4 - (0.2)(0.0709) = 0.3858$$
$$w_2^* = w_3^* = 0.2 - (0.6)(0.0709) = 0.1575$$
$$w_4^* = 0.2 + (0.4)(0.0709) = 0.2284$$

Then the minimum cost is

$$y^* = \left(\frac{40}{0.3858}\right)^{0.3858}\left(\frac{20}{0.1575}\right)^{0.1575}\left(\frac{10}{0.1575}\right)^{0.1575}$$
$$\times \left(\frac{40}{0.2284}\right)^{0.2284}\left(\frac{5}{0.0709}\right)^{0.0709} = 108.75 \tag{6-26}$$

The optimal length of the box comes from the cost of the runners and the optimal weight, w_5^*:

$$x_1^* = \frac{(108.75)(0.0709)}{5} = 1.542$$

and the optimal height is found from the cost of the sides and the optimal weight w_3^*:

$$x_3^* = \frac{(108.75)(0.1575)}{10(1.542)} = 1.107$$

while the width is obtained from the term giving the cost of the bottom and the optimal weight w_4^*:

$$x_2^* = \frac{(108.75)(0.2284)}{40(1.107)} = 0.561$$

The arithmetic–geometric mean inequality, as expressed in Eq. (6-10) may be written as

$$y(\mathbf{x}) \geq y^* = d^* \geq d(\mathbf{w}) \tag{6-27}$$

where the asterisks denote global optima—a minimum for y and a maximum for the dual function $d(\mathbf{w})$. Equation (6-27) can often be used for obtaining quick estimates of the optimal value of the objective function when only a few degrees of difficulty are present. For example, consider the problem above: a fast estimate is obtainable by neglecting one of the terms and solving the resulting zero-degree problem for the weights. These weights are then substituted into the dual function for this reduced problem to obtain a lower bound on the true optimum cost. The corresponding x_n can be used to construct an upper bound by substituting them into the objective function for the original problem.

Let us, for instance, neglect the runners' cost, since this is very likely the smallest of the component costs. Then the dual Eqs. (3-5) and (3-6) are

$$\begin{aligned} w_1 + w_2 + w_3 + w_4 &= 1 \\ -w_1 + w_2 + w_3 \qquad &= 0 \\ -w_1 + w_2 \qquad + w_4 &= 0 \\ -w_1 \qquad + w_3 + w_4 &= 0 \end{aligned} \tag{6-28}$$

which are readily solved to yield $w_1 = \frac{2}{5}$, and $w_2 = w_3 = w_4 = \frac{1}{5}$. The dual function, evaluated for these weights, is

$$d\left(\frac{2}{5}, \frac{1}{5}, \frac{1}{5}, \frac{1}{5}\right) = \left(\frac{40}{2/5}\right)^{2/5}\left(\frac{20}{1/5}\right)^{1/5}\left(\frac{10}{1/5}\right)^{1/5}\left(\frac{40}{1/5}\right)^{1/5} = 100(\$) \tag{6-29}$$

The corresponding values of the design variables are found as before, by equating each term t in the objective function of the zero-degree problem to $100w_t$.

The results are $x_1 = 2$, $x_2 = \frac{1}{2}$, and $x_3 = 1$, which, when substituted into Eq. (6-13), yield the upper bound

$$y(2, \tfrac{1}{2}, 1) = 110$$

This use of Eq. (6-27) gives good bounds on the true minimum solution:

$$100 \leq y^* \leq 110$$

In this particular problem, the actual value of y^* was nearer to the upper bound, which was only 1.25% in error. Depending on the costs involved in redesign time, it might have been felt that the design $\mathbf{x} = (2, \frac{1}{2}, 1)$ was a sufficiently good approximation to the unknown $\mathbf{x}^*$. Of course, worse bounds might have been obtained had we chosen to drop a different term from the original problem.

6-04 INEQUALITY CONSTRAINTS

Optimization problems involving inequality constraints can be treated by methods very similar to those previously developed for unconstrained problems. In this section we will deal with inequality constraints involving only posynomial terms. This restriction will be released in subsequent sections when the coefficients will be allowed both positive and negative values, and equality constraints will be permitted.

The problem that we wish to solve is given by the following formulation:

$$\text{Minimize } y_o(\mathbf{x}) = \sum_{t=1}^{T_o} c_{ot} \prod_{n=1}^{N} x_n^{a_{otn}} \tag{6-30}$$

subject to

$$y_m(\mathbf{x}) = \sum_{t=1}^{T_m} c_{mt} \prod_{n=1}^{N} x_n^{a_{mtn}} \leq 1; \qquad m = 1, 2, \ldots, M \tag{6-31}$$

Note that there are now $M + 1$ posynomial expressions: one for the objective function and M for the inequality constraints. The number of terms in each posynomial can vary and will be designated by T_m for each $m = 0, 1, 2, \ldots, M$. The coefficient of each term is now double-subscripted and each exponent is triple-subscripted to identify the posynomial term to which each belongs.

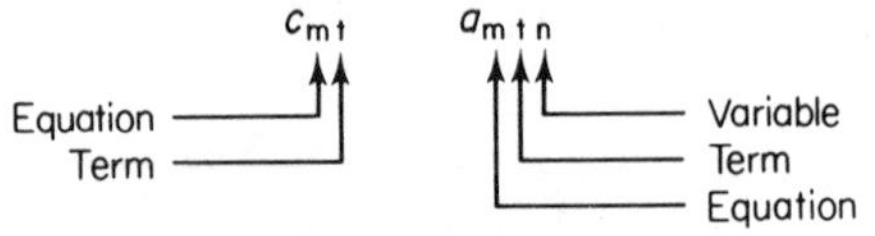

$m = 0; \quad m = 1, 2, \ldots, M; \quad n = 1, 2, \ldots, N$

The optimization problem is to minimize $y_o(\mathbf{x})$ subject to M linear or nonlinear inequality constraints, $y_m(\mathbf{x}) \leq 1$, $m = 1, 2, \ldots, M$.

Recall that the arithmetic–geometric mean inequality can be given by the following expression:

$$\delta_1 V_1 + \delta_2 V_2 + \cdots + \delta_T V_T \geq V_1^{\delta_1} V_2^{\delta_2} \cdots V_T^{\delta_T} \tag{6-32}$$

The $\delta_1, \delta_2, \ldots, \delta_T$ variables are nonnegative weights chosen such that $\delta_1 + \delta_2 + \cdots + \delta_T = 1$, and the $V_1, V_2, \ldots, V_T$ variables are arbitrary positive numbers. An equivalent expression is derived by letting $U_i = \delta_i V_i$, $i = 1, 2, \ldots, T$. This yields

$$U_1 + U_2 + \cdots + U_T \geq \left(\frac{U_1}{\delta_1}\right)^{\delta_1} \left(\frac{U_2}{\delta_2}\right)^{\delta_2} \cdots \left(\frac{U_T}{\delta_T}\right)^{\delta_T} \tag{6-33}$$

$$\delta_1 + \delta_2 + \cdots + \delta_T = 1 \tag{6-34}$$

The inequality above can be written as the generalized arithmetic–geometric mean inequality with weights that are *not* normalized. Suppose that we define such "weights" as $\omega_1, \omega_2, \ldots, \omega_T$, and represent their sum by the following quantity.

$$\lambda = \omega_1 + \omega_2 + \cdots + \omega_T \tag{6-35}$$

Hence $\delta_1, \delta_2, \ldots, \delta_T$ could be defined in the following manner:

$$\delta_i = \frac{\omega_i}{\lambda}; \qquad i = 1, 2, \ldots, T \tag{3-36}$$

Substituting Eq. (6-36) into Eq. (6-33), we obtain

$$(U_1 + U_2 + \cdots U_T)^{\lambda} \geq \left(\frac{U_1}{\omega_1}\right)^{\omega_1} \left(\frac{U_2}{\omega_2}\right)^{\omega_2} \cdots \left(\frac{U_T}{\omega_T}\right)^{\omega_T} \times \lambda^{\lambda} \tag{6-37}$$

In order to proceed, consider again the problem given by Eqs. (6-30) and (6-31), using only a single constraint for illustrative purposes.

$$\text{Minimize } y_o(\mathbf{x}) = \sum_{t=1}^{T_o} c_{ot} p_t(\mathbf{x}) \tag{6-38}$$

subject to

$$y_1(\mathbf{x}) = \sum_{t=1}^{T_1} c_{1t} Q_t(\mathbf{x}) \leq 1 \tag{6-39}$$

$$\mathbf{x} > 0$$

where

$$P_t(\mathbf{x}) = \prod_{n=1}^{N} x_n^{a_{otn}}; \qquad t = 1, 2, \ldots, T_o \tag{6-40}$$

$$Q_t(\mathbf{x}) = \prod_{n=1}^{N} x_n^{a_{1tn}}; \qquad t = 1, 2, \ldots, T_1 \tag{6-41}$$

Using the generalized arithmetic–geometric mean inequality, let us write the inequality relationships for $y_o(\mathbf{x})$ and $y_1(\mathbf{x})$.

$$[y_o(\mathbf{x})]^{\lambda_o} \geq \left[\frac{c_{01}P_1(\mathbf{x})}{\omega_{01}}\right]^{\omega_{01}} \left[\frac{c_{02}P_2(\mathbf{x})}{\omega_{02}}\right]^{\omega_{02}} \cdots \left[\frac{c_{oT_o}P_{T_o}(\mathbf{x})}{\omega_{oT_o}}\right]^{\omega_{oT_o}} \times \lambda_o^{\lambda_o} \tag{6-42}$$

$$1 \geq [y_1(\mathbf{x})]^{\lambda_1} \geq \left[\frac{c_{11}Q_1(\mathbf{x})}{\omega_{11}}\right]^{\omega_{11}} \left[\frac{c_{12}Q_2(\mathbf{x})}{\omega_{12}}\right]^{\omega_{12}} \cdots \left[\frac{c_{1T_1}Q_{T_1}(\mathbf{x})}{\omega_{1T_1}}\right]^{\omega_{1T_1}} \times \lambda_1^{\lambda_1} \tag{6-43}$$

where the $\boldsymbol{\omega}$ weights in Eq. 6-43 have been double-subscripted to relate them directly to the terms in the problem formulation. Since we are dealing with strictly positive terms, these two inequalities can be related algebraically by multiplying the objective function inequality of Eq. (6-42) by the two extreme sides of the constraint inequality of Eq. (6-43) to yield the following result.

$$[y_o(\mathbf{x})]^{\lambda_o} \geq \left[\frac{c_{01}P_1(\mathbf{x})}{\omega_{01}}\right]^{\omega_{01}} \cdots \left[\frac{c_{oT_o}P_{T_o}(\mathbf{x})}{\omega_{oT_o}}\right]^{\omega_{oT_o}} \cdots \left[\frac{c_{11}Q_1(\mathbf{x})}{\omega_{11}}\right]^{\omega_{11}} \cdots \left[\frac{c_{1T_1}Q_{T_1}(\mathbf{x})}{\omega_{1T_1}}\right]^{\omega_{1T_1}} \times \lambda_o^{\lambda_o}\lambda_1^{\lambda_1} \tag{6-44}$$

By its construction, Eq. (6-44) is valid for *any* choice of weights. It is convenient to choose the weights $\omega_{01}, \omega_{02}, \ldots, \omega_{oT_o}$ related to λ_o as normalized weights as in the unconstrained case.

$$\lambda_o = \omega_{01} + \omega_{02} + \cdots + \omega_{oT_o} \equiv 1 \tag{6-45}$$

In addition, we can also eliminate, as before, the terms

$$P_1(\mathbf{x}), P_2(\mathbf{x}), \ldots, P_{T_o}(\mathbf{x}); \qquad Q_1(\mathbf{x}), Q_2(\mathbf{x}), \ldots, Q_{T_1}(\mathbf{x})$$

by a proper choice of term exponents:

$$\sum_{m=0}^{1} \sum_{t=1}^{T_m} a_{mtn}\omega_{mt} = 0; \qquad n = 1, 2, \ldots, N \tag{6-46}$$

Equation (6-44) now becomes

$$y_o(\mathbf{x}) \geq \left(\frac{c_{01}}{\omega_{01}}\right)^{\omega_{01}} \left(\frac{c_{02}}{\omega_{02}}\right)^{\omega_{02}} \cdots \left(\frac{c_{oT_o}}{\omega_{oT_o}}\right)^{\omega_{oT_o}} \left(\frac{c_{11}}{\omega_{11}}\right)^{\omega_{11}}$$
$$\times \left(\frac{c_{12}}{\omega_{12}}\right)^{\omega_{12}} \cdots \left(\frac{c_{1T_1}}{\omega_{1T_1}}\right)^{\omega_{1T_1}} \times \lambda_1^{\lambda_1} \tag{6-47}$$

where

$$\lambda_1 = \omega_{11} + \omega_{12} + \cdots + \omega_{1T_1} \tag{6-48}$$

or, equivalently,

$$y_o(\mathbf{x}) \geq \left(\frac{c_{01}}{\omega_{01}}\right)^{\omega_{01}} \left(\frac{c_{02}}{\omega_{02}}\right)^{\omega_{02}} \cdots \left(\frac{c_{oT_o}}{\omega_{oT_o}}\right)^{\omega_{oT_o}} \left(\frac{c_{11}\lambda_1}{\omega_{11}}\right)^{\omega_{11}}$$
$$\times \left(\frac{c_{12}\lambda_1}{\omega_{12}}\right)^{\omega_{12}} \cdots \left(\frac{c_{1T_1}\lambda_1}{\omega_{1T_1}}\right)^{\omega_{1T_1}} \tag{6-49}$$

where

$$\omega_{01} + \omega_{02} + \cdots + \omega_{oT_o} = 1 \tag{6-50}$$

$$\sum_{m=0}^{1} \sum_{t=1}^{T_m} a_{mtn}\omega_{mt} = 0; \qquad n = 1, 2, \ldots, N \tag{6-51}$$

$$\lambda_1 = \sum_{t=1}^{T_1} \omega_{1t} \tag{6-52}$$

Equation (6-49) is an inequality in which the right-hand side provides a *lower bound* for the objective function $y_o(\mathbf{x})$ for any choice of $\boldsymbol{\omega}$ that satisfy Eqs. (6-50) through (6-52). Note that this inequality is of exactly the same form as Eq. (6-10). We can use the same arguments as in the unconstrained case to show that the proper choice for $\boldsymbol{\omega}$ is that set of variables which *maximizes* the right-hand side. The generalization to a geometric program with M inequality constraints is a straightforward extension of the results above. The dual geometric program for the general posynomial case is as follows.

$$\text{Maximize } d(\mathbf{w}) = \prod_{m=0}^{M} \prod_{t=1}^{T_m} \left(\frac{c_{mt}\omega_{mo}}{\omega_{mt}}\right)^{\omega_{mt}} \tag{6-53}$$

subject to

$$\sum_{t=1}^{T_o} \omega_{ot} = 1 \tag{6-54}$$

$$\sum_{m=1}^{M} \sum_{t=1}^{T_m} a_{mtn}\omega_{mt} = 0; \qquad n = 1, 2, \ldots, N \tag{6-55}$$

$$\omega_{mo} = \sum_{t=1}^{T_m} \omega_{mt}; \qquad m = 1, 2, \ldots, M \tag{6-56}$$

provided that we define $\omega_{oo} \equiv 1$ and $\omega_{mo} \equiv \lambda_m$. Note that there are $(N + 1)$ independent dual constraint equalities, and $T \equiv \sum_{m=0}^{M} T_m$ independent dual variables, one for each term of the primal problem. In the case where the number of degrees of difficulty, $T - (N + 1)$, is zero, these equalities are all that are needed to solve uniquely for the unknown dual variables. In the case where $T - (N + 1) > 0$, a true optimization problem exists. However, we have transformed a nonlinear programming problem with nonlinear constraints to a concave programming problem with linear constraints.

The only task that yet remains is to identify the exact relationship that exists between the T dual variables of Eq. (6-53) and the N primal decision variables. It is clear from the relationships already established in Eqs. (6-4) that the ω_{ot}^*, $t = 1, 2, \ldots, T_o$, variables are the true weights for the optimal primal objective function, $y_o(\mathbf{x}^*)$. It is also clear that through an extension of Eq. (6-49) to the general case, $d(\mathbf{w}^*) \equiv y_o(\mathbf{x}^*)$. Hence, for the terms of the objective function, the following relationships hold at optimality.

$$\omega_{ot}^*[y_o(\mathbf{x}^*)] = c_{ot} \prod_{n=1}^{N} x_n^{a_{otn}}; \qquad t = 1, 2, \ldots, T_o \tag{6-57}$$

Turning to the constraint terms, we see that Eq. (6-56) is a direct generalization of Eq. (6-52) for multiple constraints, where $\lambda_m \equiv \omega_{mo}$; $m = 1, 2, \ldots, M$. Recall that the weights w_{mt} are simply the values of the terms themselves for active constraints, since if a constraint is tight the terms must sum to unity. Applying Eq. (6-36) to the general case shows that the *generalized weight* for the tth term in the mth constraint is

$$w_{mt} = \frac{\omega_{mt}}{\sum_t \omega_{mt}} = c_{mt} \prod_{n=1}^{N} x_n^{a_{mtn}}; \qquad \begin{matrix} m = 1, 2, \ldots, M \\ n = 1, 2, \ldots, T_m \end{matrix} \tag{6-58}$$

If a primal constraint is *tight* at optimality, then *all* dual variables associated with that constraint will be strictly *positive* at optimality. If a primal constraint is *loose* at optimality, then *all* dual variables associated with that constraint will be *zero* at optimality. Notice that in this case, the dual objective is undefined for those dual variables that assume a value of zero. However, we might argue that if this had been known in advance of problem solution, any loose constraint could have been dropped from the problem formulation without affecting the optimal solution. This would simply mean that those terms associated with the loose constraints would not have appeared in the dual program. This situation can be remedied during problem solution provided that we define the following convention for any dual variable that is driven to zero.

$$\lim_{\omega_{mt} \to 0} \left(\frac{c_{mt}\omega_{mo}}{\omega_{mt}}\right)^{\omega_{mt}} = 1$$

To demonstrate how to use the equations just derived, a contrived example is worked below. Suppose that we wish to minimize

$$y = 0.3x_1x_3^{1.2} + 0.5x_1^{-1.5}x_2^{-1.2}x_3^{-1.4} + 0.2x_2^{1.3}$$

subject to the usual nonnegativity conditions and the inequalities

$$0.8x_1x_3 \leq 1$$
$$1.2x_1x_2^{-1} \leq 1$$

This example seems artificial, without apparent economic or industrial significance, but it will serve to illustrate the computations involved in solving the more complicated real-world problems. There are $T = 5$ terms and $N = 3$ variables, giving one degree of difficulty.

The dual problem is to find nonnegative $\boldsymbol{\omega}$, which maximize

$$d(\boldsymbol{\omega}) = \left(\frac{0.3}{\omega_{01}}\right)^{\omega_{01}}\left(\frac{0.5}{\omega_{02}}\right)^{\omega_{02}}\left(\frac{0.2}{\omega_{03}}\right)^{\omega_{03}}(0.80)^{\omega_{11}}(1.2)^{\omega_{21}}$$

and which satisfy the constraints

$$\begin{aligned}
\omega_{01} + \omega_{02} + \omega_{03} &= 1\\
\omega_{01} - 1.5\omega_{02} + \omega_{11} + \omega_{21} &= 0\\
- 1.2\omega_{02} + 1.3\omega_{03} - \omega_{21} &= 0\\
1.2\omega_{01} - 1.4\omega_{02} + \omega_{11} &= 0
\end{aligned}$$

This problem can be solved as a nonlinear programming problem with linear constraints, for which several efficient solution algorithms are available. Alternatively, any four of the $\boldsymbol{\omega}$ may be expressed in terms of the fifth, which is then substituted into the dual objective function and the resulting substituted dual maximized on this one $\boldsymbol{\omega}$. This maximization can be accomplished through various means; here we have set the first derivative to zero as in the previous example.

The solution to the present example problem is $\omega_{01}^* = 0.3339$, $\omega_{02}^* = 0.3074$, $\omega_{03}^* = 0.3587$, $\omega_{11}^* = 0.0297$, and $\omega_{12}^* = 0.0975$, giving the dual objective function value of $d^* = 0.9189$. The dual variables ω_{01}^*, ω_{02}^*, and ω_{03}^* are true weights, as in the unconstrained case. The optimal values of the primal variables are recovered in manner analogous to that employed in the unconstrained example problem. From Eq. (6-58), the second constraint may be written as

$$1.2x_1x_2^{-1} = \frac{\omega_{21}}{\omega_{20}} = 1 \tag{6-59}$$

Using Eq. (6-57) we find, from the third term in the objective function,

$$0.2x_2^{1.3} = (0.3587)(0.9189) = 0.3296$$

Hence, $x_2^* = 1.468$ and, from Eq. (6-59), $x_1^* = 1.223$. The first term in the objective function can be used to find x_3^*:

$$(0.3)(1.223)x_3^{1.2} = (0.3339)(0.9189) = 0.3068$$

so that $x_3^* = 0.8610$.

6-05 SIGNOMIAL PROGRAMMING—POSITIVE σ_{ot}

Thus far, the structure of each geometric programming problem studied has been such that every term in both the objective function and the constraints had only positive coefficients. In addition, each constraint was written as $y_m(\mathbf{x}) \leq 1, m = 1, 2, \ldots, M$. Such formulations are called *posynomial* geometric programs. In many real-world problems, however, there arise terms that have negative coefficients. If a geometric program contains one or more terms with negative coefficients, or if any constraints are of the form $y_m(\mathbf{x}) \geq 1$, such formulations are called *signomial* geometric programming problems.

Define a nonlinear objective function in N variables and T terms of the following form:

Program 1

$$\text{Minimize } y_o(\mathbf{x}) = \sum_{t=1}^{T_o} \sigma_{ot} c_{ot} \prod_{n=1}^{N} x_n^{a_{otn}} \tag{6-60}$$

constrained by M inequality constraints given by

$$y_m(\mathbf{x}) = \sum_{t=1}^{T_m} \sigma_{mt} c_{mt} \prod_{n=1}^{N} x_n^{a_{mtn}} \leq \sigma_m; \qquad m = 1, 2, \ldots, M \tag{6-61}$$

where

$$\sigma_{mt} = \pm 1; \qquad t = 1, 2, \ldots, T_m; \quad m = 0, 1, \ldots, M$$

$$\sigma_m = \pm 1; \qquad m = 1, 2, \ldots, M$$

$$c_{mt} > 0; \qquad t = 1, 2, \ldots, T_m; \quad m = 0, 1, \ldots, M$$

$$x_n > 0; \qquad n = 1, 2, \ldots, N$$

The T signum functions, σ_{mt}, $m = 0, 1, \ldots, M$; $t = 1, 2, \ldots, T_m$, are introduced in order to absorb the sign of each term, while the M signum functions σ_m, $m = 1, 2, \ldots, M$, are used to generalize the constraint right-hand sides to ± 1.

Just as in the posynomial case, our task is now to exploit the special form of this problem in order to obtain an equivalent problem with linear constraints, which is much easier to solve. The development that follows will proceed in two phases. First, the case will be explored in which all signum functions in the objective function are positive. In other words, $y_o(\mathbf{x})$ is posynomial in form while the $y_m(\mathbf{x})$ are signomial in form. The second case will be the development of the generalized geometric program with all signomial terms. When all σ_{ot} are positive, the objective function may be written as

$$y_o(\mathbf{x}) = \sum_{t=1}^{T_o} c_{ot} \prod_{n=1}^{N} x_n^{a_{otn}} \tag{6-62}$$

Note that the constraints can be written in the following form:

$$f_m(\mathbf{x}) \equiv f_m \equiv \sigma_m\left(1 - \sigma_m \sum_{t=1}^{T_m} \sigma_{mt} c_{mt} \prod_{n=1}^{N} x_n^{a_{mtn}}\right) \geq 0; \quad m = 1, 2, \ldots, M \tag{6-63}$$

Again, we introduce weights for each term in the objective function:

$$w_{ot} = \frac{c_{ot} \prod_{n=1}^{N} x_n^{a_{otn}}}{y_o(\mathbf{x})}; \qquad t = 1, 2, \ldots, T_o \tag{6-64}$$

The "weights" for the constraint terms are now the *absolute values* of the terms themselves, and if the constraint is tight (binding) at optimality, the contributions of each term (multiplied by σ_{mt}) must sum to σ_m. Hence

$$w_{mt} = c_{mt} \prod_{n=1}^{N} x_n^{a_{mtn}}; \qquad m = 1, 2, \ldots, M; \quad t = 1, 2, \ldots, T_m \tag{6-65}$$

For notational convenience, let $x_o \equiv y_o(\mathbf{x})$. Using Eqs. (6-63), (6-64), and (6-65), the original problem of minimizing $y_o(\mathbf{x})$ subject to $y_m(\mathbf{x}) \leq \sigma_m$, $m = 1, 2, \ldots, M$, now becomes

Program 2

$$\text{Minimize } x_o$$

subject to

$$x_o w_{ot} = c_{ot} \prod_{n=1}^{N} x_n^{a_{otn}}; \qquad t = 1, 2, \ldots, T_o$$

$$f_m = \sigma_m\left(1 - \sigma_m \sum_{t=1}^{T_m} \sigma_{mt} w_{mt}\right) \geq 0; \qquad m = 1, 2, \ldots, M$$

$$w_{mt} - c_{mt} \prod_{n=1}^{N} x_n^{a_{mtn}} = 0; \qquad m = 1, 2, \ldots, M; \quad t = 1, 2, \ldots, T_m$$

$$1 - \sum_{t=1}^{T_o} w_{ot} = 0$$

Define $x_n = e^{u_n}$; $n = 0, 1, \ldots, N$; the problem above now can be written as follows:

$$\text{Minimize } e^{u_o}$$

subject to

$$e^{u_o}\left[\frac{w_{ot}}{c_{ot}}\right] = \exp\left(\sum_{n=1}^{N} a_{otn}u_n\right)$$

$$f_m = \sigma_m\left(1 - \sigma_m \sum_{t=1}^{T_m} \sigma_{mt}w_{mt}\right) \geq 0; \qquad m = 1, 2, \ldots, M$$

$$\frac{w_{mt}}{c_{mt}} = \exp\left(\sum_{n=1}^{N} a_{mtn}u_n\right); \qquad \begin{array}{l} m = 1, 2, \ldots, M \\ t = 1, 2, \ldots, T_m \end{array}$$

$$1 - \sum_{t=1}^{T_o} \omega_{ot} = 0$$

By taking natural logarithms, we obtain

Program 3

$$\text{Minimize } u_o$$

subject to

$$\sum_{t=1}^{T_o} w_{ot} = 1$$

$$\sum_{n=1}^{N} a_{otn}u_n - u_o = \ln\left(\frac{w_{ot}}{c_{ot}}\right); \qquad t = 1, \ldots, T_o$$

$$\sum_{n=1}^{N} a_{mtn}u_n = \ln\left(\frac{w_{mt}}{c_{mt}}\right); \qquad \begin{array}{l} m = 1, 2, \ldots, M \\ t = 1, 2, \ldots, T_m \end{array}$$

$$f_m = \sigma_m\left(1 - \sigma_m \sum_{t=1}^{T_m} \sigma_{mt}w_{mt}\right) \geq 0; \qquad m = 1, 2, \ldots, M$$

From their respective definitions, the weights must all be nonnegative, whereas the variables $u_o, u_1, \ldots, u_N$ are unconstrained in sign.

We now write out the Lagrangian function for Program 3, employing λ_o, δ_{ot}, δ_{mt}, and λ_m as the Lagrange multipliers for the respective constraints. At an optimum, such a Lagrangian must be stationary with respect to these multipliers, the weights, and the u_n. Since the last M constraints are inequalities, they must be treated differently from the equality constraints. If any of these constraints are loose [$f_m(\mathbf{x}) > 0$] at optimality, then the corresponding Lagrange multiplier λ_m must be zero; if the constraint is tight, then $f_m(\mathbf{x}) = 0$. These two possibilities for the inequality constraints are covered by the Kuhn–Tucker *complementary slackness* conditions:

$$\lambda_m f_m(\mathbf{x}) = 0; \qquad m = 1, \ldots, M \tag{6-66}$$

In addition, since the objective is to minimize $y_o(\mathbf{x})$, and since all constraints are written in the form $f_m(\mathbf{x}) \equiv f_m \geq 0$, the λ_m are also constrained to be nonnegative ($\lambda_m \geq 0$; $m = 1, 2, \ldots, M$). Hence the terms $\lambda_m f_m(\mathbf{x})$ may be added to the Lagrangian without changing its value. The Lagrangian function for Program 3 is

$$\begin{aligned} L(\mathbf{u}, \mathbf{w}, \boldsymbol{\lambda}, \boldsymbol{\delta}) \equiv\ & u_o - \lambda_o\left(1 - \sum_{t=1}^{T_o} w_{ot}\right) \\ & - \sum_{t=1}^{T_o} \delta_{ot}\left[\ln\left(\frac{w_{ot}}{c_{ot}}\right) + u_o - \sum_{n=1}^{N} a_{otn}u_n\right] \\ & - \sum_{m=1}^{M}\sum_{t=1}^{T_m} \delta_{mt}\left[\ln\left(\frac{w_{mt}}{c_{mt}}\right) - \sum_{n=1}^{N} a_{mtn}u_n\right] \\ & - \sum_{m=1}^{M} \lambda_m\sigma_m\left(1 - \sigma_m\sum_{t=1}^{T_m}\sigma_{mt}w_{mt}\right) \end{aligned} \tag{6-67}$$

Necessary conditions for a primal stationary point are obtained through the use of the first derivatives.

$$\frac{\partial L}{\partial u_o} = 1 - \sum_{t=1}^{T_o} \delta_{ot} = 0 \tag{6-68}$$

$$\frac{\partial L}{\partial w_{ot}} = \lambda_o - \frac{\delta_{ot}}{w_{ot}}; \qquad t = 1, \ldots, T_o \tag{6-69}$$

These two equations can be combined with Eq. (6-64), which shows that the w_{ot} must sum to unity to yield the following:

$$\lambda_o = 1 \tag{6-70}$$

$$\delta_{ot} = w_{ot}; \qquad t = 1, 2, \ldots, T_o \tag{6-71}$$

Equation (6-71) yields an interesting fact: the Lagrangian multipliers (δ_{ot}) for the objective function terms are simply the corresponding weights (w_{ot}). Since $y_o(\mathbf{x})$ is posynomial, the δ_{ot} variables are *true* weights. That is, $0 < \delta_{ot} \leq 1$, for all $t = 1, 2, \ldots, T_o$, and their sum is 1.

Next consider the derivatives of the Lagrangian function with respect to the constraint term "weights," w_{mt} ($m \neq 0$):

$$\frac{\partial L}{\partial w_{mt}} = -\frac{\delta_{mt}}{w_{mt}} + \lambda_m\sigma_m^2\sigma_{mt} = 0$$

and since $\sigma_2^m \equiv 1$ regardless of the sign of σ_m:

$$\delta_{mt} = \lambda_m w_{mt}\sigma_{mt} \tag{6-72}$$

Summing both sides of Eq. (5-14) on t and multiplying through by σ_m, we find that

$$\sigma_m \sum_{t=1}^{T_m} \delta_{mt} = \lambda_m \sigma_m \sum_{t=1}^{T_m} \sigma_{mt} w_{mt} = \lambda_m; \qquad m = 1, \ldots, M \tag{6-73}$$

since the $\sigma_{mt} w_{mt}$ terms sum to σ_m for each tight constraint. Therefore, if $\lambda_m \neq 0$ (constraint is tight), using Eqs. (6-70) and (6-73), we may define

$$w_{mt} = \frac{\delta_{mt}}{\lambda_m \sigma_{mt}} = \delta_{mt}\left(\sigma_m \sigma_{mt} \sum_{t=1}^{T_m} \delta_{mt}\right)^{-1}; \qquad \begin{matrix} m = 1, \ldots, M \\ t = 1, \ldots, T_m \end{matrix} \tag{6-74}$$

so that the weights can be obtained from the Lagrange multipliers, δ_{mt}, which can be found by differentiating with respect to the u_n $(n \neq 0)$:

$$\frac{\partial L}{\partial u_n} = \sum_{m=0}^{M} \sum_{t=1}^{T_m} \delta_{mt} a_{mtn} = 0; \qquad n = 1, 2, \ldots, N \tag{6-75}$$

with the convention that $\sigma_{ot} = 1$, for $t = 1, \ldots, T_o$. These are the orthogonality conditions for the constrained case. Notice that the variables are the Lagrange multipliers δ_{mt} rather than the "weights," w_{mt}. Also observe the presence of the signum functions σ_{mt}, which, together with the exponents a_{mtn}, are known from the statement of the problem. No distinction is made in these orthogonality conditions between objective function and constraint function exponents. On the other hand, the normality conditions of Eq. (6-68) involve only multipliers for the objective function terms, which are equal to the true weights. Conversely, the w_{mt} of Eq. (6-74) are defined as a function of both the signum functions and the δ_{mt}. All that is required is that the $\sigma_{mt} w_{mt}$ values sum to $\sigma_m = \pm 1$, and there is no sign restriction on the individual $\sigma_{mt} w_{mt}$ values.

The dual problem may now be found by expressing the Lagrangian of Eq. (6-67) in terms of the u_n, the Lagrange multipliers δ_{mt}, and the constraint functions $f_m(\mathbf{x})$ using the fact that $\lambda_o = 1$, and $\lambda_m f_m(\mathbf{x}) = 0$ for $m = 1, \ldots, M$.

$$L(\boldsymbol{\delta}, \mathbf{u}, \mathbf{w}) = \sum_{m=0}^{M} \sum_{t=1}^{T_m} \delta_{mt}\left[\ln\left(\frac{c_{mt}}{w_{mt}}\right)\right] + (u_o - 1)\left(1 - \sum_{t=1}^{T_o} \delta_{ot}\right) + \sum_{n=1}^{N} u_n \sum_{m=0}^{M} \sum_{t=1}^{T_m} a_{mtn} \delta_{mt} \tag{6-76}$$

The Lagrangian above can be considered as that for *another* constrained optimization problem by regarding the variables $(u_o - 1)$ and $u_1, u_2, \ldots, u_N$ as Lagrangian multipliers for this new problem. Note that in this new formulation, the variables δ_{mt} become the *decision variables*. Using this line of thought, what does the other constrained problem look like?

The equivalent problem has for its objective function

$$z(\boldsymbol{\delta}) = \sum_{m=0}^{M} \sum_{t=1}^{T_m} \delta_{mt}\left[\ln\left(\frac{c_{mt}}{w_{mt}}\right)\right] \tag{6-77}$$

The independent dual variables δ_{mt} must satisfy a normality condition

$$\sum_{t=1}^{T_o} \delta_{ot} = 1 \tag{6-78}$$

and N orthogonality conditions

$$\sum_{m=0}^{M} \sum_{t=1}^{T_m} a_{mtn}\delta_{mt} = 0; \qquad n = 1, \ldots, N \tag{6-79}$$

where the w_{mt} are given by Eq. (6-74). Using that equation, the dual problem above can be expressed solely in terms of the decision variables, δ_{mt}:

$$z(\boldsymbol{\delta}) = \sum_{m=0}^{M} \sum_{t=1}^{T_m} \delta_{mt}\left[\ln\left(\frac{c_{mt}\sigma_m\sigma_{mt}\sum_{t=1}^{T_m}\delta_{mt}}{\delta_{mt}}\right)\right] \tag{6-80}$$

subject to

$$\sum_{t=1}^{T_o} \delta_{ot} = 1 \tag{6-81}$$

$$\sum_{m=0}^{M} \sum_{t=1}^{T_m} a_{mtn}\delta_{mt} = 0; \qquad n = 1, 2, \ldots, N \tag{6-82}$$

It is now necessary to determine the character of the solution variables, δ_{mt}. Equation (6-74) may be written as

$$w_{mt} = \delta_{mt}\left(\sigma_{mt}\sigma_m \sum_{t=1}^{T_m} \delta_{mt}\right)^{-1}$$

Hence, by Eq. (6-73),

$$\delta_{mt} = w_{mt}\sigma_{mt}\lambda_m; \qquad m = 1, \ldots, M; \quad t = 1, \ldots, T_m$$

Noting that $w_{mt} > 0$ and $\lambda_m \geq 0$, it is clear that δ_{mt} assumes the sign of σ_{mt}. Therefore, if $\sigma_{mt} = 1$, $\delta_{mt} \geq 0$; if $\sigma_{mt} = -1$, $\delta_{mt} \leq 0$. We wish now to construct a dual problem such that the solution vector will be nonnegative. Toward that end, we define new dual variables

$$\omega_{mt} = \sigma_{mt}\delta_{mt} \geq 0 \tag{6-83}$$

Substituting these new variables into Eqs. (6-80), (6-81), and (6-82), we obtain the following dual problem:

$$z(\boldsymbol{\omega}) = \sum_{m=0}^{M} \sum_{t=1}^{T_m} \sigma_{mt}\omega_{mt}\left[\ln\left(\frac{c_{mt}\sigma_{mt}\sum_t \omega_{mt}\sigma_{mt}}{\omega_{mt}}\right)\right]$$

subject to

$$\sum_{t=1}^{T_o} \omega_{ot} = 1$$

$$\sum_{m=0}^{M} \sum_{t=1}^{T_m} \omega_{mt}\sigma_{mt}a_{mtn} = 0; \qquad n = 1, 2, \ldots, N$$

$$\omega_{mt} \geq 0; \qquad m = 0, 1, \ldots, M; \quad t = 1, \ldots, T_m$$

Altogether there are $N + 1$ linear constraint equations in as many variables as there are terms T in the primal problem. Therefore, the number of degrees of difficulty is $T - (N + 1)$, provided that the equations are linearly independent. When there are no degrees of difficulty, the solution is unique, and once the optimal values ω^*_{mt} of the dual variables are known, the optimal "weights" w^*_{mt} for all terms can be computed directly from Eq. (6-74) and the relationship $\omega_{mt} = \sigma_{mt}\delta_{mt}$. When degrees of difficulty are present, however, more analysis is required to find these optimal weights.

It is usually more convenient to work with the exponential of $z(\boldsymbol{\omega})$, which is called the *dual* function, $d(\boldsymbol{\omega})$. Hence, the *dual geometric program* (DGP) is

$$d(\boldsymbol{\omega}) = \prod_{m=0}^{M} \prod_{t=1}^{T_m} \left(\frac{c_{mt}\omega_{mo}}{\omega_{mt}}\right)^{\sigma_{mt}\omega_{mt}}$$

subject to

$$\sum_{t=1}^{T_o} \omega_{ot} = 1$$

$$\sum_{m=0}^{M} \sum_{t=1}^{T_m} \sigma_{mt}a_{mtn}\omega_{mt} = 0; \qquad n = 1, 2, \ldots, N$$

$$\omega_{mo} = \sigma_m \sum_{t=1}^{T_m} \sigma_{mt}\omega_{mt}; \qquad m = 1, \ldots, M$$

$$\omega_{mt} \geq 0; \qquad \text{all } m \text{ and } t$$

The character of ω_{mo} is easily established since by Eq. (6-74)

$$w_{mt} = \frac{\omega_{mt}}{\omega_{mo}}$$

Therefore, $\omega_{mo} \geq 0$ since w_{mt} is positive (by construction) and ω_{mt} is nonnegative (by definition). Note that the ω_{mt} variables are related to the *primal*

Lagrangian multipliers λ_m by Eq. (6-72) and the variable transformation $\omega_{mt} = \sigma_{mt}\delta_{mt}$. The important dual relationships between the dual solution variables of problem DGP and the primal decision variables **x** of the (original) program 2 are easily derived.

Recall that from Eq. (6-74)

$$w_{mt} = \delta_{mt}\left(\sigma_m\sigma_{mt}\sum_{t=1}^{T_m}\delta_{mt}\right)^{-1}$$

Hence, from Eq. (6-83):

$$w_{mt} = \frac{\omega_{mt}}{\sigma_m \sum_{t=1}^{T_m} \sigma_{mt}\omega_{mt}} = \frac{\omega_{mt}}{\omega_{mo}}$$

and

$$w_{ot} \equiv \omega_{ot}$$

The primal decision variables can, therefore, be recovered through the simultaneous solution of N of the T following equations.

$$c_{ot}\prod_{n=1}^{N} x_n^{a_{otn}} = y_o(\mathbf{x})\omega_{ot}; \qquad t = 1, 2, \ldots, T_o \tag{6-84}$$

$$c_{mt}\prod_{n=1}^{N} x_n^{a_{mtn}} = w_{mt} = \frac{\omega_{mt}}{\omega_{mo}}; \qquad \begin{matrix} m = 1, 2, \ldots, M \\ t = 1, 2, \ldots, T_m \end{matrix} \tag{6-85}$$

At this point we have not yet determined whether the problem is to maximize, minimize, or find some other special point for the dual function or its logarithm. What we have established is that any set of dual variables $\boldsymbol{\omega}$ making u_o a *local minimum* (the original problem) will also be such that, at optimality,

$$z(\boldsymbol{\omega}^o) = u_o^o$$

This is true because at the local minimum, u_o^o, all the constraint terms in the Lagrangian of Eq. (6-76) are zero, and for the corresponding $\boldsymbol{\delta}^o$ variables (which determine the $\boldsymbol{\omega}^o$), the dual constraint terms disappear. When there are *no* degrees of difficulty, the normality and orthogonality conditions have a unique solution. In this case, we can obtain the minimizing solution by solving the set of linear dual constraints and calculating $y_o(\mathbf{x}^o)$ from the relation $z(\boldsymbol{\omega}^o) = u_o^o$ and the definition $u_o = \ln(y_o)$; then, since $z(\boldsymbol{\omega}^o) = \ln[d(\boldsymbol{\omega}^o)]$, we have

$$y_o(\mathbf{x}^o) = d(\boldsymbol{\omega}^o)$$

The nonnegativity of the λ_m and ω_{mt}, together with the complementary slackness requirements of Eq. (6-66), complete the Kuhn–Tucker necessary conditions for a stationary point. This stationary point will be a minimum for $y_o(\mathbf{x})$ only if the proper sufficiency conditions are satisfied.

The following problem, due to Passy, illustrates the application of the preceding results to a signomial program:

$$\text{Minimize } y_o(\mathbf{x}) = x_1$$

subject to

$$y_1(\mathbf{x}) = 5x_1^{-1}x_2 - x_1^{-1}x_3^2x_4^4 \leq 1$$
$$y_2(\mathbf{x}) = -\tfrac{5}{2}x_2x_3^{-2} + \tfrac{3}{2}x_3^{-1}x_4 \leq 1$$

For this problem

$$\begin{aligned}
&\sigma_{01} = 1; &&\sigma_{21} = -1; &&\sigma_2 = 1; &&c_{12} = 1;\\
&\sigma_{11} = 1; &&\sigma_{22} = 1; &&c_{01} = 1; &&c_{21} = \tfrac{5}{2};\\
&\sigma_{12} = -1; &&\sigma_1 = 1; &&c_{11} = 5; &&c_{22} = \tfrac{3}{2}
\end{aligned}$$

Since there are zero degrees of difficulty, the dual constraint set is all that is needed to determine the dual variables.

$$\begin{aligned}
\omega_{01} &= 1 & &[y_o(\mathbf{x})]\\
\omega_{01} - \omega_{11} + \omega_{12} &= 0 & &[x_1]\\
\omega_{11} - \omega_{21} &= 0 & &[x_2]\\
-2\omega_{12} + 2\omega_{21} - \omega_{22} &= 0 & &[x_3]\\
-4\omega_{12} + \omega_{22} &= 0 & &[x_4]
\end{aligned}$$

The unique solution is given by

$$\omega_{01}^* = 1.00; \quad \omega_{11}^* = 1.50; \quad \omega_{12}^* = 0.50; \quad \omega_{21}^* = 1.50; \quad \omega_{22}^* = 200$$

and

$$\omega_{10}^* = (1.0)[(1)(1.5) + (-1.0)(.50)] = +1.0 \geq 0$$
$$\omega_{20}^* = (1.0)[(-1)(1.50) + (1)(2.00)] = 0.50 \geq 0$$

The optimal dual solution is given by

$$d(\boldsymbol{\omega}^*) = \left\{\left(\frac{1}{1}\right)^1\left(\frac{5(1)}{1.5}\right)^{1.5}\left(\frac{1}{0.50}\right)^{-0.50}\left[\frac{(2.5)(0.5)}{1.5}\right]^{-1.5}\left[\frac{(1.5)(0.5)}{2.0}\right]^{2.0}\right\} = 0.796$$

The optimal objective functional value is given by

$$y_o(\mathbf{x}^*) = d(\boldsymbol{\omega}^*) = 0.796$$

From the primal–dual relationships of Eqs. (6-84) and (6-85) we obtain

$$x_1^* = 0.796; \qquad x_2^* = 0.239; \qquad x_3^* = 0.446; \qquad x_4^* = 1.19$$

Although these primal values have been marked with asterisks, further analysis would be required to establish this stationary point as a global minimum.

It remains to characterize the nature of the dual geometric program and to address the question of whether to maximize or minimize this (generalized) dual objective function.

The proofs of sufficiency and uniqueness are quite straightforward when all the signum functions are positive, the case developed earlier. Suppose that

$$\sigma_m = 1; \qquad m = 0, 1, \ldots, M$$

and (6-86)

$$\sigma_{mt} = 1; \qquad \begin{cases} m = 0, 1, \ldots, M \\ t = 1, \ldots, T_m \end{cases}$$

For this case of a posynomial program, the objective function $y_o(\mathbf{x})$ is strictly convex with respect to the transformed variables $\mathbf{u}$. Similarly, all the posynomial constraints are strictly convex functions of the $\mathbf{u}$. Note that all the constraints take the form:

$$y_m(\mathbf{x}) \leq 1; \qquad m = 1, \ldots, M \tag{6-87}$$

which forms a convex set with respect to $\mathbf{u}$. It is well known that a strictly convex function defined on a convex set has only one local minimum. Thus the local minimum satisfying the normality and orthogonality conditions must also be the global minimum.

As for the transformed dual function $z(\boldsymbol{\omega})$, it is strictly concave when Eqs. (6-86) hold because it is then a positively weighted sum of negative logarithms of the weights

$$w_{mt} = \frac{\omega_{mt}}{\sum_{j=1}^{T_m} \omega_{mj}}$$

Therefore, it has only one stationary point: a global maximum. Hence, the minimum of $y_o(\mathbf{x})$ can be found by maximizing the dual function $d(\boldsymbol{\omega})$ or its logarithm $z(\boldsymbol{\omega})$, and this condition is sufficient. Consequently, the dual problem for positive signum functions is to maximize $d(\boldsymbol{\omega})$, subject to the normality and orthogonality conditions [Eqs. (6-81) and (6-82)], with the ω_{mt} restricted to be nonnegative. The relationship between the primal objective function $y_o(\mathbf{x})$ and the dual objective function in this case is

$$y_o(\mathbf{x}) > y_o(\mathbf{x}^*) \equiv d(\boldsymbol{\omega}^*) > d(\boldsymbol{\omega}) \tag{6-88}$$

This relationship says that, at optimality, the value of the dual objective function will equal that of the primal objective function; furthermore, *any* solution vector $\boldsymbol{\omega}$ for the dual problem will yield a lower bound on the optimal value of y_o, and *any* solution vector $\mathbf{x}$ for the primal problem will yield an upper bound on the optimal value of y_o. Consider the following example, due to Petropoulos:

$$\text{Minimize } y_o(\mathbf{x}) = 452x_1^{-1}x_2^{-1} + 10^{-5}x_1^{2.33}x_2^{0.40}$$

subject to

$$y_1(\mathbf{x}) = 11{,}000x_1^{-1.52}x_2 \leq 1$$
$$y_2(\mathbf{x}) = 0.0193x_1x_2^{0.83} \leq 1$$
$$x_1 > 0; \qquad x_2 > 0$$

This problem has four terms ($T = 4$) and two variables ($N = 2$). Hence there is $T - (N + 1) =$ one degree of difficulty. With positive degrees of difficulty present, the constraint set constitutes necessary but not sufficient conditions for optimality. Since all signum functions are positive, let us utilize Eq. (6-88) to bound our optimal solution. If we drop the second term in the objective function (dropping *constraint* terms will *not* work), we will obtain a zero degree of difficulty problem, which has the solution

$$\hat{\omega}_{01}^* = 1.0; \qquad \hat{\omega}_{11}^* = 0.075; \qquad \hat{\omega}_{21}^* = 1.114$$

so that both constraints are tight. The optimal value of the dual objective function is given by

$$d(\hat{\boldsymbol{\omega}}^*) = \left(\frac{452}{1}\right)^1 (11{,}000)^{0.075}(0.0193)^{1.114} = 11.184$$

The optimal primal variables can be calculated from Eqs. (6-84) and (6-85):

$$\hat{x}_1^* = 174.28; \qquad \hat{x}_2^* = 0.232$$

Substitution of these values into the original objective function yields $y_o(\hat{\mathbf{x}}^*) = 12.113$.

Equation (6-88) thus produces the bounds:

$$12.113 \geq y_o(\mathbf{x}^*) = d(\boldsymbol{\omega}^*) \geq 11.184$$

Since we now have good bounds on the original objective function, it might not be worth the extra effort to obtain the true optimal $\mathbf{x}^*$. In making this decision, each problem must be judged separately, weighing the advantages of obtaining the exact optimal solution against the cost of obtaining it. For this example problem, the optimal solution is $\mathbf{x}^* = (172.30, 0.234)$, producing an optimal value for the objective function of $y_o(\mathbf{x}^*) = 12.113$, which happens to equal the upper bound in this particular problem. For most problems, we cannot, of course, expect the true optimal value of $y_o(\mathbf{x}^*)$ to be equal to one of the bounding values. Nonetheless, Eq. (6-88) offers a quick and efficient means for bounding the optimal solution, which is not obtainable through other optimization techniques.

6-06 GEOMETRIC PROGRAMS WITH MIXED CONSTRAINTS AND NEGATIVE SIGNUM FUNCTIONS

Suppose now that some or all of the signum functions σ_m are negative; in this case the constraint set will not be convex. Since the objective function is bounded below by zero and is a continuous function, it must have a minimum, provided that there exist points satisfying the constraints. Let $\mathbf{u}^*$ be a feasible local minimum for the primal problem. It has been shown that the Lagrangian function for the primal problem, when evaluated at a stationary point or constrained local optimum $\mathbf{u}^*$, is the same as that associated with finding a stationary point or constrained local optimum for the transformed dual function

$$z(\boldsymbol{\omega}) \equiv \sum_{m=0}^{M} \sum_{t=1}^{T_m} \sigma_{mt}\omega_{mt} \ln \left(c_{mt}\omega_{mt}^{-1} \sum_{j=1}^{T_m} \omega_{mj} \right) \tag{6-89}$$

with respect to the nonnegative dual variables $\boldsymbol{\omega}$, which are constrained to satisfy the normality and orthogonality dual constraint conditions.

At any dual stationary point, the complementary slackness conditions of Eq. (6-66) must hold. Then, by Eqs. (6-72) and (6-83), and the fact that the w_{mt} are positive, we have

$$f_m(\mathbf{x})\omega_{mt} = 0; \qquad t = 1, 2, \ldots, T_m; \quad m = 1, \ldots, M \tag{6-90}$$

If the mth constraint is loose ($f_m > 0$), then *all* its corresponding duals ω_{mt} ($t = 1, \ldots, T_m$) will be zero and consequently on the boundary of the set of dual variables satisfying the dual constraints. Therefore, $z(\boldsymbol{\omega})$ attains a local maximum there with respect to any nonvanishing ω_{mt}. On the other hand, if the mth constraint is tight ($f_m = 0$), then the corresponding duals ω_{mt} will be positive, placing the point inside the dual constraint set. Passy and Wilde showed that in this case $z(\boldsymbol{\omega})$ must also be stationary there with respect to the ω_{mt}.

Consequently, the dual solution vector $\boldsymbol{\omega}^o$ could produce a corresponding constrained local maximum for u_o rather than the desired minimum; it could also be a stationary point. Hence any point obtained by dual manipulations must be checked and tested to see that it has the proper character. Since u_o is convex, satisfaction of the Kuhn–Tucker nonnegativity and complementary slackness conditions is sufficient to establish the solution vector u_o^* as a local minimum ($l \min u_o$). The solution to the primal minimization problem may, therefore, be written

$$u_o^* = \min \, (l \min u_o) \tag{6-91}$$

In terms of the transformed dual function, this is interpreted as saying that

$$u_o^* = \min \, [z(\boldsymbol{\omega}^o)] \tag{6-92}$$

In other words, to guarantee a global minimum one must find the smallest of all the primal local minima, each computed in the dual space by maximizing $z(\boldsymbol{\omega})$. Passy and Wilde called this procedure *pseudominimization*; current practice is to call this "quasimaximization," since one must find all maximizing dual stationary points and then choose the minimum from among them. Although the pseudominimization technique might seem to limit the usefulness of geometric programming for solving signomial programs, in reality it only reduces the procedure to a status similar to all other nonconvex programming techniques. The authors know of no algorithm for solving nonconvex programming problems that guarantees global solutions. On the other hand, pseudominimization at least provides a systematic procedure for obtaining global solutions.

This result extends the geometric programming algorithm to sets of signomial constraints of arbitrary sense. Note, however, that an important feature of posynomial programming is lost when the signs are mixed. This is the use of the dual solution for quick estimates and successive approximation algorithms. *The important inequality given by Eq. (6-88) holds if and only if all signum functions are positive.* Algorithmic procedures for finding dual stationary points in the general signomial case are covered in the Beightler–Phillips book.

Finally, suppose that the objective function has negative terms. The prope way to handle such problems depends on whether the optimal cost $y_o(\mathbf{x}^*)$ is positive or negative. Although it is not always possible to predict which case applies, a wrong guess will be exposed in the course of the computation and can easily be corrected.

The orthogonality conditions of the dual geometric program are exactly as before; however, as shown in Beightler and Phillips, the normality condition must be replaced by a *generalized normality condition*:

$$\sum_{t=1}^{T_o} \sigma_{ot}\omega_{ot} = \sigma_o \quad (= \pm 1) \tag{6-93}$$

The value of σ_o will usually be known in advance for most problems. Since the orthogonality conditions are homogeneous, changing the sign of σ_o simply reverses the signs of all the other dual variables ω_{mt}. Hence a wrong initial guess for σ_o will only cause all the dual variables to have the wrong sign (all will be negative), but they will be correct in absolute value.

We now summarize the signomial case with a few compact formulas. Details of the developments of these formulas are to be found in the Beightler–Phillips book.

A general signomial may be written

$$y_m(\mathbf{x}) \equiv \sum_{t=1}^{T_m} \sigma_{mt} c_{mt} \prod_{n=1}^{N} x_n^{a_{mtn}}; \qquad m = 0, 1, \ldots, M \tag{6-94}$$

with

$$\sigma_{mt} = \pm 1 \tag{6-95}$$

and

$$c_{mt} > 0 \tag{6-96}$$

The primal problem is to minimize

$$y_o(\mathbf{x}) \tag{6-97}$$

subject to

$$y_m(\mathbf{x}) \leq \sigma_m \quad (\equiv \pm 1); \qquad m = 1, \ldots, M \tag{6-98}$$

and

$$x_n > 0; \qquad n = 1, \ldots, N \tag{6-99}$$

Consider now a set of signum functions σ_m ($m = 0, 1, \ldots, M$) and a set of

$$T = \sum_{m=0}^{M} T_m$$

dual variables $\boldsymbol{\omega}$ that satisfy a (generalized) normality condition,

$$\sum_{t=1}^{T_o} \sigma_{ot}\omega_{ot} = \sigma_o \tag{6-100}$$

N orthogonality conditions,

$$\sum_{m=0}^{M} \sum_{t=1}^{T_m} \sigma_{mt} a_{mtn} \omega_{mt} = 0; \qquad n = 1, \ldots, N \tag{6-101}$$

T nonnegativi'y conditions,

$$\omega_{mt} \geq 0; \qquad t = 1, \ldots, T_m; \qquad m = 0, 1, \ldots, M \tag{6-102}$$

as well as M linear inequality constraints

$$\omega_{mo} \equiv \sigma_m \sum_{t=1}^{T_m} \sigma_{mt} \omega_{mt} \geq 0; \qquad m = 0, 1, \ldots, M \tag{6-103}$$

and the definitions of the signum functions.

$$\begin{aligned} \sigma_m &\equiv \pm 1; \qquad m = 0, 1, \ldots, M \\ \sigma_{mt} &= \pm 1; \qquad \begin{cases} t = 1, \ldots, T_m \\ m = 0, 1, \ldots, M \end{cases} \end{aligned} \tag{6-104}$$

From these variables, the coefficients c_{mt} and the signum functions, form the dual function

$$d(\boldsymbol{\omega}) \equiv \sigma_o \left[\prod_{m=0}^{M} \prod_{t=1}^{T_m} \left(\frac{c_{mt} \omega_{mo}}{\omega_{mt}} \right)^{\sigma_{mt} \omega_{mt}} \right]^{\sigma_o} \tag{6-105}$$

where it is understood formally that

$$\lim_{\omega_{mt} \to 0} \left(\frac{c_{mt} \omega_{mo}}{\omega_{mt}} \right)^{\sigma_{mt} \omega_{mt}} = 1 \tag{6-106}$$

Note that as a consequence of Eqs. (6-100) and (6-103),

$$\omega_{00} \equiv 1 \tag{6-107}$$

Then for every point $\mathbf{x}^o$ where $y_o(\mathbf{x})$ is locally minimum there exists a set of dual variables $\boldsymbol{\sigma}^o$ and $\boldsymbol{\omega}^o$ satisfying Eqs. (6-100) through (6-104) and such that

$$d(\boldsymbol{\omega}^o) = y_o(\mathbf{x}^o) \tag{6-108}$$

Once the dual variables $\boldsymbol{\omega}$ and $\boldsymbol{\sigma}$ are known, the corresponding values of the primal variables $\mathbf{x}$ are found from the following relations:

$$c_{ot} \prod_{n=1}^{N} x_n^{a_{otn}} = \omega_{ot} \sigma_o y_o(\mathbf{x}^o); \qquad t = 1, \ldots, T_o \tag{6-109}$$

and

$$c_{mt}\prod_{n=1}^{N} x_n^{a_{mtn}} = \frac{\omega_{mt}}{\omega_{mo}}; \qquad \begin{matrix} t = 1, \ldots, T_m \\ m = 1, \ldots, M \end{matrix} \qquad (6\text{-}110)$$

Since there are always more primal terms than variables **x**, one can find N equations solvable for the N primals, assuming that enough primal constraints are tight at optimality (this subject is pursued in detail in Beightler and Phillips). This task is not difficult because, even though Eqs. (6-109) and (6-110) are nonlinear, each has only one term. Consequently, one can take logarithms and solve the resulting simultaneous equations, *linear* in $\log x_n$.

$$\sum_{n=1}^{N} a_{otn}(\log x_n) = \log\left[\frac{\omega_{ot}\sigma_o y_o(\mathbf{x}^o)}{c_{ot}}\right]; \qquad t = 1, \ldots, T_o \qquad (6\text{-}111)$$

$$\sum_{n=1}^{N} a_{mtn}(\log x_n) = \log\left(\frac{\omega_{mt}}{c_{mt}\omega_{mo}}\right); \qquad \begin{matrix} t = 1, \ldots, T_m \\ m = 1, \ldots, M \end{matrix} \qquad (6\text{-}112)$$

Now consider a signomial problem to illustrate the use of the equations above.

$$\text{Maximize } \hat{y}_o(\mathbf{x}) = 5x_1^2 - x_2^2x_3^4$$

subject to

$$\hat{y}_1(\mathbf{x}) = 1.5x_2^{-1}x_3 - 2.5x_1^2x_2^{-2} \geq 1$$

$$x_1, \quad x_2, \quad x_3 > 0$$

In standard geometric programming form:

$$\text{minimize } y_o(\mathbf{x}) = -5x_1^2 + x_2^2x_3^4$$

subject to

$$y_1(\mathbf{x}) = 2.5x_1^2x_2^{-2} - 1.5x_2^{-1}x_3 \leq -1$$

$$x_1, \quad x_2, \quad x_3 > 0$$

For this problem:

$$\sigma_{01} = -1.0; \qquad \sigma_{02} = 1.0; \qquad \sigma_{11} = 1.0; \qquad \sigma_{12} = -1.0;$$

$$c_{01} = 5.0; \qquad c_{02} = 1.0; \qquad c_{11} = 2.5; \qquad c_{12} = 1.5$$

Although the original objective function is to maximize (possibly profits), we will guess $\sigma_o = +1.0$ for $y_o(\mathbf{x})$, the negative value of $\hat{y}_o(\mathbf{x})$.

This problem has four terms and three variables, hence zero degrees of difficulty. From Eqs. (6-100) and (6-101):

$$-\omega_{01} + \omega_{02} = +1.0; \qquad 2\omega_{02} - 2\omega_{11} + \omega_{12} = 0;$$
$$-2\omega_{01} + 2\omega_{11} = 0; \qquad 4\omega_{02} - \omega_{12} = 0$$

These equations have the following solution.

$$\omega_{01} = -1.50; \qquad \omega_{11} = -1.50;$$
$$\omega_{02} = -0.50; \qquad \omega_{12} = -2.0$$

Note that all dual variables are negative. This signifies that we have chosen the wrong sign for σ_o. If we now set $\sigma_o = -1.0$, the optimal solution becomes:

$$\omega_{01}^* = 1.50; \qquad \omega_{11}^* = 1.50;$$
$$\omega_{02}^* = 0.50; \qquad \omega_{12}^* = 2.0$$

The additional nonnegativity constraint is given by

$$\omega_{10} = \omega_{11} - \omega_{12} \geq 0; \qquad \omega_{10} = 0.50$$

The value of the dual objective function is given by Eq. (6-105):

$$d(\boldsymbol{\omega}^*) = -1.0\left\{\left(\frac{5.0}{1.5}\right)^{-1.5}\left(\frac{1.0}{0.50}\right)^{0.50} \times \left[\frac{(0.50)(2.5)}{1.5}\right]^{1.5}\left[\frac{(0.50)(1.5)}{2.0}\right]^{-2}\right\}^{-1}$$

Hence $y(\mathbf{x}^*) = d(\boldsymbol{\omega}^*) = -0.795495$.

The optimal primal variables are now recovered as before, yielding the values

$$x_1^* = 0.4885; \qquad x_2^* = 0.446; \qquad x_3^* = 1.189$$

6-07 FUNCTIONAL SUBSTITUTIONS

Although geometric programming algorithms are very powerful in solving nonlinear problems consisting only of posynomial forms, they do not apply directly to problems containing other types of functions. Nonetheless, such functions can usually be transformed into posynomial or signomial forms through an appropriate change of variable.

Consider a nonlinear programming problem of the following form.

$$\text{Minimize } y(\mathbf{x}) = f(\mathbf{x}) + [q(\mathbf{x})]^a h(\mathbf{x}); \qquad \mathbf{x} > 0, \quad a > 0$$

where $f(\mathbf{x})$, $q(\mathbf{x})$, and $h(\mathbf{x})$ are single or multiterm functionals of posynomial or signomial form. This generalized formulation is not directly solvable using geometric programming; however, under a simple transformation it can be changed into standard geometric programming form. Let

$$P = q(\mathbf{x})$$

and replace the above problem with the following one.

$$\text{Minimize } \bar{y}(\mathbf{x}) = f(\mathbf{x}) + P^a h(\mathbf{x})$$

subject to

$$P^{-1}[q(\mathbf{x})] \leq 1$$
$$\mathbf{x}, \quad P > 0$$

The rationale used in constructing the equivalent problem with an inequality constraint is based on the following logic. Since $y(\mathbf{x})$ is to be minimized, if $q(\mathbf{x})$ is replaced by P, then it is correct to say that $P \geq q(\mathbf{x})$, realizing that in the minimization process P will remain as small as possible. Hence $P = q(\mathbf{x})$ at optimality. Note that $h(\mathbf{x})$ and/or $q(\mathbf{x})$ are permitted to be multiple-term expressions, and that the optimal (minimizing) solution to $\bar{y}(\mathbf{x})$ is obviously the same as the optimal solution to $y(\mathbf{x})$.

As a simple example of this type of functional substitution, suppose that it is desired to find nonnegative values x_1, x_2 that minimize the function

$$f(\mathbf{x}) = (x_1^{-2} + x_2)^{0.8}(x_1 + x_2^{-1})^{1.4}$$

An equivalent problem is finding nonnegative x_1, x_2, x_3, x_4 that minimize the posynomial

$$y_0(\mathbf{x}) = x_3^{0.8} x_4^{1.4}$$

and that satisfy the posynomial constraints

$$x_3 \geq x_1^{-2} + x_2$$
$$x_4 \geq x_1 + x_2^{-1}$$

This may now be written in the standard form of a posynomial program.

$$\text{Minimize } y_0(\mathbf{x}) = x_3^{0.8} x_4^{1.4}$$

subject to

$$x_1^{-2} x_3^{-1} + x_2 x_3^{-1} \leq 1$$
$$x_1 x_4^{-1} + x_2^{-1} x_4^{-1} \leq 1$$
$$x_1, \quad x_2, \quad x_3, \quad x_4 > 0$$

Since this is a zero-degree problem, the dual constraints are

$$\begin{aligned}
&\omega_{01} && && && && = 0\\
& && -2\omega_{11} && && +\omega_{21} && = 0\\
& && && \omega_{12} && && -\omega_{22} = 0\\
&0.8\omega_{01} && -\omega_{11} && -\omega_{12} && && = 0\\
&1.4\omega_{01} && && && -\omega_{21} && -\omega_{22} = 0
\end{aligned}$$

which have a unique solution:

$$\omega_{11}^* = 0.6; \quad \omega_{12}^* = 0.2; \quad \omega_{21}^* = 1.2; \quad \omega_{22}^* = 0.2; \quad \omega_{01}^* = 1.0;$$
$$\omega_{10}^* = 0.8; \quad \omega_{20}^* = 1.4$$

from which the value of the dual is easily calculated.

$$d^*(\boldsymbol{\omega}) = \left(\frac{1}{1}\right)^1 \left(\frac{0.8}{0.6}\right)^{0.6} \left(\frac{0.8}{0.2}\right)^{0.2} \left(\frac{1.4}{1.2}\right)^{1.2} \left(\frac{1.4}{0.2}\right)^{0.2} = 2.7844$$

The primal variables are recovered as before from Eqs. (6-109) and (6-110):

$$x_1 x_4^{-1} = \frac{1.2}{1.4} = 0.857; \qquad x_2^{-1} x_4^{-1} = 0.143;$$
$$x_1^{-2} x_3^{-1} = \frac{0.6}{0.8} = 0.75; \qquad x_2 x_3^{-1} = 0.25;$$
$$x_3^{0.8} x_4^{1.4} = 2.7844$$

yielding the optimal solution:

$$x_1^* = 0.0556; \quad x_2^* = 107.778; \quad x_3^* = 431.152; \quad x_4^* = 0.06488$$

Other techniques for transforming given nonlinear programming problems to equivalent geometric programs are possible, many depending on the ingenuity of the analyst. Duffin, Peterson, and Zener, for example, show that *generalized polynomials*, defined by

$$G(\mathbf{x}) = \sum_{t=1}^{T} \prod_{i=1}^{U} \frac{[p_{ti}(\mathbf{x})]^{\alpha_{ti}}}{[1 - q_{ti}(\mathbf{x})]^{\beta_{ti}}} \tag{6-113}$$

can be used both as the objective function to be minimized and as constraint functions to be satisfied by the primal variables $\mathbf{x}$. Here, $p_{ti}(\mathbf{x})$ and $q_{ti}(\mathbf{x})$ are multiterm functions, and the constants α_{ti} and β_{ti} are positive numbers. It is convenient to assume that the functions $1 - q_{ti}$ $\mathbf{x})$ and $p_{ti}(\mathbf{x})$ are positive. Otherwise, the direction of the transformed constraint inequalities would have to be "guessed" prior to problem solution. As one example of how such generalized posynomials might appear in a simple problem, consider the following:

Find positive $\mathbf{x}$ that minimize

$$G(\mathbf{x}) = f(\mathbf{x}) + \frac{q(\mathbf{x})}{[u(\mathbf{x}) - h(\mathbf{x})]^a} \tag{6-114}$$

where f, q, and h are posynomials; u is a posynomial with one term; and a is positive.

Now consider the related problem of minimizing the posynomial

$$y(\mathbf{x}, v) = f(\mathbf{x}) + \frac{q(\mathbf{x})}{v^a}$$

subject to the constraint

$$\frac{v}{u(\mathbf{x})} + \frac{h(\mathbf{x})}{u(\mathbf{x})} \leq 1$$

where v is an additional independent variable. Since $u(\mathbf{x})$ is a single-term posynomial, this problem is in standard *posynomial* programming form. It is clear that $\mathbf{x}^*$ solves this problem if, and only if, $\mathbf{x}^*$ minimizes $G(\mathbf{x})$. It also follows that the constrained minimum value for $y(\mathbf{x}, v)$ is equal to the minimum value for $G(\mathbf{x})$. Since the formulation above is *posynomial*, the solution to the equivalent problem of minimizing $y(\mathbf{x}, v)$ is a *global* minimizing solution both for $y(\mathbf{x}, v)$ and $y(\mathbf{x})$ as well. This procedure is equally applicable to signomial forms, but nonconvexity problems arise in the dual problem.

In many real-world formulations, a logarithmic term may appear in the descriptive mathematical model. Within the conventional geometric programming structure, such logarithmic terms usually cannot be substituted out; however, appropriate limiting approximations can be used to transform such terms into po ynomials. Recall from elementary calculus that the logarithm of an arbitrary real number ϕ is defined by

$$\ln(\phi) = \int_1^{\phi} \frac{dx}{x} = \int_1^{\phi} x^{-1}\, dx \simeq \int_1^{\phi} x^{\epsilon - 1}\, dx$$

for ϵ a very small positive number.
Hence $\ln(\phi) \simeq \phi^{\epsilon}/\epsilon - 1/\epsilon$, and

$$\lim_{\epsilon \to 0} \left(\frac{\phi^{\epsilon}}{\epsilon} - \frac{1}{\epsilon}\right) \longrightarrow \ln(\phi)$$

Therefore, $\ln(\phi)$ may be approximated by

$$\epsilon^{-1}(\phi^{\epsilon} - 1) \qquad \text{as } \epsilon \longrightarrow 0$$

In order to apply this approximation, consider a nonlinear programming problem of the following form.

$$\text{Minimize } Y(\mathbf{x}) = f(\mathbf{x}) + \ln [g(\mathbf{x})]$$

Suppose further that $g(\mathbf{x})$ is a single-term posynomial and $f(\mathbf{x})$ is a general geometric programming term. Using ϵ as an arbitrarily small number, this expression can be approximated by:

$$\text{minimize } Y(\mathbf{x}, \epsilon) = f(\mathbf{x}) + \epsilon^{-1}[g(\mathbf{x})]^{\epsilon} - \epsilon^{-1} \tag{6-115}$$

The problem is now in standard (unconstrained) geometric programming form (note that since ϵ^{-1} is simply a constant, it can be dropped from the equation above). Now suppose that the restriction of $g(\mathbf{x})$ as a single-term posynomial is removed. Equation (6.115) is no longer in standard form. However, using the techniques described at the beginning of this section, the problem can be stated as follows.

$$\text{Minimize } f(\mathbf{x}) + \epsilon^{-1}z^{\epsilon} - \epsilon^{-1}$$

subject to

$$z^{-1}[g(\mathbf{x})] \leq 1$$

The problem is now in standard form. Finally, note that if $f(\mathbf{x})$ and $g(\mathbf{x})$ are both posynomial, the dual geometric program can be uniquely solved for the global (minimizing) solution.

A simple example problem which illustrates this approximation is that of finding positive values x_1, x_2 that minimize

$$Y(\mathbf{x}) = x_1^{-2}x_2^{1/2} + 3x_1^{-1/2}x_2^{-3/2} + \ln (x_1^3 x_2^2)$$

Using Eq. (6.115), we minimize the expression

$$Y(\mathbf{x}, \epsilon) = x_1^{-2}x_2^{1/2} + 3x_1^{-1/2}x_2^{-3/2} + \epsilon^{-1}x_1^{3\epsilon}x_2^{2\epsilon}$$

resulting in, as the limit of $\epsilon \longrightarrow 0$, $x_1^* = 1.054$, $x_2^* = 1.423$.

Another common function that may appear in nonlinear problems is the exponential term. Again, through appropriate limiting techniques this difficulty can also be resolved within the general geometric programming framework. Consider the following nonlinear problem.

$$\text{Minimize } Y(\mathbf{x}) = f(\mathbf{x}) + ce^{g(x)}$$

where $f(\mathbf{x})$ = posynomial or signomial
$g(\mathbf{x})$ = single-term posynomial
c = arbitrary nonnegative constant

From the calculus, it is well known that

$$e^z = \lim_{\phi \to \infty} \left[\left(1 + \frac{z}{\phi}\right)^{\phi} \right]$$

Hence the problem above can be approximated to any desired accuracy by the following approximating program for sufficiently large ϕ.

$$\text{Minimize } Y(\mathbf{x}, \phi) = f(\mathbf{x}) + c\left[1 + \frac{g(\mathbf{x})}{\phi}\right]^{\phi}$$

This approximating problem is not in standard geometric programming form; however, we can form the following equivalent problem.

$$\text{Minimize } y(\mathbf{x}, \phi, \beta) = f(\mathbf{x}) + c\beta^{\phi}$$

subject to

$$\beta^{-1} + \beta^{-1}\phi^{-1}g(\mathbf{x}) \leq 1$$

Note that in the process of making the approximating transformations, the number of terms in the resultant transformed problem will be increased by one.

6-08 GENERAL NONLINEAR PROGRAMS

In a significant paper, Duffin and Peterson demonstrated that any nonlinear algebraic program of the form:

$$\text{minimize } y(\mathbf{x})$$

subject to

$$h_k(\mathbf{x}) \geq 0; \qquad k = 1, \ldots, k$$
$$\mathbf{x} \geq 0$$

where y and h_k are real-valued functions formed by addition, subtraction, or exponentiation to real powers, can be transformed to the following signomial program.

$$\text{Minimize } y_0(\mathbf{x})$$

subject to

$$y_m(\mathbf{x}) \leq \sigma_m; \qquad m = 1, \ldots, M$$
$$\mathbf{x} \geq 0$$

where the functions y_m, $m = 0, 1, M$ are signomials. This is accomplished primarily by introducting additional variables together with their defining constraints into the problem so as to break up all functions into sums or differences of individual terms.

The following example, due to Kavlie, Kowalik, and Moe, and presented as a geometric program by Reklaitis and Wilde, illustrates this procedure.

The minimum-weight design of the corrugated bulkheads for a tanker requires the solution of the following optimization problem.

$$\text{Minimize } y(\mathbf{x}) = \frac{5.885x_4(x_1 + x_3)}{x_1 + (x_3^2 - x_2^2)^{1/2}}$$

subject to

$$h_1(\mathbf{x}) = x_2x_4(0.4x_1 + \tfrac{1}{6}x_3) - 8.94[x_1 + (x_3^2 - x_2^2)^{1/2}] \geq 0$$
$$h_2(\mathbf{x}) = x_2^2x_4(0.2x_1 + \tfrac{1}{12}x_3) - 2.2[8.94x_1 + (x_3^2 - x_2^2)^{1/2}]^{4/3} \geq 0$$
$$h_3(\mathbf{x}) = x_4 - 0.0156x_1 - 0.15 \geq 0$$
$$h_4(\mathbf{x}) = x_4 - 0.0156x_3 - 0.15 \geq 0$$
$$h_5(\mathbf{x}) = x_4 - 1.05 \geq 0$$
$$h_6(\mathbf{x}) = x_3 - x_2 \geq 0 \qquad \text{and all } x_i \geq 0$$

The variables x_i correspond to the width, depth, and length of the corrugation and the thickness of the steel plate, respectively. This algebraic nonlinear programming problem can be transformed into a signomial problem by introducing several additional variables and their defining relations.

First, the objective function is transformed to

$$5.885x_4(x_1 + x_3)x_5^{-1}$$

by introducing nonnegative variables x_5 and x_6 together with the constraints

$$x_5 \leq x_1 + x_6 \leq x_1 + (x_3^2 - x_2^2)^{1/2}$$

These reduce to two equivalent constraints

$$x_5 \leq x_1 + x_6; \qquad x_6^2 \leq x_3^2 - x_2^2$$

or to two signomial constraints

$$y_1(\mathbf{x}) = -x_1x_5^{-1} - x_6x_5^{-1} \leq -1$$
$$y_2(\mathbf{x}) = -x_3^2x_6^{-2} + x_2^2x_6^{-2} \leq -1$$

Next, by introducing nonnegative variable x_7, the first constraint h_1 is replaced by the following inequalities

$$x_2x_4(0.4x_1 + \tfrac{1}{6}x_3) \geq 8.94(x_1 + x_7) \geq 8.94[x_1 + (x_3^2 - x_2^2)^{1/2}]$$

These reduce to the following equivalent constraints:

$$x_2x_4(0.4x_1 + \tfrac{1}{6}x_3) \geq 8.94(x_1 + x_7)$$
$$x_7^2 \geq x_3^2 - x_2^2$$

or, to two signomial constraints

$$y_3(\mathbf{x}) = 53.64x_1x_2^{-1}x_3^{-1}x_4^{-1} + 53.64x_2^{-1}x_3^{-1}x_4^{-1}x_7 - 2.4x_1x_3^{-1} \leq 1$$
$$y_4(\mathbf{x}) = x_3^2x_7^{-2} - x_2^2x_7^{-2} \leq 1$$

Finally, the second constraint, $h_2(\mathbf{x})$, becomes, with the introduction of the nonnegative variable x_8,

$$x_2^2x_4(0.2x_1 + \tfrac{1}{12}x_3) \geq 2.2x_8^{8/3} \geq 2.2[8.94(x_1 + x_7)]^{4/3}$$
$$\geq 2.2\{8.94[x_1 + (x_3^2 - x_2^2)^{1/2}]\}^{4/3}$$

or

$$x_2^2x_4(0.2x_1 + \tfrac{1}{12}x_3) \geq 2.2x_8^{8/3}$$
$$x_8^2 \geq 8.94(x_1 + x_7)$$

since the definition of x_7 is already given. In signomial form these two constraints become

$$y_5(\mathbf{x}) = -\tfrac{1}{11}x_1x_2^2x_4x_8^{-8/3} - \tfrac{5}{132}x_2^2x_3x_4x_8^{-8/3} \leq -1$$
$$y_6(\mathbf{x}) = 8.94x_1x_8^{-2} + 8.94x_7x_8^{-2} \leq 1$$

With these modifications, the resulting signomial program becomes:

$$\text{Minimize } y_0(\mathbf{x}) = 5.885x_1x_4x_5^{-1} + 5.885x_3x_4x_5^{-1}$$

subject to

$$y_1(\mathbf{x}) = -x_1x_5^{-1} - x_5^{-1}x_6 \leq -1$$
$$y_2(\mathbf{x}) = -x_3^2x_6^{-2} + x_2^2x_6^{-2} \leq -1$$
$$y_3(\mathbf{x}) = 53.64x_1x_2^{-1}x_3^{-1}x_4^{-1} + 53.64x_2^{-1}x_3^{-1}x_4^{-1}x_7 - 2.4x_1x_3^{-1} \leq 1$$
$$y_4(\mathbf{x}) = x_3^2x_7^{-2} - x_2^2x_7^{-2} \leq 1$$
$$y_5(\mathbf{x}) = -\tfrac{1}{11}x_1x_2^2x_4x_8^{-8/3} - \tfrac{5}{132}x_2^2x_3x_4x_8^{-8/3} \leq -1$$
$$y_6(\mathbf{x}) = 8.94x_1x_8^{-2} + 8.94x_7x_8^{-2} \leq 1$$
$$y_7(\mathbf{x}) = 0.0156x_1x_4^{-1} + 0.15x_4^{-1} \leq 1$$
$$y_8(\mathbf{x}) = 0.0156x_3x_4^{-1} + 0.15x_4^{-1} \leq 1$$
$$y_9(\mathbf{x}) = 1.05x_4^{-1} \leq 1$$
$$y_{10}(\mathbf{x}) = x_2x_3^{-1} \leq 1 \qquad \text{and all } x_i \geq 0; \quad i = 1, \ldots, 8$$

The transformation has been accomplished at the cost of introducing four additional variables and four additional constraints.

Thus it is evident that the family of signomial programs in fact encompasses all inequality constrained algebraic programs. By proceeding further with this simple device of adding artificial variables, it can, moreover, be shown that the signomial program itself can always be transformed to an equivalent problem in which all the problem functions are *posynomials* but where some of the posynomial inequalities are *reversed.* That is, the signomial program can be transformed to:

Program 4

$$\text{Minimize } y_o(\mathbf{x})$$

subject to

$$y_k(\mathbf{x}) \geq 1 \qquad k = 1, \ldots, R$$
$$y_k(\mathbf{x}) \leq 1 \qquad k = R + 1, \ldots, K$$
$$x \geq 0$$

where y_o and the y_k are posynomials.

To see this, consider any signomial constraint

$$y_m(\mathbf{x}) = \sum_{t \in P} c_{mt} \prod_{n=1}^{N} x_n^{a_{mtn}} - \sum_{t \in N} c_{mt} \prod_{n=1}^{N} x_n^{a_{mtn}}$$

where $P = \{t\colon \sigma_{mt} > 0\}$

$N = \{t\colon \sigma_{mt} < 0\}$

If $\sigma_m = +1$, then by defining x_{n+1} such that

$$\sum_{t \in P} c_{mt} \prod_{n=1}^{N} x_n^{a_{mtn}} \leq x_{n+1} \leq \sum_{t \in N} c_{mt} \prod_{n=1}^{N} x_n^{a_{mtn}} + 1$$

the constraint can be rewritten as two posynomials,

$$x_{n+1}^{-1} \sum_{t \in P} c_{mt} \prod_{n=1}^{N} x_n^{a_{mtn}} \leq 1$$

and

$$x_{n+1}^{-1} + x_{n+1}^{-1} \sum_{t \in N} c_{mt} \prod_{n=1}^{N} x_n^{a_{mtn}} \geq 1$$

if $\sigma_m = -1$, then, similarly, the constraint can be rewritten as

$$x_{n+1}^{-1} + x_{n+1}^{-1} \sum_{t \in P} c_{mt} \prod_{n=1}^{N} x_n^{a_{mtn}} \leq 1$$

and

$$x_{n+1}^{-1} \sum_{t \in N} c_{mt} \prod_{n=1}^{N} x_n^{a_{mtn}} \geq 1$$

Of course, if only a single negative or positive term occurs in a signomial, the transformation into a posynomial can often be effected without adding new variables, merely by rearranging terms.

For example, the corrugated bulkhead design problem above may be written as:

$$\text{Minimize } y_o(\mathbf{x}) = 5.885x_1x_4x_5^{-1} + 5.885x_3x_4x_5^{-1}$$

subject to the normal posynomial constraints

$$x_3^{-2}x_6^{-2} + x_2^2x_3^{-2} \leq 1$$
$$8.94x_1x_2^{-1}x_4^{-1}x_9^{-1} + 8.94x_2^{-1}x_4^{-1}x_7x_9^{-1} \leq 1$$
$$8.94x_1x_8^{-2} + 8.94x_7x_8^{-2} \leq 1$$
$$0.0156x_1x_4^{-1} + 0.15x_4^{-1} \leq 1$$
$$0.0156x_3x_4^{-1} + 0.15x_4^{-1} \leq 1$$

the reversed posynomial constraints

$$x_1x_5^{-1} + x_5^{-1}x_6 \geq 1$$
$$0.4x_1x_9^{-1} + \tfrac{1}{6}x_3x_9^{-1} \geq 1$$
$$x_3^{-2}x_7^2 + x_2^2x_8^{-2} \geq 1$$

the single-term constraints

$$4.4x_2^{-2}x_4^{-1}x_8^{8/3}x_9^{-1} \leq 1$$
$$1.05x_4^{-1} \leq 1$$
$$x_2x_3^{-1} \leq 1$$

and the nonnegativity conditions

$$x_i \geq 0; \qquad i = 1, \ldots, 9$$

Note that the transformation from signomial to reversed posynomial program has been accomplished at the cost of one additional variable and one additional constraint.

As shown in Beightler and Phillips, the real advantage of the reversed posynomial formulation is that it allows a clear view of the underlying structure of the corresponding signomial program. This view is completely obscured when the problem is viewed in its parent algebraic form.

The formulation of the reversed program 4 may be exploited by converting it into an approximating program that is also strictly posynomial but contains only prototype $[y_m(\mathbf{x}) \leq 1]$ constraints. As previously noted, a prototype posynomial geometric program can be solved for a global minimizing solution.

6-09 POSYNOMIAL SENSITIVITY ANALYSIS

This section is concerned with *postoptimality*, or *sensitivity*, analysis, in which one is interested in the magnitude of the variations in the optimal solution to a programming problem due to changes or uncertainties in the input data. In other words, we would like to know how sensitive the optimal solution is to perturbations in any of the coefficients or constraints. Furthermore, we would like to be able to do this without the necessity of re-solving the problem for a series of many different values of each parameter. This is especially important for the case of input data produced by fluctuating costs and profits, since small changes in these values can sometimes lead to surprisingly large changes in the optimal policy or solution. Furthermore, such analyses provide insight into the structure of the problem that cannot be gained by examining only the optimal solution. Our task, then, is to estimate the impact of these uncertainties on the solution that was obtained by treating all the input data as constants.

Let the mth constraint be written

$$f_m(\mathbf{x}) \equiv 1 - \sum_{t=1}^{T_m} c_{mt} \prod_{n=1}^{N} x_n^{a_{mtn}} \geq 0; \qquad m = 1, \ldots, M \tag{6-116}$$

We may now introduce slack variables v_m and write the above inequality constraints as the equations

$$f_m(\mathbf{x}) - v_m = 0; \qquad m = 1, \ldots, M \tag{6-117}$$

and the nonnegativity conditions

$$v_m \geq 0; \qquad m = 1, \ldots, M \tag{6-118}$$

Consider any feasible point $\mathbf{x}$ in F, and let exactly K of the M slack variables v_m be zero there. No generality is lost in numbering the slack variables so that the first K are zero, and they will be given the running index $k = 1, \ldots, K$.

Using the constrained derivative approach developed in Chapter 2, we may now carry out a sensitivity analysis on any posynomial program. In Beightler and Phillips, this same approach is used to develop an efficient algorithm for solving the general geometric programming problem. To illustrate the use of the constrained derivatives in performing sensitivity analyses, consider first the following simple problem, having only one constraint.

$$\text{Minimize } y = 7x_1^2 + 0.2x_1^{3.5}x_2^{2.5} + 15x_1^{-2}x_2^{-0.5}$$

subject to

$$8x_1^{-2}x_2^{-1} \leq 1$$

$$x_1, \quad x_2 > 0$$

This problem has one degree of difficulty and, for the given input data, the optimal solution is $x_1^* = 1.522$, $x_2^* = 3.458$, giving the objective function a value of $y^* = 38.980$.

For a postoptimality study, let us check the effects on y^* of a small change in the constraint. For this purpose we may employ Eq. (2-113), choosing x_1 as the state (solution) variable and treating the constraint as tight at optimality. Recall from Eq. (6-116) that the constraint must be written as

$$f_1 = 1 - 8x_1^{-2}x_2^{-1} \geq 0$$

and hence, when assumed tight, as $f_1 = 0$. Then, by Eq. (2-113),

$$\frac{\partial z}{\partial f_1} = \frac{\partial y/\partial x_1}{\partial f_1/\partial x_1} = 46.48$$

This change in f_1 is equivalent to a corresponding change in the value of the right-hand side (now zero). Thus a change of -0.01 in f_1 would produce the new constraint

$$f_1 = 1 - 7.9208x_1^{-2}x_2^{-1} = 0$$

while a change of $+0.01$ in f_1 would result in

$$f_1 = 1 - 8.0808x_1^{-2}x_2^{-1} = 0$$

Computer runs on this problem for $\Delta f_1 = -0.01$ and $+0.01$ result in

$$\frac{\Delta y^*}{\Delta f_1} = \frac{38.521 - 38.980}{-0.01} = 45.9$$

and

$$\frac{39.446 - 38.980}{0.01} = 46.6$$

Now consider the following three-variable, two-constraint problem, containing one degree of difficulty.

$$\text{Minimize } y = x_3x_1^{-1}x_2^{-1} + 25x_2^{-1}x_1$$

subject to

$$\tfrac{1}{10}x_1x_2 \leq 1$$
$$5x_1x_3^{-1} + x_2^2x_3^{-1} \leq 1$$
$$x_1, \quad x_2, \quad x_3 \geq 0$$

The optimal dual solution to this problem is

$$\omega_{01}^* = 0.55; \quad \omega_{02}^* = 0.45; \quad \omega_{11}^* = 0.033; \quad \omega_{21}^* = 0.066;$$
$$\omega_{22}^* = 0.48; \quad y^* = 10.7 = d^*; \quad x_1^* = 1.387; \quad x_2^* = 7.21;$$
$$x_3^* = 58.85$$

Note that both constraints are tight; let us now use Eq. (2-113) to perform a sensitivity analysis on the first constraint: $f_1 = x_1x_2 - 10$. Since $M = 2$ and $N = 3$, we have one decision variable: let it be x_3. Then the state variables are x_1 and x_2. In detail, then, we have

$$\frac{\partial z^*}{\partial f_1} = \frac{\begin{vmatrix} \partial y/\partial s_1 & \partial y/\partial s_2 \\ \partial f_2/\partial s_1 & \partial f_2/\partial s_2 \end{vmatrix}}{\begin{vmatrix} \partial f_1/\partial s_1 & \partial f_1/\partial s_2 \\ \partial f_2/\partial s_1 & \partial f_2/\partial s_2 \end{vmatrix}}$$

$$\frac{\partial y}{\partial s_1} = \frac{\partial y}{\partial x_1} = x_1^{-2}(x_2^{-1}x_3) + 25x_2^{-1} = -0.772$$

$$\frac{\partial y}{\partial s_2} = \frac{\partial y}{\partial x_2} = -(x_1^{-1}x_3)x_2^{-2} + (25x_1)(-x_2^{-2}) = -1.485$$

$$\frac{\partial f_2}{\partial s_1} = \frac{\partial f_2}{\partial x_1} = -5x_3^{-1} = -0.0849$$

$$\frac{\partial f_2}{\partial s_2} = \frac{\partial f_2}{\partial x_2} = -2x_2x_3^{-1} = -0.2449$$

$$\frac{\partial f_1}{\partial s_1} = \frac{\partial f_1}{\partial x_1} = -x_2 = -7.2054$$

$$\frac{\partial f_1}{\partial s_2} = \frac{\partial f_1}{\partial x_2} = -x_1 = -1.3878$$

$$\frac{\partial z^*}{\partial f_1} = \frac{\begin{vmatrix} -0.772 & -1.485 \\ -0.0849 & -0.2449 \end{vmatrix}}{\begin{vmatrix} -7.2054 & -1.3878 \\ -0.0849 & -0.2449 \end{vmatrix}} = \frac{0.063}{1.647} = 0.038$$

This result indicates that the optimal solution decreases at a rate of 0.038 as the product x_1x_2 is increased. Arbitrarily changing the constant c_{11} to $\frac{1}{12}$ decreases y^* by 12 (0.038) = 0.456, that is, from 10.7 to 10.244.

Other sensitivity analyses are possible, as well, including those which are performed in the dual space. For example, Theil has developed expressions that give the change in the optimal values of the dual variables resulting from changes in the objective function coefficients c_{0t}. Dinkel and Kochenberger have extended Theil's analysis of substitution effects to include analyses of the changes $d\omega_{mt}$ that result from changes in the constraint coefficients

c_{mt}. All this work is summarized in the book by Beightler and Phillips, which also contains numerous examples of sensitivity analyses carried out on real-world problems.

BIBLIOGRAPHY

ABRAMS, R. A., "Consistency, superconsistency, and dual degeneracy in posynomial geometric programming," *Operations Res.*, **24**, 3 (May 1976).

AVRIEL, M., and D. J. WILDE, "Optimal condenser design by geometric programming," *Ind. Eng. Chem. Des. Dev.*, **6**, 2 (April 1967).

———, and A. C. WILLIAMS, "Complimentary geometric programming," *SIAM J. Appl. Math.*, **19** (1970).

———, and A. C. WILLIAMS, "On the primal and dual constraint sets in geometric programming," *J. Math. Anal. Appl.*, **32** (1970), 684–688.

BEIGHTLER, C., and D. T. PHILLIPS, *Applied Geometric Programming* (Wiley, New York, 1976).

———, and D. T. PHILLIPS, "Geometric programming: a technical state-of-the-art survey," *AIIE Trans.*, **5**, 2 (June 1973), 97–112.

———, T. LO, and H. G. RYLANDER, "Optimal design by geometric programming," *ASME Trans.* (February 1970). Also published as ASME Paper 69-WA-DE-7.

BOUCHEY, G. D., C. S. BEIGHTLER, and B. V. KOEN, "Optimization of nuclear systems by geometric programming," *Nucl. Sci. Eng.*, **44** (1971), 267–272.

CHARNES, A., and W. W. COOPER, "Optimizing engineering designs under inequality constraints," *Northwestern Univ. ONR Research Memo 64*, August 1962.

DINKEL, J. J., and G. A. KOCHENBERGER, "A note on substitution effects in geometric programming," *Man. Sci.*, **20**, 7 (March 1974), 1141–1143.

DUFFIN, R. J., "Dual programs and minimum cost," *J. Soc. Ind. Appl. Math.*, **10** 1 (March 1962), 119–123.

———, "Cost minimization problems treated by geometric means," *Operations Res.*, **10**, 5 (September 1962), 668–675.

———, and E. L. PETERSON, "Geometric programming with signomials," *J. Opt. Theory Appl.*, **11** (1973).

DUFFIN, R. J., E. L. PETERSON, and C. ZENER, *Geometric Programming* (Wiley, New York, 1967).

ECKER, J. G., and J. R. MCNAMARA, "Geometric programming and the preliminary design of industrial waste treatment plants," *Water Resources Res.*, **7** (1971), 18–22.

FOLKERS, J. S., "Preliminary fleet design by geometric programming," *Int. Shipbuilding Progr.*, **16** (1969), 308–326.

HELLINCKX, L. J., and M. J. RIJCKAERT, "Optimal capacities of production facilities; an application of geometric programming," *Can. J. Chem. Eng.*, **50** (1972), 148–150.

KVALIE, D., J. KOWALIK, and J. MOE, "Structural optimization by means of nonlinear programming," Technical Report, Norwegian Technical Institute, Trondheim, Norway, 1966.

PASSY, URY, "Modular design: an application of structured geometric programming," *Operations Res.*, **18** (1970), 441–453.

———, "Generalized weighted mean programming," *SIAM J. Appl. Math.*, **19**, 4 (1971).

———, "Condensing generalized polynomials," *J. Opt. Theory Appl.*, **9**, 4 (1972).

———, and D. J. WILDE, "Generalized polynomial optimization," *SIAM Appl. Math.*, **15**, 5 (September 1967), 1344–1356.

PETERSON, E. L., "Geometric programming and some of its extensions," in *Optimization and Design*, M. Avriel, M. J. Rijckaert, and D. J. Wilde, eds. (Prentice-Hall, Englewood Cliffs, N.J., 1973).

———, and C. ZENER, "The rectangularity law of transformers," *Proc. Natl. Acad. Sci.* (June 1964).

PETROPOULOS, P. G., "Optimal selection of machining rate variables by geometric programming," *Int. J. Prod. Res.*, **11**, 4 (1973), 305–314.

PHILLIPS, D., and C. BEIGHTLER, "Optimization in tool engineering using geometric programming," *AIIE Trans.* (December 1970).

REKLAITIS, G. V., and D. J. WILDE, "Geometric programming via a primal auxiliary problem," *AIIE Trans.*, **6**, 4 (December 1974).

RIJCKAERT, M. J., in "Engineering applications of geometric programming," Part II, *Optimization and Design*, M. Avriel, M. J. Rijckaert, and D. J. Wilde, eds. (Prentice-Hall, Englewood Cliffs, N.J., 1974).

———, and X. M. MARTENS, "Analysis and optimization of the Williams–Otto process by geometric programming," *Amer. Inst. Chem. Engrs. J.*, **20**, 4 (1974).

THEIL, H., "Substitution effects in geometric programming," *Man. Sci.*, **19**, 1 (1972), 25–30.

ZENER, C., "A mathematical aid in optimizing engineering designs," *Proc. Natl. Acad. Sci.*, **47**, 4 (April 1961), 537–539.

———, and R. J. DUFFIN, "Optimization of engineering problems," *Westinghouse Engr.* (September 1964), 154–160.

EXERCISES

6-1. A vertical cylindrical tank is to be designed to have a total volume of 100 ft³. Find the optimal values of the diameter and the height to minimize the total surface area.

6-2.

$$\text{Minimize } y_o \equiv 1000x_1 + 4 \times 10^9 x_1^{-1} x_2^{-1}$$

subject to

$$2.5 \times 10^5 x_2 + 9000 x_1^{-1} x_2^{-1} \leq 1$$

$$x_1, \quad x_2 > 0$$

6-3. The simplest form of the economic lot size problem may be stated as follows: For a given product, the manufacturer must decide what size lots he will put into stock periodically. The total variable cost associated with the manufacture and storage of this product is given by

$$y = \frac{Q}{2}I + \frac{R}{Q}S$$

where Q = lot size (pieces per run)
I = carrying cost per piece per year
R = annual requirements
S = setup cost (dollars per run)

The first term in this objective function represents the total carrying costs, and the second term gives the total setup costs. The function to be minimized is then of the form

$$y = c_1 Q + c_2 Q^{-1}$$

(a) Find the values of the dual variables *by inspection*, and hence show that the optimum lot size should always be chosen to make the total carrying and setup costs equal. From this result find Q^* by inspection.
(b) A modification of the problem above may be made to include quantity discounts when the product is being purchased rather than produced. Accordingly, Q would be the number of pieces per order, and S the cost of processing an order. Suppose the vendor charges $P + kQ$ dollars for an order of Q pieces, where P and k are given constants. Then an additional term, $RQ^{-1}(P + kQ)$, must be added to the objective function given in part (a). Show that this new function has the same form as the previous one had, and hence find Q^* by inspection.

6-4. Consider the problem of finding positive x_n which minimize

$$y = \sum_{t=1}^{T} c_t \prod_{n=1}^{N} x_n^{a_{tn}}$$

where the exponents a_{tn} satisfy

$$\sum_{t=1}^{T} a_{tn} = 0; \qquad n = 1, \ldots, N$$

and

$$T = N + 1$$

(a) Show that the optimal value of each dual variable is $1/T$.
(b) For the special case $c_t = 1$, for $t = 1, \ldots, T$, show that $x_n^* = 1$ for $n = 1, \ldots, N$.

6-5. (Avriel and Wilde) The following problem arises in the design of horizontal vapor condensers, after some simplifying assumptions. Minimize the variable cost

$$c = \frac{\beta_1}{N^{7/6}DL^{4/3}} + \frac{\beta_2 D^{0.8}}{N^{0.2}L} + \beta_3 NDL + \frac{\beta_4 L}{D^{4.8}N^{1.8}}$$

where c = variable cost (dollars per year)
N = number of tubes
D = average tube diameter (inches)
L = tube length (feet)

and β_1, β_2, β_3, and β_4 are coefficients that vary with the fluids involved and construction costs. The first two terms represent the cost of thermal energy; the third, the fixed charges on the heat exchanger; and the fourth, the cost of pumping the cold liquid through the tubes.

(a) Verify that the optimal cost distribution is, in all circumstances, as follows: 43.3% thermal energy, 53.3% fixed charges, and 3.33% pumping cost. (This case is oversimplified; see the article for a more realistic model.)

(b) Find the minimum cost, in dollars per year, given that, for a certain seawater desalination plant using low pressure steam for heating, $\beta_1 = 1.724 \times 10^5$, $\beta_2 = 9.779 \times 10^4$, $\beta_3 = 1.57$, and $\beta_4 = 3.82 \times 10^{-2}$.

6-6. Consider the (highly simplified) problem of designing an air compressor system for compressing air from a pressure of P_o up to a value of P_N. The annual operating costs are given by

$$C_1 \sum_{j=1}^{N} \left(\frac{P_j}{P_{j-1}}\right)^{\alpha} - C_1 N$$

where C_1 and $0 < \alpha < 1$ are given constants, and N is the number of compressors. The cost of the jth compressor is a function of the operating pressure ratio and is given by

$$C_2\left(\frac{P_j}{P_{j-1}}\right)^{\beta}$$

where C_2 and $0 < \beta < 1$ are given constants.

Use geometric programming to find the optimal number of compressors to use, as well as their optimal pressure ratios.

6-7. An open cylindrical vessel is to be constructed to transport 10 ft^3 of liquid between two locations in a processing plant. The sides of the vessel can be made from material costing \$1 per ft^2, but the bottom must be made from a special alloy costing \$10 per ft^2. The cost per round trip is \$0.10 and the container will have no salvage value on completion of the operation.

(a) What are the optimal dimensions of the vessel to minimize the total cost of moving the liquid?

(b) If the round-trip cost were to double, by what factor would the total cost increase, provided that the vessel could be redesigned?

6-8.

$$\text{Minimize } y_o \equiv -5x_1^2 + x_2^2 x_3^4$$

subject to

$$\tfrac{5}{2}x_1^2 x_2^{-2} - \tfrac{3}{2}x_2^{-1}x_3 \leq -1$$

$$x_1, \quad x_2 > 0$$

6-9. Solve the following nonposynomial problem by making an appropriate change of variable: Find real numbers x_1, x_2, x_3, that minimize

$$y = (x_1 + x_2 + x_3)^{-6} e^{2x_1 - 2x_2^2 - 3x_2 + 3x_3}$$

and which satisfy

$$(x_1 + x_2 + x_3)^4 e^{-x_1 + x_2^2 + 2x_2 - 2x_3} + (x_1 + x_2 + x_3)^{-2} e^{-2x_1 + 2x_2^2} \leq 2$$

$$(x_1 + x_2 + x_3)^4 e^{x_1 - x_2^2 + x_2 - x_3} \leq 1$$

6-10.

$$\text{Minimize } Y(\mathbf{x}) = 4x_1^3 x_2^{-1} + 3e^{x_1^{-1} x_2^{-1}} + x_2$$

where x^1 and x^2 are nonnegative numbers.
(*Ans.* $x_1^* = 0.791$, $x_2^* = 2.818$.)

Optimization of Multistage Systems 7

In rivers the water that you touch is the last of what has passed and the first of that which comes: so with time present.

LEONARDO DA VINCI (1452–1519)

Like the waters of the river, decisions are often connected to each other. What we decide tomorrow and in the future depends upon choices we make today, and vice versa. Any planning problem, be it corporate expansion program, manufacturing plant layout, or family vacation trip, involves many decisions, each affecting the other. In guiding their lives, people continually balance immediate pleasures against future advantages. This chapter shows how sequences of interacting decisions can be optimized when timing is important.

Problems of this nature can be attacked by methods developed in preceding chapters, as long as there are not too many decisions. But since the computing effort grows as the square, cube, or higher power of the number of variables, the procedures already described have ceilings on the size of problem they can solve. Large systems must be decomposed into components small enough for optimization by previous methods, taking proper account of how the different parts interact.

In planning problems, the components are arranged in series, each representing the same technological complex at a different point in time. This serial structure, in which decisions at any stage affect the condition of all later ones, occurs also in processes having manufacturing units in series. Serial systems lend themselves to decomposition by the *conditional* or *partial* optimization schemes of this chapter. The earliest and best-known example of the conditional optimization approach to serial problems is Bellman's "dynamic programming," the first method to be described here. Once understood for serial systems, conditional optimization can be extended to branching or even cyclic structures by the methods of Nemhauser, Meier, Beightler, Phillips, and Wilde. Such complex arrangements are often analyzed by condensing them into serial subsystems, although excessive interconnection may render decomposition impractical.

The distinction between state and decision variables, somewhat arbitrary in the methods described previously, is more clearly marked in conditional optimization problems. A *state variable* is one which, being both an input to one component and an output from another, transmits information between stages. Only variables which can be manipulated directly qualify as decisions. Conditional optimization of a stage involves finding the optimal values of its decisions for every possible value of its input state variables. Although decisions can be guided efficiently by any appropriate optimization method, state variables must be searched exhaustively to avoid overlooking an optimal possibility. Hence the distinction between states and decisions is very important in structured decision problems.

In this world of rapidly increasing complexity, it is hard to improve one thing without affecting others. The study of conditional optimization gives insight into the wise selection of interacting alternatives for the good of the whole.

7-01 SERIAL SYSTEMS: THE INITIAL-VALUE PROBLEM

A *serial multistage system* involves a series of decisions, each affecting the circumstances under which the next in the sequence must be made. In such problems, the output from one stage becomes the input to the next and thus affects all those following. This type of system and the technique for solving it are most clearly illustrated by the sort of network problem (Beckwith) shown in Fig. 7-1. The numbers on the links give the cost of shipping one unit of a given commodity between the indicated nodes (cities). For simplicity of discussion, the nodes have been divided horizontally by dashed lines into northern (N), central (C), and southern (S) cities, and vertically by dashed lines into five east–west zones, I–V; the boundaries of these zones represent the *stages* of this problem.

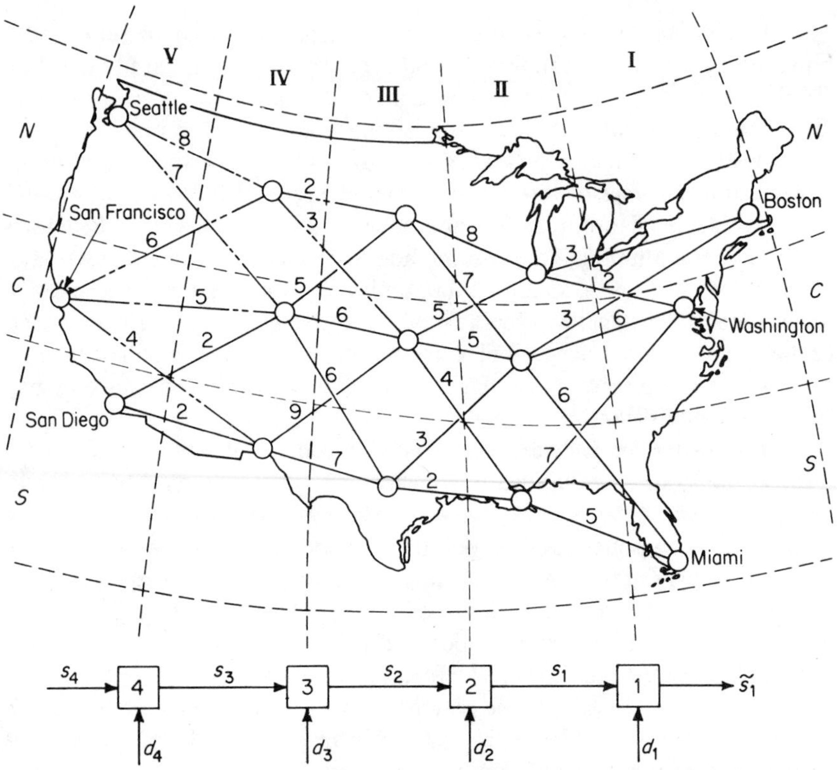

Figure 7-1. Network for first serial example problem.

An example of an *initial-value serial optimization problem* is the following. Starting in the southern city at stage 4 (San Diego), find the least-cost shipping path to any east coast (stage 1) city. Thus a path would consist of four links, where motion is restricted to easterly (NE, E, or SE) directions. This is called an *initial-value problem* because the initial condition (San Diego) is fixed, but the final condition is not. We merely wish to ship to the east coast by means of the least expensive route, whether this minimum-cost path terminates in Boston, Washington, or Miami. The restriction that the process must originate in San Diego effectively deletes some links from the system; these infeasible links are shown as double-dashed lines in Fig. 7-1.

In this problem there are just three possible inputs or outputs at each state: north (N), central (C), or south (S). The problem is solved recursively beginning with the last stage (1), computing the optimal (minimum) cost for each possible entering condition, s_1. The decisions are concerned with selecting the appropriate link leading out of the stage. The decision variables can take on at most one of three discrete values, L (left), F (forward), or R (right), depending upon whether the decision is to move northeasterly,

easterly, or southeasterly, respectively. For example, if $s_1 = C$ (central; that is, the shipment is in Cairo, Illinois), the decisions $d_1 = L, F$, and R would be associated with one-stage costs of 3, 6, and 6, respectively. Nodes representing northern and southern cities are restricted to only two possible decision choices; thus, for $s_1 = N$(Chicago), d_1 can take on only the values F and R. The solution procedure begins with the nodes in zone II, for each of which we find the least-cost path to an east-coast node. Then, the problem is solved by moving iteratively stage by stage from the east coast to the west coast. Since at this point we do not know through which nodes the optimal path will pass, we must find the best decision to make at each one.

The optimal one-stage decisions are readily seen to be $d_1^*(N) = R$, $d_1^*(C) = L$, and $d_1^*(S) = F$, producing the corresponding optimal returns (costs) of 2, 3, and 5. Next, we compute, for each node at stage 2 (zone III), the cost to a stage 1 node plus the minimum cost from that stage 1 node to the east coast. For example, if $s_2 = C$, the decision $d_2 = L$ sends the shipment to Chicago at a cost of 5 plus the 2 incurred in getting the shipment from Chicago to the east coast by way of the least-cost path. Thus, this decision results in a total cost of 7, whereas the decision $d_2 = F$ produces a total two-stage cost of $5 + 3 = 8$, and the decision $d_2 = R$ yields a cost of $4 + 5 = 9$. Hence, for $s_2 = C$, the optimal decision is $d_2 = L$. The same analysis is carried out at the remaining two stages (3 and 4). For each node at each stage we compute, for each possible decision, the sum of the link cost to the next adjoining stage plus the minimum total cost from that node to the east coast. The decision that produces the smallest such cost is then the optimal decision to make at that node, and it is recorded.

In working network problems, it is convenient to write the values of the optimal cumulative costs and decisions on the network nodes, as shown in Fig. 7-2. From this figure we see, for example, that the minimum shipment cost from Denver to the east coast is 12, and the optimal decision at that point is $d_3^*(C) = R$. The figure also shows the optimal solution to the initial value problem $s_4 = S$: the least-cost shipping path from San Diego to the east coast has a total cost of 14, and the path may be traced out by following the optimal decision at each node. The optimal set of decisions is thus L, R, L, L, as shown by the arrows in Fig. 7-2. Also shown on this figure are the solutions to the other two initial-value problems, $s_4 = C$ and $s_4 = N$, and their associated costs of 16 and 18, respectively.

The reader can verify how efficient and rapid is this computational procedure by solving the network problem given in Fig. 7-3 and checking the answers printed in the nodes. Further examples are provided in the exercises at the end of the chapter.

We may now generalize both the solution technique just employed and the type of problem for which it is applicable. The most general form of a serial multistage decision problem is shown in Fig. 7-4, in which the stages

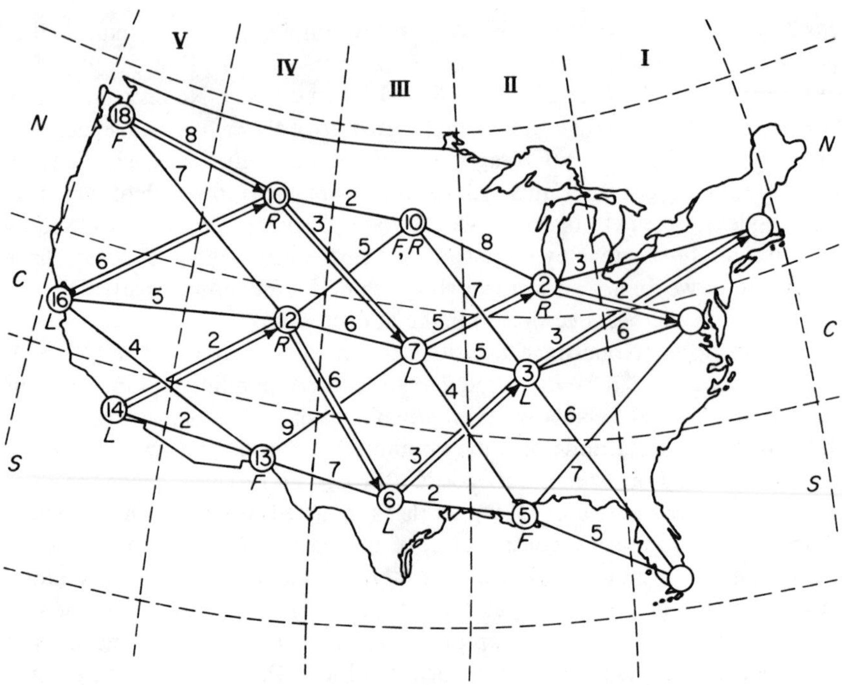

Figure 7-2. Solution to network problem of Fig. 7-1.

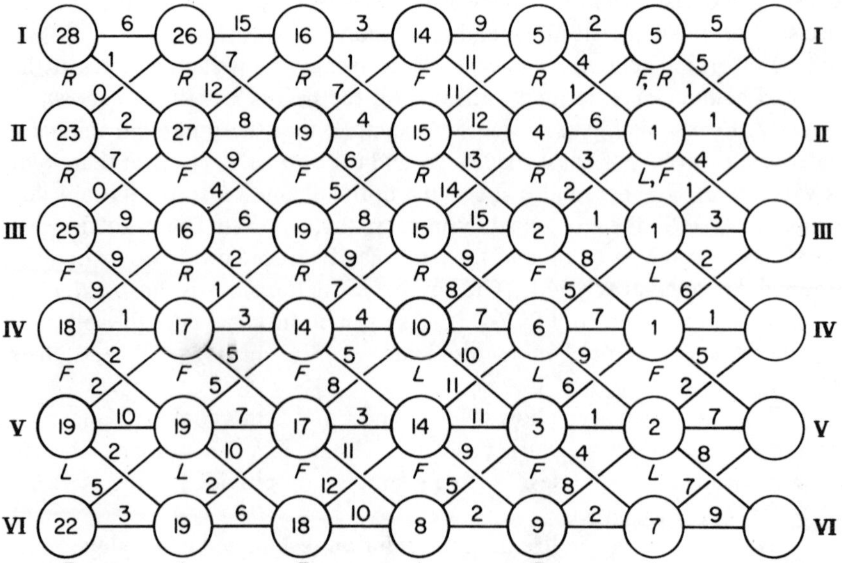

Figure 7-3. Network for second serial problem.

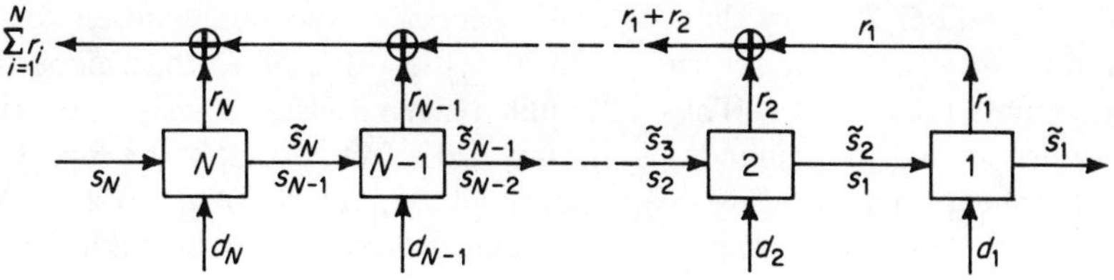

Figure 7-4. Serial system.

are represented schematically by appropriately numbered rectangles, with arrows used to indicate inputs and outputs to the various stages. Associated with the ith stage is an output r_i, called the *stage return*, which is a function of the inputs to the stage. All outputs which are not returns are called *state variables*, written $\tilde{s}_i$, where i is the index of the inputs generating the state, and the tilde ($\sim$) identifies this as an output. The transformation having the state $\tilde{s}_i$ as its output is given the name *transition function* and is written T_i. Many states act also as inputs; a state serving as an input to stage i is written s_i. All inputs which are not states are called *decision variables* and are written d_i, the index i identifying the return and transition having d_i as an input. Thus we may write

$$r_i = R_i(d_i, s_i) \tag{7-1:i}$$

$$\tilde{s}_i = T_i(d_i, s_i) \tag{7-2:i}$$

The set of all transformations carrying the same index i is called the ith *stage*, and the set of all stages is called the *system*.

Since a state variable is usually the output of some stage (say, i) and the input to at least one other (say, j), it can be represented by more than one symbol—either $\tilde{s}_i$ or s_j, for example. The identification of the several symbols with each other ($\tilde{s}_i \equiv s_j$) defines precisely how the stages i and j interact, and such a relation will be called an *incidence identity*. A system is specified completely by its stages and incidence identities. For the serial system of Fig. 7-3, the incidence identities are

$$\tilde{s}_{i+1} \equiv s_i; \qquad i = 1, 2, \ldots, N-1 \tag{7-3:i}$$

This means that the stages are numbered in the direction opposite to the flow of information shown by the arrows in Fig. 7-4. Although this backward numbering is confusing at first, it leads to simpler notation for the proofs and for the explanation of the solution procedure.

In the *initial-value problem*, the value of the initial state s_N, the total system input, is given, and the decision problem is to find the optimal sequence $d_1^*(s_N), \ldots, d_N^*(s_N)$ maximizing the *total return* R, defined by

$$R = \sum_{i=1}^{N} R_i(d_i, s_i) \tag{7-4}$$

Eqs. (7-1) and (7-2) show that for a given input s_i, the value chosen for d_i determines not only the return r_i, but also the value of T_i that maps the input s_i onto the output $\tilde{s}_i$. Thus, although a given decision d_i may maximize the return at that particular stage, it may also affect adversely the inputs to all subsequent stages, leading to a nonoptimal total return for the system. An optimal sequence, or *policy*, $\{d_i^*(s_N)\}$ can be found only by taking into account the transitions coupling the stages together.

Equations (7-1)–(7-4) can be combined to express the total return as a function of the various inputs

$$R = \sum_{i=1}^{N} R_i(T_{i+1}(T_{i+2} \cdots (T_N(s_N, d_N), d_{N-1})) \cdots d_i) \tag{7-5}$$

so that for a given value of the initial state s_N, R is a function only of the decisions d_i:

$$R \equiv R(d_i, s_N) \tag{7-6}$$

The *initial-value maximum return function* is $R^*(s_N)$, defined as the function which, for *all* values of $d_1, \ldots, d_N$ and for any *particular* value of s_N is such that

$$R^*(s_N) = \max_{d_i}[R(d_i, s_N)] \tag{7-7}$$

The functions $d_i^*(s_N)$ form an *initial-value optimal policy*, and since there are N functions of the single state variable s_N to be specified by the policy, the problem is said to be an *N-decision, one-state optimization problem*. The number of decisions plus the number of states equals the number of degrees of freedom.

Strictly speaking, s_N, not being the output of any stage, should be called a decision rather than a state. If one can choose freely among the many possible values of s_N, then s_N will be called a *choice* variable and written c_N instead of s_N. Although there is no difference between a choice variable and a decision variable as far as actual computations are concerned, the distinction in notation will be preserved for analytical reasons to be made apparent later.

Multiple decision or state variables may be distinguished from each other by adding a second subscript (d_{i1}, d_{i2}, etc.). A stage can have several transition functions (Beightler and Mitten), given double subscripts where necessary to avoid confusion. Each stage has only one return function, and the returns are always scalar. Every decision and state is, for simplicity, treated as a single variable here, although under circumstances discussed later the relations derived hold for vector variables as well.

7-02 DECOMPOSITION BY DYNAMIC PROGRAMMING

The sequential structure of a serial system can be exploited to transform the N-decision, one-state initial-value optimization problem into a set of N one-decision, one-state problems. This is accomplished by the procedure called *dynamic programming*, due to Richard Bellman.

Let S_n be the sum of the returns from stages 1 through n.

$$S_n \equiv \sum_{i=1}^{n} R_i(d_i, s_i); \qquad n = 1, \ldots, N \tag{7-8: n}$$

$$S_n = S_n(d_1, \ldots, d_n, s_n); \qquad n = 1, \ldots, N \tag{7-9: n}$$

Let state s_n be treated as a parameter, and let $f_n(s_n)$ be such that

$$f_n(s_n) \geq S_n(d_1, \ldots, d_n, s_n) \quad \text{for all } d_1, \ldots, d_n; \qquad n = 1, \ldots, N \tag{7-10: n}$$

with equality achieved at each stage for at least one set of decisions. The function $f_n(s_n)$, which is the *n-stage maximum return*, can be made to depend on the input and decision for the *previous* stage $n + 1$ by substitution of the transition function (7-2: $n + 1$) for s_n:

$$f_n(s_n) = f_n(T_{n+1}(d_{n+1}, s_{n+1})) \equiv f_n(d_{n+1}, s_{n+1}); \qquad n = 1, \ldots, N-1 \tag{7-11: n}$$

where state s_{n+1} is regarded as a parameter. Addition of the return r_{n+1} from stage $n + 1$ to both sides of inequality (7-10: n) and subsequent application of definition (7-8: n) and Eq. (7-11: n) gives

$$R_{n+1}(d_{n+1}, s_{n+1}) + f_n(d_{n+1}, s_{n+1}) \geq S_{n+1}(d_1, \ldots, d_{n+1}, s_{n+1}) \quad \text{for all } d_1, \ldots, d_n; \qquad n = 1, \ldots, N-1 \tag{7-12: n}$$

Notice that this upper bound holds for the last n decisions, but not for d_{n+1}. Moreover, the upper bound depends only upon the independent decision variable d_{n+1} and the state parameter s_{n+1}. Thus it is only a one-decision, one-state optimization problem to find the function $U_{n+1}(s_{n+1})$, defined to be such that

$$U_{n+1}(s_{n+1}) \geq R_{n+1}(d_{n+1}, s_{n+1}) + f_n(d_{n+1}, s_{n+1}) \quad \text{for all } d_{n+1}; \qquad n = 1, \ldots, N-1 \tag{7-13; n}$$

with equality achieved at each stage for at least one set of decisions. Inequalities (7-12: n) and (7-13: n) together establish $U_{n+1}(s_{n+1})$ as a least upper bound on the $(n+1)$ stage return S_{n+1} for all the last $n+1$ decisions, *including* d_{n+1}.

$$U_{n+1}(s_{n+1}) \geq S_{n+1}(d_1, \ldots, d_{n+1}, s_{n+1}) \quad \text{for all } d_1, \ldots, d_{n+1}; \quad n = 1, \ldots, N-1 \tag{7-14: n}$$

Hence, by definition, (7-10: n), $U_{n+1}(s_{n+1})$ must be the $(n+1)$-stage maximum return function $f_{n+1}(s_{n+1})$, provided that it exists.

$$U_{n+1}(s_{n+1}) \equiv f_{n+1}(s_{n+1}); \qquad n = 1, \ldots, N-1 \tag{7-15: n}$$

Therefore, one can find f_{n+1} from f_n by determining the one-parameter decision function $d^*_{n+1}(s_{n+1})$, which is such that

$$f_{n+1}(s_{n+1}) = R_{n+1}(d^*_{n+1}(s_{n+1}), s_{n+1}) + f_n(d^*_{n+1}(s_{n+1}), s_{n+1}); \quad n = 1, \ldots, N-1 \tag{7-16: n}$$

for any value of s_{n+1}. Equation (7-16: n) may also be written as

$$f_{n+1}(s_{n+1}) = \max_{d_{n+1}}[R_{n+1}(d_{n+1}, s_{n+1}) + f_n(d_{n+1}, s_{n+1})]; \quad n = 1, \ldots, N-1 \tag{7-17: n}$$

Since the return $R_1(d_1, s_1)$ depends only on one decision variable, determination of $f_1(s_1)$ is a one-decision, one-state optimization problem. With $f_1(s_1)$ known, one can find $f_2(s_2)$ and $d^*_2(s_2)$ by another one-decision, one-state optimization involving Eq. (7-16: 1). After N such optimizations, one obtains the N decision functions $d^*_1(s_1), \ldots, d^*_N(s_N)$ and the N-stage maximum return $f_N(s_N)$.

Equation (7-17) is a mathematical statement of Bellman's *Principle of Optimality* for serial multistage systems. This principle states that the optimal policy, $d^*_N(s_N), \ldots, d^*_1(s_1)$, for an N-stage system must be such that the subset of decision functions $d^*_n(s_n), \ldots, d^*_1(s_1)$ $(n = 1, \ldots, N)$ is optimal for the last n stages of the N-stage system, for *any* input s_n. Starting with $d^*_N(s_N)$, the recursive substitution of $d^*_n(s_n)$ into transition Eq. (7-2) to beget the optimal input function $s_{n-1}(s_n)$, which is, in turn, put into $d^*_{n-1}(s_{n-1})$ to give $d^*_{n-1}(s_n)$, eventually generates the entire initial value optimal policy.

In practice one rarely needs to know $R^*(s_N)$ for every possible value of s_N. Usually either s_N is specified to be a particular constant value k_N or else it can be selected at will and is therefore really a choice variable c_N amenable to search techniques. In the former case one simply determines the constant $R^*(k_N)$ and the corresponding optimal policy. The latter situation, formulated in the previous section, can be solved by dynamic programming with one modification. Since stage N has a choice c_N and a decision d_N as inputs, the optimum return R^* is the result of a two-decision, no-state optimization.

$$R^* = \max_{d_N, c_N}[R_N(d_N, c_N) + f_{N-1}(d_N, c_N)] \tag{7-18}$$

7-03 CONDITIONAL OPTIMIZATION

Now define the *maximand*

$$M_n \equiv M_n(s_n, d_n) \equiv R_n(s_n, d_n) + f_{n-1}(T_n(s_n, d_n)); \quad n = 2, \ldots, N \tag{7-19: n}$$

The maximand can be regarded as an objective function in the two variables s_n and d_n. Contours of a typical maximand are shown in Fig. 7-5. Suppose that for each possible input state s_n, the decision $d_n^*(s_n)$ is found which maximizes (in general, *optimizes*) M_n. That is,

$$M_n(s_n, d_n^*(s_n)) \geq M_n(s_n, d_n) \tag{7-20: n}$$

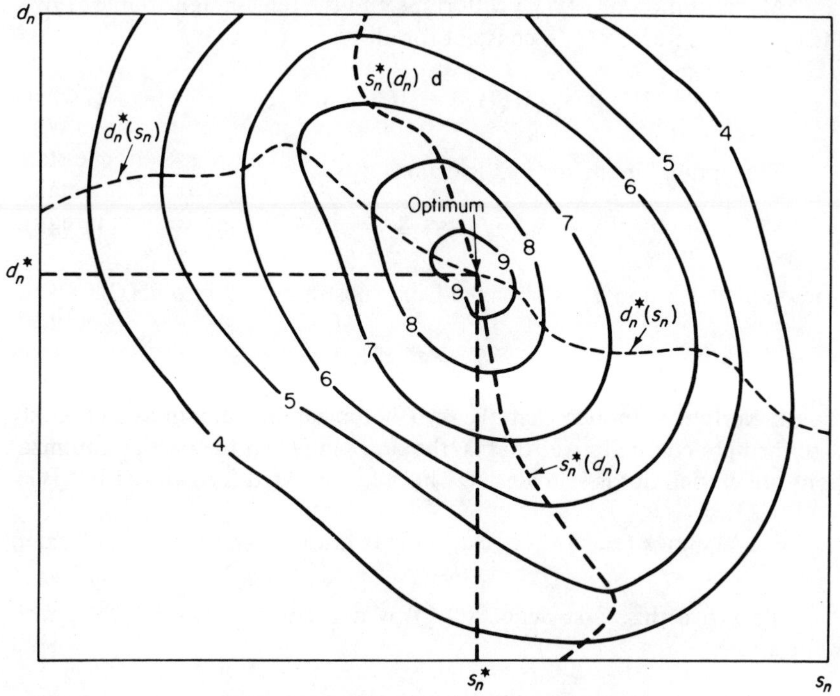

Figure 7-5. Graphic illustration of commutativity.

for every s_n. The left member of Eq. (7-20) depends only upon the input state s_n, and we symbolize this by defining

$$M_n^*(s_n) \equiv M_n(s_n, d_n^*(s_n)) \qquad (7\text{-}21\!: n)$$

We shall refer to the process of finding $d_n^*(s_n)$ and $M_n^*(s_n)$ as "conditional optimization with respect to d_n." It corresponds graphically to determining the locus of points where the maximand contour tangents are parallel to the d_n axis in Fig. 7-5. Notice that this conditional optimization produces a functional relationship, $d_n^*(s_n)$, between d_n and s_n which removes one degree of freedom, so that specification of only *one* variable (the input state s_n) completely determines the value of M_n. By analogy with the phrase "integrated out," we can say that d_n has been "optimized out" of the return function. Optimization with respect to d_n may be considered an operator mapping M_n from the two-dimensional set of all possible d_n and s_n onto the one-dimensional set of the allowable s_n. We lose no generality in assuming that we are dealing with a *maximization* problem; then we may write

$$\max_{d_n} [M_n(s_n, d_n)] \equiv M_n^*(s_n) \qquad (7\text{-}22\!: n)$$

Maximization of $M_n^*(s_n)$ with respect to the remaining variable s_n gives the optimum return, M_n^*, a constant for all s_n:

$$\max_{s_n} [M_n^*(s_n)] \equiv M_n^* \geq M_n^*(s_n) \qquad (7\text{-}23\!: n)$$

The optimal input state s_n^* is defined by

$$M_n^*(s_n^*) \equiv M_n^* \qquad (7\text{-}24\!: n)$$

The optimal decision d_n^* is obtained from substituting s_n^* into $d_n^*(s_n)$.

$$d_n^* \equiv d_n^*(s_n^*)$$

It is important to notice that the two optimization operations previously described are commutative; that is, the order in which the partial optimizations are carried out is immaterial. This may be stated formally as

$$\max_{d_n} \{\max_{s_n} [M_n(s_n, d_n)]\} = \max_{s_n} \{\max_{d_n} [M_n(s_n, d_n)]\} \qquad (7\text{-}25\!: n)$$

To prove this, first define $s_n^*(d_n)$ by a conditional optimization with respect to s_n:

$$M_n(s_n^*(d_n), d_n) \equiv M_n^*(d_n) \geq M_n(s_n, d_n) \qquad (7\text{-}26\!: n)$$

for every allowable decision d_n. Optimization of $M_n^*(d_n)$ with respect to d_n gives the optimal decision d_n^* such that

$$M_n^*(d_n^*) \geq M_n^*(d_n) \tag{7-27: n}$$

for all d_n.

Since by Eqs. (7-26) and (7-27),

$$M_n^*(d_n^*) \geq M_n^*(d_n) \geq M_n(s_n, d_n)$$

and by definition of Eq. (7-23),

$$M_n^* \geq M_n(s_n, d_n^*)$$

it follows that

$$M_n^*(d_n^*) = M_n^* = M_n^*(s_n^*)$$

which proves Eq. (7-25) and the commutativity property. This commutativity is demonstrated graphically in Fig. 7-5, which shows how both functions $s_n^*(d_n)$ and $d_n^*(s_n)$ intersect at the optimum value of the maximand.

In serial problems, one first optimizes stage 1 conditionally with respect to d_1, the maximand being the stage 1 return r_1. This gives $d_1^*(s_1)$ and $f_1(s_1)$, from which the maximand $M_2(s_2, d_2)$ can be constructed by adding the stage 2 return. A new conditional optimization determines $d_2^*(s_2)$ and $f_2(s_2)$, and the procedure is iterated until stage N has been reached.

7-04 SERIAL NETWORKS

In order to visualize clearly how much computational effort is saved by conditional optimization, consider a network of 100 nodes, arranged in 10 columns, with 10 nodes per column. Further, let each node in every column be joined by means of a link to each node in the nearest adjacent columns, for a total of 900 links. For a given initial state s_9, there are then 10^9 total paths (here a path consists of 9 links) to the final column (stage), and to find the shortest path, all 1 billion must be examined if a complete enumeration method is used. Now consider the solution by dynamic programming. Beginning with the last stage, we must examine a total of 10 links for each of the 10 possible input states s_1, or 100 links to obtain $f_1(s_1)$. For the two-stage returns, 100 values of $r_2(s_2, d_2) + f_1(T_1(s_2, d_2))$ must be calculated to find $f_2(s_2)$, 100 values of $r_3 + f_2$ will be needed to find $f_3(s_3)$, and so on.

Thus, to find the length of the shortest path, $f_9(s_9)$, only 810 partial paths need be evaluated, resulting in savings which, for hand computations, mark the difference between solving and not solving the problem. Numerous problems have appeared in applications that would defy solution by the largest electronic computers, were they to be solved by totai enumeration.

Network problems also provide an excellent transition to the more general serial optimization problem. The relation between the general initial-value problem and the network problem is best illustrated by writing the $f_i(s_i)$ and $d_i^*(s_i)$ in tabular form. This is, in fact, the way a digital computer would be programmed to solve any discrete variable serial optimization problem. For example, consider the tabular solution of the network given in Fig. 7-1. The input data to the computer (whether electronic or human) are the tables of returns and transitions for each of the four stages shown in Fig. 7-6. In general, a different transition table would be required for each stage, but in this simple problem the transitions are identical at each stage.

The computer first determines, for each row (state) in the r_1 table, the minimum return, $f_1(s_1)$, and the corresponding optimal decision, $d_1^*(s_1)$ and enters them in a tabular array as shown in Fig. 7-7(a) and (b). Next, a working

State s_1 \ Decision d_1	L	F	R
N	–	3	2
C	3	6	6
S	7	5	–

(a) Table of stage 1 returns, r_1

State s_2 \ Decision d_2	L	F	R
N	–	8	7
C	5	5	4
S	3	2	–

(b) Table of stage 2 returns, r_2

State s_3 \ Decision d_3	L	F	R
N	–	2	3
C	5	6	6
S	9	7	–

(c) Table of stage 3 returns, r_3

State s_4 \ Decision d_4	L	F	R
N	–	8	7
C	6	5	4
S	2	2	–

(d) Table of stage 4 returns, r_4

State s_i \ Decision d_i	L	F	R
N	–	N	C
C	N	C	S
S	C	S	–

(e) Transition table for output $\tilde{s}_i$ $(i = 1, \ldots, 4)$

Figure 7-6. Transition and stage return tables for network problem of Fig. 7-1.

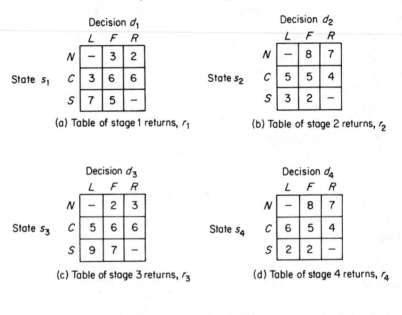

(a) $f_1(s_1)$

s_1	$f_1(s_1)$
N	2
C	3
S	5

(b) $d_1^*(s_1)$

s_1	$d_1^*(s_1)$
N	R
C	L
S	F

(c) Table of $r_2 + f_1$

s_2 \ d_2	L	F	R
N	–	10	10
C	7	8	9
S	6	7	–

(d) $f_2(s_2)$

s_2	$f_2(s_2)$
N	10
C	7
S	6

(e) $d_2^*(s_2)$

s_2	$d_2^*(s_2)$
N	F, R
C	L
S	L

(f) Table of $r_3 + f_2$

s_3 \ d_3	L	F	R
N	–	12	10
C	15	13	12
S	16	13	–

(g) $f_3(s_3)$

s_3	$f_3(s_3)$
N	10
C	12
S	13

(h) $d_3^*(s_3)$

s_3	$d_3^*(s_3)$
N	R
C	R
S	F

(i) Table of $r_4 + f_3$

s_4 \ d_4	L	F	R
N	–	18	19
C	16	17	17
S	14	15	–

(j) $f_4(s_4)$

s_4	$f_4(s_4)$
N	18
C	16
S	14

(k) $d_4^*(s_4)$

s_4	$d_4^*(s_4)$
N	F
C	L
S	L

Figure 7-7. Optimal stage returns and decisions for network problem of Fig. 7-1.

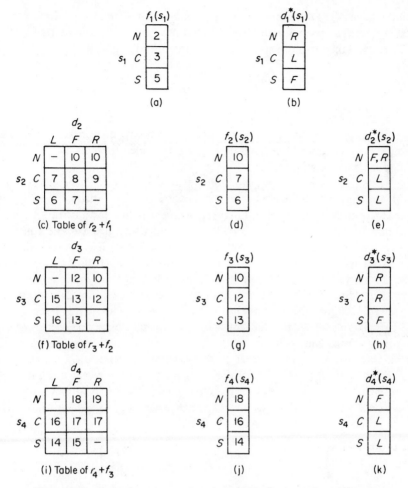

table of the two-stage returns, $r_2(s_2, d_2) + f_1(T_1(s_2, d_2))$ (abbreviated as $r_2 + f_1$), is prepared as shown in Fig. 7-7(c). The entries in this table are obtained using tables (b) and (e) of Fig. 7-6 and (a) of Fig. 7-7. For example, for $s_2 = C$, $d_2 = L$, we find the two-stage return to be the sum of 5 and 2, as obtained from the tables cited. The computer now finds $f_2(s_2)$ by locating the minimum entry in each row and records it, along with the optimal decision, $d_2^*(s_2)$, as shown in Fig. 7-7(d) and (e). This process is now continued, producing the remaining tables of Fig. 7-7. Owing to its repetitive nature, this procedure is well adapted to computer work, and the working tables may be destroyed as soon as they have been used, since only the $f_i(s_i)$ tables need be stored. [The $d_i^*(s_i)$ tables may be printed out or stored on tape for later use in tracing out the optimal path.] The reader may verify that the values in Fig. 7-7 are the same as those shown on the network of Fig. 7-1.

7-05 *DISCRETE VARIABLE PROBLEMS*

The analysis used in constructing the tables of Fig. 7-7 may now be employed to solve a more general class of problems. The analytic approach and computational procedure to be used on discrete variable problems are best illustrated by a few examples.

First, consider a highly simplified example of the replacement type of problem that occurs in almost every industrial organization. A factory is interested in developing an optimal replacement policy for a certain type of equipment which costs 12 units when new, and which has a resale value of $4 - t$, for $0 \leq t \leq 4$ (no resale value after 4 years) when it is t years old. The net profit from 1 year of operation of a t-year-old piece of equipment has been found to be $16 - t^2 (0 \leq t \leq 4)$. Assuming that this type of equipment will be discontinued after 4 years, it is desired to find the optimal replacement policy to maximize the total net profit if the present equipment is 2 years old.

In this problem, the decision variable at each stage (year) can take on only one of two values, *Keep* (K) or *Replace* (R). If n is the number of years remaining before discontinuance of the equipment, define $f_n(t)$ as the total net profit from n years of operation beginning with a t-year-old piece of equipment, when an optimal replacement policy is used. Assuming that the equipment will be sold at the end of the final year ($n = 0$), then for a one-stage process, we have

$$f_1(t) = \max \begin{cases} K: & 19 - t - t^2 \\ R: & 11 - t \end{cases}$$

The value associated with the decision to replace is the sum of the net profit from 1 year of operation with a new piece of equipment, 16, plus the resale value this year, $4 - t$, plus the resale value next year, 3, minus the cost of the new equipment, 12. By a similar analysis, we find

$$f_n(t) = \max \begin{cases} K: & 16 - t^2 + f_{n-1}(t + 1) \\ R: & 8 - t + f_{n-1}(1) \end{cases}; \qquad n = 2, 3, 4$$

Since the decisions in this problem are all limited to just two possible values, all optimal returns and associated decisions can be conveniently placed in one table, as shown in Fig. 7-8. The values in this table are computed by comparing the returns (profits) produced by the decisions K and R and retaining the larger. For example, $f_1(2)$ is the maximum of the values $19 - 2 - 4$ and $11 - 2$, associated with the decisions K and R, respectively, and

n \ t	1	2	3	4
1	17 (K)	13 (K)	8 [R]	7 [R]
2	28 (K)	23 [R]	22 [R]	21 [R]
3	38 (K)	34 (K) [R]	33 [R]	32 [R]
4	49 (K)	45 (K)	43 [R]	42 [R]

Figure 7-8. Optimal returns and decisions for the equipment replacement problem.

$f_4(2)$ is the maximum of $16 - 4 + 33$ and $8 - 2 + 38$, produced by the respective decisions, K and R. Beginning with a 2-year-old piece of equipment, the optimal policy for 4 years of operation is shown by the arrows in Fig. 7-8 as *KRKK*, for a total net profit of 45. This value can be checked by computing the individual returns for each of the four years: 12, 5, 15, and 12, plus the resale value of the 3-year-old piece of equipment at the end of the four years, 1, which sum to 45.

As a second example, consider the problem of loading a space capsule having a total capacity of W tons with scientific instruments in such a way as to maximize the total value of the payload. Note that this same type of problem occurs in numerous other contexts, such as the selection of items to go into survival kits, the selection of machinery types for factory departments, and, in general, the optimal choice of components under capacity (usually dollar) restrictions. For illustration we again select a much simplified example which, however, completely describes the computational procedure.

Four kinds of items are available for loading into the capsule, where the ith item has weight w_i (in tons, say) and value v_i, as shown in Fig. 7-9(a).

It is not so obvious what constitutes a stage in this problem, but this can be made clear by considering the solution procedure to be inductive as follows. Having found the optimal cargo selection when considering only the first $i - 1$ items and a weight restriction W_{i-1}, we can now use conditional optimization to find the optimal cargo for the first i items under the weight limitation W_i. Thus the stages are the individual items under consideration, whereas the states are the total remaining unused weight capacity.

Let $f_i(W_i)$ be the maximum total cargo value obtainable from the first i items when the weight restriction is W_i, and let n_i be the number of type i items selected. Then, if no more than one of each item is desired, or available, we have for $f_0 = 0$ and W_i a positive integer:

$$f_i(W_i) = \max_{n_i = 0, 1} [n_i v_i + f_{i-1}(W_i - n_i w_i)]; \qquad i = 1, 2, 3, 4$$

i	w_i	v_i
1	4	7
2	3	4
3	1	1
4	5	9

(a) Weights and values of the four components

$n_1 \backslash W_1$	0	1	2	3	4	5	6	7	8	9	10	11	12	13	14	15	16	17	18	19	20
0	0	0	0	0																	
1					7	7	7	7													
2									14	14	14	14	14	14	14	14	14	14	14	14	14

(b) Stage 1

$n_2 \backslash W_2$	0	1	2	3	4	5	6	7	8	9	10	11	12	13	14	15	16	17	18	19	20
0	0	0	0	0	7	7	7	7	14	14	14	14	14	14	14	14	14	14	14	14	14
1				4	4	4	4	11	11	11	11	18	18	18	18	18	18	18	18	18	18
2							8	8	8	8	15	15	15	15	22	22	22	22	22	22	22

(c) Stage 2

$n_3 \backslash W_3$	0	1	2	3	4	5	6	7	8	9	10	11	12	13	14	15	16	17	18	19	20
0	0	0	0	4	7	7	8	11	14	14	15	18	18	18	22	22	22	22	22	22	22
1		1	1	1	5	8	8	9	12	15	15	16	19	19	19	23	23	23	23	23	23
2			2	2	2	6	9	9	10	13	16	16	17	20	20	20	24	24	24	24	24

(d) Stage 3

$n_4 \backslash W_4$	0	1	2	3	4	5	6	7	8	9	10	11	12	13	14	15	16	17	18	19	20
0	0	1	2	4	7	8	9	11	14	15	16	18	19	20	22	23	24	24	24	24	24
1						9	10	11	13	16	17	18	20	23	24	25	27	28	29	31	32
2											18	19	20	22	25	26	27	29	32	33	34

(e) Stage 4

Figure 7-9. Input data and stage returns for the space capsule problem.

when $w_i \leq W_i$. [Clearly, for $w_i > W_i$, $f_i(W_i) = f_{i-1}(W_i)$.] If the n_i may be any nonnegative integer, the recurrence relation must be written as

$$f_i(W_i) = \max_{n_i=0,1,\ldots} [n_i v_i + f_{i-1}(W_i - n_i w_i)]$$

$$n_i \leq \left[\frac{W_i}{w_i}\right]$$

where the symbol $[x]$ indicates the greatest integer in x. For example, $[\frac{7}{3}] = 2$, $[\frac{1}{2}] = 0$, and $[4] = 4$. Assume that for the present problem, the decision variable n_i can take on only the values 0, 1, or 2, and consider the solution of the problem for values of the initial state, W, up to a maximum of 20.

Beginning with the last stage, that is, considering only item 1, if the remaining unused capacity after loading the other three item types is W_1, then

$$f_1(W_1) = \max_{\substack{n_1=0,1,2 \\ n_1 \leq [W_1/4]}} (7n_1)$$

$$= \begin{cases} 7\left[\dfrac{W_1}{4}\right], & W_1 \leq 8 \\ 14, & W_1 > 8 \end{cases}$$

where the restriction $W_1 \leq 8$ is derived from $[W_1/4] \leq 2$. From this equation, values of $f_1(W_1)$ can be computed as shown in Fig. 7-9(b). The optimal two-stage returns, involving items 1 and 2, are then found from the relation

$$f_2(W_2) = \max_{n_2} [4n_2 + f_1(W_2 - 3n_2)]$$

where $n_2 = 0, 1, \ldots, \min(2, [W_2/3])$ as shown by the *circled* values in Fig. 7-9(c), which contains all the two-stage returns, $r_2(W_2, n_2) + f_1(T_1(W_2, n_2))$. As an example of the procedure, consider the calculations for the two-stage input, $W_2 = 7$. For $n_2 = 0$, we find $r_2 = 0$, and $f_1(7) = 7$ [from (a)]; for $n_2 = 1$; $r_2 = 4, f_1(4) = 7$; for $n_2 = 2$: $r_2 = 8, f_1(1) = 0$. When $W_2 < 6$, the upper limit $[W_2/3] < 2$, and when $W_2 < 3$, this limit is zero, so that n_2 must be less than 2 when $W_2 < 6$, and must be zero when $W_2 < 3$, as common sense would dictate. Continuing in this manner, the optimal four-stage returns are found as shown by the circled values in Fig. 7-9(e), where, of course, $W_4 \equiv W$.

Thus, if the capsule has a weight capacity W of 13 tons, the optimal payload will have a total value of 23. From (e), we find $n_4^* = 1$, so that the state input to the remaining three stages is $W_3 = 8$, resulting in $n_3^* = 0$, which makes $W_2 = 8$, $n_2^* = 0$, and finally, $W_1 = 8$, $n_1^* = 2$. As a check, we compute $2 \cdot 7 + 0 \cdot 4 + 0 \cdot 1 + 1 \cdot 9 = 23$. If for some secondary consideration we should desire to have $n_4 = 2$, the total return will be 22, and by tracing through the remaining stages, we find that this is produced by the policy: $n_4 = 2$, $n_3 = 0$, $n_2 = 1$, $n_1 = 0$. Note that if the weight limitation is 16 tons, there are two alternative optimal policies that will produce the maximum return of 27: $n_4 = 2$, $n_3 = 2$, $n_2 = 0$, $n_1 = 1$, and $n_4 = 1$, $n_3 = 0$, $n_2 = 1$, $n_2 = 2$. Readers should convince themselves that the solution procedure used on this problem is independent of the ordering of the items by re-solving the problem after the items have been renumbered in some arbitrary way.

7-06 PROBLEMS IN CONTINUOUS VARIABLES

In many problems, the state and decision variables are not restricted to discrete values but are continuous within various ranges. In this case, Eqs. (7-1) and (7-2) are given as algebraic functions, rather than in tabular form. For example, consider the problem (L. G. Mitten) of maximizing $\sum_{i=1}^{4} r_i$, where

$$r_i = 0.5s_i - 0.2d_i, \qquad i = 1, \ldots, 4$$

and

$$\tilde{s}_i = 0.7s_i + 0.4d_i, \qquad i = 1, \ldots, 4$$

The initial value is given as $s_4 = 100$, and each decision may take on any value in the open interval $[0, s_i]$.

Then in this problem, Eq. (7-17) is simply

$$f_n(s_n) = \max_{0 \le d_n \le s_n} [0.5s_n - 0.2d_n + f_{n-1}(0.7s_n + 0.4d_n)]; \qquad n = 2, 3, 4$$

and

$$f_1(s_1) = \max_{0 \le d_1 \le s_1} (0.5s_1 - 0.2d_1) = 0.5s_1; \qquad d_1^* = 0$$

since the coefficient of d_1 is negative. The two-stage return is $0.5s_2 - 0.2d_2 + 0.5(0.7s_2 + 0.4d_2) = 0.85s_2$, which is completely independent of d_2. Hence, *any* feasible decision at stage 2 is also an optimal decision. By a similar analysis, the three-stage return is $1.095s_3 + 0.14d_3$; to maximize this, d_3 is chosen as large as possible: $d_3^* = s_3$, making $f_3(s_3) = 1.235s_3$. Continuing in this manner, we find that $d_4^* = s_4$, which produces an optimal four-stage return of $f_4(s_4) = 1.6585s_4$.

Notice that we have solved a larger problem than that given in the problem definition, for we have the optimal solution for *any* initial input s_4; thus the original problem has been embedded in a much larger problem. By tracing through the system from left to right, one finds the optimal decisions to be $d_4^* = s_4$, $d_3^* = 1.1s_4$, $0 \le d_2^* \le 1.21s_4$, and $d_1^* = 0$. The optimal four-stage return of 165.85, corresponding to the given input, $s_4 = 100$, may be verified by adding the individual returns produced by this policy.

Problems involving decisions to be made at specified time intervals usually exhibit a serial structure and are therefore amenable to the type of analysis associated with dynamic programming. The following is an example of an *allocation* type of problem. The particular formulation of the example given here is due to L. G. Mitten. Multistage allocation problems were first formulated as dynamic programming problems by Bellman in 1954.

In a certain (highly simplified) economy, the following rules apply: commodity K can be produced by either or both of two industries, A and B. Each dollar of capital invested in industry A at the beginning of the year yields, at the end of that year and each succeeding year, 2 tons of commodity K and \$0.50 capital; each dollar of capital invested in industry B at the beginning of a year yields, at the end of that year and each succeeding year, 1 ton of commodity K and \$0.90 capital. Given an initial capital of $\$s_5$, how should the available capital at the beginning of each year be allocated between the two industries so as to maximize the total amount of commodity K produced during the next 5 years?

For this program, let s_i be the amount of capital available at the beginning of the ith year, $\tilde{s}_i$ the amount of capital available at the end of the ith year, r_i the tons of commodity K produced during year i, and $0 \leq d_i \leq s_i$ the capital invested in industry A at the beginning of the ith year. Then a total of $s_i - d_i$ dollars will be invested in industry B at the start of year i, since it will always pay to invest all available funds. This problem differs somewhat from those considered previously because of the production of the commodity in years subsequent to the initiating decision. The method of solution is unchanged, however, and the accumulation feature may be treated as follows;

$$r_i = \sum_{j=i}^{5} [2d_j + 1 \cdot (s_j + d_j)] = \sum_{j=i}^{5} (s_j + d_j)$$

and the transitions are defined by

$$\tilde{s}_i = 0.5d_i + 0.9(s_i - d_i) + s_i = 1.9s_i - 0.4d_i; \qquad i \neq 5$$

since the total capital available at the end of the year is that available at the beginning of the year plus the amount of capital produced during the year. There is no accumulation of the initial capital available, however, so, for the first year,

$$\tilde{s}_5 = 0.9s_5 - 0.4d_5$$

The recurrence relations for the problem are now

$$f_n(s_n) = \max_{0 \leq d_n \leq s_n} \left[\sum_{i=n}^{5} (s_i + d_i) + f_{n-1}(\tilde{s}_n) \right]$$

where, as always, $f_0 \equiv 0$. Then, recursively, we find that

$$f_1(s_1) = \max_{0 \leq d_1 \leq s_1} \left[s_1 + d_1 + \sum_{i+2}^{5} (s_i + d_i) \right]$$

which has the value $2s_1 + \sum_{i=2}^{5} (s_i + d_i)$, for $d_1^* = s_1$, and

$$f_2(s_2) = \max_{0 \leq d_2 \leq s_2} \left[s_2 + d_2 + \sum_{i=3}^{5} (s_i + d_i) + 2(1.9s_2 - 0.4d_2) + \sum_{i=2}^{5} (s_i + d_i) \right]$$

$$= \max_{0 \leq d_2 \leq s_2} \left[5.8s_2 + 1.2d_2 + 2 \sum_{i=3}^{5} (s_i + d_i) \right]$$

which has the value $7s_2 + 2\sum_{i=3}^{5}(s_i + d_i)$ produced by the optimal decision $d_2^* = s_2$.

In this same manner, we then find that

$$f_3(s_3) = 16.5s_3 + 3\sum_{i=4}^{5}(d_i + s_i); \qquad d_3^* = s_3$$

and

$$f_4(s_4) = 35.35s_4 + 4(s_5 + d_5); \qquad d_4^* = 0$$

Then, since

$$f_5(s_5) = \max_{0 \le d_5 \le s_5} [5(s_5 + d_5) + 35.35(0.9s_5 - 0.4d_5)]$$

the optimal first-year decision is $d_5^* = 0$, so $f_5(s_5) = 36.815s_5$ tons, for any initial capital s_5.

This problem may be approached from another point of view by defining $\hat{r}_i$ as the total amount of commodity K produced *as a consequence* of the decision d_i (rather than the amount produced *during* year i). Then the recurrence relations become

$$f_n(s_n) = \max_{0 \le d_n \le s_n} [n(s_n + d_n) + f_{n-1}(1.9s_n + 0.4d_n)]; \qquad n \neq 5$$

and

$$f_5(s_5) = \max_{0 \le d_5 \le s_5} [5(s_5 + d_5) + f_4(0.9s_5 - 0.4d_5)]$$

As the reader can see, these relations require exactly the same computational work as those given before, the only difference being that the summation terms need not be carried along in the computations. Thus, with this approach we find that $f_1(s_1) = 2s_1$, $f_2(s_2) = 7s_2$, $f_3(s_3) = 16.5s_3$, $f_4(s_4) = 35.35s_4$, and $f_5(s_5) = 36.815s_5$, with the same optimal decisions as before.

Finally, consider an initial-value problem for which the stage returns are nonlinear. (All discrete variable problems are naturally nonlinear, but the computational details require somewhat different handling in the case of continuous variables.) Specifically, we wish to maximize

$$p = \sum_{i=1}^{3} r_i$$

where $r_i = 5d_i - id_i^2$, $\tilde{s}_i = s_i - 0.4d_i$, and where each decision variable is restricted to nonnegative values that cannot exceed the stage input: $0 \le d_i \le s_i$. This is the usual restriction found in allocation problems, since no more can be allocated than is available, and negative allocations have no meaning.

In this problem, unlike the linear problems previously considered, the optimal decision at a given stage is not the same function of the stage input for all inputs. Thus at the last stage, the optimal return must be written

$$f_1(s_1) = \begin{cases} 6.25; & s_1 \geq 2.5 \quad (d_1^* = 2.5) \\ 5s_1 - s_1^2; & s_1 \leq 2.5 \quad (d_1^* = s_1) \end{cases}$$

Then the two-stage return with input s_2 is

$$M_2 = \begin{cases} 5d_2 - 2d_2^2 + 6.25; & s_2 - 0.4d_2 \geq 2.5 \\ 3d_2 - 2.16d_2^2 + 5s_2 - s_2^2 + 0.8s_2d_2; & s_2 - 0.4d_2 \leq 2.5 \end{cases}$$

For the *upper* branch, we find $f_2(s_2) = 9.375$ when $s_2 \geq 3$, with the corresponding $d_2^* = 1.25$; when $s_2 < 3$, the two-stage return will be maximized by keeping d_2 at the largest possible value (consistent with $s_1 = 2.5$), so that $d_2^* = 2.5s_2 - 6.25$, producing an optimal return of $f_2(s_2) = 75s_2 - 12.5s_2^2 - 103.125$. The latter value of d_2^* becomes negative when the input falls below 2.5, so that the total restriction is $2.5 \leq s_2 \leq 3$.

For the *lower* branch, the optimal stage 2 decision is found by differentiation to be $d_2^* = 0.185s_2 + 0.695$, producing the return $f_2(s_2) = 5.556s_2 - 0.926s_2^2 + 1.042$, which is valid in the range $0.853 \leq s_2 \leq 3$ (since for s_2 below 0.853, $d_2^* = 0.185s_2 + 0.695$ would be greater than s_2). When s_2 is below 0.853, the two-stage return is again maximized by keeping d_2 at the largest possible value: $d_2^* = s_2$, for an optimal return of $f_2(s_2) = 8s_2 - 2.36s_2^2$. We have now generated two different optimal returns $f_2(s_2)$ for the range $2.5 \leq s_2 \leq 3$: $75s_2 - 12.5s_2^2 - 103.125$, and $5.556s_2 - 0.926s_2^2 + 1.042$, but since the former never exceeds the latter, it may be dropped as nonoptimal, so that in summary we have

$$f_2(s_2) = \begin{cases} 9.375; & 3 \leq s_2 \leq \infty \quad (d_2^* = 1.25) \\ 5.556s_2 - 0.926s_2^2 + 1.042; & 0.853 \leq s_2 \leq 3 \\ & \quad (d_2^* = 0.185s_2 + 0.695) \\ 8s_2 - 2.36s_2^2; & 0 \leq s_2 \leq 0.853 \quad (d_2^* = s_2) \end{cases}$$

The foregoing analyses and results are shown graphically in Fig. 7-10; note that d_2^* is piecewise continuous and linear in s_2.

The three-stage returns must now be computed separately for the three different outputs $3 \leq s_3 - 0.4d_4 \leq \infty$, $0.853 \leq s_3 - 0.4d_3 \leq 3$, and $0 \leq s_3 - 0.4d_3 \leq 0.853$, which form the input to the last two stages. The procedure is exactly like that previously given for $f_2(s_2)$, and again, when different expressions are generated for $f_3(s_3)$ over the same range of s_3, they must be compared and only the one producing the larger value retained. For example, in the range $0.853 \leq s_3 \leq 1.082$, one finds $f_3(s_3) = 44.454s_3 - 18.75s_3^2 - 19.16$, and $f_3(s_3) = 8.503s_3 - 2.096s_3^2 + 0.241$, generated by the decisions

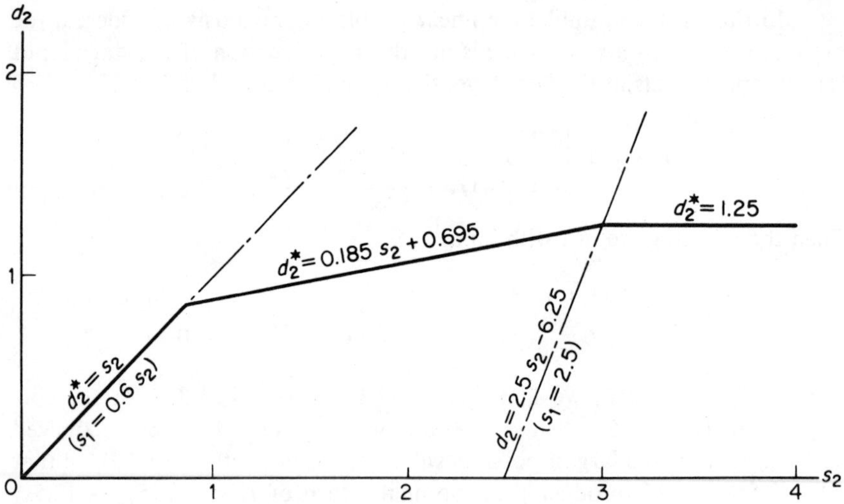

Figure 7-10. Optimal stage 2 decisions for nonlinear initial-value problem.

$d_3^* = 2.5s_3 - 2.13$ and $d_3^* = 0.28s_3 + 0.267$, respectively, where the former will be found to be nonoptimal and is, accordingly, dropped from the solution.

The complete solution to this three-stage problem is given below, and in Fig. 7-11 the optimal decisions and stage restraints are shown graphically.

$$f_3(s_3) = \begin{cases} 11.46; & 3.33 \le s_3 \le \infty \quad (d_3^* = 0.833) \\ 5.89s_3 - 0.883s_3^2 + 1.657; & 1.082 \le s_3 \le 3.33 \\ & \quad (d_3^* = 0.118s_3 + 0.442) \\ 8.503s_3 - 2.096s_3^2 + 0.241; & 0.371 \le s_3 \le 1.082 \\ & \quad (d_3^* = 0.28s_3 + 0.267) \\ 9.8s_3 - 3.849s_3^2; & 0 \le s_3 \le 0.371 \quad (d_3^* = s_3) \end{cases}$$

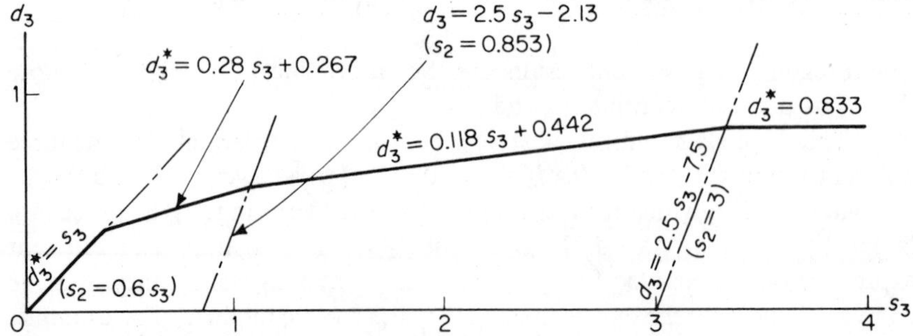

Figure 7-11. Optimal stage 3 decisions for nonlinear initial-value problem.

Notice that $f_3(s_3)$ is sectionally continuous and that it increases monotonically with s_3. That $f_3(s_3)$ is concave follows from the concavity of the transition functions and stage returns (Bellman).

The conditional optimizations of dynamic programming emphasize how different is the role of a decision variable from that of a state variable when the computations are done numerically. For every possible value of an input state, s_i, for instance, one finds the value of the decision variable d_i maximizing the function $R_i(d_i, s_i) + f_{i-1}(d_i, s_i)$. Since d_i is a system input, unaffected by any stage outputs, none of its values are important except the optimal ones, which may be found efficiently by direct-search methods. On the other hand, since s_i is an output from other parts of the system, one cannot know which of its values might be optimal until the entire problem has been solved. Hence an optimal decision $d_i^*(s_i)$ must be found for every value of s_i, which rules out using direct-search techniques on the state variable. Furthermore, $d_i^*(s_i)$ must be saved for all i and every value of s_i in order to generate the optimal policy. This storage requirement can exceed the rapid-access memory of large digital computers.

The states and decisions may in general be vectors (written in **boldface**) whose components are single variables. Except for obvious changes in notation the conditional optimization procedure is no different than before, although the number of computations certainly increases. But state dimensionality is more of a computational and storage burden than decision dimensionality.

It is important to recognize that the state variables in *any* dynamic programming problem need not be expressed in the same units from stage to stage; in fact, for a given stage, even the input and output state variables may be in different units. The same is true for the decision variables, which conceivably could differ at every stage. Nemhauser, in his dissertation, has given an excellent example of this in the form of a hypothetical plant design problem. In his example, the input to the system is pounds of raw material entering a mixer, the first manufacturing unit in the series. The decision variable at this stage is the type of mixer selected, while the output state variable is the mixing efficiency. Succeeding stages of the system consist of heaters, reactors, and separators, so that the state and decision variables are not in the same units at every stage. For example, at the first of the reactor stages, the input state is the temperature of the feed, the decision variable is the type of reactor and type of catalyst chosen, while the output state is the percent conversion achieved in the reactor.

In a dynamic programming problem, only the returns need be in the same units at each stage. In the Nemhauser problem, the measure of effectiveness upon which the best design is chosen is the maximum net profit over a 5-year period of operation. This net profit is the profit from sale of products minus the operating costs and cost of equipment. Hence all stage returns are in units of dollars. At each stage, for a given state input, the decision variable determines the contribution of that stage to the total net profit as well as its associated equipment and operating costs. The sum of all the stage returns is thus the total net profit to be maximized.

Nemhauser's design problem is unique among most published examples in showing the power of the dynamic programming solution method. It is well to keep in mind the ability of this method to handle stages which, although coupled together through the transition functions, may be expressed in totally different units.

7-07 FINAL-VALUE PROBLEM: STATE INVERSION

Consider the problem of obtaining the maximum return from a serial process as a function of $\tilde{s}_1$, called the *final state* because it is the only state not a stage input. Suppose there exist N *inverse transition functions* $\tilde{T}_i(d_i, \tilde{s}_i)$ which express the input s_i as a function of the decision and output.

$$s_i = \tilde{T}_i(d_i, \tilde{s}_i); \qquad i = 1, \ldots, N \tag{7-28: i}$$

Finding such a function for a stage is called *state inversion* because the roles of the input and output states are interchanged. The inverse transitions can be used to express the stage returns in terms of decision and *output* state.

$$\tilde{R}_i(d_i, \tilde{s}_i) = R_i(d_i, \tilde{T}_i(d_i, \tilde{s}_i)); \qquad i = 1, \ldots, N \tag{7-29: i}$$

A total return function $\tilde{R}(d_i, \tilde{s}_1)$ depending only on the decisions and the final state, $\tilde{s}_1$, can be obtained by adding the inverted stage returns and using the inverse transitions and incidence identities to eliminate states s_1 through s_N. The corresponding *final-value maximum return function*, given $\tilde{s}_1$, is

$$\tilde{R}^*(\tilde{s}_1) = \max_{d_1, \ldots, d_N} [\tilde{R}(d_i, \tilde{s}_1)] \tag{7-30}$$

The *final-value optimization problem* is to find a set of functions $d_1^*(\tilde{s}_1), \ldots, d_N^*(\tilde{s}_1)$ for which

$$\tilde{R}(d_1^*(\tilde{s}_1), \tilde{s}_1) = \tilde{R}^*(\tilde{s}_1) \tag{7-31}$$

The N functions $d_i^*(\tilde{s}_1)$ form a *final-value optimal policy*, and like the initial-value case, finding them is an N-decision, one-state problem. As before, there are $N + 1$ degrees of freedom.

When every stage has been inverted, the final-value problem can be put into the same notational form as the initial-value problem by renumbering the stages in reverse. Then the final-value problem can be decomposed into N one-decision, one-state problems by partial optimization in the manner already described.

7-08 STATE INVERSION: NETWORK ROUTING PROBLEMS

Consider the network shipping problem of Fig. 7-12, which is identical to that discussed in connection with Fig. 7-1 except that now it is the *final* state, which is fixed, and the *initial* state, which is free. In this problem, the given final condition is $\tilde{s}_1 \equiv C$; the shipment may originate in *any* west coast city, but it must terminate in Washington, D.C. As before, the solution procedure begins at the free end and proceeds recursively toward the fixed end, infeasible links being shown as double-dashed lines. Using state inversion, one begins at stage 4 and finds the optimal one-stage return (and optimal decision) for a given *output* $\tilde{s}_4$. Thus, if $\tilde{s}_4 = C$ (Denver), the optimal decision, d_4, is L (that is, *if* the shipment has arrived in Denver, we would prefer that it had been shipped *from* San Diego) and the associated return is 2. The problem is then solved just as were the initial-value problems, except that one now proceeds from left to right.

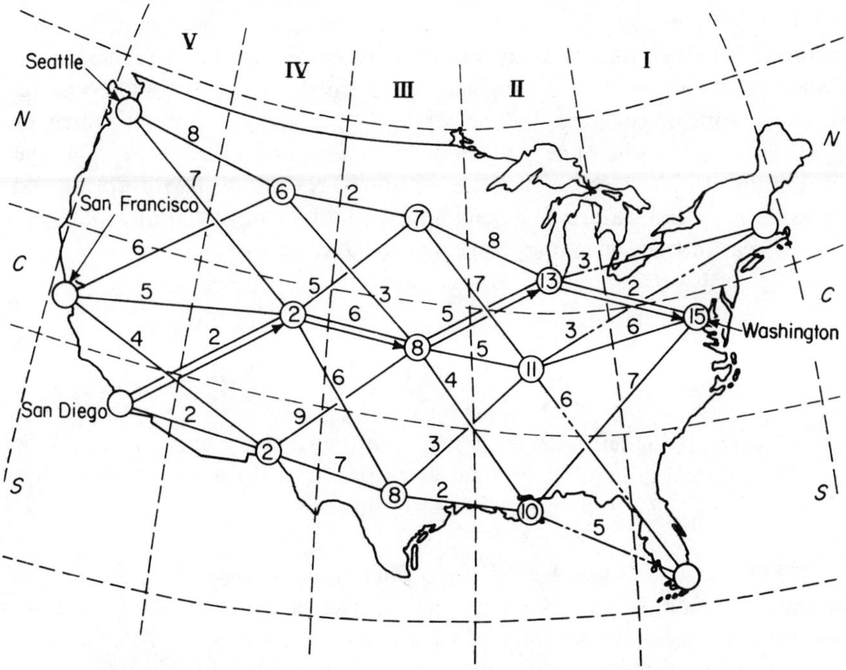

Figure 7-12. Solution of network routing problem by stage inversion.

The optimal return (least cost) is 15, and by tracing back through the path shown by the arrows, one finds that $s_4^* = S$; San Diego is the best city from which to start if Washington is to be the destination. Comparison of Figs. 7-1 and 7-12 illustrates the fundamental difference between initial- and final-value problems.

In an initial-value problem, the initial state s_N is fixed, and the final state is free to take on any value which the optimal stage decisions may produce as a consequence of optimizing the sum of the stage returns. For example, given the initial state $s_4 = C$ (San Francisco), Fig. 7-1 shows that the optimal return of 16 is achieved by selecting a path which terminates at $\tilde{s}_1 = C$ (Washington). In a final-value problem, the final state $\tilde{s}_1$ is fixed, and the initial state is free to take on whatever value results from making the optimal set of decisions. Here it is instructive to select as the final state the one which resulted when it was treated as a free end in the earlier initial-value problem: $\tilde{s}_1 \equiv C$. From Fig. 7-12 we find an optimal return of 15 produced by a path that originates not at San Francisco, but at San Diego. In other words, for a shipment that originates in San Francisco, the optimal route will terminate at Washington, whereas the optimal route that must terminate in Washington will originate in San Diego.

7-09 INITIAL-VALUE/FINAL-VALUE THEOREM

This property may now be set down formally as follows: For an initial-value problem, let the initial state be $s_N \equiv k_N$. Then the set of optimal decisions for this problem generates a final state, $\tilde{s}_1(k_N)$, and an optimal return of $f_N(k_N, \tilde{s}_1(k_N))$. For the same problem worked as a final-value problem, fix the final state at $\tilde{s}_1 \equiv \tilde{s}_1(k_N)$. Then the optimal solution to this problem will generate an initial state, $s_N(k_N)$, and an optimal return, $f_N(s_N(k_N), \tilde{s}_1(k_N))$.

Then for maximization problems, we have:

Theorem:

$$f_N(s_N(k_N), \tilde{s}_1(k_N)) \geq f_N(k_N, \tilde{s}_1(k_N))$$

Proof: If the left member is less than the right member, it cannot be an optimal policy for the final-value problem, since one could always select $s_N = k_N$ and obtain a larger return.

The converse is also true: If for a final-value problem $\tilde{s}_1 \equiv K$, and the optimal solution is $f_N(s_N(K), K)$, where $s_N(K)$ is the optimal initial state generated by the solution, then let the initial-value problem for $s_N \equiv s_N(K)$ generate $\tilde{s}_1(K)$ as the resulting final state, and $f_N(s_N(K), \tilde{s}_1(K))$ as the optimal return. Then, by the previous reasoning,

$$f_N(s_N(K), \tilde{s}_1(K)) \geq f_N(s_N(K), K)$$

7-10 STATE INVERSION: ALLOCATION PROBLEMS

As another example of a final-value problem solved by state inversion, let it be required to maximize the function $p = \sum_{i=1}^{4} r_i$, where $r_i = 0.5s_i - 0.2d_i$, $\tilde{s}_i = 0.7s_i + 0.4d_i$, $\tilde{s}_1 \equiv c_1$, and $0 \leq d_i \leq s_i$, $i = 1, \ldots, 4$. Then Eqs. (7-28) and (7-29) are $s_i = 1.43\tilde{s}_i - 0.572d_i$, and $r_i = 0.715\tilde{s}_i - 0.486d_i$; the restriction $0 \leq d_i \leq s_i$ becomes $0 \leq d_i \leq 0.91\tilde{s}_i$.

Just as in the case of the network problem solved earlier, the solution is begun at the initial stage (4), and the work is carried on iteratively to the right, ending with the final stage (1). Thus

$$f_1(\tilde{s}_4) = \max_{0 \leq d_4 \leq 0.91\tilde{s}_4} (0.715\tilde{s}_4 - 0.486d_4) = 0.715\tilde{s}_4$$

since $d_4^* = 0$. Then

$$\begin{aligned} f_2(\tilde{s}_3) &= \max_{0 \leq d_3 \leq 0.91\tilde{s}_3} [0.715\tilde{s}_3 - 0.486d_3 + 0.715(1.43\tilde{s}_3 - 0.572d_3)] \\ &= \max_{0 \leq d_3 \leq 0.91\tilde{s}_3} (1.736\tilde{s}_3 - 0.894d_3) = 1.736\tilde{s}_3 \end{aligned}$$

since $d_3^* = 0$. Continuing this process, one finds that $f_3(\tilde{s}_2) = 3.2\tilde{s}_2$ and $f_4(\tilde{s}_1) = 5.28\tilde{s}_1$, generated by the optimal decisions $d_2^* = 0$ and $d_1^* = 0$. Since $\tilde{s}_1 \equiv \tilde{c}_1$, we have $f_1(\tilde{c}_1) = 5.28\tilde{c}_1$ and, tracing back through the system, $s_1 = 1.43\tilde{c}_1$, $s_2 = 2.04\tilde{c}_1$, $s_3 = 2.92\tilde{c}_1$, and $s_4 = 4.17\tilde{c}_1$.

The interesting feature of this problem lies in the comparison of the optimal solution for the final-value problem, found previously, with that for the same problem treated earlier as an initial-value problem; namely, $d_1^* = 0$, $0 \leq d_2^* \leq s_2$, $d_3^* = s_3$, and $d_4^* = s_4$, for an optimal four-stage return of $f_4(s_4) = 1.6585s_4$. Selecting $d_2^* = 0$, we find that $\tilde{s}_1 = 0.5929s_4$, or $s_4 = 1.69\tilde{s}_1$, so $f_4(\tilde{s}_1) = 2.8\tilde{s}_1$, compared to the optimal return of $f_4(\tilde{s}_1) = 5.28\tilde{s}_1$ for the final-value problem. [The choice $d_2^* = s_2$ in the initial-value problem yields only $f_4(\tilde{s}_1) = 1.94\tilde{s}_1$.] This is an even more striking demonstration of the essential difference in character between initial- and final-value problems.

The contrast between these two types of multistage decision problems may be seen clearly in the economic context of the industry allocation problem described earlier. If one begins with a fixed amount of capital (the state variable in this problem), the optimal policy consists in first building up investment capital by investing in growth industry B in the early stages, and then in the later stages investing the accumulated capital in extractive industry A to produce the maximum total quantity of the commodity. If, however, the circumstances require that one must end with a fixed amount of capital, while being free to choose the amount of capital with which to begin, then the optimal policy will differ from that of the initial-value problem. For this final-value problem, there is no need to build up capital, since one can begin with the amount producing the maximum quantity of the commodity at each stage while finishing with the required capital (that is, invest in extractive industry A at every stage).

7-11 STATE INVERSION: DIMENSIONALITY CONSIDERATIONS

Certain complications arise in state inversion when the states are vectors with different dimensionalities. Let the dimensionality of $\mathbf{s}_i$ and $\tilde{\mathbf{s}}_i$ be M and N, respectively, and assume that the transformation $\mathbf{T}_i$ consists of N independent equations, each one expressing one of the components of $\tilde{\mathbf{s}}_i$ as a function of $\mathbf{s}_i$ and $\mathbf{d}_i$. If all N equations are not independent, we deal with the largest set of independent equations and make a corresponding adjustment in the dimensionality of the output state vector. There are three cases to be considered: $M = N$, $M > N$, and $M < N$.

1. If $M = N$, the M input state variables can be expressed in terms of the decision variables and N output state variables.
2. If $M > N$, N of the input state variables can be expressed in terms of the decision variables, N output state variables, and the remaining $M - N$ input state variables, which become *choice variables*.
3. If $M < N$, M of the equations of $\mathbf{T}_i$ can be used to express the M input variables in terms of the decisions and M of the output state variables; the remaining $M - N$ equations become constraints relating $\mathbf{d}_i$ and $\mathbf{s}_i$.

Of course, the computational feasibility of state inversion is another question. Fortunately, when state inversion is not practical, there is another way of formulating and decomposing the final-value problem.

7-12 FINAL-VALUE AND TWO-POINT BOUNDARY-VALUE PROBLEMS: DECISION INVERSION

Even when state inversion is impractical, the N-decision, one-state final-value problem can still be decomposed into N smaller problems. These subproblems each involve only one decision, but instead of a single state as in the initial-value case, each has *two* state variables. Since the effort needed to solve these two-state subproblems is considerably greater than for one-state functions, this approach should be used only for final-value serial problems when state inversion is either impossible or computationally infeasible. The main reason for studying this method is its applicability to two-point boundary-value problems arising in the optimization of systems with cycles and branches (Aris, Nemhauser, and Wilde). Moreover, the apparent doubling of the state dimensionality is not usually as disastrous as might appear at first glance, for one of the state variables can, under favorable circumstances, be treated as a choice variable to which direct-search methods may be applied.

Suppose that

$$\tilde{\mathbf{s}}_1 = \mathbf{T}_1(\mathbf{s}_1, \mathbf{d}_1) \tag{7-32}$$

can be solved for $\mathbf{d}_1$ in terms of $\mathbf{s}_1$ and $\tilde{\mathbf{s}}_1$ to give

$$\mathbf{d}_1 = \hat{\mathbf{T}}_1(\mathbf{s}_1, \tilde{\mathbf{s}}_1) \tag{7-33}$$

This mathematical interchange of the roles of $\mathbf{d}_1$ and $\mathbf{s}_1$ is called *decision inversion*, and Eq. (7-33) tells what decision is needed to transform $\mathbf{s}_1$ to $\tilde{\mathbf{s}}_1$. As with state inversion, decision inversion calls for care when there are several output and decision variables. If, for example, there are more decision variables than output state variables, there is still some freedom of choice left among the decisions that would permit limited optimization at stage 1. Even when there are exactly as many decisions as outputs, and when decision inversion is possible, any constraints on the decision variables will be translated into complicated state constraints. But although decision inversion is in principle just as difficult as state inversion, only one decision inversion is needed to solve a final-value problem which would otherwise require N state inversions.

For the serial system of Fig. 7-4, the sum of returns from stages 1 through n can be expressed in terms of the $n - 1$ independent decisions $d_2, \ldots, d_n$ and the two states s_n and $\tilde{s}_i$ by applying decision inversion Eq. (7-33) to Eq. (7-9) to obtain

$$S_n = S_n(\tilde{s}_1, d_2, \ldots, d_n, s_n); \qquad n = 1, \ldots, N \tag{7-34: n}$$

Let the two-state n-stage maximum return function $f_n(\tilde{s}_1, s_n)$ be defined by

$$f_n(\tilde{s}_1, s_n) \equiv \max_{d_2, \ldots, d_n} [S_n(\tilde{s}_1, d_2, \ldots, d_n, s_n)]; \qquad n = 1, \ldots, N \tag{7-35: n}$$

When $n = 1$, the definition degenerates into an equation involving no maximization, since the right side of Eq. (7-35: n) would have no decision variables.

$$f_1(\tilde{s}_1, s_1) = R_1(\hat{T}_1(s_1, \tilde{s}_1), \tilde{s}_1) \tag{7-36}$$

Thus Eq. (7-36) gives the optimal (and only possible) one-stage return as a function of the input and output states.

As in the initial-value problem, knowing $f_n(\tilde{s}_1, s_n)$ allows one to determine the maximum $(n + 1)$-stage return $f_{n+1}(\tilde{s}_1, s_{n+1})$ from the following relation:

$$f_{n+1}(\tilde{s}_1, s_{n+1}) = \max_{d_{n+1}} [R_{n+1}(d_{n+1}, s_{n+1}) + f_n(\tilde{s}_1, T_{n+1}(d_{n+1}, s_{n+1}))]; \qquad n = 1, \ldots, N \tag{7-37: n}$$

It is a one-decision, two-state optimization problem to find the two-state decision function $d^*_{n+1}(\tilde{s}_1, s_{n+1})$ such that

$$R_{n+1}(d^*_{n+1}(\tilde{s}_1, s_{n+1}), s_{n+1}) + f_n(\tilde{s}_1, T_{n+1}(d^*_{n+1}(\tilde{s}_1, s_{n+1}), s_{n+1})) = f_{n+1}(\tilde{s}_1, s_{n+1}); \qquad n = 1, \ldots, N-1 \tag{7-38: n}$$

From the way the final-value problem is formulated, the initial state (the "state" input to stage N) is free to take on its optimal value and is therefore a choice state, written c_N. Hence, the desired final-value maximum return function $R^*(\tilde{s}_i)$ is obtained ultimately by a two-decision one-state optimization:

$$R^*(\tilde{s}_1) = \max_{c_N, d_N} [R_N(c_N, d_N) + f_{N-1}(\tilde{s}_1, T_N(d_N, c_N))] \tag{7-39}$$

In all there is a decision inversion, only $N - 2$ one-decision, two-state optimizations, and a two-decision one-state optimization. If s_N were not a choice state, the problem would be a *two-point boundary-value problem* with the maximization in Eq. (7-39) being over d_N only. Boundary-value problems always require decision inversion; final-value problems can be solved by either state or decision inversion.

Notice that $\tilde{s}_1$ appears as one of the states in every partial optimization. When a state such as $\tilde{s}_1$ is repeated at successive stages and calculations are being done tabularly, its value should be fixed so that an ordinary one-state optimization by dynamic programming can be carried out. This completed, the maximum return and optimal policy for that value of $\tilde{s}_1$ can be stored and the intermediate calculations discarded. The computations are then repeated for another value of $\tilde{s}_1$, and so on. The advantage of multiple one-stage optimizations over a single two-state optimization is the former method's saving in storage. Moreover, if $\tilde{s}_1$ itself is free to assume any value advantageous for maximizing the total return, it can be treated as a choice state and directed by optimum-seeking methods.

7-13 DECISION INVERSION: NETWORK ROUTING PROBLEMS

The difference between state inversion and decision inversion techniques is shown clearly in the final-value network problem of Figs. 7-12 and 7-13. It will be recalled that this problem, solved by state inversion in Fig. 7-12, has the given (fixed) final state $\tilde{s}_1 \equiv C$. Using decision inversion, the solution procedure begins at stage 1 (the last stage). From Eq. (7-33) it can be seen that this is a decisionless stage, since, for given values of s_1 and $\tilde{s}_1$, d_1 is uniquely determined. Thus, with $\tilde{s}_1 \equiv C$, if $s_1 = N$, then $d_1 = R$; if $s_1 = C$, then $d_1 = F$; and if $s_1 = S$, then $d_1 = L$. The corresponding one-stage returns of Eq. (7-36) are 2, 6, and 7, respectively, as shown in Fig. 7-13. The solution now proceeds recursively to the left, just as though this were an initial-value problem; unlike the case in which state inversion is used, no more inversions are necessary. The process terminates at stage 4, having generated the optimal return, $f_4(s_4, C)$, which is clearly seen to be maximized by the choice $s_4^* = S$, leading to an optimal return of 15, produced by the same policy (path) as that of Fig. 7-12. Notice from Fig. 7-1 that the initial-value problem, $s_4 = S$, achieves a better (smaller) cost of $f_4(s_4) = 14$, with a resulting final value of $\tilde{s}_1 = N$.

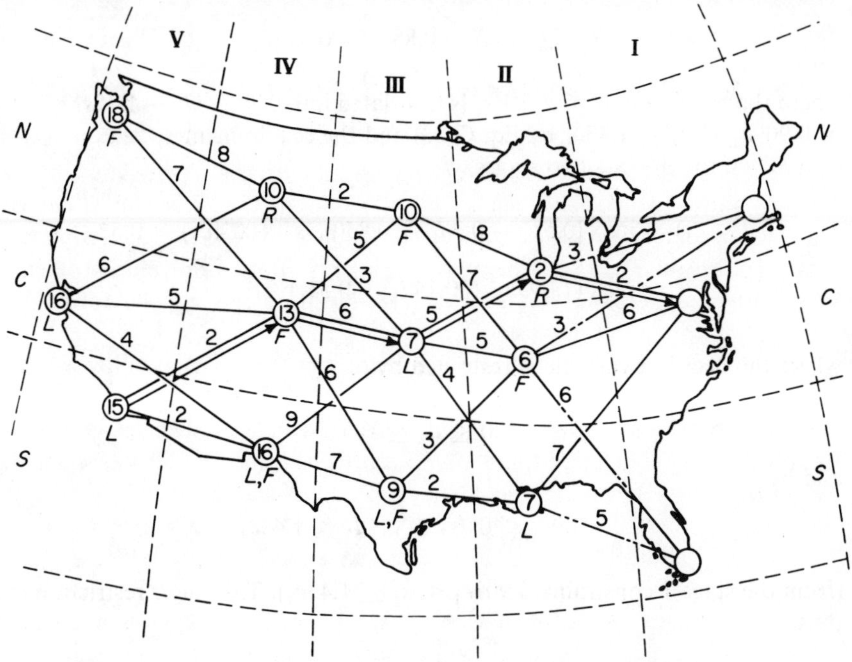

Figure 7-13. Solution of network routing problem by decision inversion.

7-14 DECISION INVERSION: CONTINUOUS VARIABLES

The analytic implications of the decision inversion method are best illustrated by problems in which d_i and s_i are continuous variables. To this end, consider the problem of maximizing $p = \sum_{i=1}^{4} r_i$, where

$$r_i = 0.5s_i - 0.2d_i$$
$$\tilde{s}_i = 0.7s_i + 0.4d_i$$
$$\tilde{s}_1 \equiv \tilde{c}_1$$
$$0 \leq d_i \leq s_i; \qquad i = 1, \ldots, 4$$

This problem was solved previously using state inversion, with the result $f_4(\tilde{c}_1) = 5.28\tilde{c}_1$, generated by the optimal decisions $d_i^* = 0$, $i = 1, \ldots, 4$.

Using decision inversion, Eq. (7-33) becomes

$$d_1 = 2.5\tilde{s}_1 - 1.75s_1$$

and the optimal return at the decisionless stage 1 is, by Eq. (7-36),

$$f_1(\tilde{s}_1, s_1) = 0.85s_1 - 0.5\tilde{s}_1$$

where the restriction $0 \leq d_1 \leq s_1$ is translated into $0 \leq 2.5\tilde{s}_1 - 1.75s_1 \leq s_1$, or $0.909\tilde{c}_1 \leq s_1 \leq 1.43\tilde{c}_1$ by Eq. (7-33) and the requirement $\tilde{s}_1 \equiv \tilde{c}_1$.

Then for the last two stages,

$$\begin{aligned} f_2(s_2, \tilde{c}_1) &= \max_{d_2} [0.5s_2 - 0.2d_2 + 0.85(0.7s_2 + 0.4d_2) - 0.5\tilde{c}_1] \\ &= \max_{d_2} (1.095s_2 + 0.14d_2 - 0.5\tilde{c}_1) \end{aligned}$$

where the decision variable is restricted by

$$0 \leq d_2 \leq s_2$$

and by

$$0.909\tilde{c}_1 \leq 0.7s_2 + 0.4d_2 \leq 1.43\tilde{c}_1$$

(from the stage 1 constraint, $0.909\tilde{c}_1 \leq s_1 \leq 1.43\tilde{c}_1$). This latter restriction is thus

$$2.27\tilde{c}_1 - 1.75s_2 \leq d_2 \leq 3.57\tilde{c}_1 - 1.75s_2$$

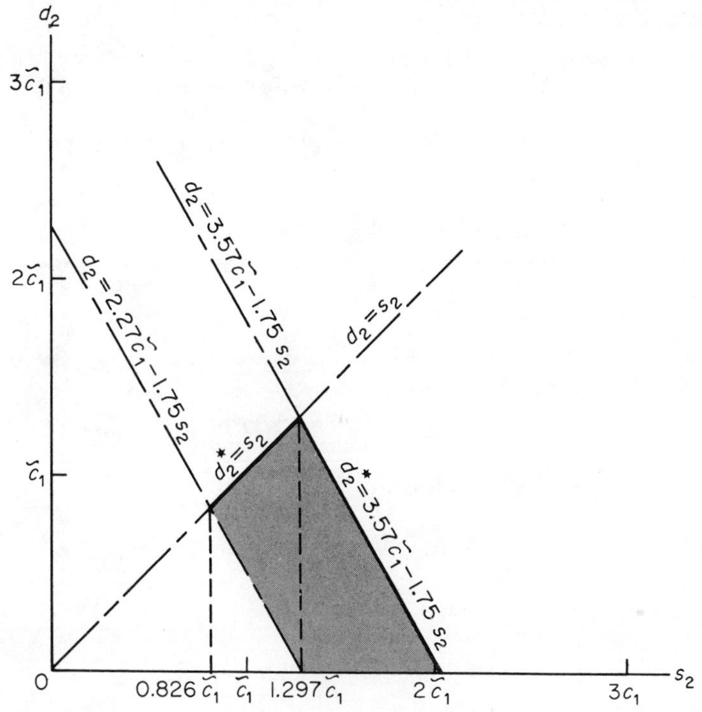

Figure 7-14. Optimal stage 2 decisions for linear problem solved by decision inversion.

and the total restraints on d_2 are shown graphically in Fig. 7-14. Since the coefficient of d_2 in the expression for $f_2(s_2, \tilde{c}_1)$ is positive, the optimal decision at this stage is to make d_2 as large as possible, consistent with the constraints. The resulting value of d_2^* is given by the solid lines in Fig. 7-14, which place a limitation on the input, s_2: $0.826\tilde{c}_1 \leq s_2 \leq 2.045\tilde{c}_1$. For $s_2 = 0.826\tilde{c}_1$, only one choice of decisions, $d_1^* = s_1$, $d_2^* = s_2$, will yield the required output, $\tilde{s}_1 = \tilde{c}_1$. Similarly, the input $s_2 = 2.045\tilde{c}_1$ requires the decisions, $d_1^* = 0$, $d_2^* = 0$ to produce the output, $\tilde{c}_1$. Between these two values of s_2, a true optimization problem exists, having the solution:

$$f_2(s_2, \tilde{c}_1) = \begin{cases} 0.850 s_2, & 1.297\tilde{c}_1 \leq s_2 \leq 2.045\tilde{c}_1 \quad (d_2^* = 3.57\tilde{c}_1 - 1.75 s_2) \\ 1.235 s_2 - 0.5\tilde{c}_1, & 0.826\tilde{c}_1 \leq s_2 \leq 1.297\tilde{c}_1 \quad (d_2^* = s_2) \end{cases}$$

achieved by substituting the optimal values of d_2 into the expression $1.095 s_2 + 0.14 d_2 - 0.5\tilde{c}_1$.

Notice that, superficially, $d_2^* \neq 0$, apparently contradicting the result obtained by state inversion. This seeming paradox occurs because we are solving here a more general problem; with state inversion, the initial state of the system is free to be directed to its optimal value; with decision inversion, we are effectively solving a problem with two fixed ends. This distinction will be made more precise after stages 3 and 4 are analyzed.

Continuing with the preceding type of analysis, one finds for the last three stages,

$$f_3(s_3, \tilde{c}_1) = \max_{d_3} (1.095s_3 + 0.14d_3), \quad \text{when} \quad 1.297\tilde{c}_1 \leq s_2 \leq 2.045\tilde{c}_1$$

and

$$f_3(s_3, \tilde{c}_1) = \max_{d_3} (1.365s_3 + 0.294d_3 - 0.5\tilde{c}_1), \quad \text{when} \quad 0.826\tilde{c}_1 \leq s_2 \leq 1.297\tilde{c}_1$$

where d_3 is restrained by the relations

$$0 \leq d_3 \leq s_3$$

and for $1.297\tilde{c}_1 \leq s_2 \leq 2.045\tilde{c}_1$,

$$3.243\tilde{c}_1 - 1.75s_3 \leq d_3 \leq 5.11\tilde{c}_1 - 1.75s_3$$

and for $0.826\tilde{c}_1 \leq s_2 \leq 1.297\tilde{c}_1$,

$$2.065\tilde{c}_1 - 1.75s_3 \leq d_3 \leq 3.243\tilde{c}_1 - 1.75s_3$$

as shown in Fig. 7-15. The optimal three-stage returns are then

$$f_3(s_3, \tilde{c}_1) = \begin{cases} 0.85s_3 + 0.715\tilde{c}_1, & 1.86\tilde{c}_1 \leq s_3 \leq 2.92\tilde{c}_1 \quad (d_3^* = 5.11\tilde{c}_1 - 1.75s_3) \\ 1.235s_3, & 1.18\tilde{c}_1 \leq s_3 \leq 1.86\tilde{c}_1 \quad (d_3^* = s_3) \\ 1.66s_3 - 0.5\tilde{c}_1, & 0.751\tilde{c}_1 \leq s_3 \leq 1.18\tilde{c}_1 \quad (d_3^* = s_3) \end{cases}$$

Continuing with this analysis, we arrive at the optimal four-stage returns:

$$f_4(s_4, \tilde{c}_1) = \begin{cases} 0.85s_4 + 1.736\tilde{c}_1, & 2.66\tilde{c}_1 \leq s_4 \leq 4.17\tilde{c}_1 \quad (d_4^* = 7.3\tilde{c}_1 - 1.75s_4) \\ 1.235s_4 + 0.715\tilde{c}_1, & 1.69\tilde{c}_1 \leq s_4 \leq 2.66\tilde{c}_1 \quad (d_4^* = s_4) \\ 1.659s_4, & 1.075\tilde{c}_1 \leq s_4 \leq 1.69\tilde{c}_1 \quad (d_4^* = s_4) \\ 2.126s_4 - 0.5\tilde{c}_1, & 0.685\tilde{c}_1 \leq s_4 \leq 1.075\tilde{c}_1 \quad (d_4^* = s_4) \end{cases}$$

where the limits on s_4 are obtained from the constraints on d_4, as shown in Fig. 7-16.

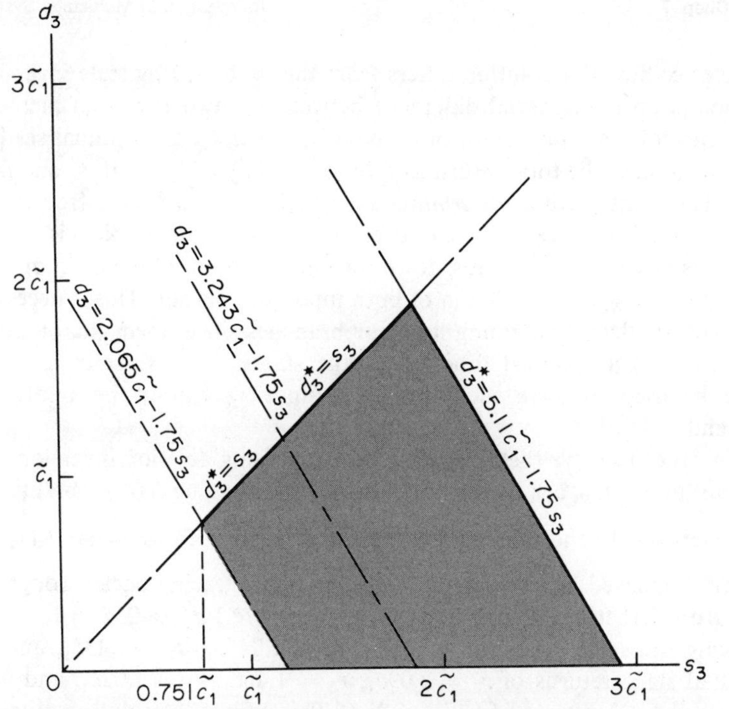

Figure 7-15. Optimal stage 3 decisions for linear problem solved by decision inversion.

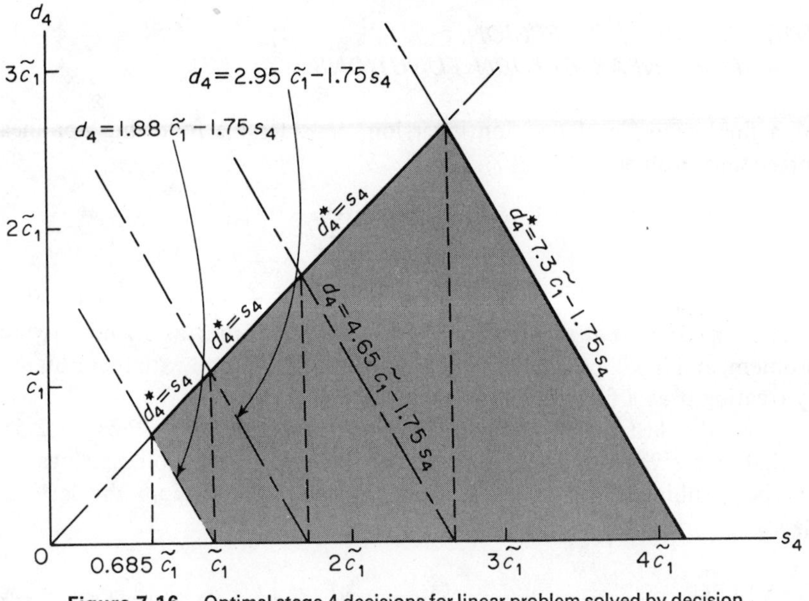

Figure 7-16. Optimal stage 4 decisions for linear problem solved by decision inversion.

Notice how this solution differs from that obtained by state inversion. This points up the essential difference between the two inversion methods. In the case of state inversion, one finds, for fixed $\tilde{c}_1$, the optimal set $\{d_i^*\}$ which maximizes the total return as a function only of $\tilde{c}_1$; that is, one finds $f_N(\tilde{c}_1)$. Then this method *determines* s_N, which is treated as a free end in working the problem. In the case of decision inversion, the problem is worked with limits on each stage input s_i down the line, finally resulting in an optimal total return $f_N(s_N, \tilde{c}_1)$, a function of both input and output. This is necessary in solving problems containing loops or branches since, then, the output is required to be a specified function of the states. Clearly, state inversion cannot be used to solve such problems, since s_N cannot be treated as a free end.

In order to solve the final-value problem using decision inversion, one computes $\max_{s_N} f_N(s_N, \tilde{c}_1)$, which is identical with the value $f_N(\tilde{c}_1)$, obtained by state inversion. In the present problem, it is easily seen that $\max_{s_4} f_4(s_4, \tilde{c}_1) = 5.28\tilde{c}_1$, achieved by selecting $s_4^* = 4.17\tilde{c}_1$. Then tracing back through the stages from left to right, one finds (for $s_4 = 4.17\tilde{c}_1$) $d_i^* = 0$, $i = 1, \ldots, 4$, producing the stage inputs $s_3 = 2.92\tilde{c}_1$, $s_2 = 2.045\tilde{c}_1$, $s_1 = 1.43\tilde{c}_1$, and the individual stage returns of $r_4 = 2.09\tilde{c}_1$, $r_3 = 1.46\tilde{c}_1$, $r_2 = 1.02\tilde{c}_1$, and $r_1 = 0.71\tilde{c}_1$, for a total return of $f_4(\tilde{c}_1) = 5.28\hat{c}_1$. This verifies the solution obtained for this problem using state inversion.

7-15 DECISION INVERSION: NONLINEAR RETURN FUNCTIONS

As a final example of decision inversion, consider the following nonlinear three-stage problem.

$$\text{Maximize } p = \sum_{i=1}^{3} r_i$$

where $r_i = 5d_i - id_i^2$, $\tilde{s}_i = s_i - 0.4d_i$, $0 \le d_i \le s_i$, and the final output state is fixed: $\tilde{s}_1 \equiv \tilde{c}_1$. This problem was solved earlier as an initial-value problem, and it will be interesting to compare this solution with that obtained by treating it as a final-value problem.

At the last stage, the decision inversion gives $d_1 = 2.5s_1 - 2.5\tilde{c}_1$, so that, for given input and output, this becomes a decisionless stage and can be combined with stage 2. Then the total return from the last two stages is

$$S_2 = 12.5s_2 - 6.25s_2^2 + 5s_2d_2 - 5\tilde{c}_1d_2 - 3d_2^2 + 12.5\tilde{c}_1s_2 - 12.5\tilde{c}_1 - 6.25\tilde{c}_1^2$$

and setting $\partial S_2/\partial d_2 = 0$, $d_2^* = 0.83s_2 - 0.83\tilde{c}_1$, within the range of s_2 where this is a feasible solution. Since $0 \le d_1 \le s_1$, one has, through the inversion relation, $\tilde{c}_1 \le s_1 \le 1.67\tilde{c}_1$, and from $s_2 - 0.4d_2 = s_1$,

$$2.5s_2 - 4.18\tilde{c}_1 \le d_2 \le 2.5s_2 - 2.5\tilde{c}_1$$

in addition to the requirement: $0 \le d_2 \le s_2$. The situation is shown graphically in Fig. 7-17, which accounts for the two-valued nature of $f_2(s_2)$:

$$f_2(s_2, \tilde{c}_1) = \begin{cases} 12.5s_2 - 12.5s_2^2 + 41.8s_2\tilde{c}_1 - 37.77\tilde{c}_1^2 - 12.5\tilde{c}_1 \\ \qquad \begin{cases} 2\tilde{c}_1 \le s_2 \le 2.77\tilde{c}_1 \\ (d_2^* = 2.5s_2 - 4.18\tilde{c}_1) \end{cases} \\ 12.5s_2 - 4.16s_2^2 + 8.33s_2\tilde{c}_1 - 4.17\tilde{c}_1^2 - 12.5\tilde{c}_1 \\ \qquad \begin{cases} \tilde{c}_1 \le s_2 \le 2\tilde{c}_1 \\ (d_2^* = 0.83s_2 - 0.83\tilde{c}_1) \end{cases} \end{cases}$$

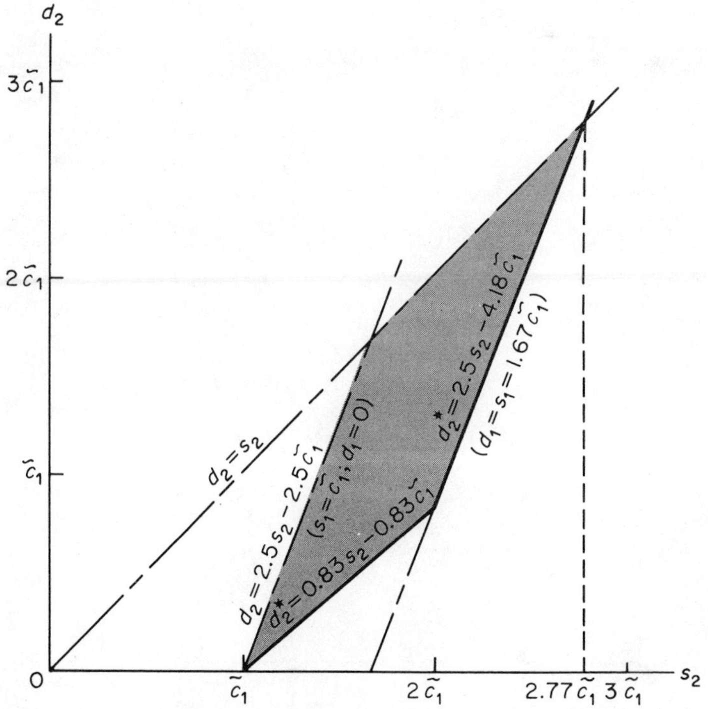

Figure 7-17. Optimal stage 2 decisions for nonlinear problem solved by decision inversion.

Continuing this analysis into the third stage, the solution to the problem is found to be

$$f_3(s_3, \tilde{c}_1) = \begin{cases} 12.5s_3 - 18.75s_3^2 + 103.875s_3\tilde{c}_1 - 161.76\tilde{c}_1^2 - 12.5\tilde{c}_1 \\ \qquad \begin{cases} 3.5\tilde{c}_1 \leq s_3 \leq 4.62\tilde{c}_1 \\ d_3^* = 2.5s_3 - 6.925\tilde{c}_1 \end{cases} \\ 12.5s_3 - 7.50s_3^2 - 25.04s_3\tilde{c}_1 - 23.76\tilde{c}_1^2 - 12.5\tilde{c}_1 \\ \qquad \begin{cases} 2.22\tilde{c}_1 \leq s_3 \leq 3.5\tilde{c}_1 \\ d_3^* = s_3 - 1.67\tilde{c}_1 \end{cases} \\ 12.5s_3 - 3.40s_3^2 + 6.82s_3\tilde{c}_1 - 3.41\tilde{c}_1^2 - 12.5\tilde{c}_1 \\ \qquad \begin{cases} \tilde{c}_1 \leq s_3 \leq 2.22\tilde{c}_1 \\ d_3^* = 0.455s_3 - 0.455\tilde{c}_1 \end{cases} \end{cases}$$

where the decision variable, d_3, is shown pictorially in Fig. 7-18. Again it can be seen that $f_3(s_3, \tilde{c}_1)$ is sectionally continuous.

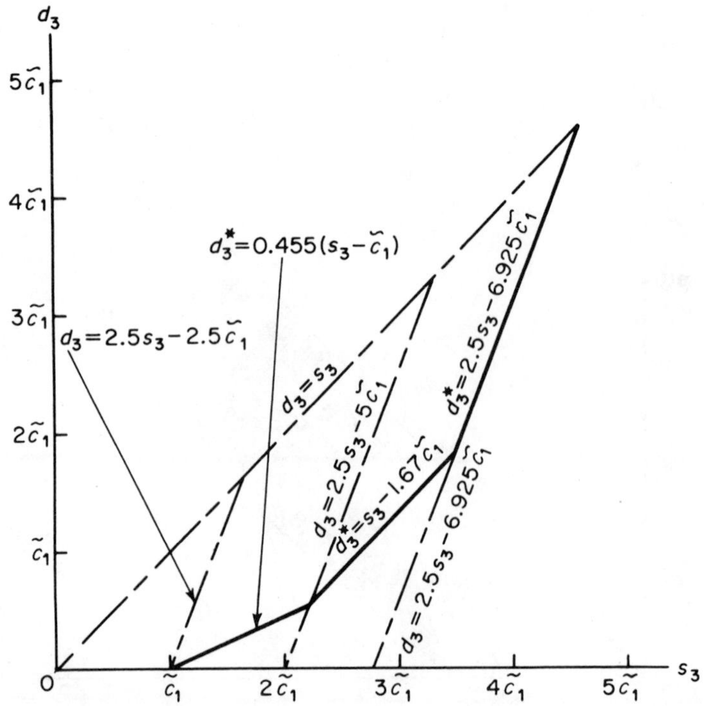

Figure 7-18. Optimal stage 3 decisions for nonlinear problem solved by decision inversion.

From the solution to the initial-value problem, notice that an input of $s_3 = 3.33$ floods the system in the sense that each stage, i, achieves its individual optimum of $6.25/i$, with $d_i^* = 2.5/i$. The final output is then 1.50, and any further input is unused, serving only to increase the output. Therefore, the optimal solution to this problem when the input is $s_3 = 3.33 + K$, for K a nonnegative number, produces an output of $\tilde{s}_1 = 1.50 + K$ and a return of $f_3(3.33 + K) = 11.46$. Thus, from the final-value problem, we would like to have

$$\frac{s_3}{\tilde{c}_1} = \frac{3.33 + K}{1.50 + K}$$

which, for $K \geq 0$, lies in the range

$$1 \leq \frac{3.33 + K}{1.50 + K} \leq 2.22$$

or $\tilde{c}_1 \leq s_3 \leq 2.22\tilde{c}_1$. For this range, we find

$$f_3(s_3, \tilde{c}_1) = 12.5s_3 - 3.40s_3^2 + 6.82s_3\tilde{c}_1 - 3.41\tilde{c}_1^2 - 12.5\tilde{c}_1$$

Now,

$$\left.\begin{aligned} \frac{\partial}{\partial s_3} f_3(s_3, \tilde{c}_1) = 0 \quad \text{implies} \quad s_3^* = \tilde{c}_1 + 1.83 \\ \frac{\partial}{\partial \tilde{c}_1} f_3(s_3, \tilde{c}_1) = 0 \quad \text{implies} \quad \tilde{c}_1^* = s_3 - 1.83 \end{aligned}\right\} s_3 = \tilde{c}_1 + 1.83$$

Then setting $s_3 = 3.33 + K$ results in

$$\tilde{c}_1 = 1.50 + K \quad \text{and} \quad f_3(3.33 + K, 1.50 + K) = 11.46$$

7-16 CLOSED-FORM SOLUTIONS

Several attempts have been made to provide closed-form solutions for $f_N(s_N)$, as well as to find the optimal decision policy without the need to solve the recursion equations of dynamic programming. Under certain conditions, problems having convex returns and linear transitions are amenable to this type of analysis. We begin with the fully linear case.

Crisp and Beightler have studied the "linear-allocation" type of problem, defined by:

$$\text{maximize } R = \sum_{n=1}^{N} r_n$$

where

$$\left.\begin{aligned} r_n &= A_n s_n + B_n d_n \\ \tilde{s}_n = s_{n-1} &= a_n s_n + b_n d_n \\ 0 \leq d_n &\leq s_n \\ a_n + b_n > 0, &\quad a_n > 0 \end{aligned}\right\} n = 1, \ldots, N$$

This is an initial-value serial system with given input $s_N > 0$.

Beightler, Johnson, and Wilde showed that this problem has the following properties:

1. The optimal decisions are $d_n^* = 0$ or $d_n^* = s_n$, for all n.
2. The optimal N-stage return $f_N(s_N) = k_N s_N$, where k_N is a constant which depends only on A_n, B_n, a_n, and b_n, $n = 1, \ldots, N$. [This is also true for any intermediate stage n: $f_n(s_n) = k_n s_n$.]

For the case in which all of the given constants, A_n, B_n, a_n, and b_n $(n = 1, \ldots, N)$ are positive, it is intuitively clear that all of the optimal decisions are $d_n^* = s_n$. Crisp and Beightler showed that for the special case

$$A_n = A; \qquad B_n = B; \qquad a_n = a; \qquad b_n = b$$

(that is, a *nonautonomous* problem), and A, B, a, and b all positive, the closed-form solution is given by

$$k_N = (A + B)\frac{1 - (a + b)^N}{1 - (a + b)} \tag{7-40}$$

All other nonautonomous problems of interest fall into the following two cases:

Case 1

$$A, B > 0, \quad a + b > 0, \quad a > 1, \quad b < 0$$

Case 2

$$a, b, A > 0, \quad B < 0, \quad a + b > 1, \quad A + B > 0$$

For case 1,

$$d_n^* = s_n \quad \text{for } n = 1, \ldots, J \quad \text{and} \quad d_n^* = 0 \quad \text{for } n = J + 1, \ldots, N$$

The value for J is found from the equation

$$J = I\left[1 + \frac{\log\left\{\dfrac{Ab + a(1 - B)}{A(a + b)}\right\}}{\log(A + B)}\right] \tag{7-41}$$

where $I[x]$ is the integer portion of x. The value of k_N is given by

$$k_N = a^{N-J}k_J + A\frac{1 - a^{N-J}}{1 - a} \tag{7-42}$$

and the value of k_J is found from Eq. (7-40), substituting in J for N.

For case 2, the optimal policy is $d_n^* = 0$, $n = 1, \ldots, j$; $d_n^* = s_n$, $n = j + 1, \ldots, N$; where j is obtained from

$$j = I\left[1 + \frac{\log\left\{1 + \frac{B(1 - a)}{bA}\right\}}{\log a}\right] \tag{7-43}$$

To obtain the optimal system return, $f_N(s_N) = k_N s_N$, the value of k_N is found from the equation

$$k_N = k_j(a + b)^{N-j} + (A + B)\frac{1 - (a + b)^{N-j}}{1 - (a + b)} \tag{7-44}$$

where k_j is given by

$$k_j = A\frac{1 - a^j}{1 - a} \tag{7-45}$$

The derivations of the equations above, along with a fuller discussion of linear allocation problems, are to be found in the paper by Crisp and Beightler.

A numerical example will illustrate the use of the equations above. Consider the following 15-stage problem, taken from the Crisp–Beightler paper.

$$\text{Maximize } R = \sum_{n=1}^{15} r_n$$

where

$$r_n = 2.2s_n - 2.0d_n$$
$$s_{n-1} = 1.1s_n + 0.1d_n$$
$$0 \le d_n \le s_n$$

for

$$n = 1, \ldots, 15 \quad \text{and} \quad s_{15} = 100$$

This problem satisfies the conditions for case 2:

$$A = 2.2, \qquad B = -2.0, \qquad a = 1.1, \qquad b = 0.1$$

The solution procedure begins by obtaining j from Eq. (7-43):

$$j = I\left[1 + \frac{\log\left\{1 + \dfrac{-2.0(-0.1)}{0.22}\right\}}{\log 1.1}\right] = 7$$

Next, k_7 is found from Eq. (7-45):

$$k_7 = (2.2)\frac{1 - (1.1)^7}{1 - 1.1} = 20.87$$

and $k_N \equiv k_{15}$ is obtained from Eq. (7-44):

$$k_{15} = (20.87)(1.2)^8 + (0.2)\frac{1 - (1.2)^8}{1 - 1.2} = 93.037$$

The optimal system return is $f_{15}(100) = 100k_{15} = 9303.7$, and the optimal decisions are $d_1^*, \ldots, d_7^* = 0$, and $d_n^* = s_n$, for $n = 8, \ldots, 15$. [Numerically, $d_n^* = 100(1.2)^{15-n}$, for $n = 8, \ldots, 15$.]

The dynamic programming recursive solution to the problem above would require 15 optimizations, one for each stage of the problem. Thus the Crisp–Beightler procedure can greatly reduce the amount of computational effort required to obtain a solution. In addition, this method gives some insight into the structure of the system under study, and permits a more direct sensitivity analysis of the input parameters.

Bagwell, Beightler, and Stark investigated the possibility of obtaining closed-form solutions to a quite general class of multistage decision problems having convex return functions and linear constraints. They were indeed able to find such solutions, but the complexity of the resulting expressions does not permit us to present them here; the interested reader is referred to their paper for the detailed results. The problem they considered is that of finding decision variables $d_1, \ldots, d_N$ which

$$\text{maximize} \sum_{n=1}^{N} [gd_n^p + h(s_n - d_n)^q]$$

and which satisfy the transition functions

$$s_{n-1} = ad_n + b(s_n - d_n); \qquad n = 2, \ldots, N$$

and the constraints

$$0 \leq d_n \leq s_n; \qquad n = 1, \ldots, N$$

where a, b, g, h, p, and q are positive constants such that

$$q > p > 1 \quad \text{and} \quad 0 < a, \quad b < 1$$

Bagwell, Beightler, and Stark showed that the optimal policy for the above problem is $d_n^* = 0$ or $d_n^* = s_n$, for $n = 1, \ldots, N$. This result is readily obtained by showing that all of the $f_n(s_n)$ are convex functions, and, in fact,

$$f_n(s_n) = \max\,[gs_n^p + f_{n-1}(as_n),\ hs_n^q + f_{n-1}(bs_n)]; \qquad n = 1, \ldots, N \qquad (7\text{-}46\colon n)$$

From the form of Eq. (7-46), it can be seen that $f_n(s_n)$ is a continuous function composed of convex segments. Let $f_n^i(s_n)$ denote the $(i - 1)$st segment, where $i = 0, 1, \ldots,$ and let

$$\lambda_n^j(s_n) = gs_n^p + f_{n-1}^j(as_n) \qquad (7\text{-}47\colon j, n)$$

$$\mu_n^j(s_n) = hs_n^q + f_{n-1}^j(bs_n) \qquad (7\text{-}48\colon j, n)$$

Equation (7-46: n) may now be written as

$$f_n^i(s_n) = \max_{s_n}\,[\lambda_n^j(s_n),\ \mu_n^j(s_n)] \qquad (7\text{-}49\colon i, n)$$

$\lambda_n^j(s_n)$ and $\mu_n^j(s_n)$ are convex functions that intersect at one unique point, $\bar{s}_n$, and furthermore,

$$\lambda_n^j(s_n) > \mu_n^j(s_n) \qquad \text{for } 0 \le s_n \le \bar{s}_n$$

and

$$\mu_n^j(s_n) > \lambda_n^j(s_n) \qquad \text{for } s_n > \bar{s}_n$$

Then the optimal policy is given by

$$d_n^* = \begin{cases} s_n, & s_n \le \bar{s}_n \\ 0, & s_n < \bar{s}_n \end{cases}$$

The authors then establish four categories in which the multistage problem may fall, depending upon the relationships among the parameters a, b, p, and q. A closed-form solution (the N-stage optimal return and the value for the changeover points $\bar{s}_n$) is given for each of these categories.

In his excellent book, Nemhauser investigated a more restrictive, but still important, class of convex problems (that is, convex returns and linear transitions) for which the state variables do not appear explicitly in any of the return or transition functions. He considers, for example, the problem:

$$\text{minimize} \sum_{n=1}^{N} d_n^p$$

subject to

$$\sum_{n=1}^{N} d_n \geq k$$

$$d_n \geq 0; \qquad n = 1, \ldots, N$$

where $p \geq 1$ and k are given numbers.

Nemhauser shows (by an induction argument) that

$$f_n(s_n) = \min_{0 \leq d_n \leq s_n} \left[d_n^p + \frac{(s_n - d_n)^p}{(n-1)^{p-1}} \right]; \qquad n = 2, 3, \ldots, N$$

with

$$f_1(s_1) = s_1^p \quad \text{and} \quad d_1^* = s_1$$

the solution is

$$d_n^* = \frac{s_n}{n} \quad \text{and} \quad f_n(s_n) = \frac{s_n^p}{n^{p-1}}$$

where $s_N = k$ and $s_{n-1} = s_n - d_n$, for all n. Hence $d_n^* = k/N$ and $f_N(k) = k^p/N^{p-1}$.

He then goes on to generalize the convex functions d_n^p to any $r(d_n)$ that is convex and monotonically increasing, by studying the problem

$$\text{minimize} \sum_{n=1}^{N} r(d_n)$$

subject to

$$\sum_{n=1}^{N} d_n \geq k$$

$$d_n \geq 0; \qquad n = 1, 2, \ldots, N$$

where $r(d_n)$ is convex and monotone-increasing. The analysis follows the same form as in the preceding problem, with an induction on n, yielding

$$f_n(s_n) = \min_{0 \leq d_n \leq s_n} \left[r(d_n) + (n-1) r\left(\frac{s_n - d_n}{n-1}\right) \right]; \qquad n = 2, 3, \ldots, N$$

Since both terms in the expression above are convex, their sum, M_n, is convex, and thus a sufficient condition for a minimum is $dM_n/dd_n = 0$, yielding

$$d_n^* = \frac{s_n}{n}$$

as before, producing the result

$$f_n(s_n) = nr\left(\frac{s_n}{n}\right)$$

So, as in the previous example, $d_n^* = k/N$; this gives the N-stage optimal return in closed form once again:

$$f_N(k) = Nr\left(\frac{k}{N}\right)$$

Clearly, further work can be done on obtaining closed-form solutions to other convex multistage problems, but the complexity of the resulting equations increases markedly. The interested reader is referred to the paper by Bagwell, Beightler, and Stark for details on how complex the closed-form solutions are for the class of problems they studied. Any further increases in problem complexity (for example, nonlinear constraints) would produce correspondingly increased complexity in the resulting equations which describe the N-stage total return. Furthermore, the optimal policy would become progressively more difficult to obtain in closed form. For most problems of general interest, one must make use of the recursive equations of dynamic programming, which means solving N one-state, one-decision problems, followed by tracing through the stages once again to obtain the optimal policy.

7-17 SEPARABLE FUNCTIONS

The dynamic programming solution method is quite well adapted for solving problems in which the return and transition functions all consist of separable functions; in other words, each term in these expressions is a function of only one variable. The individual terms may be highly nonlinear, making the problem difficult to solve through most optimization techniques, yet dynamic programming handles these problems quite well. Multistage decision problems of this form occur in practice more often than one might imagine, especially those in which the decision variables are restricted to integer values. To avoid computational difficulties, we consider here the following type of problem, having separable (generally, nonlinear) functions in the returns, and linear transitions:

$$\text{maximize } y = \sum_{j=1}^{N} g_j(x_j)$$

subject to

$$\sum_{j=1}^{N} a_j x_j \leq b$$

$$x_j \geq 0; \qquad j = 1, \ldots, N$$

where b and the a_j are given positive integers, and the $g_j(x_j)$ are given functions.

In setting up this problem in the dynamic programming functional equation format, it is helpful to think of the number b as a total quantity available for allocation to the N stages. Then, numbering the stages in the usual reverse order, the input to the last n stages, s_n, is the total unallocated quantity remaining for use in those stages. Thus the transition functions become

$$s_{n-1} = s_n - a_n x_n; \qquad n = 2, \ldots, N$$

while the return at stage n is $g_n(x_n)$. The functional equations are then

$$f_n(s_n) = \max_{0 \le x_n \le s_n/a_n} [g_n(x_n) + f_{n-1}(s_n - a_n x_n)]; \qquad n = 2, \ldots, N$$

and

$$f_1(x_1) = \max_{0 \le x_1 \le s_1/a_1} [g_1(x_1)]$$

where $s_N = b$.

Clearly, this method generalizes readily to problems for which the variables are restricted to integral values. For such problems, the functional equations remain unchanged, but the calculations are necessarily carried out in tabular form. These are precisely the types of problems for which dynamic programming is best suited, since they cannot be solved by any other optimization methods.

The following simple example will serve to illustrate the solution method.

$$\text{Maximize } y = 8x_1^2 + 4x_2^2 + x_3^3$$

subject to

$$2x_1 + x_2 + 10x_3 \le b$$

$$x_1, \quad x_2, \quad x_3 \ge 0$$

where b is a given positive number. Here

$$f_1(s_1) = \max_{0 \le x_1 \le s_1/2} (8x_1^2) = 2s_1^2 \quad \text{and} \quad x_1^* = \frac{s_1}{2}$$

Then,

$$f_2(s_2) = \max_{0 \le x_2 \le s_2} [4x_2^2 + 2(s_2 - x_2)^2] = 4s_2^2 \quad \text{and} \quad x_2^* = s_2$$

Finally,

$$f_3(b) = \max_{0 \le x_3 \le b/10} [x_3^3 + 4(b - 10x_3)^2]$$

Although the expression within the brackets, M_3, is a cubic function of x_3, it is convex in the range $0 \leq x_3 \leq b/10$, so the maximum takes place at the end points of the range. For $x_3 = 0$, $M_3 = 4b^2$, while for $x_3 = b/10$, $M_3 = b^3/1000$. Hence, for $b < 4000$, $x_3^* = 0$, resulting in $s_2^* = x_2^* = b$, and $x_1^* = 0$. When b is greater than 4000, the optimal policy is $x_3^* = b/10$, $x_2^* = x_1^* = 0$. These results are thus in the form of a parametric solution and provide a type of sensitivity analysis that is typically obtained from the use of dynamic programming.

7-18 MULTIPLE CONSTRAINTS

In general, each constraint linking the decision variables in functional form requires the addition of another state variable. Since the values of the states must be examined exhaustively, the presence of several state variables increases greatly the total amount of computations needed to solve a given problem. Bellman and Dreyfus have suggested a procedure that can be used to reduce this "curse of dimensionality," by introducing Lagrangian multipliers into the problem.

Consider the following N-stage problem having M state variables.

Program I

$$\text{Maximize } R(\mathbf{d}) = \sum_{n=1}^{N} a_{on} d_n^{\alpha_{on}}$$

subject to

$$h_i(\mathbf{d}) \equiv \sum_{n=1}^{N} a_{in} d_n^{\alpha_{in}} \leq b_i; \qquad i = 1, \ldots, M$$

where the a_{ij}, α_{ij}, and b_i are given real numbers.

By introducing $M - 1$ Lagrange multipliers, this problem may be replaced by:

Program II

$$\text{Maximize } g(\mathbf{d}, \lambda) = \sum_{n=1}^{N} \left[a_{on} d_n^{\alpha_{on}} - \sum_{i=1}^{M-1} \lambda_i a_{in} d_n^{\alpha_{in}} \right]$$

subject to

$$\sum_{n=1}^{N} a_{Mn} d_n^{\alpha_{Mn}} \leq b_M; \qquad \lambda_i \geq 0; \quad i = 1, \ldots, M$$

In program II, $M - 1$ of the state variables of program I have been replaced by $M - 1$ Lagrange multipliers. If these multipliers are each set at some fixed value, then program II becomes a standard one-state variable dynamic programming problem. In Chapter 2, using Everett's generalized

Lagrange multiplier method, it was shown that for any particular set of nonnegative multipliers λ^0 chosen, a $\mathbf{d}^*$ that maximizes $g(\mathbf{d}, \lambda^0)$ also maximizes $R(\mathbf{d})$ subject to the constraints

$$\sum_{n=1}^{N} a_{in} d_n^{\alpha_{in}} \leq h_i(\mathbf{d}^*)$$

In general, $h_i(\mathbf{d}^*)$ will not be equal to the given b_i for all i, and a new set of multipliers must be chosen. The problem is then re-solved using these new values for the λ_i, and if the new $h_i(\mathbf{d}) \neq b_i$, the entire process must be repeated. Successive values of the λ_i may be guided by using the Brooks–Geoffrion linear programming approximation method described in Chapter 2. The following example taken from Phillips, Ravidran and Solberg will illustrate the Lagrangian approach to multistage problems containing several state variables.

Consider the nonlinear, integer program

$$\text{maximize } 13x_1 - 5x_2^2 + 30.2x_2 - x_1^2 + 10x_3 - 2.5x_3^2$$

subject to

$$2x_1 + 4x_2 + 5x_3 \leq 10$$

$$x_1 + x_2 + x_3 \leq 5$$

$$x_1, \quad x_2, \quad x_3 \geq 0, \quad \text{integer}$$

The Lagrangian problem is given by

$$\text{maximize } 13x_1 - 5x_2^2 + 30.2x_2 - x_1^2 + 10x_3 - 2.5x_3^2 - \lambda(2x_1 + 4x_2 + 5x_3)$$

subject to

$$x_1 + x_2 + x_3 \leq 5$$

$$x_1, \quad x_2, \quad x_3, \quad \lambda \geq 0$$

$$x_1, \quad x_2, \quad x_3 \text{ integer}$$

As the first trial value, let us choose $\lambda = 1$; then the problem becomes

$$\text{maximize } g(\mathbf{x}, 1) = 11x_1 - 5x_2^2 + 26.2x_2 - x_1^2 + 5x_3 - 2.5x_3^2$$

subject to

$$x_1 + x_2 + x_3 \leq 5$$

The solution to this integer program (using the tabular solution technique; see the book by Phillips et al. for details) is $x_1^* = 3$, $x_2^* = 2$, $x_1^* = 0$.

The Lagrange constraint evaluated at $\mathbf{x}^*$ is

$$2x_1^* + 4x_2^* + 5x_3^* = 14$$

Now, employing the Brooks–Geoffrion method, we obtain (after two iterations): $x_1^* = 3, x_2^* = 1, x_3^* = 0$, corresponding to $\lambda^* = 4.0$, and producing a value for the lone Lagrange constraint of

$$2x_1^* + 4x_2^* + 5x_3^* = 10$$

giving the objective function a value of

$$R(3, 1, 0) = 55.2$$

This is the optimal solution to the given two-constraint problem, and illustrates the method to be employed in solving problems having any number of constraints. Notice that as a by-product of this solution technique, a number of related problems (that is, those having different right-hand-side b_i values) are solved. Thus a form of sensitivity analysis is performed on the original problem in the course of approaching the optimal solution.

As pointed out by Dreyfus, the dynamic programming and Lagrange solution techniques may be synthesized by treating $M - k$ of the original M constraints with Lagrange multipliers, adding them to the objective function, and thus produce a dynamic programming problem having k state variables. The value of k will depend upon the particular problem at hand as well as the available computational facilities. Note that if the original constrained problem consists of N variables and M constraints, then either solution method used exclusively embeds the problem into a space with $M + N$ variables—the dynamic programming technique adding M state variables and the Lagrange method adding M undetermined multipliers. Using dynamic programming, a sequence of N subproblems, having a dimensionality of $M + 1$ (M state variables and one decision variable) must be solved; using the Lagrange approach, the dimensionality of the space over which the optimizations must be performed is N (for each set of fixed values of the M multipliers). Thus, the dynamic programming method has an advantage when N is large and M quite small, while larger values of M usually make this technique computationally infeasible. Conversely, when M is large, the Lagrange approach can preserve the advantage of dynamic programming by reducing the state variable dimensionality.

7-19 INFINITE-STAGE SYSTEMS

In previous sections we have considered only those systems having a finite number of stages; we now wish to extend our results so as to be able to handle a broader range of problems. This section, which is necessarily brief, follows the development in Nemhauser's book; for a fuller treatment, the reader is referred to his book.

Systems containing an infinite number of stages occur quite naturally in practice. For example, when the stages correspond to time periods and the number of these periods, although finite, are very large, the system can be approximated by an infinite-stage model. A different type of problem is that of adjusting a control variable in an automatic control system. In this case, an analog computer adjusts the variable as a consequence of sampling the output from the process.

The examples above represent two different types of infinite-stage processes. In the first of these, there is a natural extension of an N-stage system to an infinite one, produced by letting N increase without limit. This produces the following model:

$$\text{Maximize } R = \sum_{n=1}^{\infty} r_n(s_n, d_n)$$

subject to

$$s_{n-1} = T_n(s_n, d_n); \qquad n = 1, 2, \ldots,$$

which is called a *discrete infinite-stage decision process* (Nemhauser).

Naturally, this type of model requires that R be bounded from above for all combinations of the decision variables. In problems where the stages represent time periods, the mathematical abstraction containing an infinite number of stages (infinite time horizon) can be very useful for understanding the structure of the optimal decision policy. If, as is usually the case, there is regularity in the returns and transitions, we might expect d_n^* to be independent of n. For example, suppose that f_{183} were approximately equal to f_{184}; then $d_{183}^*(s_{183})$ and $d_{184}^*(s_{184})$ will also usually be very close in value.

So, under certain circumstances, the solution to the infinite-stage problem

$$\begin{aligned} f(s) &= \lim_{n\to\infty} f_n(s_n) \\ &= \lim_{n\to\infty} \{\max_{d_n} [r(s_n, d_n) + f_{n-1}(T(s_n, d_n))\} \\ &= \max_{d} [r(S, d) + f(T(S, d))] \end{aligned} \tag{7-50}$$

where f is an *unknown* function, might serve as a good approximation to the r_n^* and d_n^* when N is large. Note that the subscripts have been omitted from r and T, so that the returns and transitions have identical functional forms.

If Eq. (7-50) can be solved for $f(S)$ and $d(S)$, then it will not be necessary to solve for the entire sequences $f_1(s_1), \ldots, f_N(s_N)$ and $d_1^*(s_1), \ldots, d_N^*(s_N)$. The optimal stage-invariant solution, given by $f(S)$ and $d^*(S)$, is called the *steady-state* solution. In many practical situations where the transient effects are negligible, only the steady-stage solution is required. Even in those cases where exact numerical results are required for $f_N(s_N)$ and all of the d_n^*, $f(S)$ and $d(S)$ will be quite valuable as approximations to these values, and will also provide insight into the structure of the system under study.

The second type of infinite-stage process may be visualized by once again considering a problem for which the stages correspond to time periods. This time, however, the horizon is finite, and the time periods are very small; that is, the time between successive decisions is negligible compared to the horizon. This problem may be approximated with a model that is obtained by permitting the size of the time periods to approach zero in the limit, so that decisions are made continuously. Hence, here is an example of an infinite-stage problem that has a *finite* time horizon! Another problem which illustrates this type of process is that of proving that the shortest distance between two points is a straight line. For simplicity, consider the case for two-dimensional Euclidean space, and let the points be designated as A and B. The problem may be formulated as follows: Let a moving point C travel from A to B along a path for which the distance traveled is a minimum. Furthermore, let point C take a very large number of very short steps; the problem then becomes one of finding the direction of each of the steps so that the sum of all the distances traveled is a minimum. In the limit, point C takes an infinite number of infinitesimal steps, and thus traces out the curve of shortest distance between A and B, which, of course, is a straight line.

Hence, in this second type of infinite-stage process, the discrete-stage problem,

$$\max_{d_n} R = \sum_{n=1}^{N} r_n(s_n, d_n) \tag{7-51}$$

subject to

$$s_{n-1} = T_n(s_n, d_n); \qquad n = 1, \ldots, N$$
$$s_N = K$$

is replaced by a continuous multistage process of the form

$$\max \hat{R} = \int_{t_1}^{t_2} F(t, S, u)\, dt$$

subject to

$$\frac{dS}{dt} = g(t, S, u); \qquad t_1 \leq t \leq t_2$$
$$S(t_1) = K$$

This latter model is called a *continuous infinite-stage decision process*. In this infinite-stage model, the stage index n of the model given by Eq. (7-51) is replaced by a continuous-stage parameter t which usually represents time or distance. Also, the finite sum of stage returns in Eq. (7-51) is replaced by the integral of stage returns, and the finite difference equations that define the stage transition functions are replaced by differential equations.

The reader who is familiar with the *calculus of variations* will see at once the close connection between that field and dynamic programming, since the calculus of variations is concerned with determining a function (extremal curve) that optimizes a given integral. In fact, dynamic programming can be used to find approximate numerical solutions to problems of this type when the classical techniques (for example, the solution of the Euler equation) cannot be carried out in closed form. Usually, the classical approach requires solving Euler's second-order nonlinear differential equation, which the optimizing function must satisfy. In general, one must resort to numerical techniques in order to solve the Euler equation, so that the dynamic programming approach of determining the unknown function numerically provides just as precise a solution as do the classical methods. Furthermore, the dynamic programming calculations are much easier to carry out than are the calculations required to solve the Euler equation. Indeed, dynamic programming will often succeed in solving a calculus-of-variations problem that cannot be solved with any accuracy by the classical approach. This is especially true when additional restrictions in the form of *transversality conditions* are present. See the book by Bellman and Dreyfus for a complete discussion of this topic.

However, we are still faced with the problem of solving Eq. (7-50), in which the unknown optimal function f appears with different arguments on both sides of the equation and, in addition, the optimizing value of d itself is unknown. In general, solving this infinite-stage optimization equation can be quite difficult if the unknown functions f and d have complicated forms. Bellman has developed a method called *approximation in policy space* for solving this type of equation.

Bellman's method begins with a guess of the optimal policy function $d_o(S)$, so that $f_o(S)$ may be found by solving

$$f_0(S) = r(S, d_0(S)) + f_0(T(S, d_0(S))) \tag{7-52}$$

In general, numerical methods must be employed to solve Eq. (7-52), but this is still far simpler than solving Eq. (7-50), which requires in addition that an optimization be performed on d. Once $f_o(S)$ is obtained, the next approximation, $d_o(S)$, is found by solving the problem

$$\max_d \,[r(S, d) + f_1(T(S, d))] \tag{7-53}$$

This procedure is now repeated by finding the function $f_1(S)$ which satisfies

$$f_1(S) = r(S, d_1(S)) + f_1(T(S, d_1(S)))$$

and this function is, in turn, used to replace f_0 in Eq. (7-53), from which one obtains $d_2(S)$. The entire process is repeated until convergence of the sequences $d_n(S)$ and $f_n(S)$ is achieved. For some classes of functions, proof of convergence is not difficult and the convergence is usually quite rapid. For functions in general, however, a rigorous proof of convergence can be difficult to obtain.

In order to illustrate the method of approximation in policy space, consider the following highly simplified problem: Find values $d_1, d_2, \ldots$ that

$$\text{minimize } R = \sum_{n=1}^{\infty} (4d_n^2 - 2s_n d_n + s_n^2)$$

and satisfy, for all n,

$$s_{n-1} = \tfrac{1}{3}s_n - \tfrac{1}{3}d_n \quad \text{and} \quad 0 \leq d_n \leq s_n$$

The infinite-stage equation is

$$f(S) = \min_{0 \leq d \leq S} [4d^2 - 2Sd + S^2 + f(\tfrac{1}{3}S - \tfrac{1}{3}d)]$$

By analyzing this problem, it is easy to show that $f_n(s_n) = ks_n^2$ (see Nemhauser), so the functional equation above is of the form

$$kS^2 = \min_{0 \leq d \leq S} [4d^2 - 2Sd + S^2 + k(\tfrac{1}{3}S - \tfrac{1}{3}d)^2]$$

As an initial guess, let us choose

$$d_o(S) = \tfrac{1}{4}S$$

Substituting this value of $d_o(S)$ into Eq. (7-52), and using the return and transition functions for this problem, results in

$$k_o S^2 = \tfrac{3}{4}S^2 + \tfrac{1}{16}k_o S^2$$

so that $k_o = \frac{4}{5}$, and hence $f_o(S) = \frac{4}{5}S^2$. After substituting this value into Eq. (7-53), the next step is to find the value of $d_1(S)$ that minimizes

$$4d^2 - 2Sd + S^2 + \tfrac{4}{5}(\tfrac{1}{3}S - \tfrac{1}{3}d)^2$$

and which satisfies

$$0 \leq d \leq S$$

This value is found to be $d_1(S) = \frac{49}{184}S = 0.266S$, which is then substituted into Eq. (7-52), which, in turn, is solved for k_1, to provide the value of $f_1(S) = k_1 S^2$. The entire process is repeated until convergence is achieved. In the present illustrative problem, convergence is quite rapid, and the infinite-stage solution is found to be

$$d(S) = 0.265S, \qquad f(S) = 0.739S^2$$

7-20 DIVERGING BRANCHES

The methods described earlier for solving initial-value serial problems can be adapted to the optimization of systems having a diverging branch as in Fig. 7-19. In such systems one of the stages (say, k) of a subsystem has, in addition

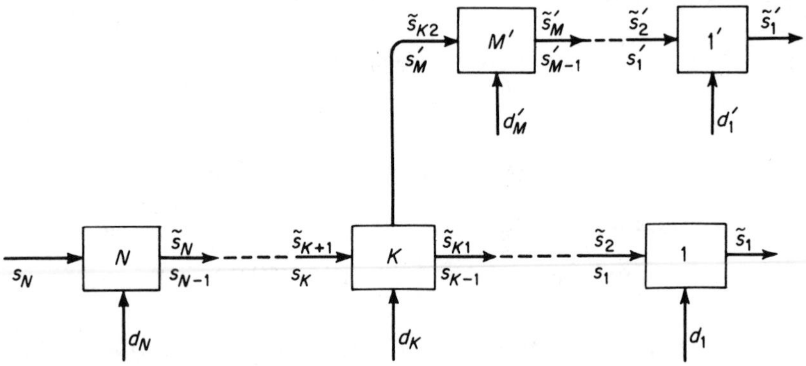

Figure 7-19. Diverging branch.

to its ordinary output state $\tilde{s}_{k_1}$, another output, $\tilde{s}_{k_2}$, which is the initial state for a different sequence of M serial stages labeled $1'$ through M' and forming a *diverging branch*. The branch transition functions are

$$\tilde{s}'_i = T'_i(d'_i, s'_i); \qquad i = 1, \ldots, M \tag{7-54: i}$$

and the incidence identities are

$$\tilde{s}'_{i+1} \equiv s'_i; \qquad i = 1, \ldots, M - 1 \tag{7-55: i}$$

Connection of the branch to the main system at stage k is represented by the additional transition function

$$\tilde{s}_{k_2} = T_{k_2}(d_k, s_k) \tag{7-56}$$

and incidence identity

$$s'_M \equiv \tilde{s}_{k_2} \tag{7-57}$$

As usual there are M return functions

$$r'_i = R'_i(d'_i, s'_i); \qquad i = 1', \ldots, M' \tag{7-58: i}$$

and it is desired to optimize the sum of the returns from all $M + N$ stages.

Let $f'_M(s'_M)$ be the *diverging branch maximum return function* defined by

$$f'_M(s'_M) \equiv \max_{d'_1, \ldots, d'_M} \left(\sum_{i=1}^{M} r'_i \right) \tag{7-59}$$

Finding this function is an initial-value problem solvable by the methods used for optimizing serial systems. By the connection Eq. (7-56) and identity (7-57) this branch return can be combined with the stage k return to give a new return function depending only on d_k and stage s_k.

$$R'_k(d_k, s_k) \equiv f'_M(T_{k_2}(d_k, s_k)) + R_k(d_k, s_k) \tag{7-60}$$

This new function can be used in place of $R_k(d_k, s_k)$ in the regular optimization plan for the main system. The replacement of $R_k(d_k, s_k)$ by $R'_k(d_k, s_k)$ is called *absorption of a diverging branch.*

It will be important in solving networks to know how to handle a final-value version of the diverging branch problem. This would involve first finding the *two-state* M-stage maximum return function $f'_M(\tilde{s}'_1, s'_M)$ of Eq. (7-35: M') by using Eq. (7-36) and iterating recursion relation (7-37: n') $M - 1$ times. This would take a decision inversion and $M - 1$ one-decision, two-state optimizations. Absorption of the branch requires that the three-variable function $R'_k(d_k, \tilde{s}'_1, s_k)$ be substituted for $R_k(d_k, s_k)$ in the main stem optimization scheme.

$$R'_k(d_k, \tilde{s}'_1, s_k) \equiv R_k(d_k, s_k) + f'_M(\tilde{s}'_1, T_{k2}(d_k, s_k))$$

As an example of a diverging branch problem, consider the system shown in Fig. 7-20. Here three networks have been linked together to form a system having one diverging branch; this system is also represented by the functional diagram just below the network. This is an initial-value problem in which it is desired to find the shortest path from stage 12 to stages 1 and 13, taking into account the length (returns) of the individual links and the returns associated with the junction stage 7.

The optimal four-stage returns, $f'_4(s_{16})$, for the upper branch are found by the usual dynamic programming methods for serial systems. The input state, s_{16}, to this branch can take on any one of five values, I–V, corresponding to the five rows of this network, as indicated on the figure. The optimal returns for this branch are easily found to be $f'_4(\text{I}) = 13$, $f'_4(\text{II}) = 17$, $f'_4(\text{III}) = 19$, $f'_4(\text{IV}) = 14$, $f'_4(\text{V}) = 20$.

The results from analysis of the six stages of the lower branch, also obtained by serial methods, are $f_6(1) = 23$, $f_6(2) = 17$, $f_6(3) = 24$, and $f_6(4) = 28$, where the values of the input state, s_6, correspond to the row numbering on Fig. 7-20. We now define the optimal (minimum) return from the 11-stage system consisting of stages 1–7 and 13–16 as

$$f_{11}(s_7) = \min_{d_7} [R_7(s_7, d_7) + f'_4(s_{16}) + f_6(s_6)]$$

where R_7, s_6, and s_{16} are functions of d_7 and s_7 as defined by the tables of Fig. 7-21. For each of the six possible values of the initial state, s_7, a decision, d_7,

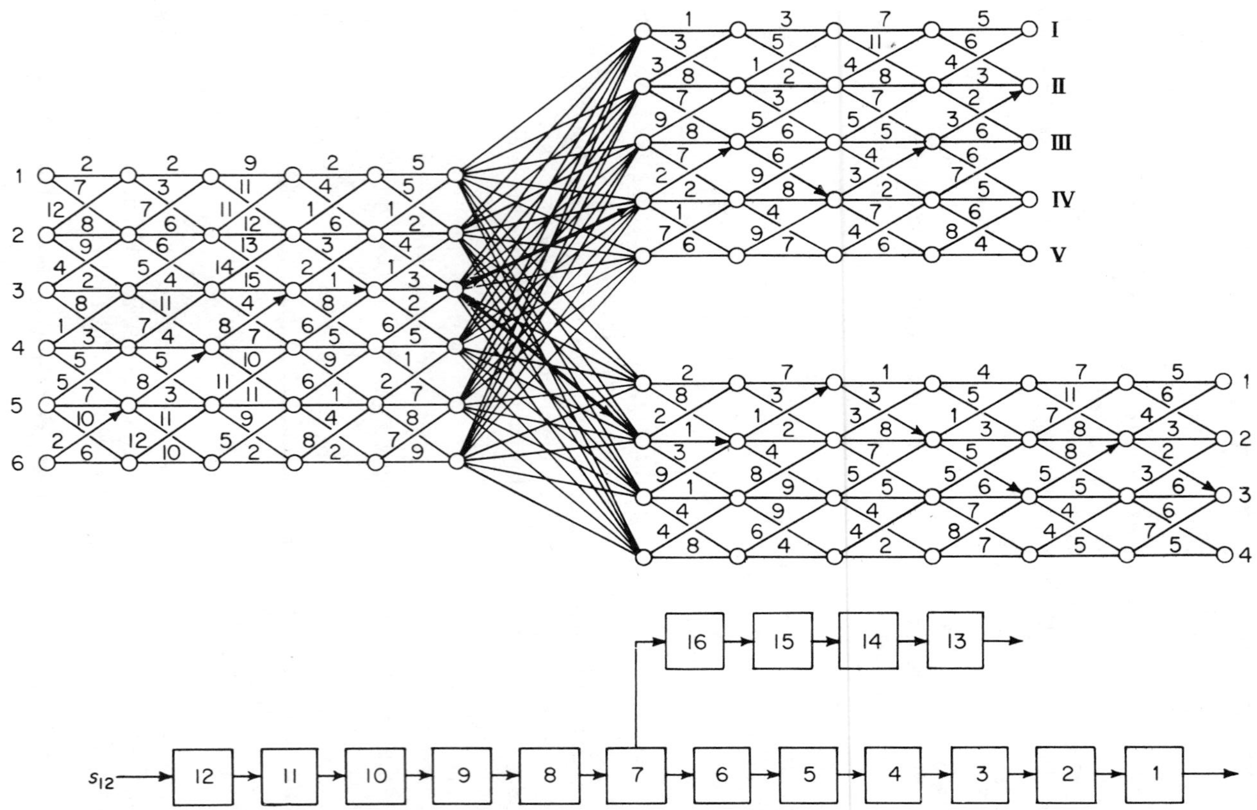

Figure 7-20. Example of diverging branch network problem.

$s_7 = 1$

s_{16} \ s_6	1	2	3	4
I	10	17	13	5
II	9	13	7	8
III	2	9	3	11
IV	8	18	(5)	6
V	3	8	2	1

$s_7 = 2$

s_{16} \ s_6	1	2	3	4
I	9	15	8	12
II	5	13	8	6
III	(2)	11	7	4
IV	10	14	7	3
V	2	11	2	1

$s_7 = 3$

s_{16} \ s_6	1	2	3	4
I	2	7	4	8
II	4	2	9	1
III	8	4	9	3
IV	2	(4)	10	3
V	7	10	2	6

$s_7 = 4$

s_{16} \ s_6	1	2	3	4
I	5	12	7	4
II	4	(6)	9	3
III	7	10	4	8
IV	15	11	7	5
V	4	4	8	2

$s_7 = 5$

s_{16} \ s_6	1	2	3	4
I	2	8	4	9
II	4	5	7	2
III	7	(1)	6	9
IV	9	8	2	4
V	5	6	1	4

$s_7 = 6$

s_{16} \ s_6	1	2	3	4
I	6	(3)	7	8
II	11	5	2	12
III	14	9	8	1
IV	2	3	4	11
V	9	10	3	5

Figure 7-21. Tabular description of stage 7 in diverging branch network problem.

must be made as to which of the 20 available combinations of output states, s_6 and s_{16}, will minimize $f_{11}(s_7)$. The tables in Fig. 7-21 are read as follows. If the value of the initial state at stage 7 is $s_7 = 5$, the return associated with the decision to go to output states $s_{16} = \text{III}$, $s_6 = 3$, is 6; to this must be added $f'_4(\text{III}) = 19$ and $f_6(3) = 24$, to yield $M_{11}(5; \text{III}, 3) = 49$. The reader may verify that the optimal returns are $f_{11}(1) = 43$, $f_{11}(2) = 44$, $f_{11}(3) = 35$, $f_{11}(4) = 40$, $f_{11}(5) = 37$, and $f_{11}(6) = 33$. The optimal decisions, d_7^*, are indicated by the circles in the tables describing stage 7 in Fig. 7-21.

The remaining stages of the system are analyzed in the usual recursive manner to find the optimal return from the process, $f_{16}(s_{12})$, for the given initial value s_{12}. If, for example, the initial state is specified as $s_{12} = 6$, the optimal return is found to be $f_{16}(6) = 57$, and the optimal decision at each stage is shown by the arrows that define the minimum path in Fig. 7-20.

7-21 DIVERGING BRANCHES: NONLINEAR RETURNS

Consider the system depicted by Fig. 7-22, in which the returns for stages 1, 2, 3, and 4 are $5d_i - id_i^2$, whereas those for stages I and II in the lower branch are $4d_i - id_i^2$. In addition, the transition functions are $s_{\text{I}} = s_{\text{II}} - 0.2d_{\text{II}}$, and

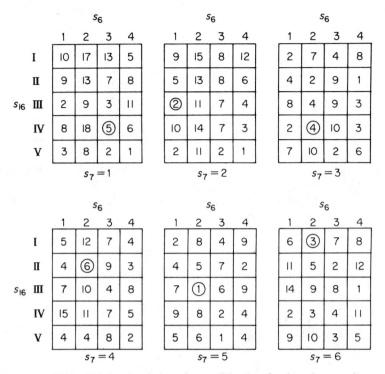

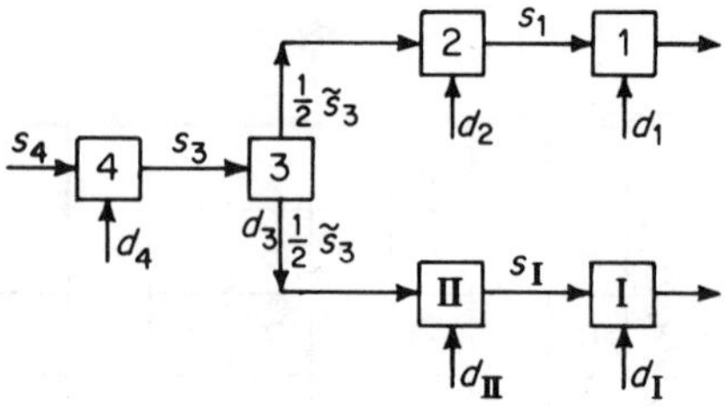

Figure 7-22. Nonlinear diverging branch example problem.

$\tilde{s}_i = s_i - 0.4d_i$, $i = 1, 2, 3, 4$. The problem is to maximize the sum of all the returns for a given system input, s_4.

Proceeding as in the nonlinear serial problem example solved earlier, one finds, for the upper branch,

$$f_2(\tilde{s}_3) = \begin{cases} 9.375, & \begin{cases} 6 \leq \tilde{s}_3 \leq \infty \\ (d_2^* = 1.25) \end{cases} \\ 2.778\tilde{s}_3 - 0.232\tilde{s}_3^2 + 1.042, & \begin{cases} 1.706 \leq \tilde{s}_3 \leq 6 \\ (d_2^* = 0.93\tilde{s}_3 + 0.685) \end{cases} \\ 4\tilde{s}_3 - 0.59\tilde{s}_3^2, & \begin{cases} 0 \leq \tilde{s}_3 \leq 1.706 \\ (d_2^* = 0.5\tilde{s}_3) \end{cases} \end{cases}$$

and, for the lower branch,

$$f_{\text{II}}(\tilde{s}_3) = \begin{cases} 6, & \begin{cases} 4.4 \leq \tilde{s}_3 \leq \infty \\ d_{\text{II}}^* = 1 \end{cases} \\ 2.157\tilde{s}_3 - 0.245\tilde{s}_3^2 + 1.255, & \begin{cases} 1.738 \leq \tilde{s}_3 \leq 4.4 \\ d_{\text{II}}^* = 0.049\tilde{s}_3 + 0.784 \end{cases} \\ 3.6\tilde{s}_3 - 0.66\tilde{s}_3^2, & \begin{cases} 0 \leq \tilde{s}_3 \leq 1.738 \\ d_{\text{II}}^* = 0.5\tilde{s}_3 \end{cases} \end{cases}$$

Then at stage 3, d_3 must be chosen so as to maximize the sum of the returns from both branches plus R_3, the return *at* stage 3. For example, in the range $1.706 \leq \tilde{s}_3 \leq 1.738$, this sum is

$$M_5(s_3, d_3) = 2.45d_3 - 3.14d_3^2 + 6.38s_3 - 0.89s_3^2 + 0.714s_3d_3 + 1.04$$

from which it follows by differentiation that

$$d_3^* = 0.1135s_3 + 0.3896$$

When this value of d_3 is substituted into the constraint

$$1.706 \le s_3 - 0.4d_3 \le 1.738$$

it is seen to be valid in the range $1.947 \le s_3 \le 1.984$. Continuing in this manner, the optimal values of d_3 for the appropriate ranges of the input s_3 are those described in Fig. 7-23. The optimal returns for the two branches follow:

$$f_5(s_3) = \begin{cases} 17.458, & \begin{cases} 6.333 \le s_3 \le \infty \\ d_3^* = \frac{5}{6} \end{cases} \\ 2.897s_3 - 0.230s_3^2 + 8.287, & \begin{cases} 4.714 \le s_3 \le 6.333 \\ d_3^* = 0.0306s_3 + 0.640 \end{cases} \\ 5.123s_3 - 0.465s_3^2 + 3.041, & \begin{cases} 1.984 \le s_3 \le 4.714 \\ d_3^* = 0.062s_3 + 0.492 \end{cases} \\ 6.656s_3 - 0.851s_3^2 + 1.519, & \begin{cases} 1.947 \le s_3 \le 1.984 \\ d_3^* = 0.1135s_3 + 0.3896 \end{cases} \\ 7.9063s_3 - 1.172s_3^2 + 0.300, & \begin{cases} 0.363 \le s_3 \le 1.947 \\ d_3^* = 0.156s_3 + 0.306 \end{cases} \\ 9.56s_3 - 3.45s_3^2, & \begin{cases} 0 \le s_3 \le 0.363 \\ d_3^* = s_3 \end{cases} \end{cases}$$

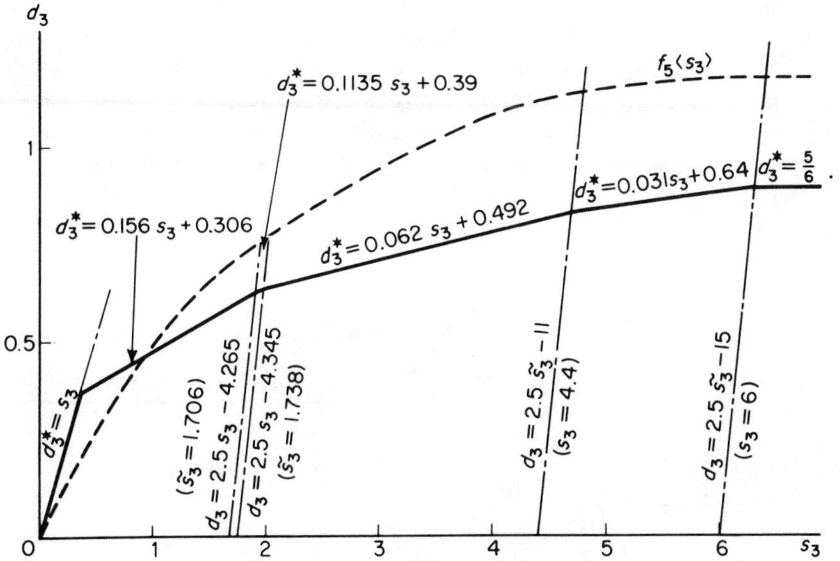

Figure 7-23. Optimal stage 3 decisions and returns for nonlinear diverging branch problem.

The remaining stage 4 is now included in the analysis by the usual serial methods of dynamic programming. The total solution to this diverging branch problem is then

$$
f_6(s_4) = \begin{cases}
19.02, & \begin{cases} 6.58 \le s_4 \le \infty \\ d_4^* = 0.625 \end{cases} \\
2.98s_4 - 0.23s_4^2 + 9.20, & \begin{cases} 4.95 \le s_4 \le 6.58 \\ d_4^* = 0.23s_4 + 0.47 \end{cases} \\
5.26s_4 - 0.46s_4^2 + 3.58, & \begin{cases} 2.17 \le s_4 \le 4.95 \\ d_4^* = 0.05s_4 + 0.362 \end{cases} \\
6.85s_4 - 0.83s_4^2 + 1.85, & \begin{cases} 2.13 \le s_4 \le 2.17 \\ d_4^* = 0.083s_4 + 0.28 \end{cases} \\
8.11s_4 - 1.11s_4^2 + 0.50, & \begin{cases} 0.47 \le s_4 \le 2.13 \\ d_4^* = 0.112s_4 + 0.219 \end{cases} \\
9.92s_4 - 3.03s_4^2 + 0.10, & \begin{cases} 0.185 \le s_4 \le 0.47 \\ d_4^* = 0.30s_4 + 0.13 \end{cases} \\
10.73s_4 - 5.24s_4^2, & \begin{cases} 0 \le s_4 \le 0.185 \\ d_4^* = s_4 \end{cases}
\end{cases}
$$

7-22 CONVERGING BRANCHES

Figure 7-24 illustrates a system with a P-stage *converging* serial branch; that is, one whose output state $\tilde{s}_{N+1}$ is an input to stage M of another N-stage system. [This choice is arbitrary; we could have just as well selected the $(N - M)$-stage system consisting of stages $M + 1$ through N as the converging branch, and stages $1, \ldots, M, N + 1, \ldots, N + P$ as the main trunk of the system.]

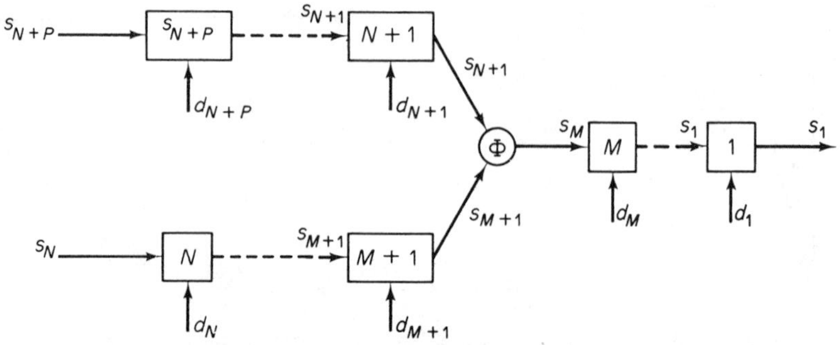

Figure 7-24. Converging branch system.

Here again, $\tilde{s}_n$ denotes the output from stage n, for all $n = 1, \ldots, N + P$, just as in serial and diverging branch systems, and the values of these output states are obtained from the given transition equations

$$\tilde{s}_n = T_n(s_n, d_n); \qquad n = 1, \ldots, N + P$$

Φ is a decisionless stage which transforms the two outputs $\tilde{s}_{N+1}$ and $\tilde{s}_{M+1}$ into the input s_M, as given by Eq. (7-61).

The stage returns are given functions, $r_n = R_n(s_n, d_n)$, just as in all previous systems studied thus far. The incidence identities are given by

$$\begin{aligned} s_n &= \tilde{s}_{n+1}; \qquad n = 1, \ldots, M - 1, M + 1, \ldots, \\ &\qquad\qquad N - 1, N + 1, \ldots, N + P - 1 \qquad\qquad (7\text{-}61) \\ s_M &= \phi_1(\tilde{s}_{M+1}) + \phi_2(\tilde{s}_{N+1}) \end{aligned}$$

where ϕ_1 and ϕ_2 are given functions. The given inputs to this initial-value problem are s_{N+P} and s_N, and the problem is to maximize the sum of returns from all $N + P$ stages. To solve it, first find $f_M(s_M)$, the optimal return from stages 1 through M, using serial dynamic programming. Next, the optimal branch return is determined, written as a function of the branch output, $\tilde{s}_{N+1}$. This is accomplished by performing a decision inversion at stage $N + 1$, to yield $f_P(s_{N+P}, \tilde{s}_{N+1})$; since s_{N+P} is a given constant, we shall not write it down explicitly, and the optimal branch return will be written as $f_P(\tilde{s}_{N+1})$. At the junction stage Φ, we are faced with a simultaneous optimization over two variables in order to obtain the total return from stages 1 through $M + 1$ and $N + 1$ through $N + P$, as shown in Eq. (7-62).

$$f_{M+P+1}(s_{M+1}) = \max_{d_{M+1};\tilde{s}_{N+1}} [R_{M+1}(s_{M+1}, d_{M+1}) + f_P(\tilde{s}_{N+1}) + f_M(s_M)] \quad (7\text{-}62)$$

where s_M may be written as

$$s_M = \phi_1(T_{M+1}(s_{M+1}, d_{M+1})) + \phi_2(\tilde{s}_{N+1})$$

Equation (7-62) requires the simultaneous selection of a value of d_{M+1} and $\tilde{s}_{N+1}$, for a given value of state s_M; this two-decision, one-state optimization step is, in general, quite difficult to carry out computationally. This is especially true in the case of tabular functions, and represented one of the major stumbling blocks to progress in solving nonserial systems until the development of the *branch compression* technique described in Section 7-25. Once the value for $f_{M+P+1}(s_{M+1})$ has been obtained, the rest of the main trunk is optimized by ordinary one-stage, one-decision dynamic programming. It is essential that the double-decision optimization be carried out at the point where the branch joins the main trunk; otherwise, the state $\tilde{s}_{N+1}$ would have to be carried as a state variable to be examined exhaustively during the optimization of every remaining stage in the main trunk.

7-23 CONVERGING NETWORK

Consider the converging branch system described by the network of Fig. 7-25 and also represented schematically by the functional diagram just below the network. The problem is to find, for a given value of the input state s_8 (s_{11} is a choice variable), the shortest path through the network (to stage 1). One begins at the free end, solving the last four stages serially to obtain the optimal returns $f_4(1) = 20$, $f_4(2) = 19$, and $f_4(3) = 17$, where the values of s_4 correspond to the row numbering in Fig. 7-25. The upper branch is treated as a final-value problem and is solved by *state* inversion (note that state inversion can be used here since s_{11} is *not* a given value; otherwise, it would be necessary to use decision inversion as developed in Section 7-22). The minimum returns for each value of s_9 are now given, along with the corre-

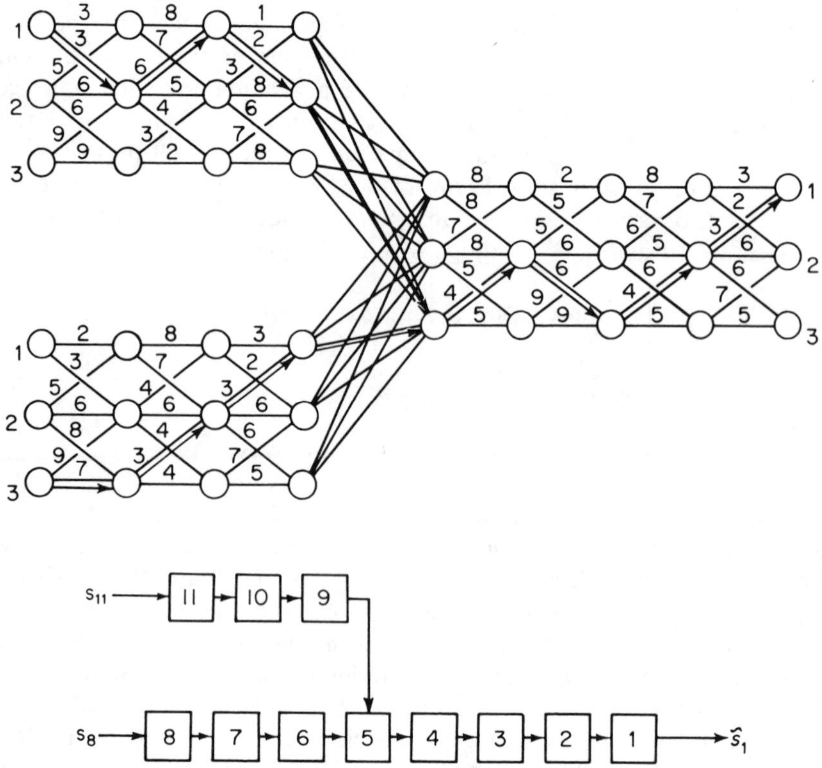

Figure 7-25. Converging branch network problem.

sponding optimal value s_{11}^*:

$$f_3'(1) = 10; \qquad s_{11}^* = 1$$
$$f_3'(2) = 11; \qquad s_{11}^* = 1$$
$$f_3'(3) = 14; \qquad s_{11}^* = 1$$

The returns for stage 5 are given in tabular form in Fig. 7-26 for the two input states $\tilde{s}_9$ and s_5, and the decision variable d_5. The values that d_5 can assume correspond to the values of the output state, $\tilde{s}_5$; in other words, these choices select the input row to the common network made up of stages 1–4. For example, if the input states to stage 5 are $\tilde{s}_9 = 3$ and $s_5 = 2$, the return associated with the decision to go to output state $\tilde{s}_5 = 2$ (that is, $d_5 = 2$) is 3.

$s_5 = 1$

$\tilde{s}_9$ \ d_5	1	2	3
1	1	7	5
2	7	7	2
3	3	3	3

$s_5 = 2$

$\tilde{s}_9$ \ d_5	1	2	3
1	8	6	7
2	5	1	7
3	7	3	6

$s_5 = 3$

$\tilde{s}_9$ \ d_5	1	2	3
1	4	7	2
2	6	7	4
3	9	5	6

Figure 7-26. Tabular description of stage 5 in converging branch network problem.

Now the two-variable optimization

$$f_8(s_9) = \min_{\tilde{s}_9, d_5} [R_5(\tilde{s}_9, s_5, d_5) + f_3'(T_4(\tilde{s}_9, d_5)) + f_4(T_5(s_5, d_5))]$$

must be carried out at stage 5; this is not difficult for this simple network problem. The results of this optimization are

$$f_8(1) = 30; \quad \text{for} \quad \tilde{s}_9^* = 2 \text{ and } d_5 = 3$$
$$f_8(2) = 31; \quad \text{for} \quad \tilde{s}_9^* = 2 \text{ and } d_5 = 2$$
$$f_8(3) = 29; \quad \text{for} \quad \tilde{s}_9^* = 1 \text{ and } d_5 = 3$$

The remaining stages of the system are optimized using the conventional serial methods. If, for example, the initial state is given as $s_8 = 3$, the optimal return is $f_{11}(3) = 43$, resulting in $s_{11}^* = 1$, $\tilde{s}_1^* = 1$, as shown by the corresponding optimal path described by the arrows in Fig. 7-25.

7-24 CONVERGING ALLOCATION: SUPERPOSITION

From the theoretical development and example problems just given, it can be seen that, for general return and transition functions, diverging branch problems can be solved with no more effort than that needed for the same size serial problem, whereas the treatment required for converging branch problems is more complicated. In allocation problems where the returns and transition functions are *linear*, however, a system having n converging branches can be solved as n serial problems, and the resulting solutions superimposed to form the solution to the original branched problem (Beightler, Johnson, and Wilde). It suffices to demonstrate this principle for a system having just one branch, as in Fig. 7-27, and transition functions:

$$\tilde{s}_i = a_i s_i + b_i d_i; \qquad i = 1, \ldots, N + P \tag{7-63; i}$$

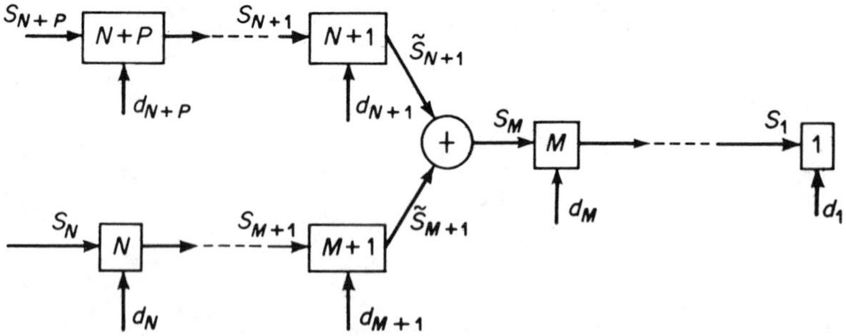

Figure 7-27. Superposition in linear converging branch problem.

where $a_i > 0$, $a_i + b_i > 0$, and the incidence identities are

$$s_i \equiv \tilde{s}_{i+1}; \qquad i = 1, \ldots, M - 1, M + 1, \ldots, N - 1, N + 1, \ldots, N + P$$

and

$$s_M \equiv \tilde{s}_{N+1} + \tilde{s}_{M+1}$$

We wish to maximize $\sum_{i=1}^{N+P} r_i$ where

$$r_i = A_i s_i + B_i d_i; \qquad i = 1, \ldots, N + P \tag{7-64: i}$$

the A_i and B_i being given constants, and the decision variables constrained by

$$0 \leq d_i \leq s_i; \qquad i = 1, \ldots, N + P \tag{7-65: i}$$

Let d_i^* be the optimal policy, and s_i^* the resulting optimal stages, $i = 1, \ldots, N + P$.

Now consider two *serial* systems derived from this branched system. Let serial problem I be

$$\max \left(\sum_{i=1}^{N} r_i \right)$$

where Eqs. (7-63), (7-64), and (7-65) hold for $i = 1, \ldots, N$, and the incidence identities are

$$s_i' \equiv \tilde{s}_{i+1}'; \qquad i = 1, \ldots, N - 1,$$

and let $d_i'^*$, $i = 1, \ldots, N$, be the optimal policy for this problem, and $\tilde{s}_i'^*$ the resulting optimal states.

Let serial problem II be

$$\max \left(\sum_{i=1}^{M} r_i \right) + \sum_{i=N+1}^{N+P} r_i$$

where Eqs. (7-63), (7-64), and (7-65) hold for $i = 1, \ldots, M$, $N + 1, \ldots, N + P$, and the incidence identities are

$$s_i'' \equiv \tilde{s}_{i+1}''; \qquad i = 1, \ldots, M - 1, N + 1, \ldots, N + P - 1$$

and

$$s_M'' \equiv \tilde{s}_{N+1}''$$

and let $d_i''^*$, $i = 1, \ldots, M$, $N + 1, \ldots, N + P$, be the optimal policy for this problem, and $\tilde{s}_i''^*$, $i = 1, \ldots, M$, $N + 1, \ldots, N + P$, the resulting optimal states.

Superposition Theorem

1. *The* qualitative *policies for all three problems are the same:*

$$\frac{d_i}{s_i} = \begin{cases} \dfrac{d_i'}{s_i'}, & \text{if } s_i' \neq 0 \\ 0, & \text{if } s_i' = 0 \end{cases} \qquad i = 1, \ldots, N \tag{7-66$'$}$$

and

$$\frac{d_i}{s_i} = \begin{cases} \dfrac{d_i''}{s_i''}, & \text{if } s_i'' \neq 0 \\ 0, & \text{if } s_i'' = 0 \end{cases} \qquad \begin{matrix} i = 1, \ldots, M, \\ N + 1, \ldots, N + P \end{matrix} \tag{7-66$''$}$$

2. *Superposition of the* quantitative *policies for problems I and II gives the quantitative policy for the branch problem:*

$$\left.\begin{aligned} s_i &= s_i' + s_i'' \\ d_i &= d_i' + d_i'' \end{aligned}\right\} \quad i = 1, \ldots, M$$

$$\left.\begin{aligned} s_i &= s_i' \\ d_i &= d_i' \end{aligned}\right\} \quad i = M+1, \ldots, N \qquad \text{(7-67: } i\text{)}$$

$$\left.\begin{aligned} s_i &= s_i'' \\ d_i &= d_i'' \end{aligned}\right\} \quad i = N+1, \ldots, N+P$$

Proof: Let

$$f_i(s_i, s_{N+1}, d_{N+1}) \equiv \max_{d_1,\ldots,d_i} \left(\sum_{j=1}^{i} r_j\right); \qquad i = 1, \ldots, N$$

where r_j is defined by Eq. (7-64: j). Then (by induction on i), for the branch problem,

$$\begin{aligned} f_i(s_i, s_{N+1}, d_{N+1}) &= \max_{0 \le d_i \le s_i} (\mu_i s_i + \lambda_i d_i + \delta_i \tilde{s}_{N+1}) \\ &= k_i s_i + \delta_i \tilde{s}_{N+1}; \qquad i = 1, \ldots, N \end{aligned} \qquad \text{(7-68)}$$

where $k_0 \equiv 0$

$$\left.\begin{aligned} \lambda_i &\equiv B_i + k_{i-1} b_i \\ \mu_i &\equiv A_i + k_{i-1} a_i \\ k_i &\equiv \max\{\mu_i, \lambda_i + \mu_i\} \end{aligned}\right\} i = 1, \ldots, N$$

$$\delta_i \equiv \begin{cases} 0, & \text{for } i = 1, \ldots, M \\ k_M, & \text{for } i = M+1, \ldots, N \end{cases}$$

Then the optimal decisions, d_i^*, are given by

$$\frac{d_i^*}{s_i^*} = \begin{cases} 0 & \text{if } \lambda_i \le 0; \\ 1 & \text{if } \lambda_i \ge 0' \end{cases} \qquad i = 1, \ldots, N$$

where

$$s_N^* \equiv s_N \quad \text{and} \quad s_i^* = a_{i+1} s_{i+1}^* + b_{i+1} d_{i+1}^*,$$

for

$$i = 1, \ldots, M-1, M+1, \ldots, N-1 \qquad \text{(7-69: } i\text{)}$$

and

$$s_M^* = a_{M+1} s_{M+1}^* + b_{M+1} d_{M+1}^* + a_{N+1} s_{N+1}^* + b_{N+1} d_{N+1}^* \qquad \text{(7-69: } M\text{)}$$

This holds for all values of s_{N+1} and d_{N+1}, and in particular when $s_{N+1} = d_{N+1} = 0$, which is the case for serial problem I, the optimal decisions and states of which are $d_i'^*$ and $s_i'^*$, respectively. Therefore,

$$\frac{d_i'^*}{s_i'^*} = \frac{d_i^*}{s_i^*}$$

as asserted in Eq. (7-66′). A similar argument can be used to prove Eq. (7-66″).

Since $s_N \equiv s_N'$, Eq. (7-67: i) for $i = M + 1, \ldots, N$, can be proved inductively using Eqs. (7-66′) and (7-69: i). The proof for $i = N + 1, \ldots, N + P$ is similar, based on the identity of s_{N+P} and s''_{N+P}.

In serial problem I, $s'_{N+1} = d'_{N+1} \equiv 0$ and Eq. (7-69: i) becomes

$$s_i'^* = a_{i+1}s_{i+1}'^* + b_{i+1}d_{i+1}'^*; \qquad i = 1, \ldots, M \qquad (7\text{-}69'\colon i)$$

Similarly for serial problem II, $s''_{M+1} = d''_{M+1} \equiv 0$, so

$$s_M''^* = a_{N+1}s_{N+1}''^* + b_{N+1}d_{N+1}''^* \qquad (7\text{-}69''\colon M)$$

and

$$s_i''^* = a_{i+1}s_{i+1}''^* + b_{i+1}d_{i+1}''^*; \qquad i = 1, \ldots, M-1 \quad (7\text{-}69''\colon i)$$

Combination of Eqs. (7-69: i), (7-69: M), (7-69′: i), (7-69″: M), and (7-69″: i) with Eqs. (7-66′) and (7-66″) gives, by induction,

$$s_i^* = s_i'^* + s_i''^*; \qquad i = 1, \ldots, M \qquad (7\text{-}67\colon i)$$

which completes the proof.

The superposition theorem also holds for more general systems. First, the transition functions may be written as inhomogeneous linear expressions containing a constant, K_i:

$$\tilde{s}_i = a_i s_i + b_i d_i + K_i$$

since adding a constant to the homogeneous linear transitions will not affect the *qualitative* policy. Second, the theorem is also valid for those systems in which the transition function at the branching junction has the more general form

$$s_M = \gamma\tilde{s}_{N+1} + \phi\tilde{s}_{M+1}$$

where γ and ϕ are any real constants. Third, the foregoing results generalize to large systems comprised of any number of linear branches, so that each branch may be analyzed independently of the others.

Generally, the method of superposition is applicable only to initial-value, linear converging branch problems or to final-value linear diverging branch problems (which are mathematically equivalent). If a nonlinear branch is adjoined to a linear system, the optimal qualitative decisions in the linear portion are unaffected by the introduction of the branch. This is clear from Eq. (7-68), which could just as well have been written

$$f_i(s_i, s_{N+1}, d_{N+1}) = k_i s_i + \delta_i[\Phi(s_{N+1}, d_{N+1})]$$

where $\Phi(s_{N+1}, d_{N+1})$ is any analytic function, without affecting the subsequent analysis and proof.

These results have an economic interpretation. Consider a firm that has worked out an optimal policy for a linear allocation problem. Even if an unknown number of mergers at arbitrary future times were to add allocation capital to the system, the original qualitative plan would still be optimal—even if the merging firms were nonlinear. Moreover, the original *quantitative* plan remains optimal until the first merger takes place. Therefore, long-range planners with linear allocation problems need never worry about their policies being upset by future mergers.

When the method of superposition is applied to linear diverging branch problems in which one is free to choose the branch inputs, the branching problem becomes a single serial system. Consider, for example, the system shown schematically in Fig. 7-28. The total return for stage $M + 1$ plus the returns for all stages to the right is

$$f_{M+P-N+1}(s_{M+1}) = \max_{d_{M+1}} (r_{M+1} + k_P s_P + k_M s_M)$$

We lose no generality in assuming that

$$s_M + s_P = \tilde{s}_{M+1}$$

since a more general relationship between these variables could be achieved by inserting a decisionless stage for that purpose at the circled junction point in Fig. 7-28. Now since the branch inputs are decision variables in this prob-

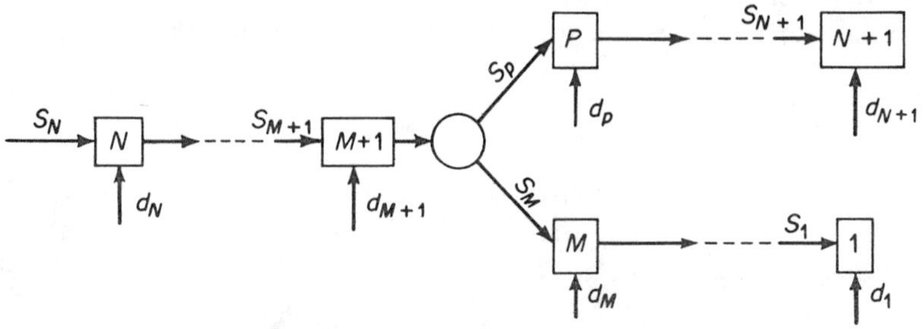

Figure 7-28. Superposition in linear diverging branch problem.

lem, one simply chooses $s_M^* = 0$ when $k_P \geq k_M$, and $s_P^* = 0$ when $k_P < k_M$. Thus, in every case, one of the branches receives no input and is effectively removed from the system, producing a simple serial structure. Note that this is a special case of the linear diverging branch problem. In general, the branch inputs may not be choice variables but might instead be determined by the expressions

$$s_P = a_{M+1}s_{M+1} + b_{M+1}d_{M+1}$$
$$s_M = \tilde{a}_{M+1}s_{M+1} + \tilde{b}_{M+1}d_{M+1}$$

where a_{M+1}, b_{M+1}, $\tilde{a}_{M+1}$, and $\tilde{b}_{M+1}$ are given constants. It is left as an exercise for the reader to show that for this general case, the superposition principle does *not* hold. In fact, it is easy to show that this principle, which is so useful in solving linear converging branches, does not apply to systems containing feedback or bypass loops (see Exercise 7-13).

7-25 BRANCH COMPRESSION PRINCIPLE

As pointed out in Section 7-22, solving the general converging branch problem presents formidable computational difficulties. The simultaneous optimization over two variables required at the junction of the converging branch with the main trunk (the decisionless stage Φ in Fig. 7-24) is the complicating factor which has prevented the solution of most nonserial multistage problems. In this section, we present a compression principle which permits converging branches to be compressed into single pseudo-stages in an equivalent serial system. This principle allows any type of system having multiple converging branches to be replaced by a single serial system that can then be optimized by dynamic programming. The resulting reduction in dimensionality makes possible the solution of complex problems which were previously not amenable to analysis.

Although branch compression is applicable to a system containing any number of converging branches, we lose no generality in considering the system shown schematically in Fig. 7-24, which has exactly one converging branch (here chosen arbitrarily as stages $N + 1$ through $N + P$). The major computational difficulty arises in attempting to maximize Eq. (7-62) over both d_{M+1} and $\tilde{s}_{N+1}$ simultaneously. This difficulty can be overcome by a suitable decomposition of the optimization operations at the branch junction point. Meier and Beightler showed that Eq. (7-62) may be decomposed into the equivalent expression

$$f_{M+P+1}(s_{M+1}) = \max_{d_{M+1}} \{R_{M+1}(s_{M+1}, d_{M+1}) + \max_{\tilde{s}_{N+1}} [f_P(\tilde{s}_{N+1}) + f_M\{\phi_1(T_{M+1}(s_{M+1}, d_{M+1})) + \phi_2(\tilde{s}_{N+1})\}]\} \tag{7-70}$$

Under the decomposition described by Eq. (7-70), the maximization on $\tilde{s}_{N+1}$ is carried out treating $\tilde{s}_{M+1}$ as a *state* variable, where $\tilde{s}_{M+1}$ is given by $\phi_1(T_{M+1}(s_{M+1}, d_{M+1}))$. This decomposition is similar to that effected by Bellman's *principle of optimality*, developed in Section 7-02, and equivalently transforms a nonserial problem into a serial problem (see the next section for the application of this principle to nonserial problems containing loops).

The remainder of the main trunk is optimized by serial dynamic programming, involving only a one-state, one-decision optimization at each stage:

$$f_{P+k}(s_k) = \max_{d_k} [R_k(s_k, d_k) + f_{P+k-1}(T_k(s_k, d_k))]; \qquad k = M+2, \ldots, N \tag{7-71}$$

Therefore, this method of decomposition permits an entire converging branch multistage problem to be solved just as one would solve a serial system. In fact, this conversion can be shown graphically by redrawing the functional diagram to exhibit the equivalent serial system, as shown in Fig. 7-29.

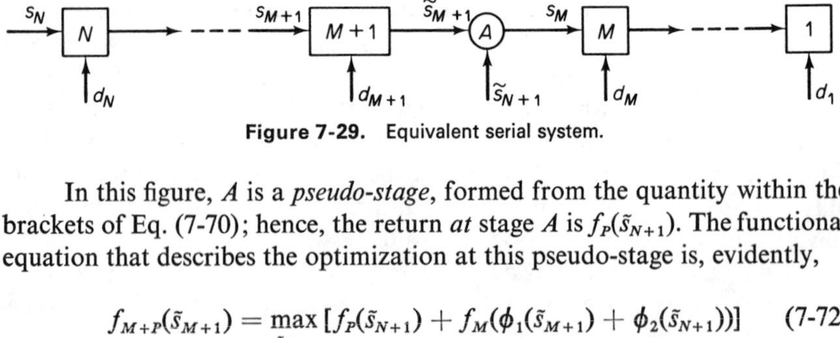

Figure 7-29. Equivalent serial system.

In this figure, A is a *pseudo-stage*, formed from the quantity within the brackets of Eq. (7-70); hence, the return *at* stage A is $f_P(\tilde{s}_{N+1})$. The functional equation that describes the optimization at this pseudo-stage is, evidently,

$$f_{M+P}(\tilde{s}_{M+1}) = \max_{\tilde{s}_{N+1}} [f_P(\tilde{s}_{N+1}) + f_M(\phi_1(\tilde{s}_{M+1}) + \phi_2(\tilde{s}_{N+1}))] \tag{7-72}$$

where $\tilde{s}_{M+1}$ is given by $T_{M+1}(s_{M+1}, d_{M+1})$.

At stage $M + 1$, then, the functional equation for this equivalent serial system is given by

$$f_{M+P+1}(s_{M+1}) = \max_{d_{M+1}} [R_{M+1}(s_{M+1}, d_{M+1}) + f_{M+P}(\tilde{s}_{M+1})] \tag{7-73}$$

where $\tilde{s}_{M+1}$ is obtained from $T_{M+1}(s_{M+1}, d_{M+1})$.

Equations (7-72) and (7-73) are completely equivalent to that of Eq. (7-63); hence the two-decision problem at the junction has been replaced by two one-decision problems, which are very much easier to solve.

Equation (7-72) defines a mathematical *compression* of the converging branch into a single pseudo-stage. This technique can be used to compress any number of converging branches into pseudo-stages and therefore will reduce any such nonserial system into a single equivalent serial system. Consider, for example, a multistage problem having K converging branches joining the main trunk at a single point. The standard analysis, exemplified by Eq. (7-70), would require that a $K + 1$-decision problem be solved; that is, the simultaneous selection of $K + 1$ variables. By compressing these branches, however, the optimization could be transformed into $K + 1$ one-decision problems. Branch compression, of course, applies no matter at what points on the main trunk the various branches join.

To fix ideas, consider the simple problem described by the functional diagram in Fig. 7-30, where the problem is to maximize the sum of the stage returns, given by $r_1 = s_1 d_1 - 4d_1^2$, $r_2 = 10d_2 - d_2^2$, and $r_3 = 2d_3 - d_3^2$. There are no restrictions on the values that the decision variables can assume.

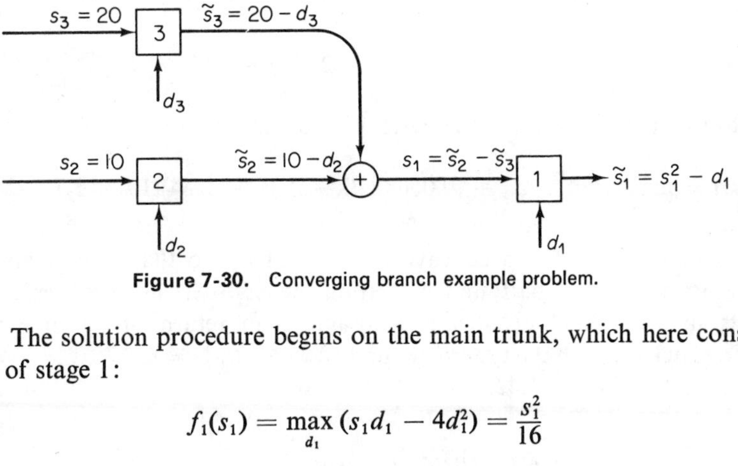

Figure 7-30. Converging branch example problem.

The solution procedure begins on the main trunk, which here consists only of stage 1:

$$f_1(s_1) = \max_{d_1} (s_1 d_1 - 4d_1^2) = \frac{s_1^2}{16}$$

where $d_1^* = s_1/8$, for any value of s_1.

Next, the converging branch, which in this simplified problem contains only stage 3, is optimized, and its total return expressed as a function of the branch output, $\tilde{s}_3$. Using decision inversion, we find that $d_3^* = 20 - \tilde{s}_3$, and the optimal branch return is $38\tilde{s}_3 - \tilde{s}_3^2 - 360$.

Figure 7-31 exhibits the functional diagram of the equivalent serial

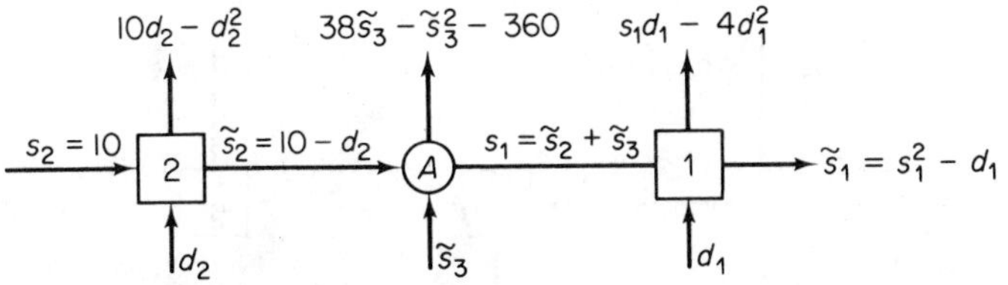

Figure 7-31. Equivalent serial system for example problem.

system for this nonserial problem, with the individual stage returns shown above the corresponding stage.

Optimization over stage 1 and the pseudo-stage A, which was formed by branch compression, is obtained by substituting the appropriate values for this problem into Eq. (7-72):

$$f_2(\tilde{s}_2) = \max_{s_3}\left[38\tilde{s}_3 - \tilde{s}_3^2 - 360 + \frac{(\tilde{s}_2 + \tilde{s}_3)^2}{16}\right]$$

For any fixed value of $\tilde{s}_2$, the expression within the brackets is a concave function of $\tilde{s}_3$; hence it is sufficient to set to zero the first partial of this expression with respect to $\tilde{s}_3$, which yields

$$\tilde{s}_3^* = \tfrac{1}{15}(\tilde{s}_2 + 304)$$

so

$$f_2(\tilde{s}_2) = 0.0668\tilde{s}_2^2 + 2.54\tilde{s}_2 + 24.7$$

The total return from the entire system is now given by

$$f_3(10) = \max_{d_2}\,[10d_2 - d_2^2 + (0.0668)(10 - d_2)^2 + (2.54)(10 - d_2) + 24.7]$$

This maximand is a concave function of d_2, so that once again the maximum value may be found by setting the first derivative to zero. This results in $d_2^* = 3.28$, producing a total system return of $f_3(10) = 67.4$. Tracing back through the system, we find $\tilde{s}_2^* = 6.72$, $\tilde{s}_3^* = 20.67$, $d_3^* = -0.67$, $s_1^* = 27.39$, and $d_1^* = 3.42$.

7-26 SYSTEMS CONTAINING LOOPS

Consider now the nonserial system shown in Fig. 7-32, which contains a feedback loop consisting of stages $A1$–$A\alpha$; such a system is said to be *cyclic* or *looped.* Cycles arise in numerous industrial "recycle" or "feedback" tech-

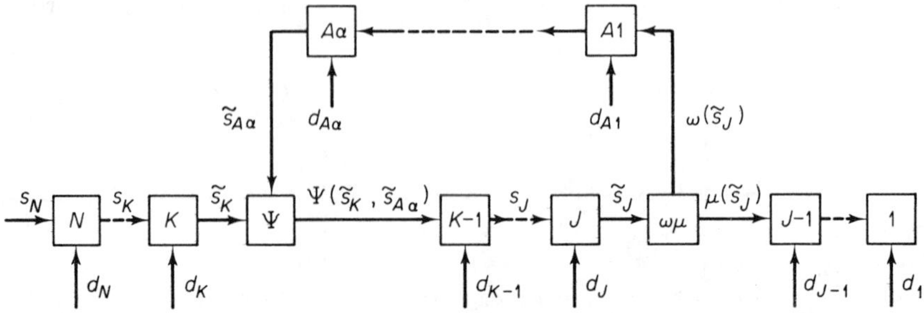

Figure 7-32. Cyclic system.

nologies, as well as in feedforward, or bypass systems. In the latter type of system, the direction of the arrows in the $A1$–$A\alpha$ loop of this figure is reversed.

In the feedback system of Fig. 7-32, s_N is a given input, and Ψ and $\omega\mu$ are decisionless stages which serve only to map their inputs onto the output states given by the real-valued functions $\Psi(\tilde{s}_K, \tilde{s}_{A\alpha})$, $\omega(\tilde{s}_J)$, and $\mu(\tilde{s}_J)$. The problem may be described mathematically as

$$\max_{\substack{d_1,\ldots,d_N,\\ d_{A1},\ldots,d_{A\alpha}}} \left[\sum_{n=1}^{N} R_n(s_n, d_n) + \sum_{i=1}^{\alpha} R_{Ai}(s_{Ai}, d_{Ai})\right], \tag{7-74}$$

where the R_j are any real-valued functions, and the transition functions

$$\begin{aligned} \tilde{s}_n &= T_n(s_n, d_n); \qquad & n &= 1, \ldots, N, \\ \tilde{s}_{Ai} &= T_{Ai}(s_{Ai}, d_{Ai}); \qquad & i &= 1, \ldots, \alpha, \end{aligned} \tag{7-75}$$

describe how each *decision* d_j maps input state s_j onto $\tilde{s}_j$ at stage j. The following incidence identities relate all of the system inputs and outputs:

$$s_{n-1} = \tilde{s}_n; \qquad n = 2, \ldots, J-1, J+1, \ldots, K-1, K+1, \ldots, N, \tag{7-76}$$

$$s_j = \tilde{s}_{j-1}; \qquad j = A2, \ldots, A\alpha. \tag{7-77}$$

No generality is lost in considering this particular system, which has exactly one feedback loop, since the analysis extends in a straightforward manner to any number of loops.

First, $f_{J-1}(\mu(\tilde{s}_J))$ is obtained as a function of $\tilde{s}_J$ by using serial dynamic programming. Then, using decision inversion, the optimal loop return, $f_\alpha(\tilde{s}_{A\alpha}, \omega(\tilde{s}_J))$, is calculated as a function of both the input $\omega(\tilde{s}_J)$ and the output $\tilde{s}_{A\alpha}$. Finally, the optimal return from the segment consisting of stages J through $K-1$ is obtained as a function of both input and output:

$$f_{K-J}(\Psi(\tilde{s}_K, \tilde{s}_{A\alpha}), \tilde{s}_J)$$

Thus the optimal return from stages $A1$–$A\alpha$ and stages 1–K is given by

$$\begin{aligned} f_{K+\alpha}(s_K) = \max_{d_K, \tilde{s}_J, \tilde{s}_{A\alpha}} & [R_K(s_K, d_K) + f_\alpha(\tilde{s}_J)) \\ & + f_{K-J}(\Psi(\tilde{s}_K, \tilde{s}_{A\alpha}), \tilde{s}_J) + f_{J-1}(\mu(\tilde{s}_J))] \end{aligned} \tag{7-78}$$

where the two-point boundary solutions f_α and f_{K-1} have been obtained by decision inversion at stages $A\alpha$ and J, respectively. Hence the optimal returns for these segments are now given as functions of the segment inputs and outputs, the decisions d_J through d_{K-1} and d_{A1} through $d_{A\alpha}$ having been optimized out. Simultaneous selection of a value for d_K, $\tilde{s}_J$, and $\tilde{s}_{A\alpha}$ would, for a given value of the state variable s_K, determine the total return for stages 1–K plus all stages in the loop. This three-decision, one-state optimization is, as one might expect, quite difficult to carry out computationally. In the past, the three-dimensional optimization process has been approached by choosing particular values for $\tilde{s}_{A\alpha}$ and $\tilde{s}_J$ and then optimizing Eq. (7-78) over d_K. Conceptually, this could be repeated for all $\tilde{s}_{A\alpha}$, $\tilde{s}_J$ pairs; in some cases, the computational effort could be reduced through the use of direct search techniques. Notice that it is essential for the triple-decision optimization to be carried out at stage K; otherwise, $\tilde{s}_{A\alpha}$ and $\tilde{s}_J$ would have to be carried as state variables to be examined exhaustively during the optimization of every stage $K + 1$ through N. Once this optimization step is carried out, the remainder of the system is optimized by serial dynamic programming.

In order to see the effects of a loop superimposed on a serial system, we begin with the simplest such problem. Let us again consider the network problem of Fig. 7-1, where now we impose the additional constraint, $s_4 = \tilde{s}_1$. One may think of this network as a complete map of the earth, as shown in Fig. 7-33, which has the network of Fig. 7-1 folded back on itself so that New York now takes the place of both Seattle and Boston, Caracas that of San Francisco and Washington, and Santiago that of San Diego and Miami. The problem is to find the optimal (initial and final) state (New York, Caracas, or Santiago) and the optimal around-the-world path, taking into account the costs of the feasible paths.

One is tempted to solve loop problems as follows: first find the optimal solution to the unconstrained problem, and then impose the loop constraint. That is, find $f_N(s_N)$, and then select as the optimal loop the path for which $s_N = \tilde{s}_1$ [or whatever equation, $\tilde{s}_N = g(\tilde{s}_1)$, forms the loop constraint]. The present problem shows clearly the error in this approach. From the solutions to the initial-value problem given in Fig. 7-1, we have $f_4(N) = 18$, $\tilde{s}_1 = C$; $f_4(C) = 16$, $\tilde{s}_1 = C$; and $f_4(S) = 14$, $\tilde{s}_1 = N$. Now, imposing the loop constraint, we have $f_4(C, C) = 16$, $s_4 = C$, $\tilde{s}_1 = C$ (that is, begin and end in Caracas, following the path through London, Cairo, and Okhotsk). This is *not* the correct solution, however, since the constraint must be made a part of the problem from the outset.

To solve this problem correctly, one begins with a decision inversion at stage 1, for a given value of the output state, and proceeds as in a final-value problem, terminating with the return $f_N(s_N, \tilde{s}_1)$. Thus one has solved a problem with two fixed ends, and may now set $s_N = \tilde{s}_1$. Solving the problem either as an initial-value problem, or as a final-value problem using *state* inversion, implicitly assumes a *free* end, which does not exist when a loop is present. In Fig. 7-33, the solution is given for $\tilde{s}_1 \equiv S$, so that $f_4(N, S) = 20$, $f_4(C, S) = 18$, and $f_4(S, S) = 15$. By repeating this procedure for the output states $\tilde{s}_1 \equiv N$ and $\tilde{s}_1 \equiv C$, one finds $f_4(N, N) = 19$, and $f_4(C, C) = 16$.

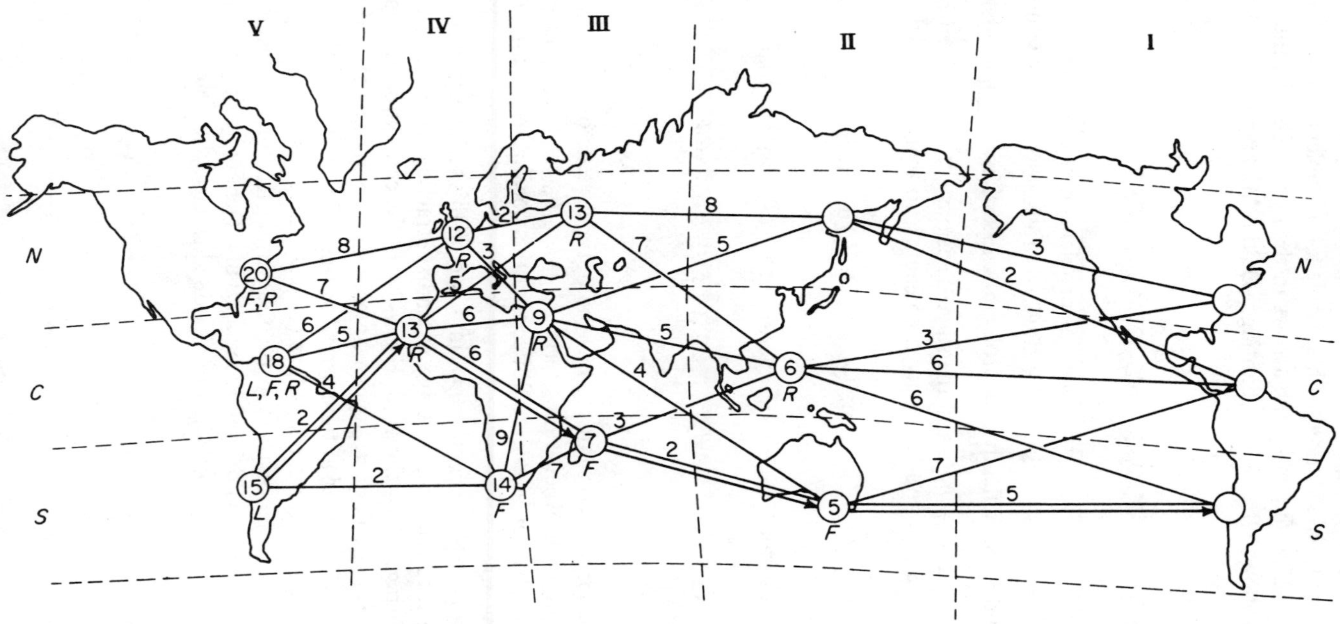

Figure 7-33. Network routing problem with loop constraint.

Therefore, the optimal solution to this simple loop problem is $f_4^*(S, S) = 15$, with $s_4 = \tilde{s}_1 = S$ as shown by the arrows in the figure. That is, the optimal loop path follows the route Santiago–Dakar–Madagascar–Melbourne–Santiago.

7-27 STAGE-LESS LOOPS AND QUASI-LOOPS

Here we consider the important class of problems first discussed in Section 7-16, in which the stage returns and transitions are *linear* in the continuous variables s_i and d_i. Specifically, the problem is to maximize $\sum_{i=1}^{P} r_i$, where $r_i = A_i s_i + B_i d_i$, $\tilde{s}_i = a_i s_i + b_i d_i$, and the stage decisions are constrained by $0 \leq d_i \leq s_i$. The A_i, B_i, a_i, and b_i are given constants, so that for a serial system,

$$f_1(s_1) = \max_{0 \leq d_1 \leq s_1} (A_1 s_1 + B_1 d_1) = k_1 s_1$$

where k_1 is a constant. Then by induction, it is easy to show that

$$f_n(s_n) = \max_{0 \leq d_n \leq s_n} [A_n s_n + B_n d_n + k_{n-1}(a_n s_n + b_n d_n)] = k_n s_n \qquad n = 2, \ldots, P \tag{7-79}$$

where k_{n-1} and k_n are constants. Thus the optimal value of d_n is either s_n or zero, depending upon whether the term $B_n + k_{n-1} b_n$, the coefficient of d_n, is respectively positive or negative. Notice that this coefficient does *not* depend upon the input s_n.

Now suppose that a *quasi-loop* is imposed on this serial system by adding a constant, βC, to the input to stage N, and subtracting a constant, C, from the output of stage M, as shown in Fig. 7-34. In the context of an allocation problem, one may think of βC as investment capital borrowed at time period N, and C as the sum repaid at time period M, including interest. [Thus $\beta < 1$, and the effective annual interest rate is $(1 - \beta)/\beta(N - M)$.]

The effect of this quasi-loop on the serial structure is best seen by introducing the constants βC and C sequentially. Denote by single primes the optimal returns which result when βC is added to the system at state N (and

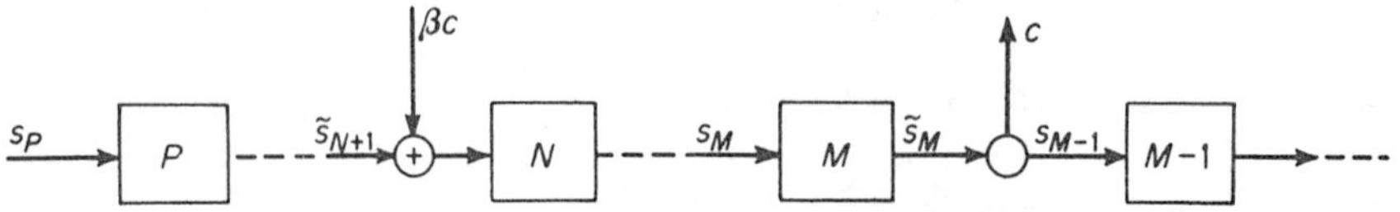

Figure 7-34. Quasi-loop imposed on serial system.

C is *not* subtracted from the output of stage M). Then by Eq. (7-79) and Fig. 7-34 it is clear that

$$f'_N = k_N(\tilde{s}_{N+1} + \beta C) \quad \text{and} \quad f'_{M-1} = k_{M-1}\tilde{s}_M\left(\frac{\tilde{s}_{N+1} + \beta C}{\tilde{s}_{N+1}}\right)$$

Denote by double primes the stage inputs and optimal returns that result when C is subtracted from the output of stage M (and βC is *still* added to the input to stage N). Then, owing to the linearity of the system, we find

$$s''_{M-1} = \tilde{s}_M\left(\frac{\tilde{s}_{N+1} + \beta C}{\tilde{s}_{N+1}}\right) - C$$

from which it follows that the optimal return from all P stages of this quasi-loop system is

$$f''_P = f_P + f'_N - f_N + f''_{M-1} - f'_{M-1} = f_P + C(\beta k_N - k_{M-1})$$

where f_P is the optimal P-stage return for the original serial system. Accordingly, β must be at least k_{M-1}/k_N to make the investment loan profitable.

Furthermore, since the coefficient of the decision variable at each stage is not a function of the input to that stage, the *qualitative* policies for the two problems are the same (Beightler, Johnson, and Wilde):

$$\frac{d''_n}{s''_n} = \begin{cases} \dfrac{d_n}{s_n} & \text{if } s_n \neq 0 \\ 0 & \text{if } s_n = 0 \end{cases} \tag{7-80}$$

This quasi-loop is the most degenerate case of a loop constraint, since the fixed amounts added to, and subtracted from, the two stages are *not* affected by the system *decisions*. We shall now consider real loop problems in which the quantities cycled are dependent upon the decisions employed at each stage, and in general the optimal loop policy will differ from the policy which is optimal when the loop constraint is removed.

As an example of a real loop constraint applied to a problem in continuous variables, consider the following: maximize $\sum_{i=1}^{4} r_i$, where $r_i = 0.5s_i - 0.2d_i$, $\tilde{s}_i = 0.7s_i + 0.4d_i$, $\tilde{s}_1 \equiv \tilde{c}_1$, and $0 \leq d_i \leq s_i$, for $i = 1, \ldots, 4$. This problem was solved earlier by decision inversion, with the resulting expression for $f_4(s_4, \tilde{c}_1)$ given in four ranges of values for s_4. If we now append the homogeneous loop constraint

$$\tilde{c}_1 = \tfrac{1}{2}s_4$$

the corresponding serial solution is $f_4(s_4, \tilde{c}_1) = 1.235s_4 + 0.715\tilde{c}_1$, since this value applies in the range $1.69\tilde{c}_1 \leq s_4 \leq 2.66\tilde{c}_1$. Substituting the loop constraint into this solution results in $f_4(\tilde{c}_1) = 3.185\tilde{c}_1$. Beginning with the given value of $d_4^* = s_4 \equiv 2\tilde{c}_1$, we find $s_3^* = (0.7)(2\tilde{c}_1) + (0.4)(2\tilde{c}_1) = 2.2\tilde{c}_1$, so from the optimal three-stage returns, we compute

$$d_3^* = 5.11\tilde{c}_1 - (1.75)(2.2\tilde{c}_1) = 1.26\tilde{c}_1$$

Using this value, it follows that

$$s_2^* = (0.7)(2.2\tilde{c}_1) + (0.4)(1.26\tilde{c}_1) = 2.044\tilde{c}_1$$

and from the value of $f_2(s_2, \tilde{c}_1)$, we have

$$d_2^* = 3.57\tilde{c}_1 - (1.75)(2.044\tilde{c}_1) = 0$$

just as was the case in the serial problem. Then, $s_1^* = (0.7)(2.044\tilde{c}_1) = 1.43\tilde{c}_1$, and by the decision inversion equation, $d_1^* = 2.5\tilde{c}_1 - (1.75)(1.43\tilde{c}_1) = 0$. Using these values of the s_i^* and d_i^*, we may verify the value of $f_4(\tilde{c}_1)$ found earlier:

$$\sum_{i=1}^{4} r_i = (0.5)(1.43\tilde{c}_1) + (0.5)(2.044\tilde{c}_1) + (0.5)(2.2\tilde{c}_1) - (0.2)(1.26\tilde{c}_1)$$
$$+ (0.5)(2\tilde{c}_1) - (0.2)(2\tilde{c}_1) = 3.185\tilde{c}_1$$

In order to illustrate once again the fallacy of not taking a constraint into account at all times in an optimization problem, consider the previous example solved as an initial-value serial problem. The solution to this serial problem was obtained earlier: $d_4^* = s_4$, $d_3^* = 1.1s_4$, $0 \leq d_2^* \leq 1.21s_4$, and $d_1^* = 0$, producing an optimal four-stage return of $f_4(s_4) = 1.6585s_4$. To be specific, let us take $d_2^* = 0$. With this solution, we find $\tilde{s}_1 = 0.5929s_4$. If the loop constraint, $s_4 = \tilde{c}_1^2 - 0.313\tilde{c}_1$, is now imposed on the preceding solution, then (with $\tilde{s}_1 \equiv \tilde{c}_1$), there results: $\tilde{c}_1^* = 2$, $s_4^* = 3.374$, for a total return of $f_4(s_4^*) = (1.6585)(3.374) = 5.596$.

The correct procedure for solving this problem, however, is to perform a decision inversion at stage 1, thus obtaining the solution in terms of both s_4 and $\tilde{c}_1$. These calculations were carried out earlier, and combining those results with the loop constraint, we find the true optimal return for this loop problem to be $f_4(s_4^*, \tilde{c}_1^*) = 23.67$, where $\tilde{c}_1^* = 4.483$, $s_4^* = 18.694$, and the optimal policy is $d_i^* = 0$, $i = 1, \ldots, 4$.

As a final example of a stage-less loop system, we shall solve a problem in which the stage returns are nonlinear. We wish to maximize $\sum_{i=1}^{3} r_i$, where $r_i = 5d_i - id_i^2$, $\tilde{s}_i = s_i - 0.4d_i$, $0 \leq d_i \leq s_i$, for $i = 1, 2, 3$, and $\tilde{s}_1 \equiv \tilde{c}_1$.

This serial problem was solved in a previous example, using decision inversion, with the solution obtained in the form $f_3(s_3, \tilde{c}_1)$. We shall employ the preceding solution in solving the loop problem that results when the constraint $s_3 = 2\tilde{c}_1$ is adjoined to this serial system.

Now consider the following context for this problem: let the input and output states be the capital (in, say, millions of dollars) available for investment in an enterprise which yields a return (in the amount of commodity K produced) of r_i during year i, when d_i millions of dollars are invested in the enterprise that year. An allocation of d_i also produces $s_i - 0.4d_i$ of new capital during the year, which results from the sale of by-products. It is desired to maximize the total production of this commodity over the next 3 years if one-half of the total capital invested now (s_3) must be available for repayment of part of the cost of obtaining the initial capital. Note that if the loop constraint were an inequality, say, $2\tilde{c}_1 \geq s_3$, it could be converted into an equality by investing the extra money at stage 1 (that is, liquidate the excess capital at that stage). The form of the solution would not change, however, and we lose no generality in using an equality constraint in this example.

For this loop, the two-point boundary serial solution which is applicable is $f_3(s_3, \tilde{c}_1) = 12.5s_3 - 3.4s_3^2 + 6.8s_3\tilde{c}_1 - 3.4\tilde{c}_1^2 - 12.5\tilde{c}_1$, since it corresponds to the range $\tilde{c}_1 \leq s_3 \leq 2.22\tilde{c}_1$. Substituting the loop constraint into this solution results in $f_3(\tilde{c}_1) = 12.5\tilde{c}_1 - 3.4\tilde{c}_1^2$, which is maximized by choosing $\tilde{c}_1^* = 1.83$; hence the optimal initial capital to invest becomes $s_3^* = 3.66$, and the total return is $f_3(1.83) = 11.5$. Tracing back through the system using the results found for the two-point boundary serial problem, we find that $d_3^* = 0.833$, $d_2^* = 1.24$, and $d_1^* = 2.5$.

7-28 COMPRESSION AND DECOMPOSITION OF LOOPS

In Section 7-26 it was shown that, in order to solve the general cyclic system of Fig. 7-32, it is necessary to carry out a double or triple optimization (depending upon whether the loop is respectively of the feedback or feedforward sense) at the junction points of the loop and the main trunk. For the feedback system of Fig. 7-32, a simultaneous optimization over three decision variables is required, as shown in Eq. (7-78). Beightler, Phillips, and Fowler have shown that these multiple-decision optimization steps can be eliminated by means of a suitable decomposition based on the compression technique described in Section 7-25.

The optimal return from stages A1–$A\alpha$ and stages 1–K, given by Eq.

(7-78), may be decomposed by the compression principle into the equivalent expression

$$f_{K+\alpha}(s_K) = \max_{d_K} \{R_K(s_K, d_K) + \max_{\tilde{s}_J} [\max_{\tilde{s}_{A\alpha}} \{f_\alpha(\tilde{s}_{A\alpha}, \omega(\tilde{s}_J)) + f_{K-J}(\Psi(\tilde{s}_K, \tilde{s}_{A\alpha}), \tilde{s}_J) + f_{J-1}(\mu(\tilde{s}_J))\}]\} \tag{7-81}$$

which consists of three one-state, one-decision optimization steps, where the *range* of $\tilde{s}_{A\alpha}$ may depend upon d_K and $\tilde{s}_J$, and the range of $\tilde{s}_J$ may depend upon d_K.

The remainder of the system, stages $K + 1$ through N, is then optimized by serial dynamic programming, again involving only a one-state, one-decision optimization at each stage:

$$f_{N+\alpha}(s_n) = \max_{d_n} [R_n(s_n, d_n) + f_{n+\alpha-1}(T_n(s_n, d_n))]; \qquad n = K + 1, \ldots, N \tag{7-82}$$

Accordingly, it can be seen that the compression method of decomposition permits any feed-back loop multistage problem to be solved as a sequence of one-state, one-decision problems, just as though the system were serial. In fact, using the pseudo-stage concept introduced in Section 7-25 this decomposition can be shown graphically by redrawing the functional diagram to exhibit the equivalent diverging branch system, as shown in Fig. 7-35. In this figure, both Φ and θ are pseudo-stages, formed from the appropriate expressions in Eq. (7-81). Thus, the return at stage θ is $f_\alpha(\tilde{s}_{A\alpha}, \omega(\tilde{s}_J)) + f_{K-1}(\Psi(\tilde{s}_K, \tilde{s}_{A\alpha}), \tilde{s}_J)$, while stage Φ has no return, but does have an associated decision variable $\tilde{s}_J$ which determines the input state values for stages 1 through J-1, as well as those for pseudo-stage θ.

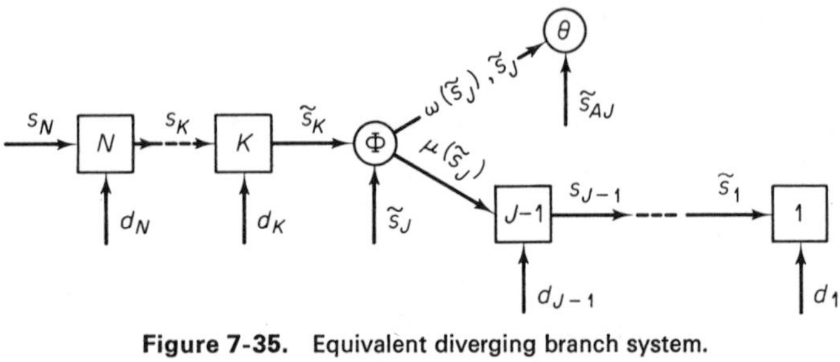

Figure 7-35. Equivalent diverging branch system.

Equation (7-81) is completely equivalent to that of Eq. (7-78), so that the three-decision problem has been replaced by three one-decision problems, which are much easier to solve. In practice the three-dimensional optimization is, in fact, often intractable. Equation (7-81) defines a mathematical compression of the feedback loop into a single diverging branch with associated pseudo-stages. This loop compression technique can be used to compress any number of feedback loops into equivalent diverging branch-pseudo-stage sets, and, therefore, will reduce any such loop system into an equivalent diverging branch system. This latter type of multistage decision system can be optimized directly by serial dynamic programming, regardless of the number of diverging branches present.

It is clear that this principle applies as well to feedforward, or bypass, loops. In such a system, the direction of the arrows in the $A\alpha$–A1 loop of Fig. 7-32 is reversed. The multidecision optimizations at the junction stages can be formulated with one less decision variable, but the compression principle is not altered. The result of the decompositions affected for feedforward loops is again a sequence of one-decision optimizations, with the attendant computational savings.

The loop compression technique will now be illustrated by solving the very simple feedback loop system shown in Fig. 7-36.

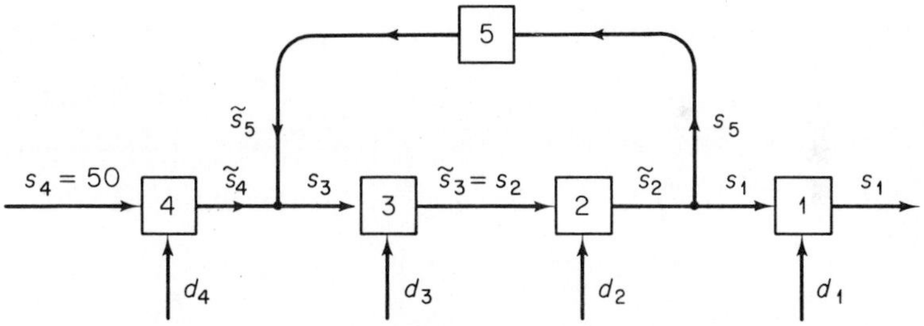

Figure 7-36. Feedback example problem.

The problem is to maximize the sum of the stage returns, r_n, subject to the following constraints and transformations:

Stage, n	*Return, R_n*	*Constraints*	*Transformations*
1	d_1	$0 \le d_1 \le s_1$	
2	$3d_2$	$0 \le d_2 \le s_2$	$s_1 + s_5 = s_2 - \frac{1}{3}d_2 = \tilde{s}_2$
3	$2d_3$	$0 \le d_3 \le \tilde{s}_4 + \tilde{s}_5$	$s_2 = \tilde{s}_4 + \tilde{s}_5 - \frac{1}{3}d_3 = \tilde{s}_3$
4	$4d_4$	$0 \le d_4 \le s_4$	$\tilde{s}_4 = s_4 - \frac{1}{3}d_4$
5	$4d_5$	$0 \le d_5 \le s_5$	$\tilde{s}_5 = s_5 - \frac{1}{3}d_5$

where $s_3 = \tilde{s}_4 + \tilde{s}_5$.

For stage 1, we have at once:

$$f_1(s_1) = s_1 = s_2 - s_5 - \tfrac{1}{3}d_2 = d_1^*$$

The optimal return from the feedback loop, consisting of stage 5, is obtained readily through decision inversion:

$$f_1(s_5, \tilde{s}_5) = 12(s_5 - \tilde{s}_5)$$

From the decision inversion,

$$d_5 = 3(s_5 - \tilde{s}_5)$$

and

$$0 \leq d_5 \leq s_5$$

implies that

$$\tilde{s}_5 \leq s_5 \leq \tfrac{3}{2}\tilde{s}_5$$

Now consider the three-stage system consisting of stages 1, 2, and 5; here there will be two state variables, $\tilde{s}_5$ and s_2. The appropriate functional equation, as obtained by loop compression, is:

$$f_3(\tilde{s}_5, s_2) = \max_{d_2} \{r_2 + \max_{\tilde{s}_5 \in T} [f_1(s_1) + f_1(s_5, \tilde{s}_5)]\}$$

where the set T is defined by

$$0 \leq d_2 \leq s_2, \qquad \tilde{s}_5 \leq s_5 \leq \tfrac{3}{2}\tilde{s}_5$$

and

$$0 \leq s_5 \leq s_2 - \tfrac{1}{3}d_2$$

In addition, the constraints imply that all the state variables be nonnegative.

At this point, we must determine the ranges of the various state and decision variables for which those variables take on their optimum values. We may then compute all the f_n for $n = 2, 3, 4, 5$. Thus

$$s_5^* = \tfrac{3}{2}\tilde{s}_5 \qquad \text{when } 0 \leq \tfrac{3}{2}\tilde{s}_5 \leq s_2 - \tfrac{1}{3}d_2$$

or

$$0 \leq d_2 \leq 3s_2 - \tfrac{9}{2}\tilde{s}_5$$

and

$$s_5^* = s_2 - \tfrac{1}{3}d_2 \qquad \text{when } \tilde{s}_5 \leq s_2 - \tfrac{1}{3}d_2 \leq \tfrac{3}{2}\tilde{s}_5$$

or

$$3s_2 - \tfrac{9}{2}\tilde{s}_5 \leq d_2 \leq 3s_2 - 3\tilde{s}_5$$

Therefore,

$$f_3(\tilde{s}_5, s_2) = 12s_2 - 12\tilde{s}_5 \qquad \text{when } \tfrac{2}{3}s_2 \leq \tilde{s}_5 \leq s_2;\ d_2^* = 0,\ s_5^* = s_2$$
$$9s_2 - \tfrac{15}{2}s_5 \qquad \text{when } \tfrac{4}{9}s_2 \leq \tilde{s}_5 \leq \tfrac{2}{3}s_2;\ d_2^* = 3s_2 - \tfrac{9}{2}\tilde{s}_5$$

and

$$s_5^* = \tfrac{3}{2}\tilde{s}_5$$
$$\tfrac{11}{3}s_2 + \tfrac{9}{2}\tilde{s}_5 \qquad \text{when } 0 \leq \tilde{s}_5 \leq \tfrac{4}{9}s_2;$$
$$d_2^* = s_2,\ s_5^* = \tfrac{3}{2}\tilde{s}_5$$

Then for stages 1, 2, 3, and 5,

$$f_4(s_3) = \max_{\tilde{s}_5} \{\max_{d_3} [2d_3 + f_3(\tilde{s}_5, s_2)]\}$$

Here, the form of $f_3(\tilde{s}_5, s_2)$ is dependent upon the range of $\tilde{s}_5$; additionally, $0 \leq d_3 \leq \tilde{s}_4 + \tilde{s}_5$.

For $0 \leq \tilde{s}_5 \leq \tfrac{4}{9}s_2$,

$$f_4(\tilde{s}_4) = \max_{\tilde{s}_5} \{\max_{d_3} [2d_3 + \tfrac{11}{3}(\tilde{s}_3 + \tilde{s}_5 - \tfrac{1}{3}d_3) + \tfrac{9}{2}\tilde{s}_5]\}$$

therefore, $d_3^* = 0$ and $\tilde{s}_5^* = \tfrac{4}{5}\tilde{s}_4$, which results in

$$f_4(\tilde{s}_4) = 8\tfrac{4}{19}\tilde{s}_4$$

By similar processes, for

$$\tfrac{4}{9}s_2 \leq \tilde{s}_3 \leq \tfrac{2}{3}s_2$$

$d_3^* = 0$ and $\tilde{s}_5^* = 2\tilde{s}_4$ in the range $\tfrac{4}{5}\tilde{s}_4 \leq \tilde{s}_5 \leq 2\tilde{s}_4$, resulting in

$$f_4(\tilde{s}_4) = 12\tilde{s}_4$$

However, $d_3^* = 3\tilde{s}_4 - \tfrac{15}{4}\tilde{s}_5$ and $\tilde{s}_5^* = \tfrac{4}{5}\tilde{s}_4$ in the range $\tfrac{8}{9}\tilde{s}_4 \leq \tilde{s}_5 \leq \tfrac{4}{5}\tilde{s}_4$, resulting in

$$f_4(\tilde{s}_4) = 10\tfrac{1}{5}\tilde{s}_4$$

Finally, for $\tfrac{2}{3}s_2 \leq \tilde{s}_5 \leq s_2$ and $\tilde{s}_5 \leq 2\tilde{s}_4$, $d_3^* = 0$, and $\tilde{s}_5^* =$ any value such that $\tfrac{2}{3}s_2 \leq \tilde{s}_5 \leq s_2$, and $f_4(\tilde{s}_4) = 12\tilde{s}_4$.

In the range, $\tilde{s}_5 \geq 2\tilde{s}_4$, $d_3^* = 3\tilde{s}_4 - \frac{3}{2}\tilde{s}_5$, $\tilde{s}_5^* = 2\tilde{s}_4$, and $f_4(\tilde{s}_4) = 12\tilde{s}_4$. From these results, $f_4(\tilde{s}_4)$ can be chosen as $12\tilde{s}_4$ to maximize the return through four stages as a function of $\tilde{s}_4$. Then

$$\begin{aligned} f_5(s_4) &= \max_{d_4} [r_4 + f_4(\tilde{s}_4)] \\ &= \max_{d_4} [4d_4 + 12(s_4 - \tfrac{1}{3}d_4)] \\ &= \max_{d_4} (12s_4) \end{aligned}$$

Since $s_4 = 50$, the total system return is

$$f_5(50) = 600$$

In summary, the solution is:

At stage: 4	$d_4^* = 0,$	$\tilde{s}_4 = s_4 = 50$	
At stage 3:	$d_3^* = 0,$	$\tilde{s}_3 = \tilde{s}_5 + 50;$	$\tilde{s}_5 = 2\tilde{s}_4$
At stage 2:	$d_2^* = 0,$	$\tilde{s}_2 = \tilde{s}_3 = 150$	
At stage 5:	$s_5^* = 100,$	$d_5 = 150;$	$\tilde{s}_5 = 150$
At stage 1:	$s_1 = 0$		

However, the optimal policy above is not unique for this simple illustrative problem. An alternative optimal policy, for example, is

$$d_1^* = 0; \quad d_2^* = 0; \quad d_3^* = 0; \quad d_4^* = 30; \quad d_5^* = 120$$

This policy produces the following values for the state variables:

$$\tilde{s}_4 = 40; \quad s_2 = 140; \quad \tilde{s}_5 = 100; \quad s_5 = 140; \quad s_1 = 0$$

so the individual stage returns are

$$R_1 = 0; \quad R_2 = 0; \quad R_3 = 0; \quad R_4 = 120; \quad R_5 = 480$$

Figure 7-37 shows the functional diagram for the diverging branch problem that results when the loop compression principle is applied to this example problem. This example was made intentionally simple to facilitate understanding of the compression method. Problems containing more stages in the loops and having more complex return and transition functions are, of course, computationally more difficult to solve. The compression technique illustrated in this problem, however, is unchanged; the strategic principles

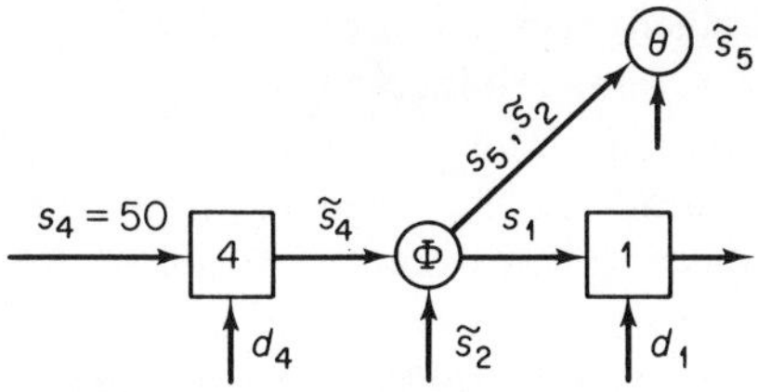

Figure 7-37. Equivalent diverging branch system for example problem.

of the method, as outlined in this section, apply without exception to all problems containing loops.

7-29 CONCLUDING SUMMARY

This chapter has shown how to optimize large systems conditionally, one stage at a time. The technique, first developed under the name *dynamic programming* for systems having the stages in series, can be extended in principle to loop and branch structures by cutting them up into serial pieces. Care must be exercised in reassembling the serial subsystems, and the extra computation at the junctions renders decomposition impractical unless branch and loop compression techniques are employed.

The guiding strategies for conditional optimization involve exploiting the information flow *between* stages, described abstractly by functional equations and diagrams. These overall strategies are the same no matter what mathematical form the stage transition and return functions take, although the conditional optimization within a stage is certainly influenced by the stage's mathematical character. Furthermore, the way information is transmitted by a state variable depends on the form of the originating stage and, at the receiving stage, affects the maximand and hence the mode of conditional optimization. Since such interactions are not easily grasped in an abstract discussion, detailed examples were given involving total enumeration, linear inequalities, and nonlinear differentiation for every system structure studied. Overall strategies are most easily grasped by studying the network examples, more subtle points concerning constraints and inversions being illustrated by the linear and nonlinear allocation problems.

The complexity of looped and branched decision processes facilitates mistakes. Optimization strategies of great plausibility which lead to inefficient or nonoptimal computations were illustrated with detailed examples so that the reader will recognize these fallacies and avoid them.

The mathematical parables of this chapter suggest guides to making intelligent decisions in complicated situations. In long-range planning when present circumstances, but not future positions, are known, one should start the analysis as far into the future as possible, working backward in time to the present. Diverging branch decision problems, which occur when a raw material is made into several products and distributed to many consumer outlets, can be analyzed in much the same way as a serial problem by working backward from distributors to the original supply. Loop systems, which correspond to planning over a seasonal or business cycle, can be analyzed through the use of loop compression, to produce an equivalent diverging branch system, which can then be solved using serial dynamic programming. Other examples of loops occur in general recycle (feedback) or bypass (feed-forward) systems which often arise in manufacturing operations. Linear allocation problems lend themselves to additive superposition of decisions in converging branch systems. The most important thing to be learned from this chapter is that information flow, as represented by functional equations and diagrams, may differ in direction from the flow of time and materials. For effective optimization of large systems, one must understand information flow and know how to reverse it when expedient.

Although conditional optimization has found many applications to serial systems, mainly involving planning over a succession of time periods, its extension to systems with more general structure has been held back by the computational complexity of nonserial problems. The recent advent of the compression techniques, which convert such problems to equivalent serial problems, should open the way to increased use of this powerful optimization technique in solving the numerous loop and branch systems which occur so frequently in practice.

BIBLIOGRAPHY

ARIS, R., *Discrete Dynamic Programming* (Blaisdell, New York, 1964).

BAGWELL, J. R., C. S. BEIGHTLER, and J. P. STARK, "On a class of convex allocation problems," *J. Math. Anal. Appl.*, **25**, 1 (January 1969).

BECKWITH, R. E., "Dynamic programming and network routing: an introduction to the technique of functional equations," *Proc. Dynamic Programming Workshop*, Purdue Univ., 1961, pp. 89–100.

BEIGHTLER, C. S., D. B. JOHNSON, and D. J. WILDE, "Superposition in branching allocation problems," *J. Math. Anal. Appl.*, **12**, 1 (1965), 65–70.

———, and L. G. MITTEN, "Design of an optimal sequence of interrelated sampling plans," *J. Amer. Stat. Assoc.*, **59** (March 1964), 96–104.

———, D. T. PHILLIPS, and S. F. FOWLER, "Compression and decomposition of feedback loops in multistage systems," *J. Math. Anal. Appl.*, **38**, 2 (1972), 497–500.

BELLMAN, R. E., *Dynamic Programming* (Princeton Univ. Press, Princeton, N.J., 1957).

———, and S. DREYFUS, *Applied Dynamic Programming* (Princeton Univ. Press, Princeton, N.J., 1962).

BOUCHEY, G. D., B. V. KOEN, and C. S. BEIGHTLER, "The optimal allocation of energy in industrial and agroindustrial complexes using dynamic programming," *Nucl. Sci. Technol.*, **41** (July 1970), 70–78.

———, B. V. KOEN, and C. S. BEIGHTLER, "Optimization of nuclear materials safeguards sampling systems by dynamic programming," *Nucl. Technol.*, **12** (September 1971), 18–25.

CRISP, R. M., and C. S. BEIGHTLER, "Closed-form solutions to certain linear allocation problems," *AIIE Trans.*, **1**, 4 (December 1969).

DENARDO, E. V., *Dynamic Programming: Theory and Application* (Prentice-Hall, Englewood Cliffs, N.J., 1975).

DREYFUS, S., "Computational aspects of dynamic programming," *Operations Res.*, **5** (1957), 409–415.

ESOGBUE, A. O., and B. R. MARKS, "The status of nonserial dynamic programming," *Man. Sci.*, **19**, 3 (November 1972).

MEIER, W. L., and C. S. BEIGHTLER, "Branch compression and absorption in nonserial multistage systems," *J. Math. Anal. Appl.*, **21**, 2 (February 1968), 426–430.

———, and C. S. BEIGHTLER, "Design of an optimum branched allocation system," *Ind. Eng. Chem.* (February 1968). Also published in *Dechema-Monographien*, **60** (1968).

———, and C. S. BEIGHTLER, "Dynamic programming in water resources development," *Proceedings of the National Symposium on the Analysis of Water-Resource Systems, Sponsored by the American Resources Association*, July 1–3, 1968.

NEMHAUSER, G. L., "A dynamic programming approach for optimal design and operation of multistage systems in the process industries," Unpublished doctoral dissertation, Northwestern Univ., 1961.

———, *Introduction to Dynamic Programming* (Wiley, New York, 1966).

PHILLIPS. D. T., C. S. BEIGHTLER, and M. W. PARKER, "Analysis of nonserial multistaged systems using compression and decomposition," *AIIE Trans.*, **7**, 4 (December 1975), 388–392.

———, and M. W. PARKER, "Dynamic programming," *Encyclopedia of Computer Science and Technology*. J. Belzer, A. Holzman, and A. Kent, eds. (Marcel Dekker, New York, 1977).

———, A. RAVINDRAN, and J. J. SOLBERG, *Operations Research: Principles and Practice* (Wiley, New York, 1976).

WILDE, D. J., "A unified approach to multivariable optimization theory," *Ind. Engrg. Chem.*, **5**, 7 (August 1965), 18–31.

YEN, J. Y., "An algorithm for finding shortest routes from all source nodes to a given destination in general networks," *Quart. Appl. Math.*, **27** (1970), 526–530.

EXERCISES

7-1. (a) Find the shortest path from each western node to an eastern node.

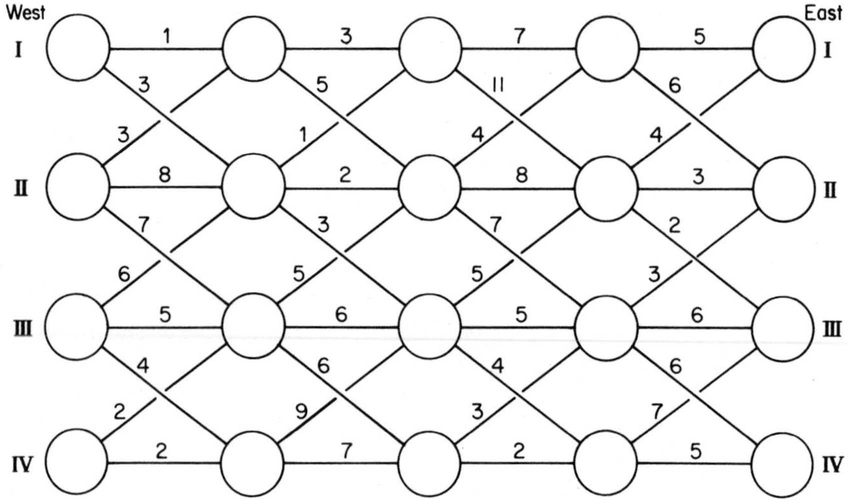

Figure E7-1(a)

(b) Find the shortest path from each eastern node to a western node.

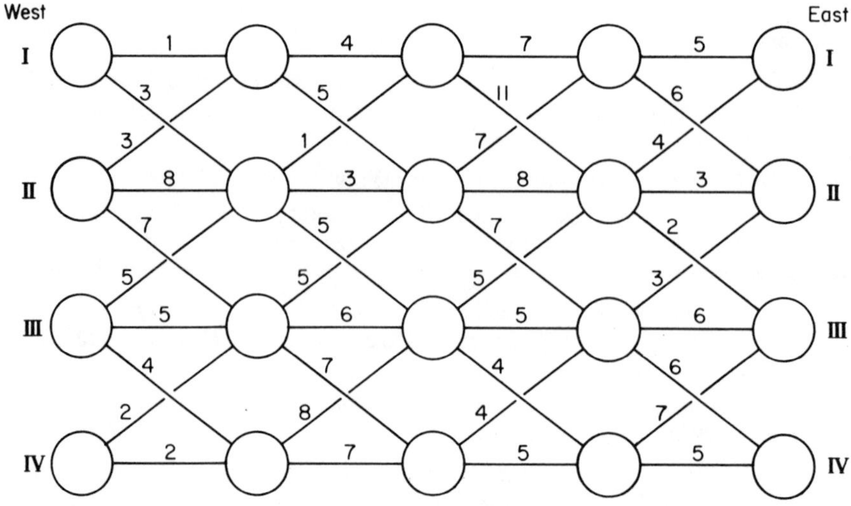

Figure E7-1(b)

7-2. Find the shortest and longest paths from O to P, where no backward movement (toward O) is permitted.

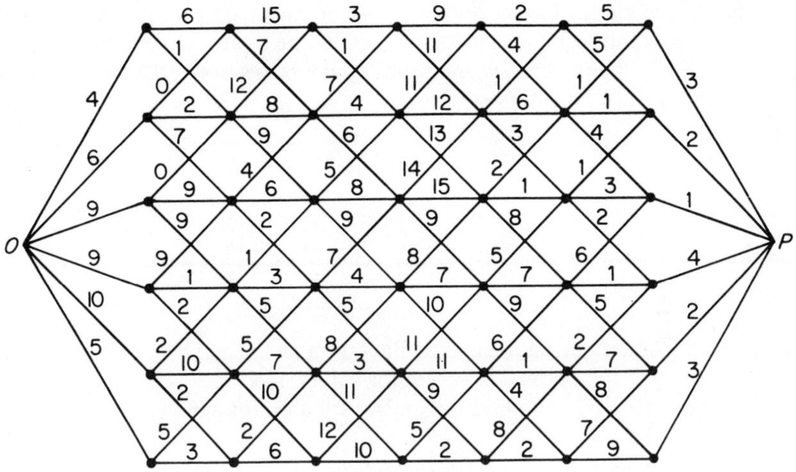

Figure E7-2

7-3. A government analyst has identified five critical subsystems which are instrumental to the mission effectiveness of a new fighter aircraft that has been designed by the Astro-Tech Aircraft Company. It has been determined that the overall effectiveness of the aircraft is equal to the product of the individual effectiveness of each of the five subsystems.

Although numerous bids have been received, Astro-Tech has identified two options for each subsystem which meet the design specifications. Only \$1000K is available for these subsystems. What is the maximum effectiveness that can be obtained for this amount?

Table of available options

Option	*UHF radio*	*TACAN*	*WPNS delivery*	*Radar*	*Navigation system*
A	Eff = 0.988 Cost = 200K	0.936 150K	0.988 300K	0.936 400K	0.988 200K
B	Eff = 0.832 Cost = 150K	0.884 100K	0.936 250K	0.884 300K	0.832 100K

(*Ans.:* Radio, A; TACAN, A; WPNS, B; radar, B; nav., B.)

7-4. Given a total resource of 10 units, and a return (profit) at stage i of $8x_i - ix_i^2$, where $x_i = 0, 1, 2, 3$ is the allocation made to the ith stage, find the optimal allocation policy to use to maximize the total return for a four-stage system if all 10 units must be allocated. That is, maximize the function $p = \sum_{i=1}^{4} (8x_i - ix_i^2)$, subject to the constraints $\sum_{i=1}^{4} x_i = 10$, and $x_i = 0, 1, 2, 3$ for $i = 1, \ldots, 4$.

7-5. Three kinds of items are available for packing into a survival kit which has a maximum volumetric capacity of V. The table gives the pertinent information on these items, where u_i and v_i are, respectively, the per-unit utility and volume of item i.

i	u_i	v_i
1	3	2
2	4	3
3	1	1

(a) Assuming that there is no limitation on the numbers of each item selected, find the optimal selection for values of V up to 10.

(b) From part (a), find the general solution for any value of V as a function of V.

7-6. A production process must be scheduled over five periods, I–V. Production during each period is restricted to integral numbers of units and the maximum production per period is four units. The following table gives the total cost of production (in hundreds of dollars) for the different numbers of units that may be produced per period. In addition to these production costs, there is an inventory cost of \$100 per unit stored per period for the units produced in periods I–IV. The inventory cost may be taken into account for complete periods only.

	Period				
Number produced	*I*	*II*	*III*	*IV*	*V*
0	2	2	3	5	3
1	3	4	4	6	8
2	7	6	8	8	10
3	10	11	13	17	15
4	11	12	14	21	18

Find the optimal number of items to be produced during each period (to minimize the total cost) if the total production requirement is (a) 18 units; (b) 15 units; (c) 13 units; (d) 10 units.

7-7. Solve Exercise 7-4 for the case in which x_i is a continuous, nonnegative variable and compare the answers with those found in Exercise 7-4.

7-8. (a) *Use dynamic programming* (in *tabular* form) to solve the following multistage decision problem.

On a certain plot of land, a farmer has a choice each year of planting corn, soybeans, oats, or hay. His profit at the end of that year depends both upon what crop he planted at the beginning of the year and also what crop was planted last year, as shown in the following figure:

Crop planted last year

Crop planted this year	Corn	Soy	Oats	Hay
Corn	3	15	9	11
Soy	2	1	4	1
Oats	4	17	6	4
Hay	1	7	5	3

Profits in $/acre yield

Figure E7-8

Find the optimal planting policy for each of the next 5 years to maximize the total profit, given that the crop planted last year was hay.

(b) Find the total return that would result from a policy of "doing the best you can at every stage": that is, a policy which chooses d_n at stage n to maximize only the return $r_n(s_n, d_n)$, for $n = 1, 2, 3, 4, 5$. Compare this total return with that produced by the optimal policy found in part (a).

7-9. (a) What are the stages, state variables, decision variables, returns, state transformations, and recursive equations for the following problem?

The accompanying graph describes the expected profits, in millions, for three divisions of a firm as a function of their next year's budget in millions. Find, to the nearest \$10 million, what the budget should be for each division in order to maximize the total expected profit from all three divisions. Also, find this maximum total expected profit. The total budget for all three divisions is \$60 million.

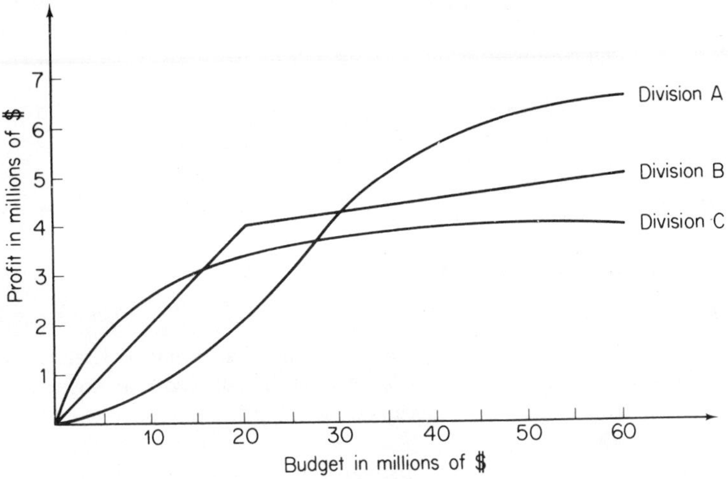

Figure E7-9

(b) *Use dynamic programming* to solve the problem in part (a).

7-10. Solve the following initial-value problem:

$$\text{maximize} \sum_{i=1}^{5} r_i$$

subject to

$$\left.\begin{aligned} r_i &= s_i + 3d_i; \\ s_{i-1} \equiv \tilde{s}_i &= 2s_i - 0.2d_i; \\ 0 \le d_i &\le s_i \end{aligned}\right\} i = 1, \ldots, 5$$

and the initial state is $s_5 = 100$.

7-11. (a) Solve the following problem using dynamic programming by writing it as a serial decision problem:

$$\text{maximize } p = x_1 x_2 x_3 x_4 x_5$$

subject to

$$x_1 + x_2 + x_3 + x_4 + x_5 = 20; \qquad x_j \ge 0; \qquad j = 1, \ldots, 5$$

(b) Generalize the method used to solve the problem in part (a) to provide a solution to

$$\text{maximize } p = \prod_{j=1}^{n} x_j$$

subject to

$$\sum_{j=1}^{n} x_j = k; \qquad x_j \ge 0; \qquad j = 1, \ldots, n$$

and from this solution show that for all $x_j \ge 0$,

$$\sqrt[n]{\prod_{j=1}^{n} x_j} \le \frac{1}{n} \sum_{j=1}^{n} x_j$$

7-12. Solve the six final-value problems, $\tilde{c}_1 = \text{I}, \ldots, \text{IV}$, described by the network of Fig. 7-1 using state inversion. Compare the solution for $\tilde{c}_1 = \text{I}$ with the initial value problem $s_6 = \text{I}$.

7-13. Solve the following problem: maximize $p = \sum_{i=1}^{3} r_i$, where $r_i = 5d_i - id_i^2$, $\tilde{s}_i = s_i - 0.4d_i$, $0 \le d_i \le s_i$ as a *final*-value problem, $\tilde{s}_1 = \tilde{c}_1$, using state inversion. Compare your answer with that given in the text where this problem is solved both as an initial-value and as a final-value problem using decision inversion.

7-14. Solve the following as a final-value problem, using state inversion: maximize $p = \sum_{i=1}^{5} r_i$, where $r_i = s_i + 3d_i$, $\tilde{s}_i = 2s_i - 0.2d_i$, $\tilde{s}_1 \equiv \tilde{c}_1$, and the decisions are restricted by $0 \le d_i \le s_i$. This problem is the final-value version of Exercise 7-10; in that problem the optimal solution produced a value of

$\tilde{s}_1^* = 2332.8$. Setting $\tilde{c}_1 = 2332.8$ in your solution, compare the resulting value of f_5 with that for the initial-value problem.

7-15. Solve Exercise 7-12 using *decision* inversion and notice the difference in the solution procedure and computational effort from that employed in state inversion for the same network.

7-16. Using state inversion, solve the two-industry problem described in the text as a final-value problem. That is, for $0 \leq d_i \leq s_i$, $i = 1, \ldots, 5$,

$$\text{maximize} \sum_{i=1}^{5} \sum_{j=1}^{5} (s_j + d_j)$$

where

$$\tilde{s}_i = 1.9s_i - 0.4d_i; \qquad i = 1, \ldots, 4$$
$$\tilde{s}_5 = 0.9s_5 - 0.4d_5$$

and the final output is fixed at $\tilde{s}_1 \equiv \tilde{c}_1$.

7-17. Find the shortest path from west to east under the loop constraint of beginning and ending at the same level (I, . . . , IV).

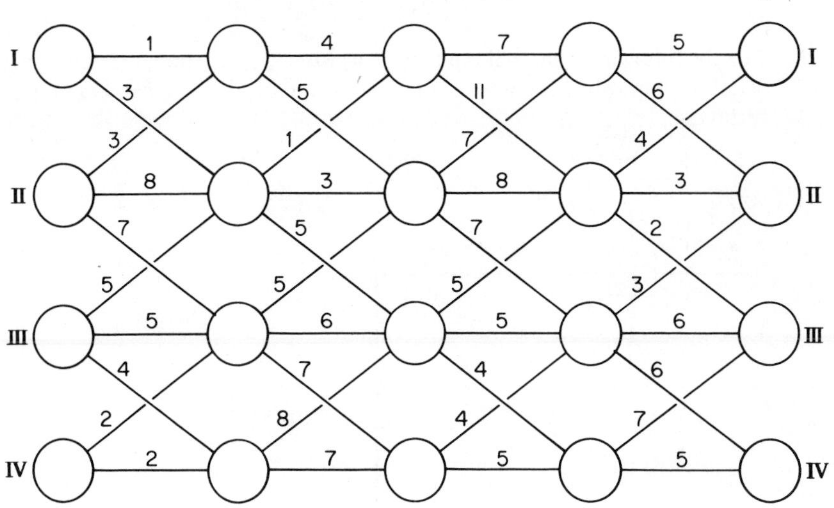

Figure E7-17

7-18. (a) Solve the following initial-value problem: maximize $p = r_1 + r_2$, where $r_n = 0.8s_n - 0.2s_n d_n$, $\tilde{s}_n = 0.7s_n + 0.5s_n d_n$, $0 \leq d_n \leq 1$, for $n = 1, 2$, and the system input is $s_2 = 100$.

(b) Consider the foregoing problem when augmented by the inhomogeneous loop constraint, $s_2 = 100 + \frac{1}{2}\tilde{s}_1$.

Show that the policy $\{d_1^* = 1, d_2^* = 1\}$ produces a larger value of p than does the policy which was optimal for the serial problem in part (a). [*Hint:* First derive the expression: $s_2 = 100/(1 - \frac{1}{2}K)$, where $K = (0.7 + 0.5d_2)(0.7 + 0.5d_1)$.]

This counterexample proves that the superposition principle of Section 7-24 is not applicable to problems containing loops.

7-19. (Nemhauser) Show that for the following problem, where $r(d_n)$ is convex and monotone-increasing, the multiplicative constraint can be replaced by $d_1 = s_1$ and the $N - 1$ transition functions

$$s_{n-1} = \frac{s_n}{d_n}; \qquad n = 2, \ldots, N$$

with $s_N \geq K$:

$$\text{minimize} \sum_{n=1}^{N} r(d_n)$$

subject to

$$\prod_{n=1}^{N} d_n \geq K$$

$$d_n \geq 0; \qquad n = 1, \ldots, N$$

Then, show that the optimal policy is given by

$$d_n^* = K^{1/N}; \qquad n = 1, \ldots, N$$

7-20. For the following nonserial multistage decision system, draw the equivalent serial system which results from application of the *branch compression* technique.

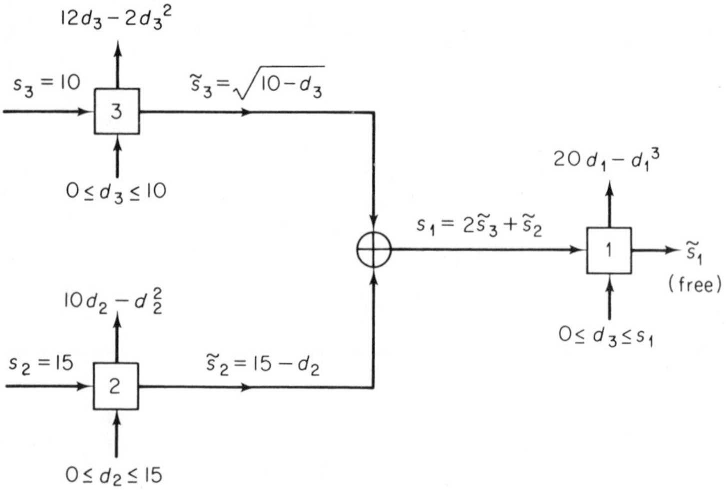

Figure E7-20

Show all restrictions on all the variables and parameters. The objective is to maximize the sum of the three stage returns: $R = 20d_1 - d_1^3 + 10d_2 - d_2^2 + 12d_3 - 2d_3^2$.

Index

M

N

O

P

Q

R

S